ANTIDESMA

IN

MALESIA AND THAINLAND

Antidesma

in
Malesia and Thailand

by Petra Hoffmann

Kew Publishing
Royal Botanic Gardens, Kew

First published in 2005 by
Royal Botanic Gardens, Kew
Richmond, Surrey, TW9 3AB, UK
www.kew.org

ISBN 1 84246 045 5

British Library Cataloguing in Publication Data
A catalogue record for this book is available from the British Library

Cover photograph: Reed S. Beaman

Production Editor: Ruth Linklater
Typesetting and page layout: Christine Beard
Design by Media Resources, Information Services Department,
Royal Botanic Gardens, Kew

Printed in the United Kingdom by Lightning Source UK Ltd

For information or to purchase all Kew titles please visit
www.kewbooks.com or email publishing@kew.org

All proceeds go to support Kew's work in saving the world's plants for life

CONTENTS

INTRODUCTION

Antidesma Burm. ex L. is a genus of dioecious shrubs and trees commonly found in the understorey of tropical rain forests as well as in more open vegetation. Its distribution ranges from West Africa to the Pacific Islands and from the Himalayas to Northern Australia, with the centre of diversity in the Malesian region. *Antidesma* is a distinct and homogeneous genus of about 100 species.

According to the most recent classification of Phyllanthaceae Martynov (Hoffmann *et al.*, in press) *Antidesma* belongs to tribe Antidesmateae Benth. subtribe Antidesmatinae Müll. Arg. Phyllanthaceae used to be considered part of Euphorbiaceae *sensu lato* (as subfamily Phyllanthoideae Kostel.) in previous classifications (Webster 1994b; Radcliffe-Smith 2001). Recent molecular studies (e.g., Savolainen *et al.* 2000; Chase *et al.* 2002; Davis *et al.* 2005) have shown that Euphorbiaceae *sensu* Webster (1994b) consist of five separate major lineages now recognised at family rank by the Angiosperm Phylogeny Group (Angiosperm Phylogeny Group 2003).

A notable character of subtribe Antidesmatinae is the elongated, U-shaped connective of the anthers, which is very rare in Phyllanthaceae (Wurdak *et al.* 2004: 1895). Besides the peculiar anthers, *Antidesma* is distinguished by racemose inflorescences, apetalous flowers with a floral disc in both sexes and its unilocular drupes with a characteristically foveolate endocarp. The endocarp sculpture and its resemblance to some genera of Icacinaceae caused the segregation of *Antidesma* as the monogeneric family Stilaginaceae by Airy Shaw (Willis 1966: 1076) which cannot be maintained in view of the molecular evidence available today.

This revision was initiated as a post-doctoral project within the research network "Botanical Diversity of the Indo-Pacific Region" of the European Community under the co-ordination of the Rijksherbarium Leiden (now National Herbarium of the Netherlands, Leiden University Branch). The network facilitated the study of some large and problematic taxa, including *Antidesma.* The genus was chosen because its great variability, and the lack of characters needed to distinguish species. The work was carried out at the Royal Botanic Gardens, Kew (K), and at the National Herbarium of the Netherlands, Leiden University Branch (L). An interval at the Natural History Museum in Paris (P) to revise *Antidesma* for Madagascar and the Comoro Islands provided valuable insights into its diversity in Africa. The study has been for the most part herbarium-based, but fieldwork was carried out in the Malay Peninsula and Borneo.

In the present revision, a total of 56 species and 13 infraspecific taxa, are recognised. This is about half of the estimated number of taxa in the entire genus. The accounts of *Antidesma* in "Flora Malesiana", "Flora of Thailand" (Hoffmann 2005) and the "Tree Flora of Sabah and Sarawak" are extracts of this revision. Only a selection of specimen citations is provided at the end of the treatment of each taxon. It excludes the type material which is cited under the taxon name and the synonyms, respectively. The revision is based on over 7600 examined specimens, c. 5000 of which were geo-referenced for the production of the maps.

ETYMOLOGY

The generic name *Antidesma* was coined by Burman (1736: 22) who explained its derivation from the Greek words αγτι ("anti-" = against) and δεσμα ("desma" = poison), referring to its reported use against bites of the hooded cobra. Burman's translation of "desma" as "venenum" is somewhat doubtful, as Greek dictionaries (e.g., Liddell & Scott 1925 – 1940: 380) give the meaning of the word as a poetic variant of "desmos" (= rope, ribbon, bond, fetter). However, there is an apparent similarity between ropes and snakes and Burman's explanation of the name is unambiguous.

The supposed efficacy of *Antidesma* against snake-bite goes back to Rheede's "Noeli-Tali" (1683: 116, t. 56), one of several elements cited by Burman under *Antidesma.* Rheede reported that the leaves contain an antidote against bites of a snake called "Heretimandel". People bitten by this snake do not die instantly; instead, their bodily tissues rot away and decompose, and they eventually die after much suffering, unless they drink a potion made from *Antidesma* leaves cooked in water with salted mango. This use for *Antidesma* leaves has not been reported since, although the plants are said to have a variety of medicinal properties.

TAXONOMIC HISTORY AND INFRAGENERIC CLASSIFICATION

The prelinnean name *Antidesma* (Burman 1737: 22, t. 10) was validated by Linné (1753: 1027; 1754: 451). By 1790, three synonyms (*Bestram* Adans., *Stilago* L., *Rhytis* Lour.) had been published but were soon found to be congeneric with *Antidesma.*

Blume (1826) was the first to describe a large number of Malesian *Antidesma* species. His ten new species account for about 40% of all specimens collected in the "Flora Malesiana" area until today. In Tulasne's excellent revision of the genus (1851), 35 species are described in great detail, 17 of them new to science. Baillon (1858: 601) treated *Antidesma* only very briefly and mainly referred to Tulasne's article. He placed *A. diandrum* (Roxb.) Spreng. (syn. *A. acidum* Retz.) in a "section A" separate from all other species with 3 – 6-merous flowers. Müller Argoviensis (1866) grouped the 63 recognised species by number of floral parts, position of the stigma, division of the calyx, shape of the bracts, inflorescence type and presence or absence of pedicels. He stated, however, that the genus was so remarkably uniform that it should not be divided into formal sections (1866: 248). Hooker (1887: 355) criticised Müller: "The stipules, the length of the pedicels of the staminate flowers, the size and form of the bracts, and the calyx-lobes are all variable. In De Candolle's Prodromus too much importance is attached to the number of stamens and the division of the calyx, whether as to depth or number of lobes, as affording sectional characters" but continued "I am not satisfied with the more restricted use I have made of the same modifications of the calyx".

The genus was last monographed by Pax & Hoffmann (1922) who accepted 146 species, including 28 newly described by them. They formally divided the genus into eight sections based on the combination of five characters: structure of the staminate disc, division of the stipules, position of the stigma, indumentum of the ovary, and number of stamens. These characters are indeed the most useful for species delimitation, but proved to be impracticable for a subgeneric classification in a dioecious group. Using their sectional key with only staminate or only pistillate material always results in a choice of three sections because two of the character states cannot be determined. In addition, there are numerous species with intermediate or variable character states which cannot be assigned to any of these sections even with complete material. In the following decades, a great number of new taxa was described, but no further attempt at a classification was made.

A major contribution to the knowledge of Asian *Antidesma* was made by H. K. Airy Shaw. In his first paper on the genus, Airy Shaw sighed "Taxonomically, *Antidesma* is a troublesome genus" (1969: 277). He nevertheless published numerous articles (e.g., 1972b, 1973) and regional treatments of Euphorbiaceae *sensu lato,* namely for Thailand (1972a), Borneo (1975), Australia (1980a), New Guinea (1980b), Sumatra (1981a), Central Malesia (1982) and the Philippines (1983). There he described 34 new species and 16 new varieties of *Antidesma,* but also synonymised a great number of names. Airy Shaw called Pax & Hoffmann's sectional classification an "artificial arrangement" (1969: 280) and wrote "Scarcely one of the characters used in their sectional and specific keys ... leads to a natural assemblage of species, although of course in some cases related species are associated. The assignment of a given species to any given Pax & Hoffmann section may therefore provide little or no information as to natural affinity." (1973: 275). In the absence of an alternative, however, Airy Shaw and other more recent authors applied the Pax & Hoffmann classification as well as possible to their newly described taxa. The most recent regional revision is that of Chakrabarty & Gangopadhyay (2000) for the Indian subcontinent, recognizing 23 species including two newly described endemics.

After several years of intensive study for this revision, no new character has been found which enables us to cleanly classify the species of *Antidesma.* The only division worthy of taxonomic recognition is that between *A. vaccinioides* Airy Shaw (Fig. 22) and all other species in the genus, based on the marginal foliar glands and articulated pedicels of *A. vaccinioides.* The species also has a peculiar leaf venation, disc structure and numerous stamens. Unfortunately, *A. vaccinioides* is only known from the staminate type specimen from Papua New Guinea. The differences could even merit distinction at the generic level, if the species was better known, and if there had not already been more than enough obscure monospecific genera in Phyllanthaceae. The only other discrete character in the whole genus is the laciniate stipules of the African *A. laciniatum* Müll. Arg., as opposed to the entire stipules in the remainder of the genus. This, however, does not appear to be a fundamental distinction, as *A. laciniatum* otherwise resembles the other African species.

However this may be, taxonomic clarity has improved by subsumption of about 50 names as new synonyms, and by the association of some ecological varieties previously recognised as separate species (the rheophytes which had previously all been treated as separate species). Some taxa exemplifying extremes of a morphological continuum are here regarded as varieties of the same species (e.g., *A. neurocarpum* Miq. var. *neurocarpum* and var. *hosei* (Pax & K.Hoffm.) Petra Hoffm.;

A. excavatum Miq. var. *excavatum* and var. *indutum* (Airy Shaw) Petra Hoffm.). Groups of similar and apparently closely related species can be identified but are too loosely defined as to deserve taxonomic recognition, such as *A. sootepense* Craib and *A. velutinum* Tul.; *A. heterophyllum* Blume, *A. japonicum* Siebold & Zucc. and *A. montanum* Blume; *A. leucopodum* Miq. and *A. polystylum* Airy Shaw; *A. concinnum* Airy Shaw, *A. excavatum*, *A. ferrugineum* Airy Shaw and *A. jucundum* Airy Shaw; *A. catanduanense* Merr., *A. celebicum* Miq. and *A. microcarpum* Elmer; *A. ghaesembilla* Gaertn., *A. myriocarpum* Airy Shaw and *A. subcordatum* Merr.; *A. tetrandrum* Blume and *A. venenosum* J.J.Sm.; as well as *A. spatulifolium* Airy Shaw and the Australian *A. parvifolium* Thwaites & F.Muell.

Beyond these informal groups, further taxonomic division based on morphology depends entirely on the choice of character. A number of species will always be polymorphic for any given character. This, and the complete lack of non-continuous characters (with the exception of those separating *A. vaccinioides* from all other species) also discouraged any thoughts of a phylogenetic analysis based on morphological characters. As a result, the arrangement of taxa in this revision is alphabetical.

SYSTEMATIC POSITION

Antidesma is conspicuous in Euphorbiaceae *sensu lato* because of its unicarpellate drupes. When A. L. de Jussieu (1789: 385) defined Euphorbiaceae as having multilocular fruits, *Antidesma* was listed as *incertae sedis* (1789: 443). Adrien de Jussieu (1824: 10) also restricted his family description to plants with 2- to multi-locular ovaries, thereby excluding *Antidesma*. In the same work (1824: 12), however, he described the genus *Thecacoris* (now known to be the closest relative of *Antidesma*), and included it in Euphorbiaceae. *Thecacoris* is distinguished from *Antidesma* only by its 3-locular, dry schizocarps. Agardh (1824: 199) placed enough importance on the fruit character to create an "ordo Stilaginae" (*Stilago* = synonym of *Antidesma*).

Antidesma was then included in Euphorbiaceae in the main works on the family for about a hundred years without reservations (Baillon 1858, Müller 1866, Pax & Hoffmann 1922). Baillon and Müller both placed it between *Thecacoris* and *Hieronyma*, Baillon in "Euphorbiacées biovulées", and Müller in tribe Phyllantheae. Pax & Hoffmann classified *Antidesma*, *Hieronyma*, *Thecacoris* and 15 other genera in Euphorbiaceae-Phyllanthoideae-Phyllantheae-Antidesmatinae.

Around the middle of the 20th century, Airy Shaw became intrigued by the unicarpellate drupes of *Antidesma* with their unusually sculptured endocarp. He recognised the family Stilaginaceae C. Agardh (Willis 1966: 1076; Airy Shaw 1969, 1972a, 1972b, 1973, 1975, 1980a, 1980b, 1981a, 1981b) with the only genus *Antidesma*, and postulated that it had no close relatives in Euphorbiaceae. Instead, he regarded it "to some extent intermediate between Icacinaceae and Euphorbiaceae" (Willis 1966: 1076). No further evidence for a relationship between *Antidesma* and Icacinaceae has been presented since, and all genera earlier placed in Icacinaceae were found to belong to three euasterid orders (Kårehed 2001). The idea of an isolated status of *Antidesma* was taken up again by Meeuse (1990: 30).

The majority of botanists, however, continued to regard *Antidesma* as a member of Euphorbiaceae-Phyllanthoideae, and followed the system of Webster (1975: 594; 1989: 4; 1994b: 52). This was supported by a number of micromorphological and anatomical studies (Punt 1962: 21; Köhler 1965: 63; Levin 1986a,b: 521 – 522; Vogel 1986: 47 – 48; Mennega 1987: 116; Stuppy 1996: 179).

These studies also provided the necessary evidence for the refinement of the systematic position of *Antidesma* in Webster's classifications of Euphorbiaceae *sensu lato* between 1975 and 1994. The most recent placement (Webster 1994b) is in Antidesmateae-Antidesmatinae together with *Hieronyma, Leptonema, Thecacoris* and the poorly known monotypic genus *Celianella* Jabl. (*Phyllanoa* Croizat was referred to Violaceae by Hayden & Hayden in 1996). One morphological synapomorphy of these genera is the peculiar anther shape. The studies cited above indicated that *Antidesma, Hieronyma* and *Thecacoris* form a particularly homogeneous group. The exclusively African genus *Thecacoris* has the dry, 3-locular, autochorous schizocarps typical of most Euphorbiaceae *sensu lato.* In neotropical *Hieronyma* the number of carpels is reduced to 2 (rarely 3) and the indumentum is lepidote. *Antidesma* can be distinguished from *Thecacoris* only with difficulty (especially in staminate material) but by the indehiscent fruits in which the number of carpels is reduced to one or rarely two. Distinctive alkaloids extracted from these three genera in the last decade, are a further putative synapomorphy (Buske *et al.* 2002).

Molecular systematics confirmed much of Webster's classification but Euphorbiaceae *sensu lato* was shown to consist of five separate major lineages (Savolainen *et al.* 2000; Chase *et al.* 2002; Davis *et al.* 2005). The Angiosperm Phylogeny Group classification (Angiosperm Phylogeny Group 2003) recognised the following euphorbiaceous lineages at family rank: Euphorbiaceae *sensu stricto* (uniovulate subfamilies Acalyphoideae [excluding tribe Galearieae and *Dicoelia*], Crotonoideae, and Euphorbioideae), Pandaceae (Acalyphoideae tribe Galearieae), Phyllanthaceae (Phyllanthoideae excluding *Centroplacus, Drypetes, Phyllanoa, Putranjiva, Sibangea* but including *Croizatia, Dicoelia* and *Tacarcuna*), Picrodendraceae (Oldfieldioideae excluding *Croizatia* and *Paradrypetes*) and Putranjivaceae (Phyllanthoideae tribe Drypeteae except *Lingelsheimia*). Tribe Antidesmateae is retained in Phyllanthaceae but major taxonomic adjustments to the composition of the tribe are necessary (Wurdack *et al.* 2004; Hoffmann *et al.*, in press; Kathriarachchi *et al.* 2005).

BIOGEOGRAPHY

Antidesma is a palaeotropical genus of about 100 species. Higher species numbers have been given in the literature, e.g., 146 (Pax & Hoffmann 1922: 107), 170 (Airy Shaw in Willis 1966: 1076; Radcliffe-Smith 1996: 105), 200 (Webster 1994b: 52) or 154 (Govaerts *et al.* 2000: 178), but there are many synonyms. In the course of the present revision, 50 new synonyms were established, presented here or in two previous publications (Hoffmann 1999a, 1999b). Even the conservative estimate given above may prove too high after a thorough study of the genus in continental Asia.

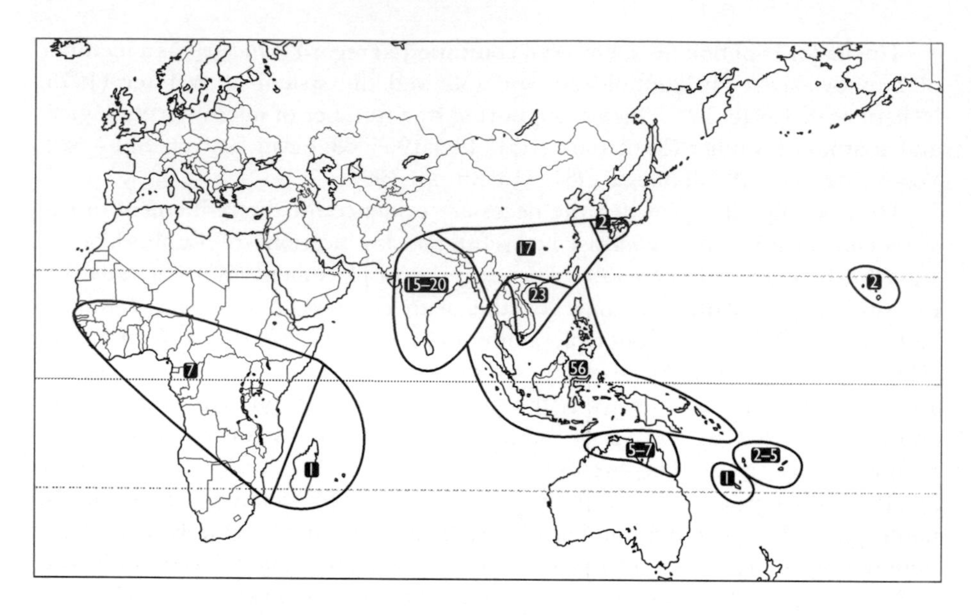

MAP 1. Worldwide distribution of *Antidesma.* Species numbers outside Malesia and Thailand compiled from literature. Widespread species are counted repeatedly.

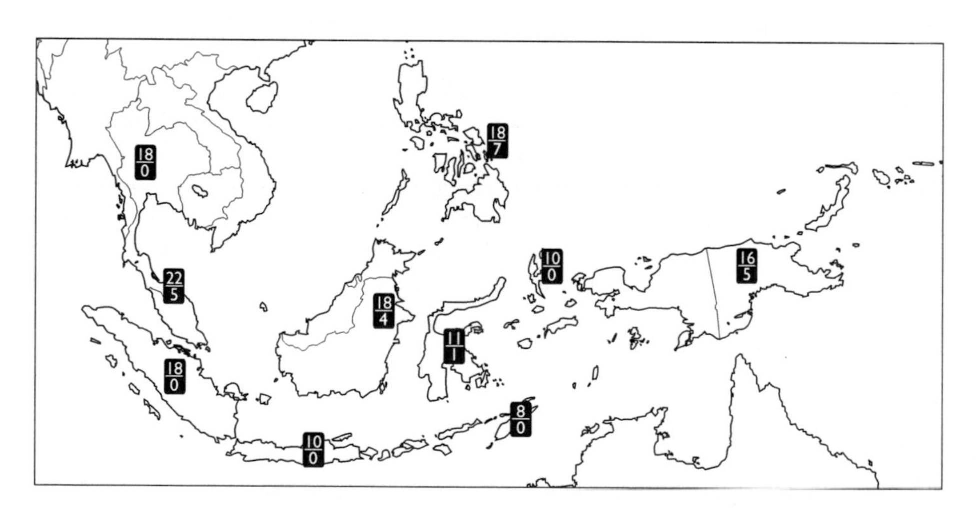

MAP 2. Diversity of *Antidesma* in Malesia and Thailand. Number above line: number of species; number below line: number of species endemic to island or group of islands.

The distribution area of *Antidesma* extends from West Africa to the Hawaiian Islands, and from Southern Japan to Northern Australia (Map 1). The centre of diversity both in terms of species number and of morphological variation is in west Malesia (38 species; 22 in Peninsular Malaysia, 18 each in Sumatra, Borneo and the Philippines). There are 27 species in south and east Malesia (16 in New Guinea, eleven in Sulawesi, ten each in Java and the Moluccas, eight in the Lesser Sunda Islands) and 18 species in Thailand (Map 2). For the entire region of Thailand and Malesia, a total of 56 species is accepted here. The term Malesia is used in this revision to signify the area comprising Indonesia, Malaysia, Singapore, Brunei Darussalam, the Philippines and Papua New Guinea.

In China, 17 species are recognised in the most recent treatment by Li Ping Tao (1994). Two of those species reach as far north-east as the Ryukyu Islands (Nansei-Shoto) in southern Japan. It is very likely that the 23 species (known synonyms not counted) listed by Gagnepain & Beille (1926: 501 – 527) for Cambodia, Laos and Vietnam will be considerably reduced by a critical revision. There are between c. 20 species in India, Bangladesh, Bhutan, Nepal, Burma and Sri Lanka (Mandal & Panigrahi 1983; Chakrabarty & Gangopadhyay 2000, synonyms not counted). Only seven species occur in continental Africa, and one in Madagascar and the Mascarenes. In the Pacific, there are two to five species in Fiji, two species in the Hawaiian Islands and one in New Caledonia. Five to seven species are found in Australia, three of which also occur in Malesia.

The species of *Antidesma* treated in this revision can be placed into ten phytogeographic groups. The majority are either Western or Eastern Malesian elements, but there are species crossing every phytogeographic border postulated by van Steenis (1950a: LXXV). This is hardly surprising given that the genus is bird-dispersed and a successful coloniser of new habitats. For a quick reference to the distribution of the individual taxa, see Table 1.

(1) WEST MALESIAN ELEMENTS. Comprising 27 taxa, this is the largest group. Eight taxa are endemic to the Philippines (*A. catanduanense*, *A. curranii* Merr., *A. digitaliforme* Tul., *A. edule* Merr. var. *edule* and var. *apoense* Petra Hoffm., *A. macgregorii* C. B. Rob., *A. microcarpum*, *A. pleuricum* Tul.), six to Borneo (*A. brachybotrys* Airy Shaw, *A. montis-silam* Airy Shaw, *A. neurocarpum* Miq. var. *linearifolium* (Pax & K. Hoffm.) Petra Hoffm., *A. polystylum*, *A. tomentosum* Blume var. *stenocarpum* (Airy Shaw) Petra Hoffm., *A. venenosum*), five to the Malay Peninsula (*A. cruciforme* Gage, *A. kunstleri* Gage, *A. orthogyne* (Hook. f.) Airy Shaw, *A. pachystachys* Hook. f., *A. pahangense* Airy Shaw) and one to Sumatra (*A. cuspidatum* Müll. Arg. var. *orthocalyx* Airy Shaw, only known from the type). Western elements common to two or more main islands or areas are *A. coriaceum* Tul., *A. cuspidatum* Müll. Arg. var. *cuspidatum*, *A. forbesii* Pax & K. Hoffm., *A. neurocarpum* var. *neurocarpum* and var. *hosei*, as well as *A. pendulum* Hook. f. *A. laurifolium* Airy Shaw shows a curious disjunction between southern Peninsular Malaysia and a pocket of evergreen rain forest in SE Thailand and Cambodia.

(2) EAST MALESIAN ELEMENTS. With 16 taxa this is the second largest phytogeographical group. Eleven taxa are endemic to New Guinea (*A. chalaranthum* Airy Shaw, *A. concinnum*, *A. contractum* J. J. Sm., *A. ferrugineum*, *A. jucundum*, *A. myriocarpum* Airy Shaw var. *myriocarpum* and var. *puberulum* Airy Shaw,

TABLE 1. Geographic distribution of *Antidesma* L. taxa covered in this revision (see p. 27 for explanation of the abbreviations of areas).

Taxon	THA	MLY	SUM	BOR	PHI	JAW	SUL	LSI	MOL	NWG	Other areas (for details see under species)
A. acidum	×					×					S and SE Asia
A. baccatum									×	×	
A. brachybotrys				×							
A. brevipes							×				
A. bunius var. *bunius*	×		×	×	×	×	×	×	×	×	S and SE Asia, Christmas Isl., Tahiti, Hawaiian Isl.
—— var. *pubescens*	×										
A. catanduanense					×						
A. celebicum							×	×	×		
A. chalaranthum										×	
A. concinnum										×	
A. contractum										×	
A. coriaceum		×	×	×							
A. cruciforme		×									
A. curranii					×						
A. cuspidatum var. *cuspidatum*		×	×	×							
—— var. *orthocalyx*			×								
A. digitaliforme					×						
A. edule var. *edule*					×						
—— var. *apoense*					×						
A. elbertii							×	×	×		
A. excavatum var. *excavatum*				(×)	(×)		×		×	×	Solomon Isl. to Samoa, Australia
—— var. *indutum*									×	×	Solomon Islands
A. ferrugineum										×	
A. forbesii	×	×	×								
A. ghaesembilla	×	×	×	×	×	×	×	×	×	×	S and SE Asia, Australia
A. helferi	×	×	×								Peninsular Burma
A. heterophyllum						×	×	×	×		
A. japonicum var. *japonicum*	×	×			×						E and SE Asia
—— var. *robustius*	×										
A. jucundum										×	
A. kunstleri		×									
A. laurifolium	×	×									Cambodia
A. leucocladon	×	×	×								
A. leucopodum	×	×	×	×	×						

Taxon	THA	MLY	SUM	BOR	PHI	JAW	SUL	LSI	MOL	NWG	Other areas (for details see under species)
A. macgregorii					×						
A. microcarpum					×						
A. minus			×			×					
A. montanum var. *montanum*	×	×	×	×	×	×	×	×	×	×	S, E and SE Asia, Australia
—— var. *microphyllum*	×										India, China, Burma, Laos, Vietnam
—— var. *salicinum*	×	×	×	×							Bangladesh, Vietnam
—— var. *wallichii*	×										Peninsular Burma
A. montis-silam				×							
A. myriocarpum var. *myriocarpum*										×	
—— var. *puberulum*										×	
A. neurocarpum var. *neurocarpum*	×	×	×	×							
—— var. *hosei*		×	×	×							
—— var. *lineariifolium*				×							
A. orthogyne	×	×									
A. pachystachys		×									
A. pahangense		×									
A. pendulum	×	×	×	×							
A. petiolatum										×	
A. pleuricum					×						
A. polystylum				×							
A. puncticulatum	×	×	×	×	×						SE Asia
A. rhynchophyllum										×	
A. riparium subsp. *riparium*			×	×	×		×				
—— subsp. *ramosum*										×	
A. sootepense	×										Burma, Laos
A. spatulifolium									×	×	
A. stipulare		×	×	×	×	×	×		×		
A. subcordatum					×			×			
A. tetrandrum			×			×		×			
A. tomentosum var. *tomentosum*	×	×	×	×	×	×	×				
—— var. *stenocarpum*				×							
A. vaccinioides										×	
A. velutinosum	×	×	×	(×)		×					Burma
A. velutinum	×	×									Burma, Cambodia
A. venenosum				×							

A. petiolatum Airy Shaw, *A. rhynchophyllum* K. Schum., *A. riparium* Airy Shaw subsp. *ramosum* Petra Hoffm., *A. vaccinioides*). *A. brevipes* Petra Hoffm. is the only species endemic to Sulawesi. More widely distributed are *A. baccatum* Airy Shaw and *A. spatulifolium* which occur also in the Moluccas, and *A. excavatum* var. *indutum* which is also found in the Solomon Islands. *A. excavatum* var. *excavatum* has the largest distribution area of all eastern Malesian elements. One collection is known each from East Kalimantan and Mindanao, respectively, whereas the vast majority of collections is from Sulawesi eastwards, stretching as far as Samoa and the Caroline Islands.

(3) SPECIES EXTENDING ACROSS THE ISTHMUS OF KRA. Nine taxa do not respect the phytogeographic demarcation line between South-East Asia and Malaysia just south of the border between Thailand and Burma postulated by van Steenis (1950a: LXXII). Their area of distribution reaches from South-East Asia into Peninsular Malaysia (*A. velutinum*), Sumatra (*A. helferi* Hook. f., *A. leucocladon* Hook. f.), Sumatra and Borneo (*A. montanum* Blume var. *salicinum* (Ridl.) Petra Hoffm.) or Sumatra, Borneo and the Philippines (*A. leucopodum*, *A. puncticulatum* Miq.). *A. japonicum* var. *japonicum* is found in southern Japan, south-eastern China, Taiwan, Burma, Indochina, Thailand, Peninsular Malaysia and Luzon and exemplifies the similarity of the upland floras of northern Luzon and eastern Asia (van Steenis 1950a: LXX). *A. montanum* Blume var. *wallichii* (Tul.) Petra Hoffm. is restricted to Peninsular Burma (Tenasserim division) and Peninsular Thailand. *A. velutinosum* Blume is the only species stretching from Burma across the Isthmus of Kra through western Malesia to Java in the south.

(4) SOUTH-EAST ASIAN ELEMENTS. The distribution of four taxa stops at the Isthmus of Kra. Of these, two are local endemics in Thailand (*A. bunius* (L.) Spreng. var. *pubescens* Petra Hoffm., *A. japonicum* Siebold & Zucc. var. *robustius* Airy Shaw). The remaining two have wider distributions, namely *A. montanum* Blume var. *microphyllum* (Hemsl.) Petra Hoffm. (India, southern China, Burma, Laos, Vietnam, Thailand) and *A. sootepense* (Burma, Laos, Thailand).

(5) SOUTH-WESTERN MALESIAN ELEMENTS. There are no taxa of *Antidesma* confined to the South Malesian province of van Steenis (1950a: LXXV), and no sharp boundary exists between this and neighbouring phytogeographic provinces. *A. tetrandrum*, for example, occurs in Sumatra, Java and Bali, whereas its presumed sister species *A. venenosum* is endemic to Borneo. *A. minus* Blume, a species common to the submontane regions of Java and Sumatra, illustrates van Steenis's statement about the similarity of the mountain flora of Java and that of most of Sumatra (van Steenis 1950a: LXX).

(6) SOUTH-EASTERN MALESIAN ELEMENTS. Three species (*A. celebicum*, *A. elbertii* Petra Hoffm., *A. heterophyllum*) are found in Sulawesi, the Lesser Sunda Islands and the Moluccas. The latter occurs also in Java.

(7) TAXA COMMON TO EASTERN AND WESTERN MALESIAN PROVINCE. *A. riparium* Airy Shaw subsp. *riparium* ranges from Sumatra over Borneo and Palawan to central and southeast Sulawesi. *A. riparium* subsp. *ramosum* is found only in Papua (Indonesia).

(8) SPECIES EXTENDING ACROSS THE MAKASSAR STRAITS (Western, Southern and Eastern Malesian province). The distribution of *A. stipulare* Blume bridges the

demarcation line of the Makassar Straits (van Steenis 1950a: LXXIV), covering the area from Peninsular Malaysia to the comparably dry Moluccas and southern Sulawesi. *A. tomentosum* Blume var. *tomentosum* has a similar but slightly more north-western distribution (Malay Peninsula, Sumatra, West Java, Borneo, Philippines, North and Central Sulawesi), occurring in more humid conditions.

(9) MONSOON FOREST ELEMENTS. *A. acidum* shows a disjunct distribution known from many other taxa including the teak tree, *Tectona grandis* L.f. (Lamiaceae). The taxa sharing this distribution pattern require a two-seasonal climate with an annual drought period. They are present in the monsoon forests of South-East Asia, avoid ever-wet West Malesia from the Isthmus of Kra southwards, but reappear in Java, parts of the Philippines, Sulawesi and the Lesser Sunda Islands. Because of the large number and the composition of taxa concerned, this cannot be explained by anthropogenic dispersal only. It probably dates back to the expansion of areas with periodic drought during the Pleistocene, which then disappeared again in the post-glacial period leaving many taxa with disjunct distribution areas. This has been discussed in detail by van Steenis (in van Meeuwen *et al.* 1960). *A. acidum* is found in the teak forests of Java and the deciduous forests of continental SE Asia, and is listed as a drought-indicating species (van Steenis & Schippers-Lammertse 1965: 71).

The rare species *A. subcordatum* is recorded only from the Philippines and the Lesser Sunda Islands. It is difficult to establish whether the gap in Sulawesi is an actual disjunction which could date back the climatic change at the beginning of the present interglacial period, or a collecting artefact.

(10) WIDESPREAD SPECIES. The three most widespread Asian species hardly show any ecological preferences. This applies particularly to the very common *A. montanum* Blume var. *montanum* which occurs from India (excluding Sri Lanka) to southern Japan, South-East Asia, the Philippines, West Malesia, Java and Sulawesi. *A. ghaesembilla* is found throughout the region including India, South China, New Guinea and northern Australia, but is absent from southern Borneo. *A. bunius* (L.) Spreng. is widely cultivated as a fruit tree, making it impossible to establish its original distribution. The species is known from Nepal and Sri Lanka to the Philippines and the Moluccas, leaving out the Malay Peninsula and most of Borneo.

ECOLOGY AND LIFE CYCLE

In Malesia and Thailand, *Antidesma* species are most commonly found as shrubs or small trees in the understorey of evergreen tropical forest. They are equally common in primary and secondary vegetation and often thrive on forest edges and river banks. *A. ghaesembilla* and the similar *A. myriocarpum* var. *myriocarpum* are the only Asian taxa with a preference for open habitats (savannah and scrubland; *A. ghaesembilla* is even reported to be fire-resistant). The available ecological data on the otherwise extremely similar *A. myriocarpum* var. *puberulum*, however, suggest that it is a substantial rain forest tree. *A. acidum, A. sootepense* and *A. velutinum* are notable for their regular occurrence in deciduous habitats.

Four South-East Asian and western Malesian species were described on the basis of their rheophytic habit and ecology. Each of these taxa has been found to correspond to a common non-rheophytic species, and each has been lowered to the level of variety (*A. montanum* var. *microphyllum* and var. *salicinum*, *A. neurocarpum* var. *linearifolium* and *A. tomentosum* var. *stenocarpum*. Further to the east, *A. riparium* shows rheophytic characteristics but apparently lacks a distinct non-rheophytic form.

Antidesma is mainly found in lowland or upland habitats. In Thailand and western Malesia, it does not seem to occur at altitudes above 2200 m (*A. leucopodum* on Mt Kinabalu, Sabah). In New Guinea, however, several species reach considerably higher altitudes (up to 3600 m for *A. excavatum*).

Plants of most *Antidesma* species seem to fruit when still young, and the often long or many-branched inflorescences produce numerous fruits. The red to black, juicy fruits are eaten and dispersed by birds. Many species thrive in a variety of habitats, and the genus is in consequence rather common and abundant throughout South-East Asia. The high colonisation potential is illustrated by a project on forest regeneration on Krakatau which investigates the re-colonization of the once sterile islands by bird-dispersed plants (Whittaker & Jones 1994). Three or four species of *Antidesma* have arrived on Krakatau so far, and *A. montanum* even dominates the forest understorey (R. Whittaker & S. Schmitt, pers. comm.).

Observations on *A. pulvinatum* Hillebr. from Hawaii (Obata 1974) confirm the impression of *Antidesma* as an early and abundantly reproducing plant. Two seedlings of 25 – 30 cm were planted on the same island but in different places. Two years later, the staminate tree had grown to 2.5 m and started flowering profusely. Another year later, the pistillate tree also started to flower at the same height. Almost all pistillate flowers developed into mature fruits, and the seeds, when planted in peat moss, began to germinate after 21 days with a germination rate of about 95%. This shows that passage through birds is not essential for seed germination at least in *A. pulvinatum*. Sosef *et al.* (1998: 76) give germination rates of 3 – 84% and germination times from 12 – 96 days for Asian species. Dry fruits of *A. bunius* were stored for up to 5 years in airtight containers without any serious loss of seed viability.

ANTIDESMA AS A PEST IN THE NEOTROPICS

In 1926, a specimen of *Antidesma ghaesembilla* (*Altson* 553) from the Botanic Gardens in Georgetown, British Guiana (now Guyana) was sent to Kew for identification. The plants had got out of control in the Botanic Gardens and its vicinity and were spreading rapidly. A number of articles were published on this subject (Anon. 1926: 305; Beckett 1927: 113; Anon. 1929: 52; Martyn 1930: 84), which described *A. ghaesembilla* as a noxious weed, menace and grave danger. It was said that it could grow anywhere, even on soil heavily impregnated with salt, that it reproduced itself, not only from millions of seeds, but also from suckers and shoots from any portion of broken root or any scattered joints however small, every such shoot growing into a robust plant, that it sends out roots 50 feet in length and to a depth of six feet. Besides the bird dispersal of the prolifically fruiting plant, children were said to be very fond of the sub-acid fruits, gathering huge quantities of them and helping to disperse the seed. Different methods of eradication such as cutting back and removal of stumps, regular cropping and even flooding of the infected area for four months were found to be inefficient. Finally, poisoning of the stumps and repeated spraying of the suckers with sodium arsenite proved to be a successful if costly method of exterminating *A. ghaesembilla* in British Guiana (Martyn 1930).

POLLINATION

Unfortunately, very little is known about the pollinators of the rather inconspicuous yellowish green flowers of *Antidesma* but they can be assumed to be entomogamous (macro fossil record in Willemstein 1987). Only two species are noted for their floral scents. *A. ghaesembilla*, with fragrant flowers, is the only species in open vegetation, predominantly savannahs. *A. bunius*, with unpleasant-smelling flowers, is probably fly-pollinated. There is no conclusive information on the other species.

The whole genus (and related *Thecacoris* and *Hieronyma*) displays an interesting mechanism of pollen presentation. The pollen sticks together and is squeezed out of the anther cells in strands (like toothpaste out of a tube). The significance of this for pollination is so far unknown.

INTERACTIONS WITH ARTHROPODS

Antidesma species are often associated with ants. The ants live in or on a variety of plant organs, but no connection with the domatia on the undersurface of the leaves could be established. Some collections possess hollow twigs with access holes (e.g., *A. petiolatum*, *Jacobs* 9490 and *Frodin et al.* 2483, both from New Guinea) which were once inhabited by ants. In some cases the ants build their nests on the abaxial leaf surfaces, apparently starting by galling the midvein. This was observed in herbarium material of *A. bunius* (trees VIII.F. 13 and 17), *A. montanum* (trees VIII.B.58a and 118, all in Bogor Botanical Garden) and *A. tomentosum* (*S* 14664 from Sarawak). In two species with foliaceous stipules, the chambers formed between the stipules in the leaf axils are inhabited by ants and their progeny (field observation in Peninsular Malaysia on *A. pachystachys* (*Hoffmann* 8) and *A. stipulare* (*Hoffmann* 13)). The glandular-fimbriate hairs at the margins of the sepals of several species (especially in *A. montanum* and relatives) might serve as feeding bodies for ants. Apart from the sepals, these marginal glands are also found on bracts and stipules. There is, however, no evidence to support this suggestion.

The galls frequently found in *Antidesma* are described under the individual species. In *A. tomentosum*, insects induce branching of the inflorescences (simple in unaffected plants). In one specimen with galled leaves, *SAN* 21270 (K), the inflorescences consist of up to 15 branches, which bear numerous bracts but do not develop buds. Very small ants were found in these galls. Several species of *Antidesma* develop domatia on the undersurface of the leaves. For more details see p. 19.

BREEDING SYSTEM

The vast majority of collections (7628 out of 7630) examined for this study are dioecious. Only two monoecious specimens were found: one of *A. excavatum* var. *excavatum* from Papua New Guinea (*UPNG (Vians & Nagari)* 4853 (K, L), and one of *A. montanum* var. *montanum* from Peninsular Malaysia (*SF (Ngadiman)* 36630, (L)). Sexual dimorphism was not observed in any of the taxa.

Staminodes occur occasionally in many genera of Phyllanthaceae. In the genus *Aporosa* (Antidesmatoideae-Scepeae), *A. hermaphrodita* Airy Shaw from New Guinea and *A. heterodoxa* Airy Shaw from the Solomon Islands have been described as obligatory hermaphrodites (Airy Shaw 1971, 1980a: 30, pl. 2, Fig. 1). In addition to those species, *A. brevicaudata* Pax & K. Hoffm. and *A. egreria* Airy Shaw have also been found to occasionally have functional bisexual flowers (Schot 2004: 39, 205). In *Antidesma*, however, the occurrence of staminodes is very rare. Two sheets of the type collection of *A. acuminatum* Wight from Calcutta Botanical Garden (*Hb. Wight* s.n. (K), Fig. 4A; cf. also Wight 1853: t. 1991 = synonym of *A. montanum* Blume), have apparently hermaphrodite flowers: a normally developed gynoecium and a whorl of stamens, the anther cells of which contain pollen. The stamens are aberrant, with unusually large, apiculate connectives and almost petaloid filaments, unlike those of the normal unisexual staminate flowers drawn in Wight's plate. Wight appears to have described *A. acuminatum* only on account of the apparently complete flowers; the specimen is otherwise typical *A. montanum*. More commonly observed than monoecy or staminodes were 1–2 additional ovaries in otherwise unremarkable individual flowers of several species.

A pistillate specimen of *A. velutinosum*, *Kerr* 11995 (K) from Thailand (Fig. 4B) possesses a small foliaceous, lanceolate organ in each flower in the position of the stamens in staminate flowers, i.e., antisepalous with their bases fully enclosed by the disc. The only other vestigial stamens or petals found in the course of this study are those in the collection of *Puasa-Angian* 10497 (K) from Sabah, representing *A. montanum*. These structures are subulate, 0.5 mm long and have no apical swellings or anthers.

CHROMOSOME NUMBERS

According to Hans (1970; 1973: 609), three African species and the two Asian species *A. ghaesembilla* and *A. diandrum* (= *A. acidum*)are diploid with n = 13. This is the prevalent base number in the "lower Phyllanthoideae" (Webster 1994a: 15). For the latter species there was also one count of n = 13 + 2B published by Mehra & Gill (1968: 574).

On the other hand, *A. acuminatum* (a synonym of *A. montanum*) and *A. bunius* have been found to be polyploid. In the former species both diploid (n = 13) and hexaploid (n = 39) races have been observed, whereas *A. bunius* has been reported to be 18-ploid (n = 117) (Hans 1970). The comparison of the two cytotypes of *A. acuminatum* showed that "the hexaploid race is found to be larger in size, have larger stipules, leaves and stomata; longer inflorescences and larger pollen grains", and the cytological evidence suggests an allopolyploid origin of the hexaploid

(Hans 1970: 325). The record of n = 117 for *A. bunius* might be the highest in Euphorbiaceae *sensu lato*. Unfortunately, the voucher specimens for the 18-ploid (*Hans* K113, K114) could not be examined. Because of the large discrepancy, Hans (1970: 325) suspected that an earlier count of n = 13 for the same species was made using incorrectly identified material. However this may be, polyploidy certainly accounts partly for the high infraspecific morphological variation in some species of *Antidesma*.

All records of polyploidy are based on collections from Northern India, where the genus reaches the north-western limit of its distribution, and both species concerned are among the most widespread, variable and ecologically tolerant species. It would be most interesting to compare these with chromosome counts of species from the centre of diversity in Malesia.

Hybridisation might be another factor responsible for the great amount of variation in the genus (see also Chakrabarty & Gangopadhyay 2000: 4), however, no studies on the reproductive biology of *Antidesma* seem to have been carried out so far. In *A. digitaliforme* and *A. japonicum* (both again from the northern part of the generic distribution area), many staminate flowers were found to be aberrant, with a reduced number of stamens, irregularly fused filaments and deformed anthers. This might be caused by genetic problems but no evidence exists to support this.

USES

The principal use of *Antidesma* for humans is as a fruit tree. The most commonly cultivated species is *A. bunius*. *A. excavatum* seems to be cultivated as a minor fruit tree in and around the Solomon Islands. Other species of which the fruits are eaten locally include *A. acidum, A. edule, A. ghaesembilla, A. montanum, A. pleuricum, A. sootepense, A. tetrandrum, A. tomentosum* and *A. velutinosum*.

The leaves and shoots of *A. bunius* and *A. ghaesembilla* are eaten, raw or steamed, as *lalab*, or mixed with rice or other food, to give it a sour taste similar to tamarind (Ochse 1931: 262 – 265; Gruèzo 1991: 78). In India and Thailand, the young leaves of *A. acidum* are used in curries and as a vegetable.

The wood is used for temporary construction, poles, posts, fence posts and small objects like walking sticks and tool handles (Sosef *et al.* 1998: 75). *Antidesma* yields a medium-weight to heavy hardwood which is also valued as fuel. Species described as having very hard and fine-grained wood include *A. cuspidatum, A. excavatum, A. neurocarpum, A. petiolatum, A. stipulare, A. tetrandrum, A. tomentosum* and *A. venenosum* (herbarium labels and M. Coode, pers. comm.). The commercial exploitation of the wood is limited by the modest size of most trees (usually 5 – 35 cm in diameter). The taller species, however, can grow up to 30 m and reach a diameter of up to 1 m.

Secondary uses include forage and dye. *A. bunius* is highly valued as feed for village ruminants (Lowry *et al.* 1992: 15; t'Mannetje & Jones 1992: 246). The fruits of *A. bunius, A. excavatum* and *A. stipulare* provide a good purple dye (Lemmens & Wulijarni-Soetjipto 1991: 144, and herbarium labels).

According to De Padua *et al.* (1999: 24), *A. bunius* can be found depicted in reliefs on the walls of ancient temples in Java, such as those of Borobudur and

Prambanan, because of its use in traditional medicine. The name *Antidesma* refers to the efficacy of the plant against snake bites (see p. 2). A range of medicinal uses are mentioned in Burkill (1935: 184 – 187). The medicinal properties of *Antidesma* are likely to be linked to the organic compounds discussed below. Specific uses as folk remedies reported on herbarium labels are described under each taxon.

PHYTOCHEMISTRY

A number of organic compounds have been reported in *Antidesma*. Triterpenoids (Hegnauer 1966: 118; 1989: 443; Rivzi *et al.* 1980) which have shown diuretic activity in rats (Rivzi *et al.* 1980), benzopyranones and ferulic acid derivatives (Buske *et al.* 1997) and also cyclopeptide alkaloids (Arbain & Taylor 1993) have been found. Recently, a novel-type quinoline alkaloid has been isolated (Buske *et al.* 1999; Bringmann *et al.* 2000a). Its synthesis follows an unprecedented biochemical pathway derived directly from glycine (Bringmann *et al.* 2000b). This substance was named antidesmone and has shown strong fungitoxic activity. It is very similar to hyeronimone extracted from the genus *Hieronyma* in the same subtribe (Tinto *et al.* 1991). Initial research focused on *A. membranaceum* Müll. Arg. from Africa, but antidesmone has subsequently been found in all examined species of *Antidesma*, *Hieronyma* and *Thecacoris*, whereas *Uapaca* (Uapaceae), *Aporosa* and *Maesobotrya* (both Scepeae) tested negative (Buske *et al.* 2002). This is a further supposed synapomorphy of these three genera; further genera of tribe Antidesmateae remain to be investigated.

THREAT AND CONSERVATION

As a genus, *Antidesma* is commonly found in a wide variety of habitats. Some species, above all *A. montanum* and *A. ghaesembilla*, are both widespread and common, and thrive in habitats disturbed by human activity. Other species are more local and fruit less abundantly. Continuing habitat destruction in the region gives reason for concern over the conservation of species of more restricted distribution, lower reproductive rate and more specialised ecology.

IUCN Red List Categories (version 3.1, IUCN 2001) are here assigned to all taxa of *Antidesma* in Malesia and Thailand. As the present revision is based mainly on the study of herbarium material, the data available to analyse the conservation status are limited to criteria B1 (extent of occurrence) and D (estimated population size). The extent of occurrence (area contained within the shortest continuous imaginary boundary which can be drawn to encompass all the known, inferred or projected sites of present occurrence of a taxon) was computed using ArcView GIS 3.2a.

The number of herbarium specimens studied in the course of an exhaustive taxonomic revision can be used as a guide for estimating population size. Further considerations for the application of criterion D included habitat information from herbarium label data and personal communications with colleagues as well as from fieldwork, the collection date of the majority of specimens, the average collection density in the region, and the general threat to the habitat of the species in its distribution area. Bird dispersal is seen as a factor that reduces the threat to a certain extent.

Of the 69 taxa treated in this revision, 21 were assigned the category Least Concern (LC), 31 were classified as Near Threatened (NT), 6 as Vulnerable (VU) and two as Endangered (EN) and as Critically Endangered (CR), respectively. The remaining 7 taxa are Data Deficient (DD). For the definition of the individual categories, see IUCN 2001.

VERNACULAR NAMES

In Thailand, the name for all *Antidesma* species is "mao". In Java and surrounding islands, *A. bunius* is grown as a fruit tree and is called "buni", "wuni", "huni" or variants thereof. Similarly, in the Philippines, the species is known as "bignay". The vernacular names of other species are often related to these names of *A. bunius.* In Malaysia, Borneo and Sumatra, where *A. bunius* is almost unknown, the commonest names for different *Antidesma* species are: "benai, beranai, berenai, bernai, bernei, borni, brenai, brunei, bunai, burnai", also: "ampenai, empanai, emparanai, empenai, emponai, mempena or mempenai"; also: "benai" in Mindanao. These names maybe derived from the country Brunei, or again somehow from "buni". In Kalimantan, the genus is called "uhai", a name used for several genera of small Euphorbiaceae *sensu lato* (P. Keβler, L, pers. comm.). Some species with small, abundant fruits are called "ki seueur" in Sundanese (seueur = many; relating to the many fruits).

TAXONOMICALLY SIGNIFICANT CHARACTERS

The lack of discrete characters in a genus of about 100 species presents great difficulties for both classification and identification. Most characters are found "mostly", "often" or "usually". The combination of a few key characters will, however, narrow down the possibilities considerably. These are, at a glance, petiole length, stipule size, position, length and branching of the inflorescences and, in pistillate material, the position of the style and the shape and indumentum of the fruits. A handlens is usually necessary to observe the division of the calyx and, in staminate material, the shape and indumentum of the disc.

Stipules

The stipules of *Antidesma* are variable in shape and size (Fig. 1). Especially conspicuous are the foliaceous stipules of some species which are highly diagnostic when present. When absent, however, they cannot be relied upon for identification as even species such as *A. stipulare* (where stipules measure up to 6 × 3 cm) can also have small, subulate stipules. Laciniate stipules occur only in the African species *A. laciniatum.*

The stipules are often caducous, and the bracts of inflorescences or generative buds can easily be mistaken for stipules. It is important to ascertain that there are no stipule scars before assuming that any structures are stipules.

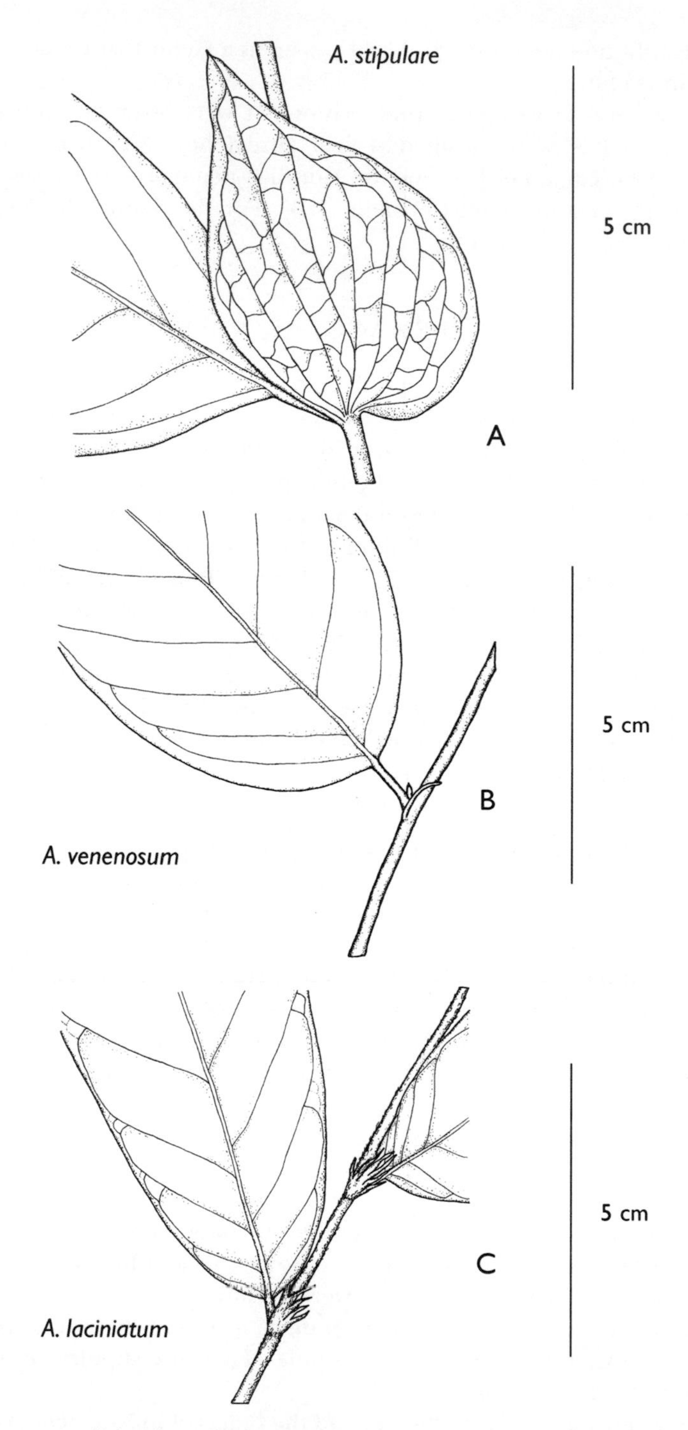

FIG. 1. Key characters of *Antidesma* I: stipules. **A** foliaceous stipule of *A. stipulare*. **B** subulate stipule of *A. venenosum*. **C** laciniate stipule of the African *A. laciniatum*. All other species of the genus have entire stipules.

Domatia

The term is used here in its descriptive sense as defined by Jacobs (1966: 275) as "spatial differentiations at the under surface of leaves, bound to nerves, and virtually always in or near their axils. They consist of a pit, a pocket, a hair-tuft, or a dome with an opening at its top, and these elements singly or variously combined. They are produced with limited regularity by ligneous plants belonging to various taxa of dicotyledons, in the rank of genus and lower, on maturing adult-stage leaves, and without any apparent cause or function." The term does not imply or deny involvement of these structures in plant-animal interactions. It has in fact been argued more recently (O'Dowd & Willson 1989) that leaf domatia are indeed mite-shelters in which mites reproduce and develop.

Of the 56 *Antidesma* species treated here, eleven have been found to develop domatia. The domatia are shaped like pockets or pits, with or without hairtufts (Fig. 2B, 21B), and may be present in some vein axils while they are absent from others. If present, they are a useful diagnostic character. Leaf domatia in *A. excavatum* and *A. celebicum* from Sulawesi are glabrous, whereas those in *A. excavatum* and relatives from New Guinea usually have hairy margins. Domatia were found in very young leaves (*A. celebicum*, *Burley et al.* 3558 (K)) as well as in a living specimen of *A. celebicum* (*Rastini* 173) at the Botanical Garden Bogor in Java where the species does not naturally occur. No arthropods were found in association with these specimens.

In the type specimen of *A. oligonervium* Lauterb. (*Schlechter* 14311 (K & other herbaria); synon. nov. of *A. excavatum* var. *excavatum*) the pocket domatia are particularly deep and noticeable even on the adaxial leaf surfaces. The material is rather poor, but remains of small arthropods are present in these domatia. This is most pronounced at the basal pair of nerves, thus deforming the leaf base, and may have been the reason for describing *A. oligonervium* as a new species with steeply ascending lowermost nerves. The Hawaiian species *A. pulvinatum* clearly has inhabited acaridomatia (Pemberton & Turner 1989; pers. obs., e.g. *Degener 8137* [K]).

Jacobs (1966: 275) and Brouwer (1983: 9) noted that domatia occur predominantly in humid habitats. In *Antidesma*, however, the occurrence of domatia is restricted to three South-East Asian species which are often found in deciduous forest (*A. acidum*, *A. sootepense* and *A. velutinum*), one Philippine endemic (*A. digitaliforme*) and otherwise only in Eastern or South-Eastern Malesian elements (*A. baccatum*, *A. celebicum*, *A. chalaranthum*, *A. excavatum*, *A. myriocarpum*, *A. rhynchophyllum* and *A. spatulifolium*). Domatia are also found in *A. messianianum* Guillaumin from New Caledonia (Schmid & McPherson 1991: 10, pl. 2.2). No domatia are known from the more humid part of the distribution area, e.g., the Malay Peninsula south of Kedah, Sumatra or Borneo.

Inflorescences

Inflorescences in *Antidesma* are axillary (Fig. 19), but the subtending leaves may either be reduced, or develop at a later stage, which makes the inflorescences appear terminal (Fig. 23). In other cases the inflorescences develop only after abscission of the subtending leaves, thus becoming cauline, or borne on the

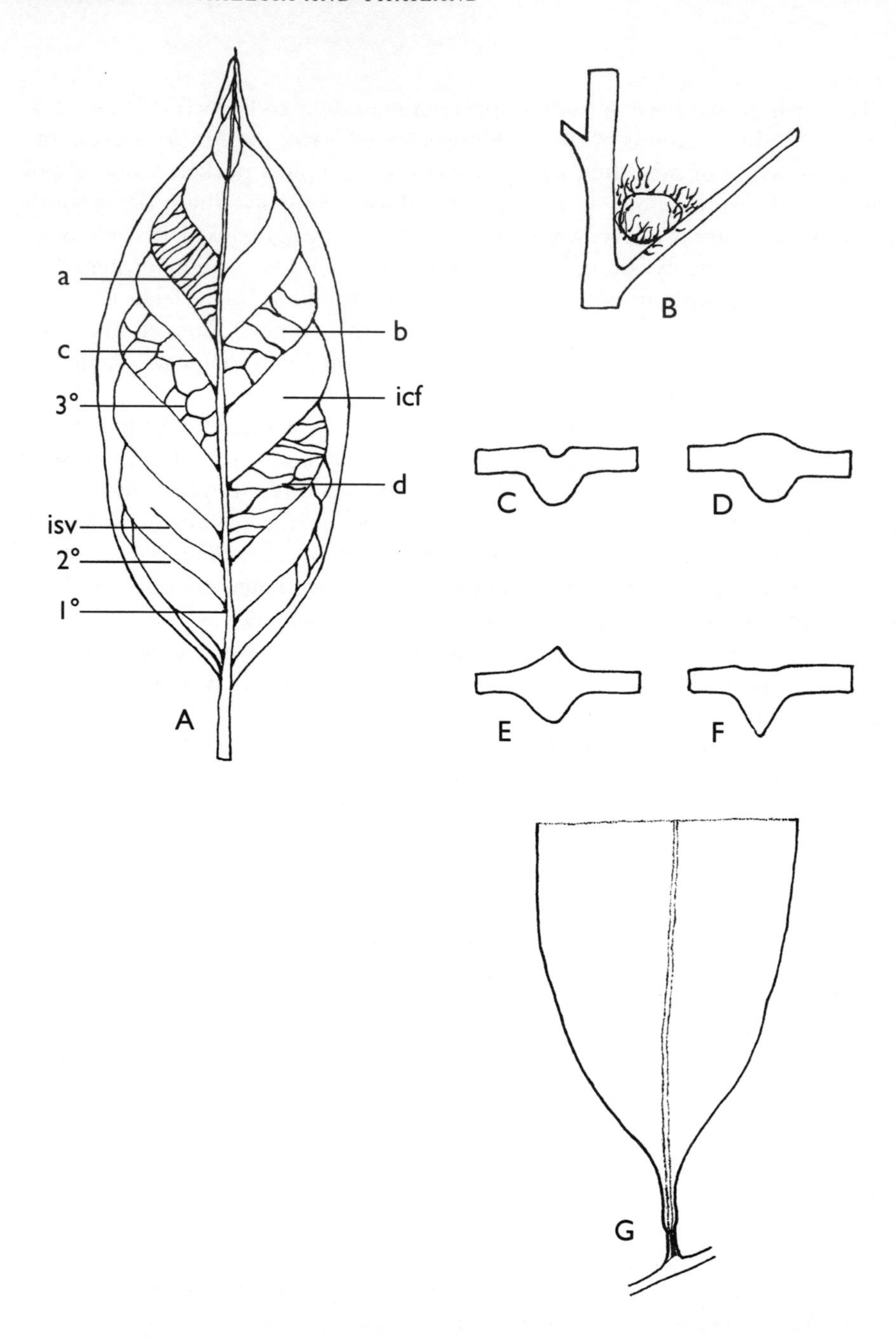

FIG. 2. Key characters of *Antidesma* II: leaves. **A** leaf venation: 1° midvein (primary vein), 2° secondary vein, 3° tertiary vein, isv intersecondary vein, icf intercostal field; a–d course of tertiary veins: **a** percurrent, **b** weakly percurrent, **c** reticulate, **a** close together, **b** widely spaced, **a** – **c** oblique, **d** perpendicular. **B** domatium in axil between midvein and secondary vein. **C** – **F** midvein in cross section: **C** impressed adaxially, **D** slightly raised adaxially, **E** sharply raised adaxially, **F** flat adaxially, **C** – **E** rounded abaxially, **F** sharply keeled abaxially. **G** strongly decurrent leaf base ("pseudo-petiole" of Airy Shaw).

branches (Fig. 17). Although transitional states are common, the position of the inflorescence is a useful diagnostic character for many species.

Calyx

The degree of fusion of the sepals, ranging from free (Fig. 8, 9, 14) to highly connate (Fig. 7, 16, 19, 20, 23), is one of the most important diagnostic characters in the genus. It is unfortunate that this can only be observed with a handlens or under a dissecting microscope, and often requires flowers from herbarium specimens to be softened first. The margin of the individual sepals is often scarcely visible in dry material. This is most probably also why many earlier descriptions describe the sepals as more highly connate than is the case here.

Disc

The staminate disc of *Antidesma* exhibits great morphological plasticity. The three principal types are shown in Fig. 3A – C, but transitions and variations were observed (e.g., Fig. 12H, 19D, 22D). More rarely, the disc lobes seem to be fused at least partly intrastaminally, and extra- or intrastaminal fusion can occur to different degrees in the same flower. The disc of pistillate flowers is uniformly annular.

The presence and distribution of hairs on the disc is another valuable diagnostic character. The three possible constellations are shown in Fig. 3D – F. Fairly long hairs from the inner base of the calyx (Fig. 3D) often obscure an otherwise glabrous disc, and can lead to misidentification. In some species, such as *A. leucopodum*, the disc indumentum is variable and of no help for species identification. The disc indumentum is usually the same in both staminate and pistillate flowers.

Peculiar apical impressions can be observed in taxa with a lobed staminate disc. This is particularly conspicuous in *A. curranii, A. digitaliforme, A. forbesii, A. tetrandrum* and *A. venenosum* (Fig. 23C, H). These impressions result from the spatial conditions in the bud, where the anthers rest on top of the disc lobes. At anthesis, each disc lobe bears an imprint of the right cell of the anther to its left and the left cell of the anther to its right.

Androecium

The number of stamens varies within the same inflorescence, e.g., three to six in *A. tomentosum* and four to eight in *A. velutinosum*. It is therefore of limited use for identification. All other androecial characters are generic and do not assist in distinguishing species.

The enlarged, U-shaped connective is a distinct feature of *Antidesma, Hieronyma* and *Thecacoris* but rare in the rest of Phyllanthaceae. Similar anthers are known from *Claoxylon* and *Lobanilia* in Euphorbiaceae *sensu stricto*, but there the "U" is formed by the elongated anther cells, while the connective is not enlarged. They are also called "rabbit's ears" and fit the name rather well. The anthers of *Antidesma* and its relatives, on the other hand, are much more like Mickey Mouse's ears. Another genus with unusual anther shape is *Acalypha* L., where the anthers are vermiform (often in long spirals) and pendulous. Anthers more similar to

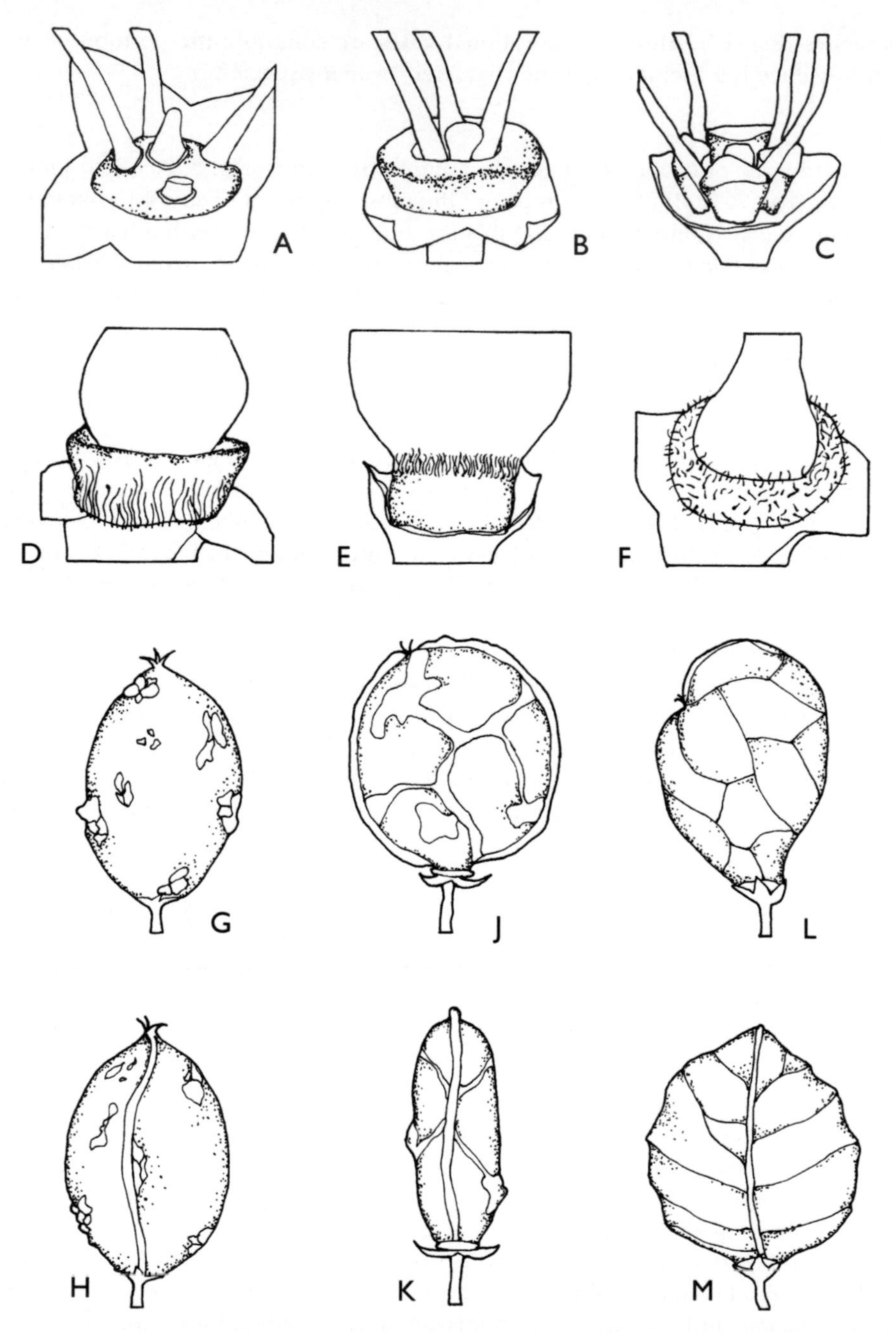

FIG. 3. Key characters of *Antidesma* III: flowers and fruits. **A** – **C** shape of staminate disc: **A** cushion-shaped, enclosing the bases of the filaments and pistillode; **B** extrastaminal-annular; **C** free, alternistaminal lobes. **D** – **F** disc indumentum: **D** glabrous (but hairs at base of calyx); **E** hairy at margin; **F** hairy all over. **G** – **L** fruit shape and position of style: **G** – **H** terete, style terminal; **J** – **K** laterally compressed, style subterminal; **L** – **M** dorsiventrally compressed, style lateral (**G**, **J**, **L** lateral view; **H**, **K**, **M** dorsal view).

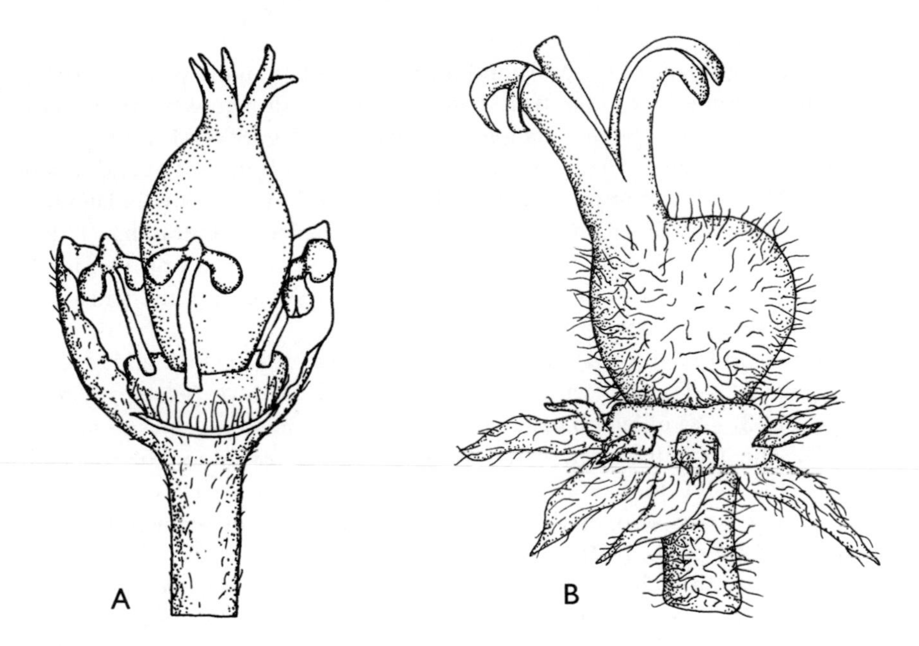

FIG. 4. Aberrant flowers in *Antidesma*. **A** Type of *A. acuminatum* (= *A. montanum*) with apparently hermaphrodite flowers. From *Hb. Wight* s.n. (K). **B** Pistillate flower of *A. velutinosum* with antisepalous "ligulae". From *Kerr* 11995 (K).

those of *Antidesma* are also found in *Flacourtia rukam* Zoll. & Moritzi (Salicaceae *sensu lato*) and related taxa (S. Zmarzty, K, pers. comm.)

The anther sacs of *Antidesma* are pendant and extrorse in bud, so that the connectives are the first parts revealed at anthesis, and the anther sacs are the last parts to remain covered. The peculiar shape of the anthers could be interpreted as having a protective function for the tender young anther sacs, as the perianth is much reduced. After anthesis, the anther sacs are raised, and point towards the pollinators. The pollen is presented from the anther cells as a compact filiform mass, as if it is being squeezed out of a tube of toothpaste. These "pollen strands" can still be observed on herbarium sheets.

Pollen

Pollen of eleven species of *Antidesma* from the whole range of geographic distribution was examined by Selling (1947: 199), Erdtman (1952: 167), Punt (1962: 22), Köhler (1965: 36) and Muller *et al.* (1989: 60, 72 – 74) and found to be uniformly 3-colporate and perprolate to prolate with long, narrow colpi and a pertectate to semitectate exine. These studies also showed that pollen of *Hieronyma* and *Thecacoris* is extremely similar to that of *Antidesma*. From this we can be reasonably certain that *Antidesma* pollen does not yield characters of taxonomic or phylogenetic value at the specific or generic level.

Ovary and fruit

The general shape of the ovary and fruit, the position of the style relative to the calyx, the symmetry of the fruit base and the indumentum are important diagnostic characters. Some examples are shown in Fig. 3G – M. Both the ovary and fruit can be terete (fruit roughly of the same width in ventral and in lateral view, Fig. 3G – H), or compressed either laterally (fruit much wider in lateral than in ventral view, Fig. 3J – K) or dorsiventrally (fruit wider in ventral than in lateral view, Fig. 3L – M). Like all other characters in the genus, it is only consistent in part of the species. In *A. neurocarpum* and a few other species, the fruits can be terete, or laterally or dorsiventrally compressed.

The position of the style can be terminal (calyx at 6 o'clock and style at 12 o'clock, Fig. 3G – H), subterminal (calyx at 6 o'clock and style between 1:30 and 2 o'clock, Fig. 3J – K) or lateral (calyx at 6 o'clock and style at 2 o'clock or more, Fig. 3L – M). The fruit base can either be symmetrical (extension of the pedicel dividing the fruit in lateral view into roughly equal parts, Fig. 3G, J) or asymmetrical (extension of the pedicel dividing the fruit in lateral view into unequal parts, Fig. 3L, 10D).

The ovary and fruit are sometimes covered with white pustules (e.g., Fig. 3G – K, 13E). These are particularly conspicuous in *A. montanum*, but are also found in many other species. They are not an artefact of drying, as the collectors reported fruits with "pigmentation in reticulate pattern" in a specimen with pustulate dry fruits (*McDonald & Afriastini* 3305 from Java), whereas a live specimen in Bogor Botanical Garden (*Hoffmann* 14) with smooth fruits and without pigmentation pattern showed no white pustules in the dry state. White pustules can also be present on the calyx (Fig. 14D) and on both leaf surfaces. They do not seem to be of a crystalline nature and become mucilaginous in water but do not dissolve.

TAXONOMIC TREATMENT

Antidesma *Burm. ex L.*

Sp. Pl.: 1027 (1753); Gen. Pl., ed. 5: 451 (1754); Burman, Thes. Zeylan.: 22, t. 10 (1736), prelinn.; L., Fl. Ceyl.: 169 (1747), prelinn.; Lam., Encycl. 1: 206 (1783); Tul., Ann. Sci. Nat. Bot., Sér. 3: 182 (1851); Baill., Étude Euphorb.: 601 (1858); Müll. Arg. in DC., Prodr. 15(2): 247 (1866); Benth., Fl. Austral. 6: 84 (1873); Benth. & Hook. f., Gen. pl. 3: 284 (1880); Hook. f., Fl. Brit. India 5: 354 (1887); Hutch. in Dyer, Fl. Trop. Afr. 6(1): 642 (1912); Pax & K. Hoffm. in Engl., Pflanzenr. 81: 107 (1922); Ridl., Fl. Malay Penins. 3: 229 (1924); Gagnep. in Lecomte, Fl. Indo-Chine 5(5/6): 501 (1926); Leandri, Notul. Syst. (Paris) 6: 23 (1937), Leandri in Humbert, Fl. Madag. 111(1): 14 (1958); Hutch. & Dalziel, Fl. Trop. W. Afr., ed. 2, 1(2): 374 (1958); Backer & Bakh. f., Fl. Java 1: 457 (1964); Chun & Chang, Fl. Hainan. 2: 118 (1965); Airy Shaw, Kew Bull. 26. 351 (1972); Whitmore, Tree Fl. Malaya 2: 54 (1973); Airy Shaw, Kew Bull., Addit. Ser. 4: 207 (1975); Kew Bull. 35: 692 (1980), Kew Bull., Addit. Ser. 8: 208 (1980), Kew Bull. 36: 358 (1981); A. C. Sm., Fl. Vitiensis Nova 2: 445 (1981); Airy Shaw, Kew Bull. 37: 5 (1982); Coode in Bosser *et al.*, Fl. Mascareignes 160: 106 (1982); Airy Shaw, Euphorb. Philipp.: 4 (1983); Mandal & Panigrahi, J. Econ. Taxon. Bot. 4: 255 (1983); Grierson & Long, Fl. Bhutan 1(3): 786 (1987); Radcl.-Sm. in Polhill, Fl. Trop. East Africa: Euphorb. 2: 572 (1988); Wagner *et al.*, Man. Fl. Pl. Hawaii 1, rev. ed.: 598 (1999); Schmid & McPherson, Fl. Nouv.-Caléd. 17: 9 (1991); Li Ping Tao, Fl. Reip. Pop. Sin. 44(1): 52 (1994); G. L. Webster, Ann. Missouri Bot. Gard. 81: 52 (1994); Léonard in Fl. Afr. Centr., Euphorb. 2: 16 (1995); Radcl.-Sm. in Pope, Fl. Zambesiaca 9(4): 105 (1996); Philcox in Dassan. & Clayton, Rev. Handb. Fl. Ceyl. 11: 275 (1997); Khan & Khan, Bangladesh J. Pl. Taxon. 5(1): 85 (1998); Petra Hoffm., Kew Bull. 54: 347 (1999), Kew Bull. 54: 877 (1999), Thai For. Bull. (Bot.) 28: 139 (2000); Gardner *et al.*, Forest Trees of Northern Thailand: 299 (2000); Govaerts, Frodin & Radcl.-Sm., World Checklist Bibliogr. Euphorb. 1: 178 (2000); Chakrabarty & Gangopadhyay, J. Econ. Taxon. Bot. 24: 1 (2000), Petra Hoffm. in Chayamarit & van Welzen, Fl. Thailand 8(1): 51(2005).

Bestram Adans., Fam. Pl. 2: 354 (1763).

Stilago L., Mant. Pl.: 16 (1767).

Rhytis Lour., Fl. Cochinchin.: 660 (1790).

Minutalia Fenzl, Flora 27: 312 (1844), nom. nud.

Antidesma sect. *Euantidesma* Müll. Arg., Flora 47: 519 (1864), nom. inval.

Type: *A. alexiteria* L. (South India and Sri Lanka)

Dioecious plants with a simple indumentum (in *A. pachystachys* and *A. pahangense* indumentum partly stellate). Shrubs and trees, up to 30 m, clear bole up to 17 m, diameter up to 1 m, bole sometimes fluted or with buttresses, sometimes with more than one stem, sometimes reported to be climbers. *Bark* usually brown or grey, smooth, often longitudinally fissured, sometimes roughened, flaky or pustular, without exudate (except in *A. excavatum* and *A. tetrandrum*). *Wood* hard to very hard, rarely reported to be soft. *Twigs* sometimes hollow and inhabited by ants. *Petioles* terete or channelled adaxially, sometimes basally and distally pulvinate and geniculate, up to 45 mm long (*A. petiolatum*). *Stipules* present, caducous or persistent, sometimes foliaceous, up to 60 × 30 mm, entire (only the African *A. laciniatum* with laciniate stipules). *Leaves* evergreen or deciduous, alternate, simple, eglandular except in *A. vaccinioides*; blades symmetrical, entire, pinnately veined, sometimes decurrent, 1.5 – 60 × 0.4 – 30 cm, membranaceous to coriaceous, discolorous, domatia in the axils between the midvein and secondary veins present or absent. *Inflorescences* raceme-like, often branched, sometimes fasciculate, 0.5 – 35 cm long, axillary or cauline, erect to pendulous, *infructescences* usually pendulous. *Bracts* 1 per flower. *Pedicels* short to absent, articulated only in *A. vaccinioides*. *Flowers* usually light yellowish green, often turning partly or completely red when mature. *Sepals* 3 – 8, imbricate, fused to varying degrees. *Petals* absent. *Disc* in staminate flowers cushion-shaped (enclosing the bases of the stamens and pistillode, Fig. 3A), extrastaminal-annular (Fig. 3B), or consisting of free alternistaminal lobes (Fig. 3C), in pistillate flowers annular (Fig. 3D – F). *Stamens* 2 – 13, usually antisepalous, filaments free, anthers extrorse in bud, versatile, connective elongated, U-shaped, thecae resembling swollen ends of the U, raised at anthesis, longitudinally dehiscing. *Pistillode* present or absent. *Ovary* 1-locular, 1-carpellate (2-locular and 2-carpellate only as an exception). *Stigmas* 2 – 16, acute. *Ovules* 2 per locule, anatropous, with individual obturators (tissue bridging the gap between pollen transmitting tissue and micropyle). Immature *fruits* green, maturing unevenly depending on position in the infructescence and exposure to the sun, through white, yellow, orange, pink, red and purple to black, sometimes glaucous or mottled with grey, globose to ellipsoid or flattened, smooth, shiny, fleshy; mesocarp soft, juicy, edible, sour, sour-sweet, sweet or bitter-sweet. *Fruits* drupaceous, ellipsoid to lenticular, often laterally compressed, 1.5 – 2.5 × 1.5 – 2 mm (*A. venenosum*) to 22 × 13 mm (*A. gillespieanum* A. C. Sm. in Fiji), base symmetrical to asymmetrical, style terminal to lateral. *Endocarp* lignified, foveolate, the sculpture usually obvious through the dried mesocarp. *Seeds* ecarunculate, 1 per fruit (rarely two, then fruits twice as big). *Endosperm* in mature seeds present; cotelydons thin, flat, several times wider and longer than the radicle.

Good photographs of Antidesma can be found in Gardner *et al.* 2000: 299 – 301.

KEY TO THE SPECIES IN WEST MALESIA AND THAILAND (EXCLUDING NEW GUINEA)

This key includes all taxa found in Thailand and Malesia, excluding New Guinea and the Solomon Islands. Ideally, specimens should either be in fruit or bear mature staminate flowers. Pistillate flowering or sterile material may be difficult and sometimes impossible to determine with this dichotomous key. Some structures can only be seen in sufficient detail under a dissecting microscope. Dried flowers may have to be softened. All measurements are given in the dry state; fruit length includes calyx and stigmas. Terminology of leaf venation follows

Hickey (1979, see also Fig. 2A). Some descriptive terms used for flowers and fruits are explained in Fig. 3.

SUM = Sumatra, MLY = Malay Peninsula, BOR = Borneo, JAW = Java, PHI = Philippines, SUL = Sulawesi, LSI = Lesser Sunda Islands, MOL = Moluccas, NWG = New Guinea, THA = Thailand. For distribution data outside this area see Table 1 or species descriptions.

Species in parentheses key out in this place only as an exception.

Key for staminate material

1. Leaf apex rounded or retuse, more rarely obtuse · · · · · · · · · · · · · · 2
   Leaf apex acute, acuminate or caudate · 4
2. Leaves base cordate to rounded, very rarely acute. Stamens 4 – 6. Sepals free, pubescent outside. SUM, MLY, BOR, JAW, PHI, SUL, LSI, MOL, NWG, THA · **21. A. ghaesembilla**
   Leaf base acute or obtuse, mostly cuneate. Stamens 2 – 4. Sepals free or fused, glabrous to pilose outside · 3
3. Sepals fused for ⅔ of their length. Leaves obovate to elliptic-oblong, (2 –)5 – 10(– 18) × (1 –)2.5 – 4(– 8) cm. Petioles 1 – 2 mm wide. Inflorescences 5 – 14 cm long. Flowers on pedicels 1 – 1.5 mm long. Pistillode absent. Disc cushion-shaped (Fig. 3A). Stamens 2(– 3). JAW, THA · **1. A. acidum**
   Sepals free or nearly so. Leaves spathulate, (1.5 –)2 – 5(– 7) × (0.7 –)1 – 2.5(– 3.5) cm. Petioles less than 1 mm wide. Inflorescences c. 3 cm long. Flowers sessile. Pistillode present. Disc consisting of 2 – 4 free lobes (Fig. 3C). Stamens 2 – 4. MOL, NWG · · · **48. A. spatulifolium**
4(1). Midvein in dry material sharply raised adaxially (Fig. 2E), distinctly perceptible to the touch · 5
   Midvein in dry material impressed to slightly raised adaxially (Fig. 2D), not distinctly perceptible to the touch · 7
5. Leaves 6 – 7.5 cm long. THA · · · · · · · **24b. A. japonicum** var. **robustius**
   Leaves 13 – 45 cm long · 6
6. Young twigs, petioles and leaves hairy, glabrescent. Stipules usually foliaceous, more than 2 mm wide, persistent. Leaves 15 – 45 × 5 – 18 cm. MLY · **38. A. pachystachys**
   Young twigs, petioles and leaves glabrous. Stipules linear, up to 1.5 mm wide, caducous. Leaves 13 – 27 × 3.5 – 8 cm. MLY, THA · **27. A. laurifolium**
7(4). Inflorescences at least partly cauline · 8
   Inflorescences axillary (sometimes aggregated at the end of the branch) · 15
8. Petioles (at least of some leaves) more than 10 mm long, no wider than 2 mm in the middle, basally and distally usually slightly pulvinate and sometimes geniculate · 9
   Petioles shorter than 10 mm (if longer, then more than 2 mm wide), basally and distally not pulvinate or geniculate · · · · · · · · · · · · · · 10

9. Disc hairy all over. All leaf veins, including the very fine veinlets, clearly visible abaxially in dry material. Leaves drying yellowish brown to yellowish green. Pubescence of inflorescence whitish. Inflorescences simple, fascicled. SUM, MLY, BOR, PHI, THA · · **44. A. puncticulatum**

Disc glabrous (sometimes with hairs originating from the base of the disc). Only the coarse leaf veins visible abaxially in dry material. Leaves drying grey to reddish brown. Pubescence of inflorescence ferrugineous. Inflorescences simple or branched. SUM, MLY, BOR · **11. A. coriaceum**

10(8). Tertiary veins close together, more than 10 between every two secondary veins (Fig. 2Aa). Petioles 1.5 – 4 mm wide. Inflorescences (4 –)15 – 24 cm long. Leaves (7 –)13 – 25(– 50) × (2.5 –)4 – 8(– 15) cm. Stipules 8 – 20 × 1 – 4 mm · 11

Tertiary veins widely spaced, less than 9 between every two secondary veins (Fig. 2Ab). Petioles up to 2 mm wide. Inflorescences 3 – 9 cm long. Leaves (5 –)8 – 14(– 19) × (2 –)3 – 5(– 9.5) cm. Stipules 2 – 10 × 0.5 – 1.5 mm · 12

11. Indumentum of leaves abaxially sparse, indumentum of inflorescence axes and calyx moderately dense. SUM, MLY, BOR, PHI, THA · **29. A. leucopodum**

Indumentum of leaves abaxially usually dense, indumentum of inflorescence axes and calyx extremely dense. BOR · **43. A. polystylum**

12(10). Abaxial leaf surfaces hairy all over. Pistillode hairy · · · · · · · · · · · · · 13

Abaxial leaf surfaces glabrous. Pistillode glabrous or absent · · · · · · 14

13. Indumentum ferrugineous. Inflorescences consisting of 4 – 11 branches. Pedicel absent. Flowers 1 – 2 × 0.7 – 1.5 mm. Stamens exserted 0.5 – 1.5 mm from the calyx. Disc extrastaminal-annular (Fig. 3B); but may sometimes appear cushion-shaped), glabrous (sometimes with hairs originating from the base of the disc; Fig. 3D). MLY, THA · **55. A. velutinum**

Indumentum whitish to ochraceous. Inflorescences simple or once-branched. Pedicels 0.2 – 1 mm long. Flowers 2 – 3 × c. 2 mm. Stamens exserted 1.5 – 2 mm from the calyx. Disc cushion-shaped, enclosing the bases of the filaments and the pistillode (Fig. 3A), glabrous at the sides, hairy on top (Fig. 3E). LSI, MOL, SUL · · · · · · · **17. A. elbertii**

14(12). Sepals free or nearly so. Disc hairy all over. Pistillode glabrous. SUM, MLY, THA · **22. A. helferi**

Sepals fused for $^2/_3$ – $^3/_4$ of their length. Disc glabrous (sometimes with hairs originating from the base of the disc). Pistillode absent. PHI · **30. A. macgregorii**

15(7). Stipules foliaceous, more than 4 mm wide · · · · · · · · · · · · · · · · · · · 16

Stipules not foliaceous, up to 4 mm wide · · · · · · · · · · · · · · · · · · · 24

16. Petioles (at least of some leaves) more than 10 mm long, no more than 2 mm wide in the middle, basally and apically slightly pulvinate and geniculate · 17

Petioles up to 10 mm long (if longer, more than 2 mm wide), not pulvinate or geniculate · 18

17. Stipules spathulate, cordate or kidney-shaped, apex rounded to acute, base symmetrical. Leaves (6 –)10 – 14(– 22) cm long, up to 3 times as long as wide, chartaceous to membranaceous. Bracts c. 0.5 mm long. Sepals fused for $^2/_3$ of their length. SUM, JAW, LSI · · · · · · **51. A. tetrandrum**

Stipules ovate, apex caudate, base asymmetrical. Leaves 13 – 21 cm long, 3 – 4 times as long as wide, subcoriaceous to chartaceous. Bracts 0.7 – 1 mm long. Sepals fused only at the base. MLY · · · **26. A. kunstleri**

18(16). Indumentum, especially of young parts and inflorescences, ferrugineous. Leaves drying reddish brown · · · · · · · · · · · · · · · · 19

Indumentum whitish or absent. Leaves drying olive-green to reddish brown · 22

19. Inflorescences 15 – 20 cm long. Flowers 2 – 3 × 2 – 3 mm. Stamens 1.5 – 3 mm long, exserted 1 – 2.5 mm from the calyx. Sepals fused for $^2/_3$ – $^3/_4$ of their length. Disc ferrugineous-pubescent. Mostly rheophytic. BOR, PHI (Palawan), SUL · · · · · · **46a. A. riparium** subsp. **riparium**

Inflorescences 2 – 14 cm long. Flowers 1 – 1.5 × c. 1.5 mm. Stamens 0.8 – 1 mm long, exserted 0.5(– 1) mm from the calyx. Sepals free. Disc glabrous to pubescent · 20

20. Stipules subulate to falcate, rarely lanceolate with an asymmetrical base, strongly parallel-veined, coriaceous, not more than 10 mm wide, at least twice as long as wide. Petioles 2 – 5 mm wide. Leaves (8 –)15 – 30(– 60) cm long. Tertiary veins percurrent, close together (Fig. 2Aa). Inflorescences 6 – 14 cm long, always simple; flowers set densely, touching each other. Disc glabrous or hairy. Pistillode present, 0.2 – 0.4 mm long, pubescent. SUM, MLY, BOR, JAW, PHI, SUL, LSI, MOL · **52. A. tomentosum**

Stipules cordate to lanceolate, base more or less symmetrical, reticulately veined, chartaceous, up to 18 mm wide, hardly longer to several times as long as wide. Petioles c. 1 mm wide. Leaves (5 –)6 – 14(– 20) cm long. Tertiary veins weakly percurrent (Fig. 2Ab) to reticulate (Fig. 2Ac). Inflorescences 2 – 6 cm long, simple or branched, very slender; flowers spaced more widely, not touching each other. Disc glabrous. Pistillode absent or up to 0.1 mm long, glabrous to sparsely pilose · · · · · · · · · 21

21. Leaves more than 2 cm wide, less than 5 times as long as wide. SUM, MLY, BOR, THA · · · · · · · · **36a. A. neurocarpum** var. **neurocarpum**

Leaves less than 2 cm wide, more than 5 times as long as wide. Rheophyte. BOR · · · · · · · · · **36c. A. neurocarpum** var. **linearifolium**

22(18). Young twigs light grey. Disc pilose. Flowers 2 – 3 × 1.5 – 2 mm. Stamens 2 – 2.5 mm long, exserted 1.5 – 2 mm from the calyx. Leaves drying greyish green to reddish brown. SUM, MLY, THA · · · **22. A. helferi**

Young twigs brown. Disc glabrous, rarely pilose. Flowers 1 – 2.5 × 1 – 1.5 mm. Stamens 1 – 2 mm long, exserted 0.5 – 1.5 mm from the disc. Leaves drying olive-green · 23

23. Leaves (9 –)20 – 40 cm long, 3.2 – 9.2 times as long as wide, midvein slightly raised adaxially. Inflorescences simple, solitary, very slender. Petioles 1 – 3 mm wide. Stamens exserted c. 0.5 mm from the calyx. Stipules, petioles, abaxial leaf surfaces and inflorescence axes usually glabrous. Sepals entire or erose, never glandular-fimbriate. SUM, MLY, BOR, JAW, PHI, SUL, LSI, MOL · · · · · · · · · · **49. A. stipulare**

Leaves 8 – 20 (– 25) cm long, 2.5 – 3.5 times as long as wide, midvein impressed adaxially. Inflorescences consisting of 10 or more branches. Petioles 1 – 2 mm wide. Stamens exserted 1 – 1.5 mm from the calyx. Stipules, petioles, abaxial leaf surfaces and inflorescence axes usually hairy. Sepals often glandular-fimbriate. THA · **33d. A. montanum** var. **wallichii**

24(15). Bracts more than 1.5 mm long, conspicuous even in very young inflorescences. Leaves abaxially hairy · 25

Bracts up to 1.5 mm long. Leaves abaxially glabrous to hairy · · · · · 27

25. Indumentum, especially of young parts, densely hirsute, whitish to ochraceous, never ferrugineous. Leaves drying olive-green. Disc glabrous. Flowers 2 – 3 × 2 – 3 mm. Stamens exserted c. 1 mm from the calyx. Sepals 5 – 7. Inflorescences 5 – 10 cm long. Most leaves 12 – 18 × 4 – 6 cm. SUM, MLY, BOR (Anambas & Natuna Isl.), JAW, THA · **54. A. velutinosum**

Indumentum, especially of young parts, ferrugineous. Leaves drying reddish brown. Disc hairy at the margin; if glabrous, then flowers not larger than 1.5 × 1.5 mm and stamens exserted only 0.5 mm from the calyx. Sepals 4 – 6. Inflorescences 6 – 25 cm long. Most leaves 15 – 30 × 5 – 12 cm · 26

26. Indumentum of all parts rather long and shaggy, sometimes with some stellate hairs. Pedicels 0.5 – 1.5 mm long. Stamens exserted 1 – 1.5 mm from the calyx. Inflorescences 11 – 25 cm long. Sepals fused to halfway. Tertiary veins reticulate (Fig. 2Ac) to weakly percurrent (Fig. 2Ab), widely spaced. Disc hairy at the margin. Usually above 1000 m altitude. MLY · **39. A. pahangense**

Indumentum of all parts short and velvety, without stellate hairs. Flowers sessile. Stamens exserted c. 0.5 mm from the calyx. Inflorescences 6 – 14 cm long. Sepals free or nearly so. Tertiary veins percurrent, close together (Fig. 2Aa). Disc hairy or glabrous. Indifferent to altitude. SUM, MLY, BOR, JAW, PHI, SUL, LSI, MOL · · · (52. *A. tomentosum*)

27(24). Petioles (at least of some leaves) more than 10 mm long, no more than 3 mm wide in the middle, basally and apically usually slightly pulvinate and sometimes geniculate · 28

Petioles up to 10 mm long (if longer, more than 3 mm wide, and leaf blades more than 25 cm long, basally and distally never pulvinate or geniculate) · 43

28. Abaxial leaf surfaces hairy all over · 29

Abaxial leaf surfaces glabrous, or hairy along the major veins only · · 33

29. Disc glabrous (sometimes with hairs originating from the base of the disc; Fig. 3D) · 30
Disc hairy all over (Fig. 3F), or at least at the margin (Fig. 3E) · · · · 31

30. Indumentum of leaves not distinctly velvety to the touch abaxially. Petioles 5 – 13 × 0.8 – 1 mm. Most leaves 10 – 13 × 3.5 – 5 cm. BOR · **56. A. venenosum**
Indumentum of leaves distinctly velvety to the touch abaxially. Petioles 12 – 35 × 1.2 – 2 mm. Most leaves 16 – 21 × 6 – 8 cm. MLY · **12. A. cruciforme**

31(29). Sepals fused for c. 1/2 of their length, unequal. Stipules 1 – 2 mm wide. Petioles 1.5 – 3 mm wide. Leaves (8 –)1 – 17(– 23) × (3 –)5 – 8(– 12) cm. PHI · **16a. A. edule** var. **edule**
Sepals free, equal. Stipules 0.5 – 1 mm wide. Petioles 0.5 – 1.5 mm wide. Leaves (2 –)4 – 11(– 16) × (2 –)3 – 5(– 9) cm · · · · · · · · · · · · · · · 32

32. Inflorescences consisting of up to 9 branches, indumentum very dense, making it hard to see floral details. Leaves (1.3 –)1.8 – 2.4(– 4) times as long as wide. PHI, LSI · · · · · · · · · · · · · · · · · **50. A. subcordatum**
Inflorescences consisting of 10 – 20 branches, indumentum of average density to sparse, so that floral details are easily seen. Leaves (1 –)1.3 – 1.7(– 2.25) times as long as wide. SUM, MLY, BOR, JAW, PHI, SUL, LSI, MOL, NWG, THA · · · · · · · · · · · · · · · · · · · (21. *A. ghaesembilla*)

33(28). Disc cushion-shaped, enclosing the bases of the filament and pistillode (Fig. 3A); (observe carefully especially when disc is hairy, dried flowers may have to be softened). Flowers on pedicels 0.3 – 1 mm long. SUM, MLY, BOR · · · · · · · · · · · · · · · · · · · **14. A. cuspidatum**
Disc extrastaminal-annular (Fig. 3B), or consisting of free lobes (Fig. 3C). Flowers sessile (pedicels up to 0.2 mm long in *A. curranii*) · · 34

34. Disc hairy all over (Fig. 3F), or at least at the margin (Fig. 3E) · · · · 35
Disc glabrous (sometimes with hairs originating from the base of the disc; Fig. 3D) · 40

35. Disc consisting of free lobes (Fig. 3C). Sepals free or fused to halfway · 36
Disc extrastaminal-annular (Fig. 3B). Sepals fused for 1/2 of their length or more · 37

36. Sepals free, pubescent outside. Inflorescences consisting of (3 –)10 – 20 branches. SUM, MLY, BOR, JAW, PHI, SUL, LSI, MOL, NWG, THA · (21. *A. ghaesembilla*)
Sepals fused for 1/2 of their length or more, glabrous outside. Inflorescences simple, or consisting of up to 6 branches. PHI · **13. A. curranii**

37(35). Young twigs brown. Leaves sometimes with domatia in the axils between the midvein and secondary veins abaxially (Fig. 2B); base acute to rounded · 38
Young twigs usually very light brown to whitish. Leaves without domatia; base obtuse to truncate, rarely acute · 39

38. Inflorescences 2 – 8 cm long, consisting of 1 – 5(– 10) branches. Petioles 1 – 2(– 3) mm wide. Leaves coriaceous to subcoriaceous. Occasionally cultivated. BOR, PHI, SUL, MOL, NWG ········ **18. A. excavatum**
Inflorescences 5 – 14 cm long, consisting of 2 – 25 branches. Petioles 0.8 – 1.2 mm wide. Leaves membranaceous to subchartaceous. SUL, LSI, MOL ···························· **7. A. celebicum**
39(37). Inflorescences consisting of 5 – 50 branches. Leaves chartaceous to membranaceous. PHI ····················· **42. A. pleuricum**
Inflorescences consisting of 2 – 10 branches. Leaves chartaceous to subcoriaceous. PHI ···················· **31. A. microcarpum**
40(34). Indumentum of young twigs, stipules, petioles and inflorescences ferrugineous-hispid. Sepals free or almost so. Leaves coriaceous, drying grey to reddish brown adaxially, reddish brown abaxially. SUM, MLY, BOR ························ **11. A. coriaceum**
Indumentum whitish. Sepals fused for ½ of their length or more. Leaves membranaceous to chartaceous, drying olive-green ····· 41
41. Stipules 1 mm long. Inflorescences 2 – 4 cm long, consisting of 1 – 4 branches. Young twigs first dark brown, then light grey to whitish. Leaf base acute to obtuse. Tertiary leaf veins reticulate (Fig. 2Ac). SUM, MLY, THA ························ **20. A. forbesii**
Stipules 5 – 25 mm long. Inflorescences 4 – 11 cm long, consisting of up to 35 branches. Young twigs brown. Leaf base rounded to truncate, more rarely obtuse. Tertiary leaf veins percurrent (Fig. 2Aa) to weakly percurrent (Fig. 2Ab) ························ 42
42. Stipules (2 –)5 – 20 mm wide. Petioles (7 –)10 – 15(– 35) mm long. SUM, JAW, LSI ······················ **51. A. tetrandrum**
Stipules c. 0.5 mm wide. Petioles 5 – 10(– 13) mm long. BOR ······· ································ **56. A. venenosum**
43(27). Stamens 2, rarely 3. Pistillode absent ······················ 44
Stamens 3 – 5. Pistillode present or absent ················· 45
44. Pedicels 1 – 1.5 mm long. Inflorescences 5 – 14 cm long. Petioles 2 – 7 mm long, hairy. Leaves chartaceous, abaxial leaf surfaces usually ± hairy, drying yellowish green. Domatia present (Fig. 2B). Stipules usually persistent, 3 – 7 × 1 – 2 mm. JAW, THA ······ **1. A. acidum**
Pedicels 0.1 mm long. Inflorescences 2 – 3 cm long. Petioles 1.5 – 2 mm long, glabrous. Leaves coriaceous, glabrous, drying reddish brown. Domatia absent. Stipules caducous, 1.5 × 0.6 mm. SUL ···· ···································· **4. A. brevipes**
45(43). Midvein sharply keeled abaxially (Fig. 2F). Leaves coriaceous. BOR ·· ································ **3. A. brachybotrys**
Midvein rounded abaxially (Fig. 2C – E). Leaves membranaceous to coriaceous ·································· 46
46. Disc glabrous (sometimes with hairs originating from the base of the disc; Fig. 3D) ································ 47
Disc hairy all over (Fig. 3F), or at least at the margin (Fig. 3E) ···· 69
47. Flowers 3 – 4 × c. 3 mm, sessile, with cheesy smell (of bat's droppings),

attracting flies. Stamens exserted 1.5 – 2 mm from the calyx. Anthers c. 0.5 × 0.7 mm. Pistillode present · 48

Flowers 1 – 2.5 × 1 – 2 mm, sessile or pedicellate, without cheesy smell. Stamens exserted 0.5 – 2 mm from the calyx. Anthers 0.2 – 0.5 × 0.3 – 0.6 mm. Pistillode present or absent · 49

48. Indumentum lacking or whitish. Abaxial leaf surfaces glabrous except sometimes along the midvein. Often cultivated. SUM, MLY, BOR, JAW, PHI, SUL, LSI, MOL, NWG, THA · · · **5a. A. bunius** var. **bunius**

Indumentum ferrugineous. Abaxial leaf surfaces pubescent all over. THA · **5b. A. bunius** var. **pubescens**

49(47). Disc extrastaminal-annular (Fig. 3B) · 50

Disc consisting of free lobes (Fig. 3C; short lobes may appear to be fused extrastaminally in *A. pendulum*); or disc cushion-shaped, enclosing the bases of the filament and, if present, pistillode (Fig. 3A; observe carefully; dried flowers may have to be softened · · · · · · · · · · · · · 51

50. Stipules 3 – 7 × 0.5 – 1 mm. Petioles 1.2 – 1.5 mm wide. Indumentum ferrugineous. Leaves pubescent all over abaxially, distinctly velvety to the touch, up to 18 cm long, always drying reddish brown. Inflorescences 4 – 7 cm long, axillary or more rarely cauline. MLY, THA · **55. A. velutinum**

Stipules 1 – 3 × 0.2 – 0.5 mm. Petioles 0.7 – 1(– 1.2) mm wide. Indumentum ochraceous. Leaves pubescent all over or only along the veins abaxially, up to 13 cm long, drying olive-green, greyish or reddish brown. Inflorescences 4 – 11 cm long, terminal or more rarely axillary. THA · **47. A. sootepense**

51(49). Disc consisting of free lobes (Fig. 3C) · 52

Disc cushion-shaped, enclosing the bases of the filament and, if present, pistillode (Fig. 3A) · 55

52. Petioles 1.5 – 6 mm wide, glabrous. Leaves (13 –)25 – 50 × (4 –)7 – 14 cm, chartaceous to coriaceous, usually drying yellowish-brown. Inflorescences c. 15 cm long, simple, solitary or fascicled in pairs, axes glabrous. Pistillode glabrous. Stipules usually persistent, 0.5 – 3 mm wide. Midvein usually slightly raised adaxially. SUM, MLY, BOR, THA · **40. A. pendulum**

Petioles 0.5 – 1.2 mm wide, hairy. Leaves 3 – 13(– 18) × 1 – 5(– 7) cm, chartaceous to membranaceous, drying olive-green, reddish or greyish brown. Inflorescences 2 – 11 cm long, simple to much-branched, solitary, axes hairy. Pistillode pubescent or glabrous. Stipules early-caducous, 0.2 – 0.5 mm wide. Midvein impressed adaxially · · · · · · 53

53. Inflorescences consisting of up to 35 branches. Disc longer than the calyx. Stipules 5 – 10 mm long. Leaf base rounded to truncate, rarely obtuse. BOR · **56. A. venenosum**

Inflorescences simple or consisting of 2 – 4 branches. Disc as long as or shorter than the calyx. Stipules 1 – 3 mm long. Leaf base acute to obtuse, rarely rounded · 54

54. Inflorescences 2 – 4 cm long. Young twigs first dark brown, then light grey to whitish, glabrous to puberulent. Stamens exserted c. 0.5 mm from the calyx. Leaves glabrous or slightly hairy only along the midvein, higher venation not conspicuously tessellated (Fig. 9A). SUM, MLY, THA ······························ **20. A. forbesii**

Inflorescences 4 – 11 cm long. Young twigs brown, (usually densely) ochraceous-tomentose. Stamens exserted 1 – 2 mm from the calyx. Leaves hairy abaxially at least along the veins, higher venation finely tessellated (Fig. 9D). THA ················ **47. A. sootepense**

55(51). Sepals fused for $^{3}/_{4}$ of their length or more. Disc small, much shorter than the calyx. Pistillode absent. Leaves completely glabrous, drying dark reddish brown to purplish grey. Young twigs light brown to whitish grey. PHI ······················ **30. A. macgregorii**

Sepals free to fused for c. $^{1}/_{2}$ of their length. Disc obvious, often protruding from the calyx. Pistillode present or absent. Leaves glabrous to hairy, drying olive-green, reddish or greyish. Young twigs dark brown to whitish grey ···························· 56

56. Indumentum, especially of young parts and inflorescences, ferrugineous to ochraceous ···························· 57

Indumentum whitish or absent ·························· 62

57. Petioles 2 – 5 mm wide. Leaves (8 –)15 – 30(– 60) cm long. Tertiary veins percurrent (Fig. 2Aa), close together. Inflorescences 6 – 14 cm long, always simple, usually dense. Pistillode present, 0.2 – 0.4 mm long, pubescent ····································· 58

Petioles c. 1 mm wide. Leaves (3 –)6 – 14(– 20) cm long. Tertiary veins weakly percurrent (Fig. 2Ab) to reticulate (Fig. 2Ac). Inflorescences 2 – 6 cm long, simple or branched, usually lax. Pistillode absent or only to 0.1 mm long, glabrous to sparsely pilose ············· 59

58. Plant rheophytic, leaves more than 6 times as long as wide. Petioles 1 – 3 mm wide. Inflorescences 4 – 6 cm long. BOR ················ ······················ **52b. A. tomentosum** var. **stenocarpum**

Plant not rheophytic, leaves less than 6 times as long as wide. Petioles 2 – 5 mm wide. Inflorescences 6 – 14 cm long. SUM, MLY, BOR, JAW, PHI, SUL, LSI, MOL, THA ·· **52a. A. tomentosum** var. **tomentosum**

59(57). Stipules 0.5 – 1 mm wide. SUM, MLY, BOR ·················· ························· **36b. A. neurocarpum** var. **hosei**

Stipules (1 –)1.5 – 18 mm wide ························· 60

60. Plant rheophytic, leaves more than 5 times as long as wide ······· 61

Plant not rheophytic, leaves less than 5 times as long as wide. SUM, MLY, BOR, THA ············ **36a. A. neurocarpum** var. **neurocarpum**

61. Leaves 1.2 – 4 cm wide. Petioles 1 – 3 mm wide. Pedicel absent. Stamens exserted c. 0.5 mm from the calyx. Pistillode 0.2 – 0.4 mm long. BOR ····················· **52b. A. tomentosum** var. **stenocarpum**

Leaves 0.7 – 1.8 cm wide. Petioles c. 1 mm wide. Pedicels 0.5 – 1 mm long. Stamens exserted 0.8 – 1 mm from the calyx. Pistillode up to 0.1 mm long. BOR ·········· **36c. A. neurocarpum** var. **linearifolium**

62(56). Young twigs light brown to whitish. Leaves drying reddish brown to grey, often with dense intersecondary and perpendicular tertiary veins conspicuous near the midvein · 63

Young twigs brown. Leaves drying olive-green, without conspicuously perpendicular intersecondary and tertiary veins near the midvein · 64

63. Higher order veins in dry material distinctly prominent on both leaf surfaces. Stipules caducous, subulate. Leaves drying reddish brown. Inflorescences 4 – 7 cm long, simple, slender. Usually between 400 – 1400 m. SUM, JAW · **32. A. minus**

Higher order veins in dry material much less prominent adaxially. Stipules usually persistent, lanceolate, linear or subulate, often falcate. Leaves drying reddish brown to greyish. Inflorescences 2 – 6 cm long, simple or branched, not particularly slender. Indifferent to altitude. SUM, MLY, BOR · · · · · · · · **36b. A. neurocarpum** var. **hosei**

64(62). Leaves (9 –)20 – 40 × (1.5 –)5 – 7(– 10) cm, predominantly oblong, midvein slightly raised adaxially. Petioles 1 – 3 mm wide. Stipules 5 – 60 × 1 – 30 mm. Flowers sessile. Sepals entire or erose, never glandular-fimbriate. Stamens exserted c. 0.5 mm from the calyx. Inflorescences simple. Stipules, petioles and inflorescence axes usually glabrous. SUM, MLY, BOR, JAW, PHI, SUL, LSI, MOL · · · · · · · **49. A. stipulare**

Leaves 3 – 15(– 30) × 0.5 – 6(– 12) cm, predominantly elliptic, midvein impressed adaxially. Petioles 0.5 – 2 mm wide. Stipules 2 – 7(– 13) × 0.5 – 1.5(– 3) mm. Flowers sessile or pedicellate. Sepals often glandular-fimbriate. Stamens exserted 1 – 2 mm from the calyx. Inflorescences simple or branched. Stipules, petioles and inflorescence axes usually hairy · 65

65. Plant rheophytic, leaves 4 – 10 times as long as wide or narrower than 2 cm · 66

Plant not rheophytic, leaves (1.2 –)2.5 – 3.5(– 6.6) times as long as wide · 67

66. Most leaves 5 – 11 × 1 – 2 cm. Leaf apex acuminate. SUM, MLY, BOR, THA · **33c. A. montanum** var. **salicinum**

Most leaves 3 – 6 × 0.4 – 1.2 cm. Leaf apex acute, sometimes with a rounded apiculum. THA · · · · **33b. A. montanum** var. **microphyllum**

67(65). Inflorescences, pedicels and flowers slender. Petioles 0.5 – 1 mm wide. Sepals free. Leaves often conspicuously shiny. MLY, PHI, THA · **24a. A. japonicum** var. **japonicum**

Inflorescences, pedicels and flowers not conspicuously slender. Petioles 0.7 – 2 mm wide. Sepals free, to fused to halfway. Leaves not conspicuously shiny · 68

68. Without differential characters in staminate material. Very common. SUM, MLY, BOR, JAW, PHI, SUL, LSI, MOL, THA · **33a. A. montanum** var. **montanum**

Rare. JAW, LSI, SUL · **23. A. heterophyllum**

69(46). Disc extrastaminal-annular, lobed with inflexed lobes or hardly lobed, longer than or as long as the calyx (Fig. 5F; observe carefully, dried flowers may have to be softened). Sepals fused for at least $^{2}/_{3}$ of their length. Inflorescences 2 – 7 cm long. Occasionally cultivated. BOR (East Kalimantan), PHI, SUL, MOL, NWG · · · · · · **18. A. excavatum**

Disc consisting of free, alternistaminal lobes which may appear to be fused extrastaminally (Fig. 3C; observe carefully, dried flowers may have to be softened) or cushion-shaped, enclosing the bases of the filaments and pistillode (Fig. 3A). Sepals free or fused. Inflorescences 2 – 20 cm long · 70

70. Disc consisting of free lobes (Fig. 3C) · 71

Disc cushion-shaped, enclosing the bases of the filaments and pistillode (Fig. 3A) · 76

71. Inflorescences c. 15 cm long. Pistillode glabrous. Petioles 1.5 – 6 mm wide. Stipules 6 – 25 × 0.5 – 3 mm, usually persistent. Leaves (13 –)25 – 50 × (4 –)7 – 14 cm. SUM, MLY, BOR, THA · · · · **40. A. pendulum**

Inflorescences 2 – 12 cm long. Pistillode hairy. Petioles up to 1.5 mm wide. Stipules 0.5 – 6(– 10) × 0.5 – 1.5 mm, usually caducous. Leaves 2 – 15(– 20) × 1 – 5(– 9) cm · 72

72. Sepals free, pubescent outside. Leaf base rounded to cordate, rarely acute · 73

Sepals fused for $^{1}/_{2}$ of their length or more, glabrous to sparsely pilose outside. Leaf base acute to rounded · 74

73. Inflorescences simple or consisting of up to 9 branches, indumentum very dense, making it hard to recognise floral details. Leaves (1.3 –)1.8 – 2.4(– 4) times as long as wide. PHI, LSI · · · · · · · **50. A. subcordatum**

Inflorescences consisting of 10 – 20 branches, indumentum less dense, allowing easy recognition of floral details. Leaves (1 –)1.3 – 1.7(– 2.25) times as long as wide. SUM, MLY, BOR, JAW, PHI, SUL, LSI, MOL, NWG, THA · (21. *A. ghaesembilla)*

74(72). Midvein slightly raised to flat adaxially. Flowers 2 – 3 mm long. Stamens exserted 1.5 – 2 mm from the calyx. Calyx cupular to cylindrical, lobes broadly rounded. Leaves membranaceous to chartaceous. Inflorescences 6 – 12 cm long. SUM, MLY, THA · · **28. A. leucocladon**

Midvein impressed adaxially. Flowers 1.5 – 2 mm long. Stamens exserted c. 1 mm from the calyx. Calyx bowl-shaped to cupular, lobes acute to rounded. Leaves chartaceous to subcoriaceous. Inflorescences 2 – 8(– 12) cm long · 75

75. Calyx 0.5 – 0.8 mm long, sepals fused for $^{2}/_{3}$ of their length or more, margin entire. Stipules c. 0.5 × 0.5 mm. Petioles 3 – 8(– 10) mm long, not pulvinate. Leaves(1 –)2 – 3.5(– 5.5) cm wide, usually shiny. PHI · **15. A. digitaliforme**

Calyx 0.8 – 1 mm long, sepals fused for $^{1}/_{2}$ – $^{2}/_{3}$ of their length, margin erose. Stipules 3 – 5 × 1 – 1.5 mm. Petioles 3 – 15 mm long, basally and distally sometimes slightly pulvinate. Leaves (2 –)3 – 5(– 6.5) cm wide, usually dull. PHI · **13. A. curranii**

76(70). Indumentum, especially of young parts, inflorescences and disc, ferrugineous to ochraceous · 77
Indumentum whitish or absent · 80

77. Inflorescences 4 – 6 cm long. Leaves 2 – 4 times as long as wide. Calyx conical. Disc about as wide as long, ± flat on top, constricted at the base. LSI, MOL, SUL · **17. A. elbertii**
Inflorescences 6 – 20 cm long (if shorter, leaves at least 6 times as long as wide). Calyx cupular. Disc wider than long, ± hemispherical, not constricted at the base · 78

78. Sepals fused for $^2/_3$ of their length or more. Inflorescences 15 – 20 cm long. Flowers 2 – 3 × 2 – 3 mm, sessile or pedicels up to 1 mm long. Stamens exserted 1 – 2.5 mm from the calyx. Tertiary veins usually at right angles to the midvein (Fig. 2Ad). Mostly rheophytic. BOR, PHI (Palawan), SUL · · · · · · · · · · · · · · **46a. A. riparium** subsp. **riparium**
Sepals free or nearly so. Inflorescences 4 – 14 cm long. Flowers c. 1.5 × 1.5 mm, sessile. Stamens exserted c. 0.5 mm from the calyx. Tertiary veins at right angles to the secondary veins, and oblique to the midvein (Fig. 2Aa – c) · 79

79. Plant rheophytic, leaves more than 6 times as long as wide. Petioles 1 – 3 mm wide. Inflorescences 4 – 6 cm long. BOR · **52b. A. tomentosum** var. **stenocarpum**
Plant not rheophytic, leaves less than 6 times as long as wide. Petioles 2 – 5 mm wide. Inflorescences 6 – 14 cm long. SUM, MLY, BOR, JAW, PHI, SUL, LSI, MOL, THA · · **52a. A. tomentosum** var. **tomentosum**

80(76). Stamens exserted c. 0.5 mm from the calyx. Leaves (9 –)20 – 30 (– 40) cm long, major veins slightly raised adaxially. Stipules lanceolate to linear, 5 – 60 × 1 – 30 mm. Inflorescences simple. Flowers sessile, c. 1 mm long. Leaves (except for some hairs along the major veins abaxially), inflorescence axes and calyx glabrous. SUM, MLY, BOR, JAW, PHI, SUL, LSI, MOL · · · · · · · · · · · · · · · · · · · (49. *A. stipulare)*
Stamens exserted (0.5 –)1 – 2 mm from the calyx. Leaves (1.5 –)5 – 20(– 30) cm long, major veins impressed adaxially. Stipules linear, 1.5 – 8(– 13) × 0.5 – 1.5(– 3) mm. Inflorescences simple or branched. Flowers sessile or pedicels up to 2 mm long, 1.5 – 3 mm long. Leaves, inflorescence axes and calyx glabrous or hairy · · · · · · · · · · · · · · 81

81. Young twigs light grey, glabrous or nearly so. Leaves drying greyish green to reddish brown, chartaceous to coriaceous. Anthers 0.5 – 1 × 0.5 – 1 mm. Sepals never glandular-fimbriate. Flowers 2 – 3 mm long. SUM, MLY, THA · **22. A. helferi**
Young twigs brown, ± densely hairy. Leaves usually drying olive-green, membranaceous to chartaceous, only the larger leaves sometimes coriaceous. Anthers 0.2 – 0.5 × 0.2 – 0.5 mm. Sepals often glandular-fimbriate. Flowers 1 – 3 mm long · 82

82. Abaxial leaf surfaces spreading-hirsute all over, especially along the veins · 83
Abaxial leaf surfaces glabrous except along the major veins · · · · · · 84

83. Flowers 1.5 – 2 mm long. Calyx cupular, sepals free or nearly so, almost orbicular. Disc wider than long, ± hemispherical, not constricted at the base, densely hirsute. Stamens exserted c. 1 mm from the calyx. MLY, THA ··························· **37. A. orthogyne**

Flowers 2 – 3 mm long. Calyx conical, sepals fused for $^1/_3$ – $^2/_3$ of their length, deltoid. Disc about as wide as long, ± flat on top, constricted at the base, glabrous at the sides, hairy on top. Stamens exserted 1.5 – 2 mm from the calyx. LSI, MOL, SUL ·········· **17. A. elbertii**

84(82). Plant not rheophytic, leaves (1.2 –)2.5 – 3.5(– 6.6) times as long as wide. SUM, MLY, BOR, JAW, PHI, SUL, LSI, MOL, THA ···························· (33a. *A. montanum* var. *montanum*)

Plant rheophytic, leaves 4 – 10 times as long as wide or narrower than 2 cm ··· 85

85. Most leaves 5 – 11 × 1 – 2 cm. Leaf apex acuminate. SUM, MLY, BOR, THA ····················· (33c. *A. montanum* var. *salicinum)*

Most leaves 3 – 6 × 0.4 – 1.2 cm. Leaf apex acute, sometimes with a rounded apiculum. THA ···· (33b. *A. montanum* var. *microphyllum)*

KEY FOR PISTILLATE MATERIAL

1. Leaf apex rounded or retuse, more rarely obtuse ··············· 2

Leaf apex acute, acuminate or caudate ······················ 4

2. Leaves base cordate to rounded, very rarely acute. Sepals free, pubescent outside. SUM, MLY, BOR, JAW, PHI, SUL, LSI, MOL, NWG, THA ···························· **21. A. ghaesembilla**

Leaf base acute or obtuse, mostly cuneate. Sepals fused or free, glabrous outside ·· 3

3. Sepals fused for $^2/_3$ of their length. Leaves obovate to elliptic-oblong, (2 –)5 – 10(– 18) × (1 –)2.5 – 4(– 8) cm. Petioles 1 – 2 mm wide. Fruiting pedicels 1.5 – 3 mm long. JAW, THA ······· **1. A. acidum**

Sepals free or nearly so. Leaves spathulate, (1.5 –)2 – 5(– 7) × (0.7 –)1 – 2.5(– 3.5) cm. Petioles less than 1 mm wide. Fruiting pedicels 0.2 – 1 mm long. MOL, NWG ··················· **48. A. spatulifolium**

4(1). Midvein in dry material sharply raised adaxially (Fig. 2E), distinctly perceptible to the touch. Inflorescences simple ·············· 5

Midvein in dry material impressed to slightly raised adaxially (Fig. 2C – D), not distinctly perceptible to the touch. Inflorescences simple or branched ·· 7

5. Leaves 6 – 7.5 cm long. Ovaries glabrous. Fruits 5 – 6(– 8) × 4 – 5(– 6) mm. Infructescences 4 – 8 cm long. THA ·································· **24b. A. japonicum** var. **robustius**

Leaves 13 – 45 cm long. Ovaries pubescent. Fruits 8 – 16 × 5 – 9 mm. Infructescences 7 – 50 cm long ·························· 6

6. Young twigs, petioles and leaves hairy, glabrescent. Stipules usually foliaceous, more than 2 mm wide. Leaves 15 – 45 × 5 – 18 cm. Sepals fused to halfway. Disc glabrous to pilose at the margin. Inflorescences

15 – 20 cm long, infructescences 25 – 50 cm long. Fruits glabrous or with stellate hairs, 12 – 16 mm long, style subterminal. MLY ········ **38. A. pachystachys**

Young twigs, petioles and leaves glabrous. Stipules linear, up to 1.5 mm wide, caducous. Leaves 13 – 27 × 3.5 – 8 cm. Sepals fused only at the base. Disc glabrous. Inflorescences c. 3 cm, infructescences up to 15 cm long. Fruits pilose, with simple hairs, 8 – 11 mm long, style terminal. MLY, THA ········ **27. A. laurifolium**

7(4). Inflorescences at least partly cauline ········ 8

Inflorescences axillary (sometimes aggregated at the end of the branch) ········ 15

8. Petioles (at least of some leaves) more than 10 mm long, no wider than 2 mm in the middle, basally and distally usually slightly pulvinate and sometimes geniculate ········ 9

Petioles shorter than 10 mm (if longer, then more than 2 mm wide), basally and distally not pulvinate or geniculate ········ 10

9. Disc hairy. All leaf veins, including the very fine veinlets, in dry material clearly visible abaxially. Leaves drying yellowish brown to yellowish green. Pubescence of inflorescence whitish. Inflorescences simple, fascicled. SUM, MLY, BOR, PHI, THA ········ **44. A. puncticulatum**

Disc glabrous (sometimes with hairs originating from the base of the disc). Only the coarse leaf veins in dry material visible abaxially. Leaves drying grey to reddish brown. Pubescence of inflorescence ferrugineous. Inflorescences simple or branched. SUM, MLY, BOR ········ **11. A. coriaceum**

10(8). Tertiary veins close together, more than 10 between every two secondary veins (Fig. 2Aa). Infructescences 8 – 25(– 69) cm long, up to 13 per fascicle. Petioles 1.5 – 4 mm wide. Leaves (7 –)13 – 25(– 50) × (2.5 –)4 – 8(– 15) cm. Stipules 8 – 20 × 1 – 4 mm ········ 11

Tertiary veins widely spaced, less than 9 between every two secondary veins (Fig. 2Ab). Infructescences 4 – 15 cm long, up to 7 per fascicle. Petioles up to 2 mm wide. Leaves (5 –)8 – 14(– 19) × (2 –)3 – 5(– 9.5) cm. Stipules 2 – 10 × 0.5 – 1.5 mm ········ 12

11. Fruits 4 – 8(– 10) mm long, lenticular to ellipsoid, rarely ovate and beaked, usually laterally compressed, areolate. Stigmas 4 – 6. Ovaries hairy to nearly glabrous. Infructescences 10 – 25(– 69) cm long, flexible. Indumentum of leaves sparse abaxially, indumentum of inflorescence axes and calyx moderately dense. SUM, MLY, BOR, PHI, THA ········ **29. A. leucopodum**

Fruits 10 – 14 mm long, elongate-ellipsoid, more or less tetragonal, not or hardly compressed, reticulate. Stigmas 6 – 16. Ovaries densely hairy. Infructescences 8 – 13 cm long, robust. Indumentum of leaves usually dense abaxially, indumentum of inflorescence axes and calyx extremely dense. BOR ········ **43. A. polystylum**

12(10). Abaxial leaf surfaces hairy all over ········ 13

Abaxial leaf surfaces glabrous 14

13. Indumentum ferrugineous. Fruits 4 – 5 × c. 3 mm. Style terminal, usually wide. Inflorescences 2 – 4 cm long, infructescences 3 – 8 cm long. Disc glabrous (sometimes with hairs originating from the base of the disc). MLY, THA .... **55. A. velutinum**

Indumentum whitish to ochraceous. Fruits 6 – 8 × 4 – 6 mm. Style lateral or subterminal, very distinct, thin. Inflorescences (4 –)7 – 8 cm long, infructescences 8 – 23 cm long. Disc glabrous, or glabrous only at the sides, hairy on top. LSI, MOL, SUL ..... **17. A. elbertii**

14(12). Fruits 6 – 8 × 4 – 6 mm, style subterminal to lateral, rarely terminal. Sepals free or nearly so. Disc hairy all over. Stigmas (2 –)3 – 4, long (up to 1.5 mm) and conspicuous. SUM, MLY, THA ... **22. A helferi**

Fruits 7 – 10 × 5 – 7 mm, style terminal to slightly subterminal. Sepals fused for 2/3 – 3/4 of their length. Disc glabrous (sometimes with hairs originating from the base of the disc). Stigmas 5 – 8(– 10), not conspicuously long. PHI .... **30. A. macgregorii**

15(7). Stipules foliaceous, more than 4 mm wide 16

Stipules not foliaceous, up to 4 mm wide 24

16. Petioles (at least of some leaves) more than 10 mm long, no more than 2 mm wide in the middle, basally and apically slightly pulvinate and geniculate 17

Petioles up to 10 mm long (if longer, more than 2 mm wide), not pulvinate or geniculate 18

17. Stipules spathulate, cordate or kidney-shaped, apex rounded to acute, base symmetrical. Leaves (6 –)10 – 14(– 22) cm long, up to 3 times as long as wide, chartaceous to membranaceous. Bracts c. 0.5 mm long. Sepals fused for 2/3 of their length. Inflorescences usually consisting of 5 – 8 branches, rarely simple. SUM, JAW, LSI .... **51. A. tetrandrum**

Stipules ovate, apex caudate, base asymmetrical. Leaves 13 – 21 cm long, 3 – 4 times as long as wide, subcoriaceous to chartaceous. Bracts 0.7 – 1 mm long. Sepals fused only at the base. Inflorescences simple or once-branched. MLY .... **26. A. kunstleri**

18(16). Indumentum, especially of young parts and inflorescences, ferrugineous. Leaves drying reddish brown 19

Indumentum whitish or absent. Leaves drying olive-green to reddish brown 22

19. Sepals fused for 2/3 – 3/4 of their length. Disc ferrugineous-pubescent. Fruits terete or hardly compressed. Mostly rheophytic. BOR, PHI (Palawan), SUL .... **46a. A. riparium** subsp. **riparium**

Sepals free or nearly so. Disc glabrous to ferrugineous-pubescent. Fruits laterally or dorsiventrally compressed, or terete 20

20. Stipules subulate to falcate, rarely lanceolate with an asymmetrical base, strongly parallel-veined, coriaceous, no wider than 10 mm, at least twice as long as wide. Petioles 2 – 5 mm wide. Leaves (8 –)15 – 30(– 60) cm long. Tertiary veins percurrent, close together (Fig. 2Aa). Inflorescences 3 – 30 cm long; flowers set densely, touching each

other. Infructescences 7 – 65 cm long. Ovaries ovoid to falcate, densely appressed-pubescent. Disc glabrous or hairy. Fruits laterally compressed. SUM, MLY, BOR, JAW, PHI, SUL, LSI, MOL · **52. A. tomentosum**

Stipules cordate to lanceolate, base more or less symmetrical, reticulately veined, chartaceous, up to 18 mm wide, hardly longer to several times as long as wide. Petioles c. 1 mm wide. Leaves (5 –)6 – 14(– 20) cm long. Tertiary veins weakly percurrent (Fig. 2Ab) to reticulate (Fig. 2Ac). Inflorescences 1.5 – 6 cm long, lax; flowers widely spaced, not touching each other. Infructescences 4 – 16 cm long. Ovaries globose, glabrous, rarely pilose. Disc glabrous. Fruits laterally or dorsiventrally compressed · 21

21. Leaves more than 2 cm wide, less than 5 times as long as wide. SUM, MLY, BOR, THA · · · · · · · · **36a. A. neurocarpum** var. **neurocarpum**

Leaves less than 2 cm wide, more than 5 times as long as wide. Rheophyte. BOR · · · · · · · · · **36c. A. neurocarpum** var. **linearifolium**

22(18). Fruits terete, (3 –)4 – 6(– 8) mm long. Stipules, petioles and abaxial leaf surfaces usually hairy. Sepals often glandular-fimbriate. Inflorescences simple to consisting of up to 13 branches. THA · **33d. A. montanum** var. **wallichii**

Fruits laterally compressed, 6 – 18 mm long. Stipules, petioles and abaxial leaf surfaces usually glabrous. Sepals never glandular-fimbriate. Inflorescences simple or consisting of up to 3 branches · · · · · · · · 23

23. Fruits 6 – 8 × 4 – 6 mm. Inflorescences 4 – 10 cm long, infructescences 4 – 15 cm long. Fruiting pedicels 0.5 – 2.5 mm long. Leaves drying greyish green to reddish brown, (5 –)9 – 14(– 19) cm long, 1.9 – 4.3 times as long as wide. Young twigs light grey. Stipules up to 5 mm wide. Disc and ovaries always hairy. SUM, MLY, THA · · · · · · **22. A. helferi**

Fruits 8 – 18 × 6 – 11 mm. Inflorescences 10 – 30 cm long, infructescences 10 – 50 cm long. Fruiting pedicels 1 – 14 mm long. Leaves drying olive-green, (9 –)20 – 30(– 40) cm long, 3.2 – 9.2 times as long as wide. Young twigs brown. Stipules up to 30 mm wide. Disc and ovaries usually glabrous. SUM, MLY, BOR, JAW, PHI, SUL, LSI, MOL · **49. A. stipulare**

24(15). Bracts more than 1.5 mm long, conspicuous even in very young inflorescences, less so in old infructescences. Leaves abaxially hairy · 25

Bracts up to 1.5 mm long. Leaves abaxially hairy or glabrous · · · · · 27

25. Indumentum, especially of young parts, densely hirsute, whitish to ochraceous, never ferrugineous. Leaves drying olive-green. Fruits lenticular, obliquely ellipsoid or bean-shaped, 4 – 7 × 4 – 6 mm, with a distinctly asymmetrical base and lateral styles, laterally compressed. Sepals 5 – 8. Infructescences 6 – 15 cm long. Most leaves 12 – 18 × 4 – 6 cm, tertiary veins percurrent, close together (Fig. 2Aa). Disc glabrous. SUM, MLY, BOR (Anambas & Natuna Isl.), JAW, THA · **54. A. velutinosum**

Indumentum, especially of young parts, ferrugineous. Leaves drying reddish brown. Fruits ellipsoid or obliquely ovate (mango-shaped) (7 –)9 – 10(– 20) × 5 – 9 mm, with a symmetrical or asymmetrical base and terminal to lateral styles, not, laterally or dorsiventrally compressed. Sepals 4 – 6. Infructescences 10 – 65 cm long. Most leaves 15 – 30 × 5 – 12 cm, tertiary veins reticulate (Fig. 2Ac) to percurrent (Fig. 2Aa), close together or widely spaced. Disc hairy or glabrous · 26

26. Indumentum of all parts rather long and shaggy; sometimes, especially on fruits, with some stellate hairs. Fruits ellipsoid, not or laterally or dorsiventrally compressed. Tertiary veins reticulate (Fig. 2Ac) to weakly percurrent (Fig. 2Ab), widely spaced. Disc hairy at the margin. Usually above 1000 m altitude. MLY · · · · · · · · · · **39. A. pahangense**

Indumentum of all parts rather short and velvety; without stellate hairs. Fruits obliquely ovate (mango-shaped), always laterally compressed. Tertiary veins percurrent, close together (Fig. 2Aa). Disc hairy or glabrous. Indifferent to altitude. SUM, MLY, BOR, JAW, PHI, SUL, LSI, MOL · (52. *A. tomentosum*)

27(24). Petioles (at least of some leaves) more than 10 mm long, no wider than 3 mm in the middle, basally and apically usually slightly pulvinate and sometimes geniculate · 28

Petioles up to 10 mm long (if longer, more than 3 mm wide, and leaf blades more than 25 cm long, basally and distally never pulvinate or geniculate) · 47

28. Abaxial leaf surfaces hairy all over · 29

Abaxial leaf surfaces glabrous, or hairy along the major veins only · · 33

29. Disc glabrous (sometimes with hairs originating from the base of the disc). Fruits less than 3 mm or more than 7 mm long, glabrous · · 30

Disc hairy all over or at least at the margin. Fruits 3 – 6 mm long, pilose or glabrous · 31

30. Fruits 1.5 – 2.5 mm long, style distinctly lateral. Inflorescences branched. Leaf indumentum abaxially not distinctly velvety to the touch. Petioles 5 – 13 × 0.8 – 1 mm. Most leaves 10 – 13 × 3.5 – 5 cm. BOR · **56. A. venenosum**

Fruits 8 – 10 mm long, style terminal or subterminal. Inflorescences simple. Leaf indumentum abaxially distinctly velvety to the touch. Petioles 12 – 35 × 1.2 – 2 mm. Most leaves 16 – 21 × 6 – 8 cm. MLY · **12. A. cruciforme**

31(29). Fruits with lateral style and asymmetrical base, bean-shaped to obliquely lenticular. Petioles 1.5 – 3 mm wide. Stipules 1 – 2 mm wide. Leaves (8 –)13 – 17(– 23) × (3 –)5 – 8(– 12) cm. PHI · **16a. A. edule** var. **edule**

Fruits with terminal style and symmetrical base, lenticular or ellipsoid. Petioles 0.5 – 1.5 mm wide. Stipules 0.5 – 1 mm wide. Leaves (2 –)4 – 11(– 16) × (2 –)3 – 5(– 9) cm · 32

32. Sepals fused to halfway. Ovaries (and to a lesser extent also the fruit) elongate-ellipsoid. Inflorescences simple. Leaves (1.3 –)1.8 – 2.4(– 4) times as long as wide. PHI, LSI · · · · · · · · · · · · · **50. A. subcordatum**
Sepals free. Ovaries ovoid to globose, fruits lenticular, more rarely ellipsoid. Inflorescences consisting of 3 – 20 branches. Leaves (1 –)1.3 – 1.7(– 2.25) times as long as wide. SUM, MLY, BOR, JAW, PHI, SUL, LSI, MOL, NWG, THA · · · · · · · · · · · · · · · · · · · (21. *A. ghaesembilla*)

33(28). Fruits with a distinctly lateral style (Fig. 3L). Fruiting pedicels 0.5 – 5 mm long · 34
Fruits with a terminal (Fig. 3G) to subterminal (Fig. 3J) style. Fruiting pedicels 0 – 2 mm long · 42

34. Fruits 1.5 – 2.5 × 1 – 2 mm. BOR · · · · · · · · · · · · · · **56. A. venenosum**
Fruits 3 – 15 × 2 – 12 mm · 35

35. Sepals free. SUM, MLY, BOR · · · · **14a. A. cuspidatum** var. **cuspidatum**
Sepals fused for at least half their length · · · · · · · · · · · · · · · · · · · 36

36. Disc glabrous (sometimes with hairs originating from the base of the disc; Fig. 3D) · 37
Disc hairy all over (Fig. 3F), or at least at the margin (Fig. 3E) · · · · 39

37. Leaves ovate to elliptic, base acute to obtuse. Petioles 0.5 – 1 mm wide. Young twigs first dark brown, then light grey to whitish. Inflorescences 2 – 5 cm long, simple, solitary or 2 per fascicle. Fruits 5 – 8 × 4 – 6 mm. Ovaries glabrous to sparsely pilose. SUM, MLY, THA · **20. A. forbesii**
Leaves usually oblong, base rounded to truncate, rarely obtuse. Petioles 0.7 – 2 mm wide. Young twigs brown. Inflorescences 5 – 10 cm long, usually consisting of 5 – 8 very regular branches, solitary. Fruits 3 – 5 × 2 – 3 mm (unknown for *A. cuspidatum* var. *orthocalyx*) · · · · · · · 38

38. Inflorescences 5 – 6 cm long. Ovaries glabrous. Stipules (2 –)5 – 20 mm wide. Fruits 3 – 5 × 2 – 3 mm. SUM, JAW, LSI · · · **51. A. tetrandrum**
Inflorescences 8 – 10 cm long. Ovaries pilose. Stipules and fruits unknown. SUM · · · · · · · · · · · · · **14b. A. cuspidatum** var. **orthocalyx**

39(36). Exocarp in dry material loosely enfolding the endocarp, thick, neither areolate nor reticulate, without ridges or keels, brittle, minutely scaly (Fig. 5C). Fruits globose, not or hardly compressed. Infructescences consisting of 1 – 4 branches, axes robust and rigid. MOL, NWG · **2. A. baccatum**
Exocarp in dry material closely enfolding the endocarp, thin, reticulate, areolate or wrinkled (fleshy when fresh), usually keeled at the suture, not brittle or scaly (Fig. 5D). Fruits laterally compressed. Infructescences consisting of up to 10 branches, axes less robust and rigid · 40

40. Petioles 0.8 – 1.2(– 2) mm wide. Leaves chartaceous to membranaceous, without domatia, base obtuse to truncate, rarely acute. Young twigs often very light brown to whitish. Fruits 3 – 4 mm across; ovaries and fruits hairy. PHI · **42. A. pleuricum**

Petioles 1 – 2(– 3) mm wide. Leaves coriaceous to subcoriaceous, sometimes with domatia, base acute to rounded. Young twigs brown. Fruit 3 – 9 mm across; ovaries and fruits glabrous to sparsely pilose · 41

41. Bracts 0.4 – 0.8 × 0.3 – 0.6 mm. Infructescences 4 – 17 cm long, consisting of 2 – 10 branches, rarely simple. Domatia in the axils between the midvein and secondary veins abaxially (Fig. 2B) usually present. Occasionally cultivated. BOR, PHI, SUL, MOL, NWG · **18. A. excavatum**

Bracts 1 – 1.5 × 0.7 – 1 mm. Infructescences 4 cm long, simple. Domatia absent. PHI · **16b. A. edule** var. **apoense**

42(33). Sepals free, pubescent or hispid outside. Indumentum of inflorescence axes usually ferrugineous · 43

Sepals fused for ½ of their length or more (fruiting calyx splits irregularly in *A. curranii*), glabrous to pilose outside. Indumentum of inflorescence axes usually whitish · 44

43 Ovaries and fruits glabrous. Fruits (3 –)4 – 8(– 13) × 4 – 6 mm. Petioles 1 – 2 mm wide. Stigmas 4 – 10. Infructescences consisting of 1 – 10 branches. Leaves (6 –)10 – 15(– 21) cm long, coriaceous, drying grey to reddish brown adaxially, reddish brown abaxially. SUM, MLY, BOR · **11. A. coriaceum**

Ovaries and fruits hairy. Fruits 3 – 4(– 5) × 2.5 – 3(– 3.5) mm. Petioles 0.7 – 1 mm wide. Stigmas 2 – 4. Infructescences consisting of (1 –)10 – 20 branches. Leaves (2 –)4 – 6(– 16) cm long, chartaceous to coriaceous, drying olive-green. SUM, MLY, BOR, JAW, PHI, SUL, LSI, MOL, NWG, THA · (21. *A. ghaesembilla*)

44(42). Fruits 4 – 5 × 3 – 4 mm. Calyx in flower fused to halfway, splitting irregularly in fruit, 1 – 1.5 mm long, much longer than the disc. Inflorescences simple or consisting of up to 6 branches, solitary. PHI · **13. A. curranii**

Fruits 2 – 4 × 1 – 3 mm. Calyx in flower fused for (½ –) ⅔ – ⅘ of its length, not split in fruit, 0.5 – 1 mm long, extending to about the same length as the disc. Inflorescences consisting of 2 – 20 branches, often 2 – 3 per fascicle · 45

45. Calyx bowl-shaped, 0.5 – 0.7 × c. 1 mm. Fruits distinctly laterally compressed, 3 – 4 × c. 3 mm. Fruiting pedicels 1 – 2 mm long. Leaves membranaceous to subchartaceous, usually with domatia in the axils between the midvein and secondary veins abaxially (Fig. 2B), base acute to obtuse, more rarely rounded. Young twigs brown. SUL, LSI, MOL · **7. A. celebicum**

Calyx urceolate to cupular, 0.5 – 1 × 0.7 – 1 mm. Fruits not or slightly compressed, 2 – 3(– 4) × 1 – 2.5 mm. Fruiting pedicels 0.2 – 2 mm long. Leaves chartaceous to subcoriaceous, without domatia, base often obtuse to rounded. Young twigs light brown to whitish · · · 46

46. Disc ferrugineous-pubescent, hairs as long as to slightly longer than the disc itself. Fruits with terminal style, 2 – 3(– 4) × 1 – 2 mm. Fruiting pedicels 0.2 – 0.5(– 1.5) mm long. PHI ······ **31. A. microcarpum**
- Disc sparsely whitish-pilose, hairs much shorter than the disc. Fruits with subterminal style, 3.5 × 2 – 2.5 mm. Fruiting pedicels 1 – 2 mm long. PHI ······························ **6. A. catanduanense**

47(27). Disc glabrous (sometimes with hairs originating from the base of the disc; Fig. 3D) ·································· 48
- Disc hairy all over (Fig. 3F), or at least at the margin (Fig. 3E) ···· 77

48. Fruits 1.5 – 2.5 mm long, glabrous, style lateral. Sepals fused for c. $^2/_3$ of their length. BOR ······················· **56. A. venenosum**
- Fruits more than 3 mm long, glabrous or hairy, style terminal to lateral. Sepals free or fused ································ 49

49. Sepals fused for more than half of their length ················ 50
- Sepals free to fused for up to half of their length (if fused to the middle, then both leads are correct) ······················ 60

50. Fruits roughly bean-shaped, with a lateral style and an asymmetrical base, 5 – 8 × 4 – 6 mm. Leaves chartaceous to submembranaceous. SUM, MLY, THA ··························· **20. A. forbesii**
- Fruits ellipsoid, with a terminal to slightly subterminal (sometimes lateral in *A. stipulare*) style and a symmetrical base, 3 – 18 × 4 – 11 mm. Leaves chartaceous to coriaceous ··················· 51

51. Fruits (3 –)4 – 6(– 8) × 3 – 4(– 6) mm, laterally compressed or terete. Leaves with or without domatia, pilose at least along the major veins ·· 52
- Fruits 5 – 18 × 4 – 11 mm, laterally compressed. Leaves without domatia, completely glabrous or pubescent ······················ 55

52. Inflorescences 2 – 3 cm long, infructescences 2 – 5(– 8) cm long. Leaves with domatia between the midvein and secondary veins abaxially. Calyx urceolate, lobed for up to $^1/_3$ of its length. JAW, THA ······ ···································· **1. A. acidum**
- Inflorescences 3 – 10 cm long, infructescences 6 – 20 cm long. Leaf domatia present or absent. Calyx urceolate or cupular, lobed for $^1/_4$ to almost all of its length ··························· 53

53. Calyx urceolate, lobed for $^1/_4$ – $^1/_2$ of its length, margin never glandular-fimbriate. Leaves often with domatia, ochraceous-pilose at least along the veins abaxially, higher venation finely tessellated (Fig. 9D). Fruits without or with inconspicuous white pustules when dry. THA ····· ································· **47. A. sootepense**
- Calyx cupular, lobed for c. $^1/_2$ of its length, margin often glandular-fimbriate. Leaves without domatia, usually glabrous except along the major veins, higher venation not conspicuously tessellated (Fig. 9A). Fruits usually with conspicuous white pustules when dry ······· 54

54. Plant rheophytic, leaves 4 – 10 times as long as wide. SUM, MLY, BOR, THA ······················ (33c. *A. montanum* var. *salicinum*)

Plant not rheophytic, leaves (2 –)2.5 – 3.5(– 6.6) times as long as wide. SUM, MLY, BOR, JAW, PHI, SUL, LSI, MOL, THA · · · · · · · · · · · · · (33a. *A. montanum* var. *montanum*)

55(51). Midvein sharply keeled abaxially (Fig. 2F). Fruits 10 – 15 × 7 – 11 mm, thinly appressed-puberulent. Leaves coriaceous. BOR · · · · · · · · · · · **3. A. brachybotrys**

Midvein rounded abaxially (Fig. 2C – E). Fruits 5 – 18 × 4 – 11 mm, glabrous or hairy. Leaves coriaceous · 56

56. Leaves drying dark reddish brown or purplish grey to greyish brown, usually elliptic. Young twigs and old petioles whitish. Stigmas 5 – 10. Inflorescences 2 – 4 cm long, infructescences 3 – 14 cm long. Fruiting pedicels 0 – 1.5 mm long. PHI · · · · · · · **30. A. macgregorii**

Leaves drying olive-green to yellowish brown, usually oblong. Young twigs brown. Stigmas 3 – 5(– 6). Inflorescences (4 –)8 – 35 cm long, infructescences 10 – 75 cm long. Fruiting pedicels 0 – 14 mm long · 57

57. Sepals fused for $^2/_3$ – $^3/_4$ of their length; calyx in flower and fruit cupular, 1 – 1.5 mm long, with long ferrugineous hairs at the base inside. Fruits 5 – 11 × 4 – 7 mm. Stipules early-caducous, 4 – 6 mm long. Leaves (5 –)10 – 18(– 32) cm long · 58

Sepals fused for c. $^1/_2$ of their length; calyx in flower and fruit bowl-shaped to cupular, up to 1 mm long, indumentum short or absent, whitish to ochraceous. Fruits 8 – 18 × 6 – 11 mm. Stipules usually persistent, 5 – 60 mm long. Leaves (9 –)20 – 30(– 60) cm long · · 59

58. Abaxial leaf surfaces glabrous except sometimes along the midvein. Ovaries glabrous. Often cultivated. SUM, MLY, BOR, JAW, PHI, SUL, LSI, MOL, NWG, THA · · · · · · · · · · · · · · · **5a. A. bunius** var. **bunius**

Abaxial leaf surfaces pubescent all over. Ovaries sparsely pilose. THA · **5b. A. bunius** var. **pubescens**

59(57). Fruiting pedicels 0 – 1(– 3) mm long. Fruits ovate to ellipsoid or lenticular, often apiculate, with a terminal to subterminal style. Ovaries hairy. Leaves usually drying yellowish brown. SUM, MLY, BOR, THA · **40. A. pendulum**

Fruiting pedicels 1 – 14 mm long. Fruits obliquely ovate (mango-shaped), with a lateral to subterminal style. Ovaries glabrous, very rarely hairy. Leaves drying olive-green. SUM, MLY, BOR, JAW, PHI, SUL, LSI, MOL · **49. A. stipulare**

60(49). Indumentum, especially of young parts and inflorescences, ferrugineous to ochraceous · 61

Indumentum whitish or absent · 67

61. Fruits 4 – 5 × 3 mm. Fruiting pedicels c. 1 mm long. Calyx urceolate, sepals fused for c. $^1/_2$ of their length. Ovaries almost cylindrical. Leaves ferrugineous-pubescent all over abaxially, distinctly velvety to the touch. Leaves up to 18 cm long. MLY, THA · · · **55. A. velutinum**

Fruits 5 – 20 × 3 – 9 mm. Fruiting pedicels 1 – 23 mm long. Calyx cupular, sepals free to fused for c. $^1/_2$ of their length, calyx not urceolate. Ovaries falcate, ovoid or globose. Leaves abaxially pubescent to glabrous. Leaves up to 60 cm long · · · · · · · · · · · · 62

62. Fruits obliquely ovate (mango-shaped) to falcate (sometimes elongate-ellipsoid in rheophytes), always laterally compressed, hairy (old fruits sometimes almost glabrous). Ovaries ovoid to falcate, densely appressed-pubescent. Inflorescences 3 – 30 cm long, infructescences 7 – 65 cm long. Petioles (1 –)2 – 5 mm wide. Leaves (8 –)15 – 30(– 60) cm long. Tertiary veins percurrent, close together (Fig. 2Aa) · · · · 63

Fruits ellipsoid, globose, lenticular, ovate or obovate, compressed or terete, glabrous or hairy. Ovaries globose to lenticular, glabrous or spreading-hirsute. Inflorescences 1.5 – 8 cm long, infructescences 3 – 24 cm long. Petioles 1 – 2 mm wide. Leaves (3 –)6 – 14(– 20) cm long. Tertiary veins weakly percurrent (Fig. 2Ab) to reticulate (Fig. 2Ac), widely spaced · 64

63. Plant rheophytic, leaves more than 6 times as long as wide. Petioles 1 – 3 mm wide. Inflorescences 3 – 5(– 20) cm long, infructescences 7 – 20 cm long. Fruits 7 – 9 × 3 – 6 mm. BOR · **52b. A. tomentosum** var. **stenocarpum**

Plant not rheophytic, leaves less than 6 times as long as wide. Petioles 2 – 5 mm wide. Inflorescences (5 –)10 – 30 cm long, infructescences 10 – 30 (– 65) cm long. Fruits (7 –)9 – 20 × 5 – 9 mm. SUM, MLY, BOR, JAW, PHI, SUL, LSI, MOL, THA · **52a. A. tomentosum** var. **tomentosum**

64(62). Petioles 1 – 2 mm wide. Leaves membranaceous to chartaceous, spreading-hirsute all over abaxially, drying olive-green to reddish green. Fruits ellipsoid to lenticular, laterally compressed, usually conspicuously beaked, pilose, rarely nearly glabrous, usually conspicuously and coarsely white-pustulate. Inflorescences (4 –)7 – 8 cm long. Ovaries spreading-hirsute. Calyx 1 × 1.5 – 2 mm. LSI, MOL, SUL · **17. A. elbertii**

Petioles c. 1 mm wide. Leaves chartaceous to coriaceous, glabrous or pilose only along the major veins (rarely all over) abaxially, drying dark reddish brown. Fruits ellipsoid, globose, ovoid or obovoid, laterally or dorsiventrally compressed, or terete, not beaked, glabrous, rarely thinly puberulent, rarely white-pustulate. Inflorescences 1.5 – 6 cm long. Ovaries glabrous, rarely pilose. Calyx 0.3 – 0.8 × 0.8 mm · 65

65. Stipules 0.5 – 1 mm wide. SUM, MLY, BOR · **36b. A. neurocarpum** var. **hosei**

Stipules 1.5 – 18 mm wide · 66

66. Plant rheophytic, leaves more than 5 times as long as wide. Infructescences 3 – 5 cm long. BOR · · · · · · · **36c. A. neurocarpum** var. **linearifolium**

Plant not rheophytic, leaves less than 5 times as long as wide. Infructescences 6 – 9 cm long. SUM, MLY, BOR, THA · **36a. A. neurocarpum** var. **neurocarpum**

67(60). Young twigs light brown to whitish. Leaves drying reddish brown to grey, often with conspicuous dense intersecondary and perpendicular tertiary veins near the midvein · 68

Young twigs brown. Leaves drying olive-green, without conspicuous perpendicular intersecondary and tertiary veins near the midvein · 69

68. Fruiting pedicels 0 – 1 mm long. Higher order veins in dry material distinctly prominent on both leaf surfaces. Stipules caducous, subulate. Leaves drying reddish brown. Fruits ellipsoid, distinctly laterally compressed, with a subterminal to terminal style. Usually at 400 – 1400 m altitude. SUM, JAW · · · · · · · · · · · · · · · · **32. A. minus**

Fruiting pedicels 1 – 18 mm long. Higher order veins in dry material much less prominent adaxially. Stipules usually persistent, lanceolate, linear or subulate, often falcate. Leaves drying reddish brown to greyish. Fruits ellipsoid, globose, ovate or obovate, dorsiventrally or laterally compressed, or terete, with a subterminal to lateral style. Indifferent to altitude. SUM, MLY, BOR · **36b. A. neurocarpum** var. **hosei**

69(67). Fruits 8 – 18 × 6 – 11 mm, reticulate, laterally compressed. Sepals never glandular-fimbriate. Flowers 4 – 5-merous. Inflorescences simple, 8 – 35 cm long, infructescences 10 – 75 cm long. Stipules 5 – 60 × 0.5 – 30 mm. Leaves (9 –)20 – 50 × (1.5 –)5 – 9(– 14) cm, midvein flat to slightly raised, rarely slightly impressed adaxially. Stipules, petioles and inflorescence axes usually glabrous · · · · · · · · · · · · · · · · · · 70

Fruits 3 – 8 × 2.5 – 6 mm, areolate, not or laterally, or slightly dorsiventrally compressed. Sepals often glandular-fimbriate. Flowers 3 – 5(– 7)-merous. Inflorescences simple or branched, 2 – 10 cm long, infructescences 2.5 – 20 cm long. Stipules 1.5 – 7(– 13) × 0.5 – 1.5(– 3) mm. Leaves 1.5 – 15(– 30) × 0.4 – 6(– 12) cm, midvein impressed adaxially. Stipules, petioles and inflorescence axes usually hairy · 71

70. Fruiting pedicels 0 – 1(– 3) mm long. Fruits ovate to ellipsoid or lenticular, often apiculate, with a terminal to subterminal style. Ovaries hairy. Leaves usually drying yellowish brown. SUM, MLY, BOR, THA · **40. A. pendulum**

Fruiting pedicels 1 – 14 mm long. Fruits obliquely ovate (mango-shaped), with a lateral to subterminal style. Ovaries glabrous, very rarely hairy. Leaves drying olive-green. SUM, MLY, BOR, JAW, PHI, SUL, LSI, MOL · **19. A. stipulare**

71(69). Plant rheophytic, leaves 4 – 10 times as long as wide or narrower than 2 cm. Fruits terete · 72

Plant not rheophytic, leaves (1.2 –)2.5 – 3.5(– 6.6) times as long as wide. Fruits compressed or terete · 73

72. Most leaves 5 – 11 × 1 – 2 cm. Leaf apex acuminate. SUM, MLY, BOR, THA · **33c. A. montanum** var. **salicinum**

Most leaves 3 – 6 × 0.4 – 1.2 cm. Leaf apex acute, sometimes with a rounded apiculum. THA · · · · **33b. A. montanum** var. **microphyllum**

73(71). Fruits not or hardly laterally compressed, with a terminal style · · · · 74

Fruits laterally compressed, with a lateral to terminal style · · · · · · · 75

74. Calyx urceolate, lobed for 1/4 – 1/2 of its length, margin never glandular-fimbriate. Leaves often with domatia, ochraceous-pilose at least along the veins abaxially, higher venation finely tessellated (Fig. 9D). Fruits without or with inconspicuous white pustules when dry. THA · **47. A. sootepense**

Calyx cupular, lobed for about half of its length, margin often glandular-fimbriate. Leaves without domatia, usually glabrous except along the major veins, higher venation not conspicuously tessellated (Fig. 9A). Fruits usually with conspicuous white pustules when dry. SUM, MLY, BOR, JAW, PHI, SUL, LSI, MOL, THA · **33a. A. montanum** var. **montanum**

75(73). Abaxial leaf surfaces spreading-hirsute all over. Calyx hirsute outside. Fruits pilose to nearly glabrous, with a lateral style, usually with conspicuous white pustules when dry. Infructescences (4 –)8 – 24 cm long. LSI, MOL, SUL · **17. A. elbertii**

Abaxial leaf surfaces glabrous, or pilose only along the major veins. Calyx glabrous outside. Fruits glabrous, with a subterminal to terminal style and only occasionally with rather fine white pustules when dry. Infructescences 4 – 9 cm long · · · · · · · · · · · · · · · · · · 76

76. Leaves chartaceous to subcoriaceous, often conspicuously shiny, narrowly elliptic-oblong to ovate or obovate, most leaves 6 – 10 × 2 – 3.5 cm. Petioles 0.5 – 1 mm wide. Petioles, inflorescences and pedicels slender. Fruiting pedicels (2 –)3 – 6 mm long. Sepals free. MLY, PHI, THA · · · · · · · · · · · · · **24a. A. japonicum** var. **japonicum**

Leaves membranaceous to chartaceous, not conspicuously shiny, oblong, more rarely elliptic or obovate, most leaves 10 – 15 × 3 – 5 cm. Petioles 1 – 2 mm wide. Petioles, inflorescences and pedicels not particularly slender. Fruiting pedicels 1 – 3 mm long. Sepals free to fused for up to half of their length. JAW, LSI, SUL · · · · · · · · · **23. A. heterophyllum**

77(47). Sepals fused for more than half of their length · · · · · · · · · · · · · · · · 78

Sepals free to fused for up to half of their length · · · · · · · · · · · · · · 95

78. Infructescences up to 10 cm long, simple or branched · · · · · · · · · · 79

Infructescences more than 10 cm long, always simple · · · · · · · · · · · 90

79. Style in flower and fruit distinctly lateral (Fig. 3L) · · · · · · · · · · · · · 80

Style in flower and fruit terminal (Fig. 3G) to subterminal (Fig. 3J) · · 83

80. Fruits beaked, longer than wide (including the beak). Leaves membranaceous to chartaceous. Midvein usually slightly raised, rarely flat, never impressed adaxially. Infructescences 7 – 10 cm long. SUM, MLY, THA · **28. A. leucocladon**

Fruits not beaked, as long as wide or wider than long. Leaves chartaceous to coriaceous. Midvein impressed, flat or rarely slightly raised adaxially. Infructescences 2 – 7(– 10) cm long · · · · · · · · · 81

81. Exocarp in dry material loosely enfolding the endocarp, thick, neither areolate nor reticulate, without ridges or keels, brittle, minutely scaly. Fruits 5 – 12 mm across, globose, not compressed. MOL, NWG · · · · · · · · · · **2. A. baccatum**

Exocarp in dry material closely enfolding the endocarp, thin, reticulate, areolate or wrinkled (fleshy when fresh), not brittle or scaly. Fruits 3 – 9 mm across, laterally compressed · · · · · · · · · · 82

82. Bracts 0.4 – 0.8 × 0.3 – 0.6 mm. Infructescences 4 – 17 cm long, consisting of 2 – 10 branches, rarely simple. Domatia in the axils between the midvein and secondary veins abaxially (Fig. 2B) usually present. Occasionally cultivated. BOR, PHI, SUL, MOL, NWG · · · · · · · · · · **18. A. excavatum**

Bracts 1 – 1.5 × 0.7 – 1 mm. Infructescences c. 4 cm long, simple. Domatia absent. PHI · · · · · · · · · · **16b. A. edule** var. **apoense**

83(79). Fruits 10 – 11 mm long, beaked. Leaves membranaceous to chartaceous. Midvein usually slightly raised, rarely flat, never impressed adaxially. Infructescences 7 – 10 cm long. SUM, MLY, THA · · · · · · · · · · **28. A. leucocladon**

Fruits up to 8 mm long, not beaked. Leaves chartaceous to subcoriaceous. Midvein impressed to flat adaxially. Infructescences 2 – 9 cm long · 84

84. Fruits pilose. Calyx pubescent outside. Leaf base cordate to rounded. Sepals 5 – 6. PHI, LSI · · · · · · · · · · **50. A. subcordatum**

Fruits glabrous. Calyx glabrous outside. Leaf base acute to obtuse, rarely rounded. Sepals 4 – 5 · · · · · · · · · · 85

85. Calyx urceolate, lobed for no more than 1/3 of its length. Leaves sometimes with domatia in the axils between the midvein and secondary veins abaxially (Fig. 2B) · · · · · · · · · · 86

Calyx cupular, usually lobed for more than 1/3 of its length. Leaf domatia absent · · · · · · · · · · 87

86. Fruiting pedicels 1.5 – 3 mm long. Leaves obovate to elliptic-oblong, margins basally straight to convex, abaxial leaf surfaces pilose to pubescent at least around the domatia, very rarely completely glabrous. Stipules usually persistent, 3 – 7 × 1 – 2 mm. Ovaries and fruits laterally compressed or not. JAW, THA · · · · · · · **1. A. acidum**

Fruiting pedicels 0.5 – 1.3 mm long. Leaves elliptic to ovate, margins often basally concave, abaxial leaf surfaces glabrous except for sometimes some short hairs along the major veins. Stipules early-caducous, c. 0.5 × 0.5 mm. Ovaries and fruits laterally compressed. PHI · · · · · · · · · · **15. A. digitaliforme**

87(85). Fruits laterally compressed. Infructescences 3 – 8 cm long. Disc indumentum ferrugineous. Fruiting pedicels 0 – 2 mm long · · · 88

Fruits terete. Infructescences 6 – 20 cm long. Disc indumentum whitish. Fruiting pedicels 1 – 4 mm long · · · · · · · · · · 89

88. Petioles 3 – 15 mm long. Leaves drying olive-green. Calyx in fruit 1 – 1.5 × 1.5 – 2 mm. Fruiting pedicels 0.5 – 2 mm long. PHI · · · **13. A. curranii**

Petioles 1 – 3 mm long. Leaves drying reddish brown. Calyx in fruit c. 0.4 × 1 mm. Fruiting pedicels 0 – 0.4 mm long. SUL · · · · **4. A. brevipes**

89(87). Plant rheophytic, leaves 4 – 10 times as long as wide. SUM, MLY, BOR, THA · (33c. *A. montanum* var. *salicinum)*

Plant not rheophytic, leaves (2 –)2.5 – 3.5(– 6.6) times as long as wide. SUM, MLY, BOR, JAW, PHI, SUL, LSI, MOL, THA · (33a. *A. montanum* var. *montanum)*

90(78). Fruits 8 – 18 × 6 – 11 mm, always laterally compressed. Indumentum absent or whitish, never ferrugineous. Leaves drying olive-green to yellowish brown. Young twigs brown · 91

Fruits 3 – 8 × 2.5 – 7 mm, not or hardly compressed. Indumentum ferrugineous or whitish. Leaves drying reddish brown or olive-green. Young twigs whitish grey to brown · 93

91. Sepals fused for c. $^3/_4$ of their length, apex acute. Disc hirsute. Infructescences 14 – 28 cm long. BOR (Sabah) · **34. A. montis-silam**

Sepals fused for c. $^1/_2$ of their length, apex obtuse to rounded. Disc sparsely pilose. Infructescences 10 – 75 cm long · · · · · · · · · · · · · 92

92. Fruiting pedicels 0 – 1(– 3) mm long. Fruits ovate to ellipsoid or lenticular, often apiculate, with a terminal to subterminal style. Ovaries hairy. Inflorescences and infructescences robust. Leaves usually drying yellowish brown. SUM, MLY, BOR, THA · · · · · · · · · **40. A. pendulum**

Fruiting pedicels 1 – 14 mm long. Fruits obliquely ovate (mango-shaped), with a lateral to subterminal style. Ovaries glabrous, very rarely hairy. Inflorescences and infructescences rather slender. Leaves drying olive-green. SUM, MLY, BOR, JAW, PHI, SUL, LSI, MOL · (49. *A. stipulare)*

93(90). Infructescences 18 – 45 cm long. Style lateral to subterminal, rarely terminal. Calyx lobed for no more than $^1/_3$ of its length, margin entire. Indumentum ferrugineous to ochraceous, rarely absent. Leaves drying reddish brown, tertiary veins usually perpendicular (at right angle to the midvein). Young twigs first dark brown, then light grey to whitish, or whitish from the beginning. Mostly rheophytic. BOR, PHI (Palawan), SUL · · · · · · **46a. A. riparium** subsp. **riparium**

Infructescences 6 – 20 cm long. Style terminal, rarely subterminal. Calyx usually lobed for more than $^1/_3$ of its length, margin often glandular-fimbriate. Indumentum absent or whitish, never ferrugineous. Leaves drying olive-green, tertiary veins oblique in relation to the midvein. Young twigs brown · 94

94. Plant rheophytic, leaves 4 – 10 times as long as wide. SUM, MLY, BOR, THA · (33c. *A. montanum* var. *salicinum)*

Plant not rheophytic, leaves (2 –)2.5 – 3.5(– 6.6) times as long as wide. SUM, MLY, BOR, JAW, PHI, SUL, LSI, MOL, THA · (33a. *A. montanum* var. *montanum)*

95(77). Indumentum, especially of young parts and inflorescences, ferrugineous to ochraceous · 96

Indumentum whitish or absent · 100

96. Fruits 6 – 20 × 4 – 8 mm, with a lateral to subterminal style. Petioles 1 – 5 mm wide. Infructescences (4 –)7 – 65 cm long. Leaves (5 –)9 – 30(– 55) cm long, drying reddish brown to olive-green · · · · · · · · · · · · · · 97

Fruits 3 – 5 × 3.5 – 4 mm, with a terminal to slightly sublateral style. Petioles 0.7 – 1.2 mm wide. Infructescences 3 – 7 cm long. Leaves (2 –)4 – 10(– 16) cm long, drying olive-green · · · · · · · · · · · · · · · · · · 99

97. Indumentum pale yellow to ochraceous. Ovaries globose to lenticular, spreading-hirsute, with a thin, conspicuous style. Fruits 6 – 8 × 4 – 6 mm, ellipsoid to lenticular, with a lateral style, often conspicuously beaked. Fruiting pedicels 1 – 3 mm long. Infructescences (4 –)8 – 24 cm long. Leaves 5 – 13(– 20) cm long, drying greyish, reddish or olive-green. Stipules 3 – 7 × 0.5 – 1.2 mm, not falcate, coriaceous or strongly veined. Petioles 1 – 2 mm wide. LSI, MOL, SUL · · · · · **17. A. elbertii**

Indumentum ferrugineous to ochraceous. Ovaries ovoid, slightly falcate, appressed-pubescent, with an inconspicuous style. Fruits (7 –)9 – 20 × 5 – 9 mm, obliquely ovate (mango-shaped), with a lateral to subterminal style. Fruiting pedicels 0 – 23 mm long. Infructescences 7 – 65 cm long. Leaves (8 –)15 – 60 cm long, drying reddish brown. Stipules (3 –)6 – 35 × 1 – 10 mm, often falcate, coriaceous, strongly veined. Petioles (1 –)2 – 5 mm wide · 98

98. Plant rheophytic, leaves more than 6 times as long as wide. Petioles 1 – 3 mm wide. Inflorescences 3 – 5(– 20) cm long, infructescences 7 – 20 cm long. Fruits 7 – 9 × 3 – 6 mm. BOR · **52b. A. tomentosum** var. **stenocarpum**

Plant not rheophytic, leaves less than 6 times as long as wide. Petioles 2 – 5 mm wide. Inflorescences (5 –)10 – 30 cm long, infructescences 10 – 30 (– 65) cm long. Fruits (7 –)9 – 20 × 5 – 9 mm. SUM, MLY, BOR, JAW, PHI, SUL, LSI, MOL, THA · · · **52a. A. tomentosum** var. **tomentosum**

99(96). Sepals free, pubescent outside, c. 0.7 mm long. Fruits pilose. Inflorescences consisting of (1 –)10 – 20 branches. Leaves pubescent at least along the major veins abaxially. Common. SUM, MLY, BOR, JAW, PHI, SUL, LSI, MOL, NWG, THA · · · · · · (21. *A. ghaesembilla*)

Sepals fused for ½ of their length or more (splitting irregularly in fruit), glabrous outside, 1 – 1.5 mm long. Fruits glabrous. Inflorescences simple or once-branched, more rarely consisting of up to 6 branches. Leaves glabrous or nearly so. Rare. PHI · · · · · · · · · · · **13. A. curranii**

100(95). Fruits dorsiventrally compressed, 7 – 8 × 4 – 5 mm, pilose. Disc, ovaries and abaxial leaf surfaces (at least along the veins) densely hirsute. MLY, THA · **37. A. orthogyne**

Fruits laterally compressed or terete, 3 – 18 × 2.5 – 11 mm, glabrous or hairy. Disc, ovaries and leaves glabrous or hairy · · · · · · · · · · · · 101

101. Fruits 8 – 18 × 6 – 11 mm, laterally compressed. Inflorescences 8 – 35 cm long, infructescences 10 – 75 cm long, simple. Leaves and inflorescence axes glabrous · 102

Fruits 3 – 8 × 2.5 – 6 mm, laterally compressed or terete. Inflorescences 2 – 10 cm long, infructescences 3 – 20 cm long, simple or branched. Leaves and inflorescence axes glabrous or hairy ············ 103

102. Fruiting pedicels 0 – 1(– 3) mm long. Fruits ovate to ellipsoid or lenticular, often apiculate, with a terminal to subterminal style. Ovaries hairy. Leaves usually drying yellowish brown. SUM, MLY, BOR, THA ······························ **40. A. pendulum**

Fruiting pedicels 1 – 14 mm long. Fruits obliquely ovate (mango-shaped), with a lateral to subterminal style. Ovaries glabrous, very rarely hairy. Leaves drying olive-green. SUM, MLY, BOR, JAW, PHI, SUL, LSI, MOL ························· (49. *A. stipulare)*

103(101). Fruits terete, with a terminal style. Sepals often glandular-fimbriate. Leaves membranaceous to chartaceous, only larger leaves sometimes coriaceous ································· 104

Fruits laterally compressed, with a terminal to lateral style. Sepals never glandular-fimbriate. Leaves chartaceous to coriaceous ······· 106

104. Plant rheophytic, leaves 4 – 10 times as long as wide or narrower than 2 cm ······································· 105

Plant not rheophytic, leaves (1.2 –)2.5 – 3.5(– 6.6) times as long as wide. SUM, MLY, BOR, JAW, PHI, SUL, LSI, MOL, THA ·· (33a. *A. montanum* var. *montanum)*

105. Most leaves 5 – 11 × 1 – 2 cm. Leaf apex acuminate. SUM, MLY, BOR, THA ······················ (33c. *A. montanum* var. *salicinum)*

Most leaves 3 – 6 × 0.4 – 1.2 cm. Leaf apex acute, sometimes with a rounded apiculum. THA ···· (33b. *A. montanum* var. *microphyllum)*

106(103). Fruits 6 – 8 × 4 – 6 mm, with a lateral to subterminal, rarely terminal style. Leaves drying greyish, reddish or olive-green. Infructescences 4 – 24 cm long ································· 107

Fruits 3 – 5 × 2.5 – 4 mm, with a terminal to slightly subterminal style. Leaves drying olive-green. Infructescences 3 – 7 cm long ····· 108

107. Stipules, petioles and leaves glabrous. Young twigs light grey. Leaves chartaceous to coriaceous. SUM, MLY, THA ········ **22. A helferi**

Stipules, petioles and abaxial leaf surfaces hirsute. Young twigs brown. Leaves membranaceous to chartaceous. LSI, MOL, SUL ·· **17. A. elbertii**

108(106). Sepals free, pubescent outside, c. 0.7 mm long. Fruits pilose. Inflorescences consisting of (1 –)10 – 20 branches. Leaves pubescent at least along the major veins abaxially. SUM, MLY, BOR, JAW, PHI, SUL, LSI, MOL, NWG, THA ·············· (21. *A. ghaesembilla)*

Sepals fused for 1/2 of their length or more (splits irregularly in fruiting stage), glabrous outside, 1 – 1.5 mm long. Fruits glabrous. Inflorescences simple or once-branched, more rarely consisting of up to 6 branches, sometimes fascicles of 2 inflorescences. Leaves glabrous or nearly so. PHI ························· **13. A. curranii**

KEY TO THE SPECIES IN NEW GUINEA AND THE SOLOMON ISLANDS

Key for staminate material

1. Leaves with 1 – 2 pairs of marginal glands at the base (Fig. 22B), up to 2.2 cm long. Stamens 7 – 13. Pedicels articulated · · · · **53. A. vaccinioides**
   Leaves eglandular, up to 30 cm long. Stamens 2 – 6. Pedicels not articulated · 2
2. Leaf apex retuse, rounded or obtuse · 3
   Leaf apex acute or acuminate · 4
3. Leaves usually oblong, rounded to cordate, rarely acute basally, (2 –)3 – 4.5(– 9) cm wide. Petioles 4 – 17 × 0.7 – 1 mm. Stipules 3 – 6 mm long. Inflorescences 4 – 8 cm long, consisting of 10 – 20 branches. Stamens 4 – 6 · **21. A. ghaesembilla**
   Leaves spathulate, cuneate-acute basally, (0.7 –)1 – 2.5(– 3.5) cm wide. Petioles 1.5 – 4 × 0.5 – 0.7 mm. Stipules 1 – 3 mm long. Inflorescences c. 3 cm long, simple. Stamens 2 – 4 · · · · · · · · · · **48. A. spatulifolium**

4(2). Midvein raised adaxially. Sepals free or nearly so. Stamens 2 – 3. Disc glabrous · **10. A. contractum**
   Midvein impressed to flat adaxially. Sepals fused or free. Stamens 2 – 6. Disc hairy or glabrous · 5

5. Flowers on pedicels 2 – 4 mm long. Calyx twice as long as wide. Pistillode c. 1.5 × 0.5 – 1 mm · · · · · · · · · · · · · · · **8. A. chalaranthum**
   Flowers sessile or pedicels up to 2 mm long. Calyx shorter than to hardly longer than wide. Pistillode up to 1 × 0.5 mm · · · · · · · · · · · · · · · · 6
6. Stipules 3 – 4 mm wide · · · · · · · · · · **46b. A. riparium** subsp. **ramosum**
   Stipules up to 3 mm wide · 7
7. Flowers 3 – 4 × c. 3 mm, sessile, with cheesy smell (of bat's droppings), attracting flies. Inflorescences (6 –)15 – 25 cm long. Stamens exserted 1.5 – 2 mm from the calyx. Anthers c. 0.5 × 0.7 mm. Disc glabrous. Often cultivated · **5. A. bunius**
   Flowers 1 – 3 × 1 – 3 mm, sessile or pedicellate, without cheesy smell. Inflorescences 1.5 – 13 cm long. Stamens exserted 0.5 – 1.5 mm from the calyx. Anthers 0.2 – 0.5 × 0.2 – 0.7 mm. Disc glabrous · · · · · · · 8
8. Disc cushion-shaped, enclosing the bases of the filament and pistillode (Fig. 3A; observe carefully especially when disc is hairy; dried flowers may have to be softened) · 9
   Disc extrastaminal-annular (Fig. 3B), or consisting of free lobes (Fig. 3C) · 10
9. Sepals fused for $^2/_3$ – $^3/_4$ of their length, never glandular-fimbriate. Disc ferrugineous- to ochraceous-pubescent on top. Stamens 2 – 3. Pistillode absent. Leaves with hairtuft domatia in the axils between the midvein and secondary veins abaxially (Fig. 2B). Inflorescences 1.5 – 5 cm long · **45. A. rhynchophyllum**

Sepals free, to fused to halfway, often glandular-fimbriate. Disc usually glabrous, indumentum whitish. Stamens 3 – 5. Pistillode present. Leaves without domatia. Inflorescences 3 – 13 cm long .... **33. A. montanum**

10(8). Sepals free. Flowers 2 – 3 × 2 – 3 mm, sessile. Inflorescences consisting of 10 – 20 branches .... (21. *A. ghaesembilla)*

Sepals fused for at least 1/2 of their length. Flowers 1 – 2 × 1 – 2 mm, sessile or pedicellate. Inflorescences simple, or consisting of up to 10 branches 11

11. Petioles 1 – 3 × c. 0.5 mm. Leaves 2.2 – 8.5 × 1 – 3 cm, usually ovate, chartaceous, without domatia in the axils between the midvein and secondary veins abaxially (Fig. 2B). Plant slender in all parts ... 12

Petioles more than 3 mm long and/or more than 0.5 mm wide. Leaves 4 – 30 × 2 – 13.5 cm, usually elliptic, chartaceous to coriaceous, with or without domatia. Plant not particularly slender 13

12. Inflorescences 4 – 6 cm long. Pedicel absent. Flowers 3-merous. Disc consisting of 3 free alternistaminal lobes (Fig. 3C), ferrugineous-pubescent on top. Indumentum appressed or lacking .... **9. A. concinnum**

Inflorescences c. 2 cm long. Pedicels c. 1.5 mm long. Flowers 4-merous. Disc extrastaminal-annular, glabrous (Fig. 3B). Indumentum spreading .... **25. A.** cf. **jucundum** (paratype *Hoogland* 4621)

13(11). Disc consisting of free lobes (Fig. 3C; observe carefully; dried flowers may have to be softened). Flowers 1 – 2 × 1 – 1.5 mm 14

Disc extrastaminal-annular (Fig. 3B). Flowers 1.5 – 2 × 1 – 2 mm ... 16

14. Petioles 15 – 45 × 1.5 – 2 mm. Stamens 2 – 4. Stipules ovate, 3.5 – 6 × 2 – 2.5 mm. Leaves (10 –)15 – 29 × (5 –)7.5 – 13.5 cm, tertiary venation well-differentiated from finer venation, not tessellated (Fig. 9A); hairtuft domatia present (Fig. 2B) .... **41. A. petiolatum**

Petioles 5 – 18 × 0.5 – 1.5 mm. Stamens 4 – 5. Stipules linear, c. 10 × 1.5 mm. Leaves 4 – 11(– 16) × 2 – 5(– 6.5) cm, tertiary venation hardly differentiated from finer venation, giving it a tessellated appearance (Fig. 9D); domatia absent 15

15. Leaves glabrous, or puberulous only along the midvein adaxially and along the major veins abaxially ... **35a. A. myriocarpum** var. **myriocarpum**

Leaves hairy all over abaxially .. **35b. A. myriocarpum** var. **puberulum**

16(14). Leaves glabrous, or hairy only along the midvein adaxially and along the major veins abaxially. Indumentum lacking or whitish to pale yellow. Flowers sessile. Occasionally cultivated .... **18a. A. excavatum** var. **excavatum**

Leaves hairy all over abaxially. Indumentum pale yellow to ferrugineous. Flowers sessile or pedicellate 17

17. Flowers sessile .... **18b. A. excavatum** var. **indutum**

Flowers on pedicels (0.5 –)1 – 2 mm long .... **19. A. ferrugineum**

KEY FOR PISTILLATE MATERIAL

1. Leaves with 1 – 2 pairs of marginal glands at the base (Fig. 22B), up to 2.2 cm long. Pistillate plants unknown · · · · · · · · **53. A. vaccinioides**
   Leaves eglandular, up to 30 cm long · 2
2. Leaf apex retuse, rounded or obtuse. Fruits with terminal style · · · · 3
   Leaf apex acute or acuminate. Fruits with lateral to terminal style · · 4
3. Leaves usually oblong, base rounded to cordate, rarely acute, (2 –)3 – 4.5(– 9) cm wide. Petioles 4 – 17 × 0.7 – 1 mm. Stipules 3 – 6 mm long. Infructescences 4 – 7 cm long, usually much-branched, consisting of (1 –)10 – 20 branches. Fruits pilose, 3 – 4(– 5) × 2.5 – 3(– 3.5) mm · **21. A. ghaesembilla**
   Leaves spathulate, base cuneate-acute, (0.7 –)1 – 2.5(– 3.5) cm wide. Petioles 1.5 – 4 × 0.5 – 0.7 mm. Stipules 1 – 3 mm long. Infructescences 1.5 – 3 cm long, simple. Fruits glabrous, 4 – 6 × 3 – 5 mm · **48. A. spatulifolium**
4(2). Midvein raised adaxially. Sepals free or nearly so. Disc glabrous. Fruits and ovaries glabrous, with terminal style · · · · · · · **10. A. contractum**
   Midvein impressed to flat adaxially. Sepals fused to free. Disc hairy or glabrous. Fruits and ovaries glabrous or hairy, with lateral to terminal style · 5
5. Stipules 3 – 4 mm wide. Fruits, ovaries and disc pilose · **46b. A. riparium** subsp. **ramosum**
   Stipules up to 3 mm wide. Fruits, ovaries and disc glabrous or hairy · 6
6. Calyx 1 – 1.5 × c. 1.5 mm, sepals fused for $^{2}/_{3}$ – $^{3}/_{4}$ of their length. Fruits 5 – 11 mm long, always laterally compressed. Inflorescences (4 –)8 – 17 cm long, infructescences 10 – 17 cm long. Disc glabrous. Leaves glabrous or nearly so, without domatia. Often cultivated · **5. A. bunius**
   Calyx 0.4 – 1 × 0.4 – 1.5 mm, sepals free or fused. Fruits 2 – 5 mm long and laterally compressed, or up to 8 mm long and terete. Inflorescences 1.5 – 10 cm long, infructescences 2 – 20 cm long. Disc hairy or glabrous. Leaves glabrous or hairy, with or without domatia · · · · · · · · · · · · · · 7
7. Fruits with a terminal (Fig. 3G) to subterminal (Fig. 3J) style · · · · · · 8
   Fruits with a distinctly lateral style (Fig. 3L) · · · · · · · · · · · · · · · · · 14
8. Petioles 15 – 45 × 1.5 – 2 mm. Leaves 10 – 30 cm long. Fruits 2 – 3 mm long · **41. A. petiolatum**
   Petioles 2 – 18 × 0.5 – 2 mm. Leaves 2 – 30 cm long. Fruits 2.5 – 15 mm long · 9
9. Sepals free or nearly so · 10
   Sepals fused for half or more of their length · · · · · · · · · · · · · · · · · 11
10. Fruits terete, (3 –)4 – 8(– 8) mm long. Fruiting pedicels 1 – 4 mm long. Sepals often glandular-fimbriate. Petioles 1 – 2 mm wide. Leaves 2 – 6.6 times as long as wide, base acute to obtuse, rarely rounded. Inflorescences 4 – 10 cm long; infructescences 6 – 20 cm long, simple or consisting of up to 5 branches. Sepals free, to fused to half of their

length, glabrous to pilose outside. Ovaries and fruits usually glabrous ························· **33. A. montanum**

Fruits laterally compressed, 3 – 4(– 5) mm long. Fruiting pedicels 0 – 2 mm long. Sepals never glandular-fimbriate. Petioles 0.7 – 1 mm wide. Leaves 1 – 2.25 times as long as wide, base rounded to cordate, rarely acute. Inflorescences 2 – 3 cm long; infructescences 3 – 7 cm long, consisting of (1 –) 10 – 20 branches. Sepals free, pubescent outside. Ovaries pubescent, fruits pilose ··········· (21. *A. ghaesembilla)*

11(9). Infructescences 6 – 20 cm long. Fruiting pedicels 1 – 4 mm long. Fruits (3 –) 4 – 6(– 8) × 2.5 – 4(– 6) mm. Leaves without domatia in the axils between the midvein and secondary veins abaxially (Fig. 2B). Sepals often glandular-fimbriate. Petioles 1 – 2 mm wide · (33. *A. montanum)*

Infructescences 2 – 6 cm long. Fruiting pedicels 0 – 1(– 2.5) mm long. Fruits 2.5 – 4(– 6) × 2 – 3 mm. Leaves with hairtuft domatia. Sepals never glandular-fimbriate. Petioles 0.5 – 1.5 mm wide ········· 12

12. Leaf base acute-cuneate. Leaves usually dull, tertiary venation well-differentiated from finer venation, not tessellated (Fig. 9A). Petioles 2 – 7 mm long. Stipules 1 – 3 × 0.2 – 0.7(– 1) mm. Inflorescences 1.5 – 3 cm long. Disc with long hairs (about as long as the disc) at the margin ························· **45. A. rhynchophyllum**

Leaf base rounded to cordate, rarely obtuse. Leaves usually shiny, tertiary venation hardly differentiated from finer venation, giving it a tessellated appearance (Fig. 9D). Petioles (5 –) 7 – 18 mm long. Stipules c. 10 × 1.5 mm. Inflorescences 4 – 7 cm long. Disc glabrous or with short hairs (much shorter than the disc) outside and at the margin ···································· 13

13. Leaves glabrous, or puberulous only along the midvein adaxially and along the major veins abaxially ······················· ····················· **35a. A. myriocarpum** var. **myriocarpum**

Leaves hairy all over abaxially ·· **35b. A. myriocarpum** var. **puberulum**

14(7). Petioles 1 – 3 × 0.5 – 1 mm. Leaves 2.2 – 8.5 × 1 – 3 cm, usually ovate, chartaceous, without domatia in the axils between the midvein and secondary veins abaxially (Fig. 2B). Fruits 4 – 6 mm across ····· 15

Petioles (2 –) 3 – 30 × 1 – 3 mm. Leaves 4 – 30 × 2 – 13.5 cm, usually elliptic, chartaceous to coriaceous, with or without domatia. Fruits 2.5 – 15 mm across ······························ 16

15. Indumentum appressed or lacking ············· **9. A. concinnum**

Indumentum spreading ····················· **25. A. jucundum**

16(14). Exocarp in dry material loosely enfolding the endocarp, thick, neither areolate nor reticulate, without ridges or keels, brittle, minutely scaly. Fruits globose, not compressed. Infructescences simple, more rarely consisting of of 2 – 4 branches ················ **2. A. baccatum**

Exocarp in dry material closely enfolding the endocarp, thin, reticulate, areolate or wrinkled (fleshy when fresh), not brittle or scaly. Fruits laterally compressed. Infructescences simple or branched ····· 17

17. Petioles 3 – 6 × 1 – 1.5 mm. Leaves 5.5 – 10 × 2.5 – 5 cm. Inflorescences, infructescences and pedicels slender. Fruiting pedicels 3 – 4 mm long .... **8. A. chalaranthum**

Petioles (2 –)6 – 30 × 1 – 3 mm. Leaves (5 –)7 – 30 × (2 –)3 – 13.5 cm. Inflorescences, infructescences and pedicels not particularly slender. Fruiting pedicels 0.2 – 3(– 5) mm long 18

18. Leaves glabrous, or hairy only along the midvein adaxially and along the major veins abaxially. Indumentum lacking or whitish to pale yellow. Fruits glabrous to sparsely pilose. Occasionally cultivated .... **18a. A. excavatum** var. **excavatum**

Leaves hairy all over abaxially. Indumentum pale yellow to ferrugineous. Fruits pilose 19

19. Fruits (3 –)5 – 8 × (3 –)5 – 6 mm. Calyx bowl-shaped to cupular, sepals fused for $^2/_3$ – $^4/_5$ of their length. Disc at anthesis longer than or as long as the calyx. Fruiting pedicels 0.3 – 5 mm long .... **18b. A. excavatum** var. **indutum**

Fruits 2.5 – 3.5 × 2 – 3 mm. Calyx campanulate with a truncate base, sepals fused for $^2/_3$ – $^4/_5$ of their length. Disc at anthesis shorter than the calyx. Fruiting pedicels 0.2 – 1.5 mm long .. **19. A. ferrugineum**

MULTIPLE ACCESS IDENTIFICATION AID

This identification aid is an alternative to the dichotomous keys above. It may allow easier identification if the plant material stands out by conspicuous characters, or serve as a starting point if it lacks parts required in the dichotomous keys.

To use this multi-access key, choose distinct characters present in the material to be identified from the list of characters below and note the numbers listed after each of them. The number(s) common to all chosen characters represent(s), ideally, the taxon in question or, more often, a set of taxa. Among these, the final identification is hopefully possible using the descriptions, distribution data (Table 1), dichotomous keys, illustrations or a reference specimen. If none or hardly any of the characters listed below do apply to the material to be identified, it is necessary to use the dichotomous keys.

The numbers after the characters refer to the taxa as follows:

1. *A. acidum*
2. *A. baccatum*
3. *A. brachybotrys*
4. *A. brevipes*
5a. *A. bunius* var. *bunius*
5b. *A. bunius* var. *pubescens*
6. *A. catanduanense*
7. *A. celebicum*
8. *A. chalaranthum*
9. *A. concinnum*
10. *A. contractum*
11. *A. coriaceum*
12. *A. cruciforme*
13. *A. curranii*
14a. *A. cuspidatum* var. *cuspidatum*
14b. *A. cuspidatum* var. *orthocalyx*
15. *A. digitaliforme*
16a. *A. edule* var. *edule*
16b. *A. edule* var. *apoense*
17. *A. elbertii*

18a. *A. excavatum* var. *excavatum*
18b. *A. excavatum* var. *indutum*
19. *A. ferrugineum*
20. *A. forbesii*
21. *A. ghaesembilla*
22. *A. helferi*
23. *A. heterophyllum*
24a. *A. japonicum* var. *japonicum*
24b. *A. japonicum* var. *robustius*
25. *A. jucundum*
26. *A. kunstleri*
27. *A. laurifolium*
28. *A. leucocladon*
29. *A. leucopodum*
30. *A. macgregorii*
31. *A. microcarpum*
32. *A. minus*
33a. *A. montanum* var. *montanum*
33b. *A. montanum* var. *microphyllum*
33c. *A. montanum* var. *salicinum*
33d. *A. montanum* var. *wallichii*
34. *A. montis-silam*
35a. *A. myriocarpum* var. *myriocarpum*
35b. *A. myriocarpum* var. *puberulum*
36a. *A. neurocarpum* var. *neurocarpum*
36b. *A. neurocarpum* var. *hosei*
36c. *A. neurocarpum* var. *linearifolium*
37. *A. orthogyne*
38. *A. pachystachys*
39. *A. pahangense*
40. *A. pendulum*
41. *A. petiolatum*
42. *A. pleuricum*
43. *A. polystylum*
44. *A. puncticulatum*
45. *A. rhynchophyllum*
46a. *A. riparium* subsp. *riparium*
46b. *A. riparium* subsp. *ramosum*
47. *A. sootepense*
48. *A. spatulifolium*
49. *A. stipulare*
50. *A. subcordatum*
51. *A. tetrandrum*
52a. *A. tomentosum* var. *tomentosum*
52b. *A. tomentosum* var. *stenocarpum*
53. *A. vaccinioides*
54. *A. velutinosum*
55. *A. velutinum*
56. *A. venenosum*

LIST OF CHARACTERS

Stipules foliaceous, more than 4 mm wide: 22, 26, 29, 33d, 36a, 36c, 46, 49, 51, 52.
Petioles more than 20 mm long: 2, 11, 12, 14, 16a, 18, 31, 38, 40, 41, 42, 44, 51.
Petioles more than 15 mm long: 2, 5, 6, 7, 11, 12, 14, 16a, 18, 19, 20, 21, 26, 31, 35, 38, 40, 41, 42, 44, 50, 51, 52a.
Petioles more than 2 mm wide: 3, 10, 16a, 18, 27, 29, 34, 38, 39, 40, 43, 46, 49, 52.
Leaf apex rounded to obtuse: 1, 21, 48.
Leaf blades more than 25 cm long: 5, 10, 12, 18, 27, 29, 33a, 34, 37, 38, 39, 40, 41, 43, 46, 49, 52, 54.
Leaf blades less than 1.5 times as long as wide: 16, 21, 24, 48, 50, 53.
Leaf blades more than 5 times as long as wide: 5, 10, 27, 29, 33, 36c, 46, 49, 52.
Leaf blades drying reddish: 2, 3, 4, 5, 8, 10, 11, 12, 14, 15, 16, 17, 18, 19, 22, 27, 29, 30, 32, 36, 39, 46, 47, 52, 55.
Abaxial leaf surfaces hairy all over: 1, 5b, 8, 9, 12, 16a, 17, 18b, 19, 21, 25, 29, 33, 35b, 36, 37, 38, 39, 43, 46, 47, 50, 52, 54, 55, 56.
Glands in lower third of abaxial leaf surfaces (Fig. 22B): 53.
Midvein sharply keeled abaxially, perceptible to the touch (Fig. 2F): 3.
Major veins sharply raised adaxially (Fig. 2E): 10, 24b, 27, 38, 40, 46, 49.

Ultimate venation finely tessellated (Fig. 9D): 44, 47, 55.
Domatia present (Fig. 21B): 1, 2, 7, 8, 15, 18, 35, 45, 47, 48, 55.
Plant cauli- or ramiflorous: 8, 10, 11, 17, 19, 22, 29, 30, 39, 41, 43, 44, 55.
Bracts more than 1.5 mm long (Fig. 15F): 39, 52, 54.
Glandular-fimbriate margin of calyx, bracts or stipules (Fig. 14D): 2, 14, 17, 23, 24, 33, 37, 48.
Sepals three: 3, 5, 9, 10, 14, 18, 23, 24, 29, 30, 33, 36, 41, 43, 44, 45, 47, 53, 55.
Sepals five: 1, 5, 10, 11, 14, 15, 16, 17, 18, 19, 20, 21, 22, 23, 24, 29, 30, 32, 33, 34, 35, 36, 38, 39, 40, 41, 42, 44, 45, 46, 47, 48, 49, 50, 51, 52, 54, 55, 56.
Sepals six or more: 14, 15, 16, 19, 21, 22, 24, 33, 35, 36, 39, 46, 48, 50, 51, 52, 54.
Sepals free or nearly so (Fig. 15F): 10, 11, 12, 13, 14a, 17, 21, 22, 23, 24, 26, 27, 29, 32, 33, 36, 37, 39, 40, 43, 45, 48, 50, 52, 54, 55.
Sepals fused to about half of their length (Fig. 14D): 4, 13, 16, 17, 19, 20, 23, 25, 29, 31, 33, 35, 36, 38, 39, 40, 41, 42, 44, 45, 47, 49, 50, 55, 56.
Sepals fused for $^2/_3$ or more of their length (Fig 7D): 1, 2, 3, 4, 5, 6, 7, 8, 9, 13, 14b, 15, 16, 17, 18, 19, 20, 25, 28, 30, 31, 34, 35, 38, 40, 41, 42, 45, 46, 47, 49, 50, 51, 53, 56.
Staminate pedicels articulated (Fig. 22C): 53.
Staminate inflorescences more than 20 cm long: 5, 29, 38, 39, 40.
Staminate inflorescences more than 10 cm long: 1, 5, 7, 14, 15, 18, 28, 29, 31, 33, 38, 39, 40, 42, 46, 47, 49, 51, 52a, 54.
Staminate flowers 3 mm long or longer: 5, 8, 14, 17, 21, 22, 28, 43, 46, 50, 53.
Staminate pedicels more than 0.5 mm long: 1, 8, 10, 14, 16, 17, 19, 22, 23, 24, 25, 29, 32, 33, 35, 36b, 36c, 39, 43, 45, 46, 47, 53, 54.
Staminate pedicels 0 – 0.1 mm long: 4, 5, 7, 9, 11, 13, 15, 16, 18, 20, 21, 22, 23, 28, 30, 31, 33, 35, 36, 38, 40, 42, 43, 44, 46, 47, 48, 49, 50, 51, 52, 54, 55, 56.
Staminate disc glabrous, sometimes with long hairs from the base (Fig. 7B): 5, 10, 11, 14, 20, 23, 24, 25, 29, 30, 32, 33, 36, 40, 47, 49, 51, 52, 53, 54, 55, 56.
Staminate disc hairy, at least at the margin (Fig. 12H, 16C): 1, 4, 7, 8, 9, 13, 14, 15, 16, 17, 18, 19, 21, 22, 28, 29, 31, 35, 37, 38, 39, 40, 41, 42, 43, 44, 45, 46, 48, 49, 50, 52.
Staminate disc cushion-shaped (Fig. 3A): 1, 4, 5, 10, 14, 16, 17, 18, 22, 23, 24, 29, 30, 31, 32, 33, 36, 37, 39, 43, 45, 46, 49, 52, 54.
Staminate disc extrastaminal-annular (Fig. 3B): 5, 7, 8, 15, 18, 19, 25, 28, 31, 38, 41, 42, 47, 53, 55.
Staminate disc consisting of free lobes (Fig. 3C): 5, 8, 9, 11, 13, 15, 16, 20, 21, 28, 29, 35, 40, 41, 44, 47, 48, 50, 51, 56.
Stamens 2: 1, 4, 10, 41, 45, 48.
Stamens 3: 1, 4, 5, 8, 9, 10, 11, 14, 18, 19, 23, 24, 29, 30, 31, 33, 36, 41, 42, 43, 44, 45, 46, 47, 48, 52, 55.
Stamens 4: 5, 7, 8, 11, 13, 14, 15, 16, 17, 18, 19, 20, 21, 22, 23, 24, 25, 26, 28, 29, 30, 31, 32, 33, 35, 36, 37, 38, 40, 41, 42, 43, 44, 46, 47, 48, 49, 50, 51, 52, 54, 55, 56.
Stamens 5: 5, 11, 14, 15, 16, 19, 20, 21, 24, 29, 33, 35, 36, 38, 39, 40, 42, 46, 49, 50, 51, 52, 54, 56.

Stamens 6 or more: 15, 16, 21, 33, 39, 42, 52, 53, 54.

Pistillode absent (Fig. 6C, 19D, 22E): 1, 4, 10, 29, 30, 32, 36, 43, 45, 53.

Pistillode glabrous (Fig. 13A, 23C): 10, 14, 22, 23, 24, 25, 32, 33, 35, 36, 37, 40, 47, 49.

Pistillode hairy (Fig. 7B): 5, 7, 8, 9, 11, 13, 14, 15, 16, 17, 18, 19, 20, 21, 26, 28, 29, 31, 33, 35, 36, 37, 38, 39, 41, 42, 44, 46, 47, 48, 49, 50, 51, 52, 54, 55, 56.

Pistillate disc glabrous, sometimes with long hairs from the base (Fig. 3D): 3, 5, 10, 11, 12, 14, 17, 20, 21, 23, 24, 26, 27, 29, 30, 32, 33, 35a, 36, 38, 40, 47, 49, 51, 52, 54, 55, 56.

Pistillate disc hairy, at least at the margin (Fig. 3E–F): 1, 2, 4, 6, 7, 8, 9, 13, 14a, 15, 16, 17, 18, 19, 21, 22, 25, 28, 29, 31, 35b, 37, 38, 39, 40, 41, 42, 43, 44, 45, 46, 48, 49, 50, 52, 54.

Gynoecium with terminal style (Fig. 7D): 1, 3, 4, 5, 7, 10, 11, 12, 13, 21, 22, 23, 24, 27, 28, 29, 30, 31, 32, 33, 35, 36, 37, 41, 43, 44, 45, 47, 50, 55.

Gynoecium with subterminal style (Fig. 20A): 5, 6, 7, 11, 12, 13, 14, 15, 17, 21, 22, 23, 24, 26, 28, 29, 30, 32, 33, 34, 36, 37, 38, 39, 40, 41, 45, 46, 49, 51, 52, 54.

Gynoecium with lateral style (Fig. 23L): 2, 8, 9, 14, 16, 17, 18, 19, 20, 22, 23, 25, 28, 34, 36, 38, 42, 45, 46, 56.

Stigmas more than 6: 4, 11, 25, 30, 38, 43, 47, 50, 51, 52, 55.

Ovary and fruit glabrous: 1, 2, 4, 5a, 6, 7, 8, 10, 11, 12, 13, 14a, 15, 16, 18a, 20, 22, 23, 24, 26, 29, 30, 31, 32, 33, 35, 36, 38, 40, 41, 44, 45, 46, 47, 48, 49, 51, 54, 55, 56.

Ovary and fruit hairy: 2, 3, 5b, 7, 9, 12, 14, 15, 16, 17, 18, 19, 20, 21, 22, 25, 26, 27, 28, 29, 30, 31, 34, 35, 36, 37, 38, 39, 40, 42, 43, 45, 46, 49, 50, 52, 54, 55.

Infructescences more than 30 cm long: 29, 38, 39, 40, 46, 49, 52a.

Infructescences more than 15 cm long: 5, 10, 12, 14, 17, 18, 29, 33a, 33d, 34, 36a, 38, 39, 40, 46, 49, 52.

Fruiting pedicel less than 1 mm long: 2, 4, 10, 11, 12, 13, 15, 16, 18, 19, 21, 22, 24b, 27, 29, 30, 31, 32, 34, 35, 40, 41, 43, 44, 45, 48, 50, 52b, 55, 56.

Fruiting pedicel at least 4 mm long: 5, 8, 10, 14, 18, 20, 24a, 29, 33, 36a, 36b, 38, 39, 46, 47, 49, 51, 52a, 54.

Mature fruits less than 5 mm long: 1, 4, 6, 7, 9, 11, 13, 15, 16, 18, 19, 21, 25, 29, 31, 33, 35, 41, 42, 45, 47, 48, 50, 51, 54, 55, 56.

Mature fruits 5–10 mm long: 1, 2, 3, 4, 5, 8, 9, 10, 11, 12, 13, 14, 15, 16, 17, 18, 20, 21, 22, 23, 24, 25, 27, 28, 29, 30, 32, 33, 34, 36, 37, 39, 40, 43, 44, 45, 46, 47, 48, 49, 50, 51, 52, 54, 55.

Mature fruits more than 10 mm long: 2, 3, 5, 11, 18, 27, 28, 32, 34, 36, 38, 40, 43, 49, 52.

Fruit terete (Fig. 3G–H): 1, 2, 11, 29, 31, 33, 35, 36, 39, 41, 43, 45, 46, 47.

Fruit laterally compressed (Fig. 3J–K): 1, 2, 3, 4, 5, 6, 7, 8, 9, 10, 11, 12, 13, 14, 15, 16, 17, 18, 19, 20, 21, 22, 23, 24, 25, 27, 28, 29, 30, 31, 32, 33, 34, 35, 36, 38, 39, 40, 41, 42, 44, 48, 49, 50, 51, 52, 54, 55, 56.

Fruit dorsiventrally compressed (Fig. 3L–M): 33, 36, 37, 39.

Fruit with terminal style (Fig. 3G–H): 1, 3, 4, 5, 7, 10, 11, 12, 13, 21, 22, 23, 24, 27, 29, 30, 31, 32, 33, 35, 37, 39, 40, 41, 43, 45, 46, 47, 48, 50, 55.

Fruit with subterminal style (Fig. 3J–K): 1, 3, 5, 6, 7, 11, 12, 13, 15, 21, 22, 23, 24, 28, 29, 30, 32, 33, 34, 36, 37, 38, 39, 40, 41, 44, 45, 46, 47, 49, 50, 52, 55.

Fruit with lateral style (Fig. 3L–M): 2, 8, 9, 14, 16, 17, 18, 19, 20, 22, 25, 28, 34, 36, 39, 42, 44, 46, 49, 51, 52, 54, 56.

Fruit with asymmetrical base (Fig. 3L, 10D): 7, 11, 12, 14, 16, 17, 18, 19, 20, 22, 23, 24, 25, 29, 32, 34, 36, 37, 39, 40, 44, 45, 46, 51, 52, 54, 56.

1. Antidesma acidum *Retz.*, Observ. Bot. 5: 30 (1788), "acida"; Müll. Arg. in DC., Prodr. 15(2): 249 (1866); C. E. C. Fischer, Bull. Misc. Inform., Kew 1932(2): 65 (1932); Airy Shaw, Kew Bull. 26: 352 (1972); Grierson & Long, Fl. Bhutan 786, fig. 48 h – j (1987). Type: India orientalis, *Koenig* s.n. (LD!, neotype, designated by Fischer 1932: 65, as "type"). — Note: Although Fischer cited the specimen as type, there is no definite proof that it is actually part of the original material seen by Retzius, as it does not bear his handwriting and shows some features suggesting that it is a later addition to Herb. Retzius (Lassen, Curator of LD, in litt.). Fischer's typification is therefore better regarded as a neotype designation.

Stilago diandra Roxb., Pl. Coromandel 2: 35, pl. 166 (1802). — *A. stilago* Poir., Encycl. suppl. 1: 403 (1811), p.p., nom. illeg. according to Art. 11.4. (Greuter *et al.* 2000). — *A. diandrum* (Roxb.) Spreng., Syst. Veg. 1: 826 (1824), nom. illeg. as later homonym of Roth's name (see below), which was published without reference to *Stilago diandra* Roxb. Therefore the Roxburgh type cannot be used in *Antidesma*. — *A. diandrum* var. *genuinum* Müll. Arg. in DC., Prodr. 15(2): 267 (1866), nom. inval. Type: *Roxburgh* s.n. (BM!, lectotype, here designated). — Notes: More authentic material in BM! and BR (*fide* Merrill 1952). There is another specimen in P ex BM, labelled: "Stilago diandra Roxb.!, India, Roxburgh, specimen typicum!". No more authentic material was discovered by Forman (1997: 533). Roxburgh's plate is a copy of a plate of *Icones Roxburghianae* 107 at Kew. The actual specimen after which the plate was drawn in India is unlikely to have been preserved.

Antidesma diandrum Roth, Nov. Pl. Sp.: 369 (1821). Type: India, Circars, 3 July 1798, *Heyne* s.n. (K!, lectotype, here designated). — Note: The specimen indicates that this name is not based on the type of Roxburgh's *Stilago diandra*. It bears the handwriting of one of the "Tranquebar Five" (most probably Roth's) and the name followed by "nob.".

S. lanceolaria Roxb., Fl. Ind. ed. 2, vol. 3: 760 (1832); Hort. Bengal.: 71 (1814), nom. nud. — *A. lanceolarium* (Roxb.) Wall., Num. List: no. 7284 (1832); Wight, Icon. Pl. Ind. Orient. 3(1): 4, t. 766 (prob. 1844). — Note: Airy Shaw (1972a: 353) cited Wight as author of the new combination. Stafleu & Cowan (1983: 957, 1988: 40) gave the publication date of Roxburgh's basionym as "Oct.-Dec. 1832", and that of the respective pages in Wallich's Catalogue as "1832". It is likely that Wallich based his name on Roxburgh's basionym and in the absence of evidence to the contrary, Wallich should be credited with the authorship of this combination. There are three sheets of No. 7284 in K-W representing *A. acidum*, all of which seem to have come from different gatherings. One additional specimen numbered 7284? represents *A. bunius* (L.) Spreng. — *A. lanceolatum* Tul., Ann. Sci. Nat. Bot., Sér. 3: 195 (1851), nom. illeg. according to Art. 52 (Greuter *et al.* 2000). — *A. lanceolatum* Tul. var. *genuinum* Müll. Arg. in DC., Prodr. 15(2): 266 (1866), nom. inval. Type: *Icones Roxburghianae* No. 2561 (no. 2554 on drawing) at Kew (lectotype, here designated). — Note: Iconotype chosen in lack of a Roxburgh herbarium specimen (nothing found in Merrill 1952 or Forman 1997).

A. wallichianum C. Presl, Epimel. Bot.: 235 (1849). Type: India or., Silhet, *Francis de Silva in Hb. Wallich* 7285 B (PRC!, holotype; G!, K!, K-W!, L!, isotypes).

A. diandrum var. *lanceolatum* Tul., Ann. Sci. Nat., Bot., Sér. 3: 199 (1851). Type: India or., *Wallich* 7284 (CGE!, lectotype, here designated; CGE!, G!, isolectotypes). — Note: Tulasne stated explicitly that this name is not based on Roxburgh's name: "Antidesma lanceolarium Wall., herb. n. 7284, non Roxb.", and cited "herb. Lindlaei".

A. diandrum var. *ovatum* Tul., Ann. Sci. Nat. Bot., Sér. 3: 198 (1851). Type: India or., *Hb. Hamilton [F. Buchanan-Hamilton]* s.n. (CGE!, lectotype, here designated).

A. diandrum var. *parvifolium* Tul., Ann. Sci. Nat. Bot., Sér. 3: 198 (1851). — *A. parviflorum* Ham. in sched., Pax & K. Hoffm. in Engl., Pflanzenr. 81: 143 (1922), nom. nud., pro. syn., surely a typing error for "parvifolium". Type: India or., *Hb. Hamilton [F. Buchanan-Hamilton]* s.n. (CGE!, lectotype, here designated).

A. diandrum f. *javanicum* J. J. Sm. in Koord. & Valeton, Meded. Dept. Landb. Ned.-Indië 10: 275 (1910). Type: Java, *Hb. Koorders*, not located. — Note: No material of *A. acidum* in BO is annotated as f. *javanicum*, although many specimens are annotated by J. J. Smith. No material of *A. acidum* with Smith's handwriting has been found in L.

Shrub to tree (*fide Bhargava* 4215: large climber), up to 10 m, diameter up to 10 cm, usually branched from the base. Twigs brown. Bark brown or grey, thin, smooth or roughened, cracked or flaking; inner bark pink. *Young twigs* terete, pilose to pubescent, brown. *Stipules* usually persistent, linear, apically obtuse to rounded, 3 – 7 × 1 – 2 mm, slightly pilose to (especially apically) densely pubescent. *Petioles* flat to channelled adaxially, 2 – 7 × 1 – 2 mm, pilose to densely pubescent. *Leaf blades* obovate to elliptic-oblong, apically long to very shortly acuminate to rounded, basally acute or obtuse, mostly cuneate, (2 –)5 – 10(– 18) × (1 –)2.5 – 4(– 8) cm, (1.6 –)2 – 2.5(– 4.5) times as long as wide, eglandular, chartaceous, glabrous or rarely slightly pilose adaxially, pilose to pubescent at least in the axils between the midvein and secondary veins, rarely completely glabrous abaxially, dull on both surfaces, midvein flat to impressed adaxially, tertiary veins reticulate, drying yellowish green, domatia present. *Staminate inflorescences* 5 – 14 cm long, axillary, simple or branched twice at the base, axes glabrous to pubescent. *Bracts* orbicular to lanceolate, 0.6 – 1 × c. 0.6 mm, glabrous, margin fimbriate. *Staminate flowers* 2 – 2.5 × 1 – 1.5 mm. *Pedicels* 1 – 1.5 mm long, very thin, not articulated, glabrous. *Calyx* c. 0.5 × 0.8 mm, globose to cupular, sepals 4, fused for c. $^2/_3$ of their length, irregularly shaped, glabrous outside, pubescent inside with hairs often exceeding the calyx, margin erose. *Disc* cushion-shaped, enclosing the bases of the filaments, pubescent. *Stamens* 2, rarely some 3-androus flowers among the 2-androus ones, 1.5 – 2 mm long, exserted 1.5 – 2 mm from the calyx, anthers c. 0.3 × 0.6 mm. *Pistillode* absent. *Pistillate inflorescences* 2 – 3 cm long, axillary, simple to branched twice at the base, axes glabrous to pubescent. *Bracts* orbicular to lanceolate, 0.5 – 0.7 × c. 0.5 mm, glabrous, margin slightly erose, sometimes fimbriate. *Pistillate flowers* c. 2 × 1 mm. *Pedicels* 0.2 – 1 mm long, glabrous. *Calyx* c. 1 × 1 mm, urceolate, sepals 4(– 5), fused for c. $^2/_3$ of their length, apically acute, glabrous outside, pubescent inside with hairs often exceeding the calyx, margin erose to entire. *Disc* shorter than the sepals, glabrous inside, glabrous to pilose outside. *Ovary* ovoid, glabrous, style usually terminal, stigmas 3 – 4. *Infructescences* 2 – 5(– 8) cm long. *Fruiting pedicels* 1.5 – 3 mm long, glabrous. *Fruits* ellipsoid, terete or laterally compressed, basally symmetrical, with a terminal to slightly subterminal style, 4 – 6 × 3 – 4 mm, glabrous, reticulate when dry.

DISTRIBUTION. India (including Andaman & Nicobar Islands), Nepal, southern China (Yunnan), Bangladesh, Burma, Laos, Vietnam, Cambodia, Thailand, Indonesia (Java only). Absent from Malesia, except Java. Absent from Sri Lanka, but the endemic *A. walkeri* (Tul.) Pax & K. Hoffm. is very similar. *A. acidum* shows a classical monsoon forest disjunction which is discussed in detail on p. 7. Map 3.

HABITAT. Continental Asia: in dry deciduous, deciduous and evergreen forest; at forest edges, in open spaces and bamboo thickets; in open or half-shady habitats; associated with dipterocarps, pine, oak; secondary, often disturbed, much degraded or frequently burnt vegetation. Java: in teak forest. On sandy gravel in the Andaman and Nicobar Islands; on red volcanic soil in Java; on sand, silt or red lateritic soil, over limestone, granite and shale-granite in Thailand and Vietnam. 0 – 1600 m altitude.

USES. The fruits are eaten, and the young leaves are used in curry and as vegetable in India and Thailand.

CONSERVATION STATUS. Least Concern (LC). The extent of occurrence ranges over 10,000,000 km^2 and the species is very common. 210 specimens examined.

VERNACULAR NAMES. India: kati khatai, khujueva or khujuura, nakham tenga (Assamese); Burma: ihaka, kin-malin; Thailand: haw cha (Karen), kho cho la, mao sai or mug mao, nong sa (Karin); Java: ande-ande, ondeh ondeh, onjam, kenjam.

ETYMOLOGY. The epithet refers to the sour taste of the fruits (Latin, acidus = acid).

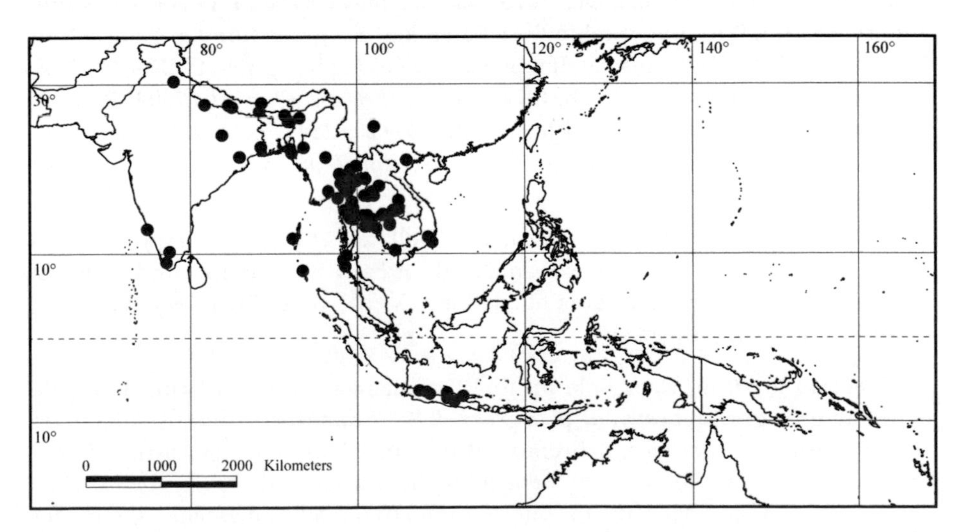

MAP 3. Distribution of *A. acidum*.

KEY CHARACTERS. Flowers 2-staminate; pistillode absent; sepals highly fused; leaves often obovate, usually with domatia.

SIMILAR SPECIES. *A. montanum* has longer pistillate inflorescences and infructescences, less highly fused sepals without glandular margins, and terete fruits.

SELECTED SPECIMENS. INDIA. Tamil Nadu, Madura distr., Pulney Hills, Periakulam road, 2000 ft, *Anglade* 532 (G); Andaman and Nicobar Islands, South Andamans, Poona Nallah, inland forest, sandy gravel soil, c. 100 m, 24 July 1976, *Bhargava* 4215 (L); NW India, Dehra Dun, Birani Naddi, Sept. 1882, *Duthie* 2407 (G); Mangalor, 1849, *Hohenacker* Pl. Indiae or. (Terr. Canara) 167 (B, G, L, MEL). NEPAL. Chitwan distr., NE section of the Park, S of Rapti R., Narayani Zone, c. 83 km W of Kathmandu, above Paidi Choki, Churia Hills, along E boundary, 27°30'N, 84°35'E, hill forest with *Semecarpus* and *Shorea*, > 500 m, 9 June 1976, *Troth* 875 (B). CHINA. Yunnan prov., Szernao forest, 5000 ft, *Henry* 13032 (L). BANGLADESH. Chittagong distr., Chunati Forest Range, Harbhang, 9 June 1979, *Khan, Huq & Rahman* 5467 (L). LAOS. Vientiane, 8 June 1952, *Vidal* 1790 (P). CAMBODIA. Ruins of Angkor, outer temple area, 16 Oct. 1966, *Schwabe* s.n. (B). BURMA. Tenasserim, Paratoba 2500 ft, 26 Jan. 1877, *Gallatly* 176 (L). VIETNAM. Annam, Phanrang prov., Cana, 28 Nov. 1923, *Poilane* 8843 (K, L, P). THAILAND. Northern, Chiang Mai prov., 10 km S of Bo Luang along the Om Koi trail, 18°04'N, 98°22'E, upper mixed deciduous forest, 1000 m, 2 July 1968, *Larsen, Santisuk & Warncke* 1951 (K, L); North-eastern, Nakhon Phanom prov., Down Tan, in mixed deciduous forest, c. 100 m, 15 May 1932, *Kerr* 21470 (K, L); Eastern, Buri Ram prov., Khao Phanom Rung, near Buri Ram, 14°32'N, 102°56'E, edge of sparse secondary forest along road, sunny to half shaded, c. 370 m, 3 Oct. 1984, *Murata et al.* T 37361 (AAU, L); South-western, Kanchanaburi prov., between Huay ban kao and Kritee, 15°00'N, 98°50'E, disturbed mixed deciduous forest, on limestone, 450 m, 2 July 1973, *Geesink & Phengkhlai* 6075 (K, L); Central, Saraburi prov., Na Pra Larn distr., Ban Khrua, dry limestone hill, 100 – 200 m, 6 Oct. 1979, *Shimizu, Toyokuni et al.* T 17944 (L); South-eastern, Prachin Buri prov., Krabin Buri, Ban Keng, evergreen forest, c. 25 m, 9 Nov. 1930, *Kerr* 19821 (K); Peninsular, Ranong prov., Khlong Nang Yon, 9°45'N, 98°40'E, secondary vegetation, 0 m, 25 Feb. 1974, *Geesink, Hiepko & Phlengklai* 7602 (B, K, L). JAVA. Semarang, Karangasem, 14 June 1897, *Koorders* 28977 (L); Indramajoe res., Forestry, Plosokerep, c. 30 m, 4 Jan. 1936, *Van Steenis* 7498 (A, BO, L).

2. Antidesma baccatum *Airy Shaw*, Kew Bull. 23: 287 (1969). Type: Western New Guinea, Fak-Fak Division, Kowap, North of Fak-Fak, young secondary forest on limestone with thin clay cover, alt. 410 m, 27 Feb. 1962, *BW (Vink)* 12193 (K!, holotype; A!, B!, L!, isotypes).

Tree, up to 20 m, clear bole up to 12 m, diameter up to 12 cm. Bark pale ochreous-grey, greyish-brown or pale green, 0.2 – 2 mm thick, smooth, sometimes slightly pustular; inner bark ochreous-red to ochre, blaze cream-brown, c. 5 mm thick; sapwood pale yellow, c. 0.5 cm thick, heartwood red. *Young twigs* slightly striate to angular, glabrous to pilose to ferrugineous-pubescent; sometimes individual trichomes very thick, forming a crust that rubs off in patches as the twig grows older. *Stipules* early-caducous, deltoid, apically acute, 1 – 4 × 0.8 – 1.5 mm, pubescent. *Petioles* channelled to nearly flat adaxially, basally and distally sometimes slightly pulvinate and geniculate, 3 – 15(– 25) × 1 – 2 mm, indumentum as on young twigs, becoming glabrous when old. *Leaf blades* oblong to elliptic, apically acuminate to caudate-mucronate, with a rounded to acute apiculum, basally obtuse to acute, up to 7 mm long decurrent, (5 –)10 – 15(– 22) × (2 –)4 – 6.5(– 9.5) cm, (1.7 –)2.3 – 2.7(– 3.1) times as long as wide, eglandular, chartaceous to coriaceous, glabrous, except sometimes for short hairs along the major veins abaxially or/and along the midvein adaxially, shiny to dull on both surfaces, midvein flat, shallowly impressed or slightly raised adaxially, tertiary veins reticulate to weakly percurrent, widely spaced, drying olive-green to reddish brown, domatia sometimes present though not very pronounced. *Staminate plants* unknown. *Pistillate inflorescences* axillary, simple or consisting of up to 4 branches, robust, axes pilose to pubescent. *Bracts* deltoid to irregularly shaped, apically acute to lacerate, 0.5 – 1 × 0.5 – 1 mm, glabrous, margin fimbriate, sometimes glandular. *Pistillate flowers* not known. *Calyx* in young fruits c. 1 × 1.5 – 2 mm, shallowly bowl-shaped with a truncate base, sepals 4, fused for 3/4 to all of their length, apically rounded to acute, sinuses wide, rounded, glabrous, margin entire. *Disc* extending to the same length as or exserted from the sepals, but indumentum always exserted from the calyx, ferrugineous-tomentose at the margin, otherwise glabrous, hairs as long as or slightly shorter than the disc. *Ovary* lenticular, very sparsely pilose, style lateral, stigmas 3 – 5(– 6). *Infructescences* (2 –)3 – 6.5 cm long, axes c. 1 mm wide. *Fruiting pedicels* (0.5 –) 1 – 2.5 mm long, pilose to nearly glabrous. *Fruits* globose, not or maybe slightly laterally compressed, basally symmetrical, with a distinctly lateral style, 5 – 12 × 5 – 12 mm, glabrous or rarely sparsely pilose to sparsely tomentose, minutely scaly, sometimes white-pustulate, exocarp loosely enfolding the endocarp, neither areolate nor reticulate, without keels or ridges, brittle when dry. Fig. 5A–C.

DISTRIBUTION. Moluccas (Seram, Obi Island); Indonesia: Papua (Vogelkop Peninsula and Fak-Fak division), and Papua New Guinea (Central, Gulf, Madang, Morobe and Western provinces). Map 4.

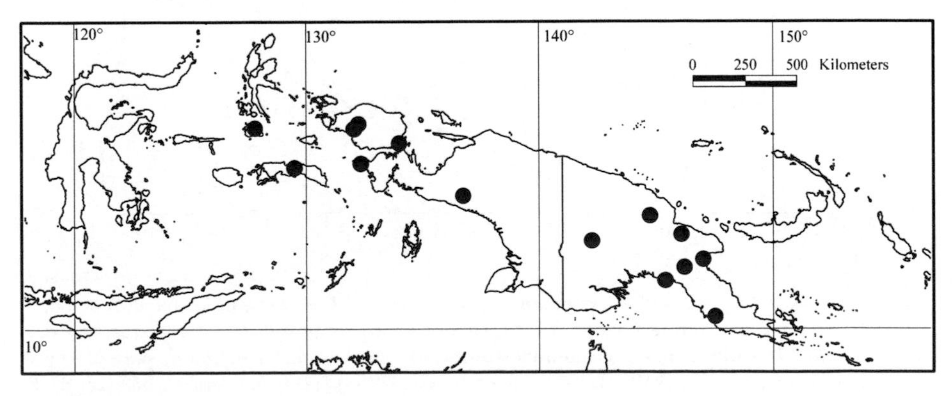

MAP 4. Distribution of *A. baccatum.*

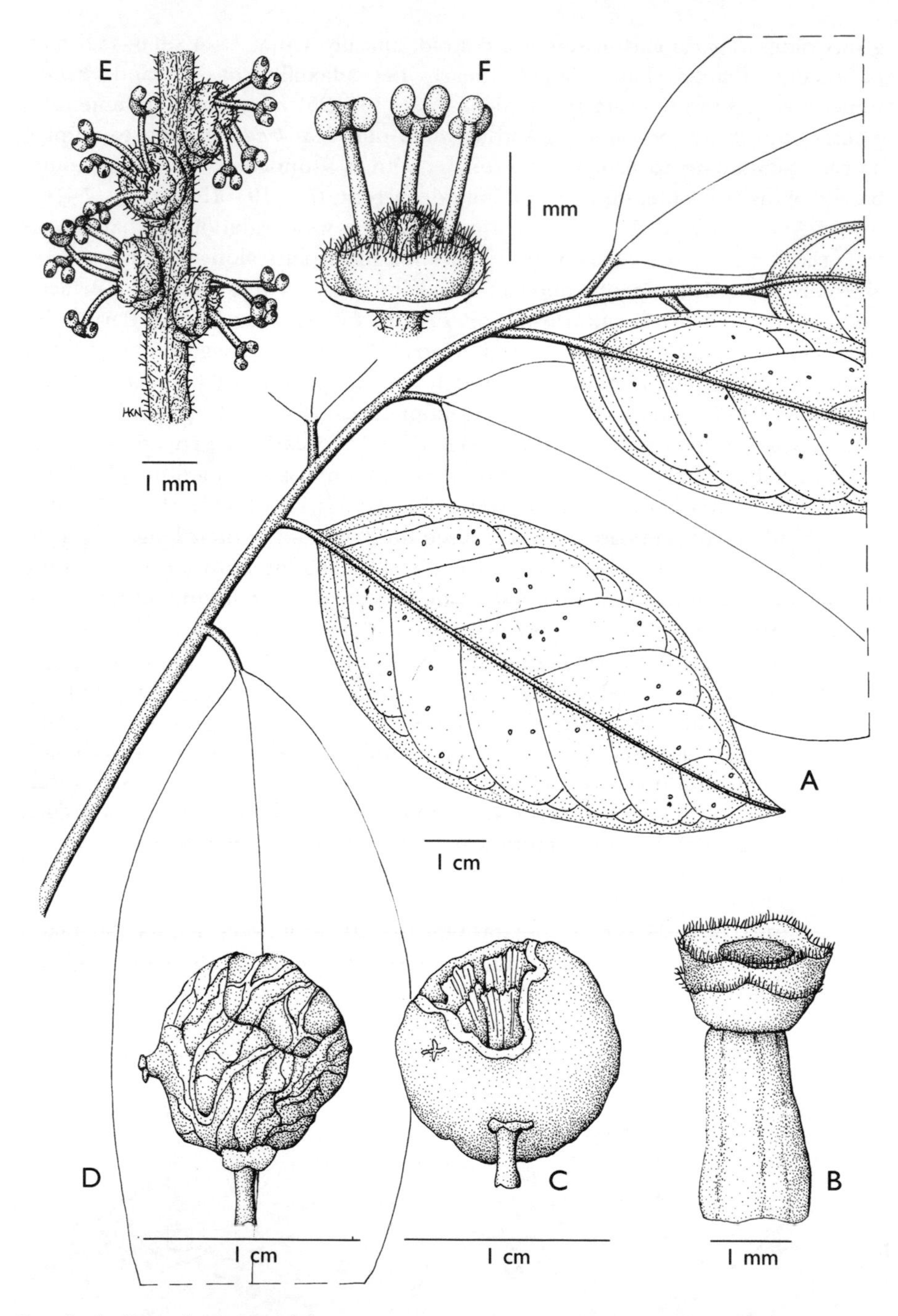

FIG. 5. *Antidesma baccatum* and *A. excavatum* var. *excavatum*. **A – C** *A. baccatum*. **A** habit; **B** pedicel, calyx and disc in fruiting stage; **C** fruit, part of exocarp removed. **D – F** *A. excavatum* var. *excavatum*. **D** fruit; **E** part of staminate inflorescence; **F** staminate flower, part of calyx removed. **A** from *de Vogel* 4196 (L); **B – C** from *Schodde* 4226 (L); **D** from *Hartley* 11963 (L); **E – F** from *NGF (McDonald)* 8221 (L). Drawn by Holly Nixon.

HABITAT. In (rain) forest; in primary or secondary, sometimes much disturbed vegetation; associated with *Nothofagus* and *Castanopsis.* On clay soil over limestone. The species seems to have an affinity to limestone but collections are too few to be certain. 10 – 1650 m altitude.

CONSERVATION STATUS. Near Threatened (NT). Extent of occurrence is very large (over 600,000 km^2), but only 16 specimens have been examined. This species is likely to be locally common but there are too few collections to classify it as Least Concern.

VERNACULAR NAMES. Indonesia: Papua: offu, po kamit; PNG: suo (Samotang).

ETYMOLOGY. The epithet refers to the large pulpy fruits (Latin, baccatus = berry-like, pulpy).

KEY CHARACTERS. Large, globose, glabrous fruits; dried exocarp loosely enfolding the endocarp, smooth (neither areolate nor reticulate), without ridges, thick, brittle.

SIMILAR SPECIES. *A. excavatum* has more abundantly branched inflorescences and fruits with a "regular" exocarp (Fig. 5). — *A. riparium* has shorter petioles, larger stipules, longer infructescences and fruits with a "regular" exocarp and dorsal and ventral ridges.

NOTE. *Schodde (& Craven)* 5025 and 5034, fruiting collections from Aseki valley, Morobe province, Papua New Guinea, differ from typical specimens in their percurrent and perpendicular tertiary leaf venation (cf. Fig. 2A).

SELECTED SPECIMENS. MOLUCCAS. Obi Island, Anggai, Gunung Batu Putih, 1°24'S, 127°48'E, much depleted very open forest with much secondary growth, hill ridge, rather sloping, with deep clayey soil, 200 m, 20 Nov. 1974, *De Vogel* 4200 (K, L); Seram, Seram Utara distr., Manusela National Park, along a trail between Kanikeh and Selumena, 3°06'S, 129°30'E, 620 – 820 m, 8 Jan. 1985, *Kato, Sunarno & Akiyama* C 3360 (L). INDONESIA: PAPUA. McCluer Gulf, Anakasi near Babo, c. 50 m, 1941, Expedition Lundquist, *Aet* 128 (L); Beriat, c. 12 km S of Teminaboean, secondary forest, clayey soil, c. 10 m, 19 April 1958, *BW (Versteegh)* 4902 (L); Central Irian Jaya prov., Freeport Mining Concession, Golf course at Kuala Kencana, 04°24'28"S, 136°52'21"E, edge of primary humid forest along 18th fairway, 20 Nov. 2000, Cowry *et al.* 5245 (K, MO); Vogelkop Peninsula, N of Ayawasi, 1°14'S, 132°12'E, secondary forest, 7 Nov. 1995, *Polak* 917 (L). PAPUA NEW GUINEA. Morobe distr., Menyama subdistr., Tawa village near Aseki, *Nothofagus*-dominated montane forest, near ridge top, 1650 m, 16 May 1968, *NGF (Streimann & Kairo)* 27636 (K, L); Madang distr., E Madang distr., between Wamundi and Budemu, S slopes of the Finisterre Range, upper Gusap valley, rain forest on ridge saddle, c. 4150 ft, 24 Oct. 1964, *Pullen* 6033 (L); Gulf distr., E base of Ihu hill, c. 1 mile N of Ihu, Vailala R., primary forest, c. 40 ft, 13 Jan. 1966, *Schodde & Craven* 4226 (K, L, US); Morobe distr., Aseki valley, c. 3 miles SE of Aseki, secondary ridge forest, c. 4000 ft, 25 April 1966, *Schodde & Craven* 5025 (K, L, US).

3. Antidesma brachybotrys *Airy Shaw,* Kew Bull. 26: 457 (1972). Type: Borneo, Sarawak, Ulu Sungei Sekaloh, Niah R., mixed dipterocarp forest on clay loam soil, alt. 90 m, 26 Nov. 1966, *(S) Anderson, Sonny Tan & E. Wright* 26069 (K!, holotype; A!, L!, SAR!, SING, isotypes).

Shrub or tree, up to 5 m, diameter up to 5 cm. Twigs grey or greyish green. Bark grey or grey-brown, smooth; inner bark pale yellow; wood reddish or pale yellow, very hard and dense. *Young twigs* terete, puberulent, soon becoming glabrous, brown. *Stipules* caducous, narrowly deltoid to linear, apically acute, 5 – 8 × 1 – 1.5 mm, pilose to pubescent. *Petioles* broadly channelled to flat adaxially, 2 – 10(– 13) × (1 –) 1.5 – 3 mm, puberulent, soon becoming glabrous. *Leaf blades* elliptic to slightly obovate, apically acuminate-mucronate, basally acute to obtuse, (9 –)15 – 20(– 24) × (3.5 –)5 – 7(– 8) cm, (2.3 –)2.8(– 3.4) times as long as wide, eglandular, coriaceous, glabrous, or slightly puberulent only along the midvein abaxially, moderately shiny to dull on both surfaces, midvein flat to broadly impressed adaxially, sharply keeled abaxially, tertiary veins reticulate, widely spaced, drying reddish brown, more rarely yellowish brown, domatia absent. *Staminate plants* unknown. *Pistillate inflorescences* 2.5 – 4 cm long, axillary, simple, solitary or rarely 2 per fascicle, axes ochraceous- to ferrugineous-pubescent. *Bracts* linear to deltoid, apically usually acute, 0.5 – 0.7 × 0.3 – 0.5 mm, pilose to pubescent. *Pistillate flowers* c. 2 × 1 mm. *Pedicels* 0 – 0.5 mm long, puberulent. *Calyx* c. 1 × 1 mm, cupular, sepals 3 – 4, fused for c. $^2/_3$ of their length, apically acute, sinuses wide, shallow, sometimes irregular, sparsely pilose to puberulent outside, glabrous inside, margin entire. *Disc* shorter than the sepals, glabrous. *Ovary* ellipsoid, densely appressed-pubescent, style terminal, thick, stigmas 6, flatly spread out. *Infructescences* 5 – 11 cm long, robust. *Fruiting pedicels* stout, 1 – 3 mm long, pilose to pubescent. *Fruits* ellipsoid, distinctly laterally compressed, basally symmetrical, with a terminal, more rarely slightly subterminal style, 10 – 15 × 7 – 11 mm, thinly appressed-puberulent, not pustulate, coarsely reticulate when dry.

DISTRIBUTION. Borneo: East Sarawak (Bintulu, Kapit and Miri divisions), Brunei and north-eastern Kalimantan Barat. The two specimens from Johore, Peninsular Malaysia (*SF (Corner)* 28984 and 29433) cited in the protologue are erroneously assigned to this species and represent *A. neurocarpum*. Map 5.

HABITAT. In mixed dipterocarp forest, hill dipterocarp forest and mixed lowland forest, often by streams. On sandy soil over Belait sandstone in Brunei; on clay loam soil in Sarawak. 30 – 350 m altitude.

CONSERVATION STATUS. Near threatened (NT). The extent of occurrence (c. 25,000 km^2) is too large to justify a higher threat category. 15 specimens examined.

ETYMOLOGY. The epithet refers to the inflorescence (Greek, brachys = short, botrys = bunch, raceme).

KEY CHARACTERS. Midvein sharply keeled abaxially; fruits thinly but consistently puberulent; disc glabrous; short, rigid inflorescences; leaves large, strongly coriaceous.

SIMILAR SPECIES. *A. helferi* has less highly fused, rounded sepals, a pilose disc, thinner petioles and smaller leaves. — *A. macgregorii* has whitish twigs as well as smaller stipules, leaves and fruits. — *A. brachybotrys* differs from *A. montis-silam* in its puberulent fruits, its more coriaceous, reddish drying leaves, its glabrous disc

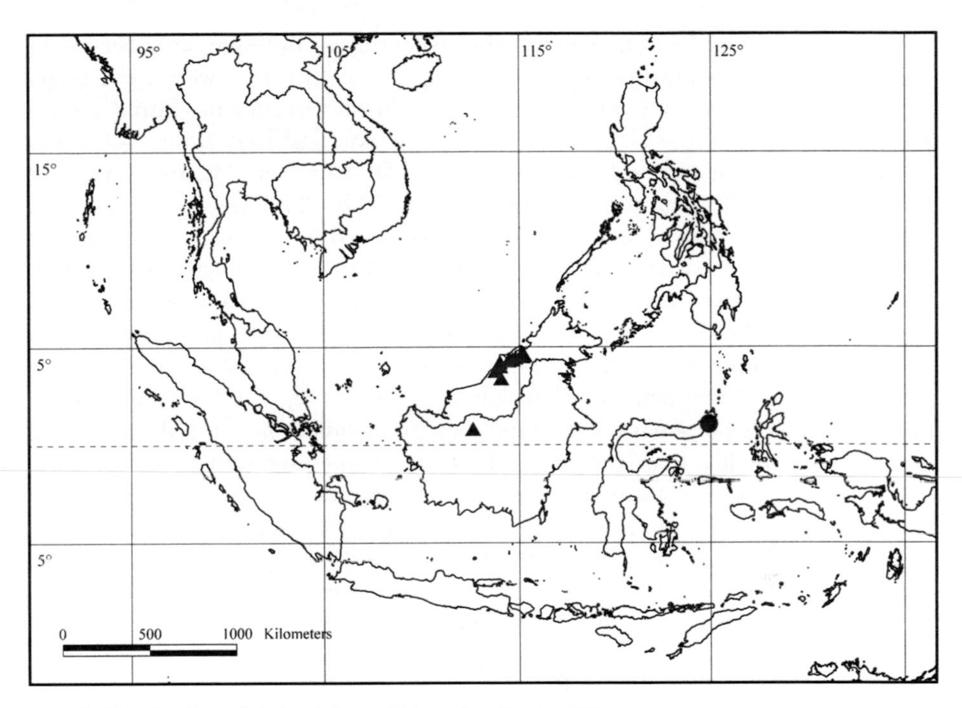

MAP 5. Distribution of *A. brachybotrys* (▲) and *A. brevipes* (●).

and its shorter, more rigid inflorescences. — It is distinguished from *A. neurocarpum* by its larger, consistently puberulous and laterally compressed fruits, and its wider petioles. — *A. pendulum* has longer inflorescences and infructescences, and larger, yellowish drying leaves. — *A. tomentosum* has longer, laxer infructescences and a percurrent tertiary leaf venation.

SELECTED SPECIMENS. BORNEO. BRUNEI. Belait distr., Labi subdistr., Sungai Rampayoh, 2.5 km from road at Waterfall No. 1, 4°00'N, 114°00'E, mixed dipterocarp forest, river bank, valley bottom, sandy soil, 30 m, 9 Jan. 1994, *Coode* 7779 (K); Labi, Bukit Teraja, N of Rest House, 4°18'N, 114°26'E, hill dipterocarp forest, by stream and on sides of valley and along ridge, 350 m, 17 Oct. 1991, *Simpson* 2091 (K, KEP); Labi Hills Forest Reserve, Bukit Telingan, mixed dipterocarp forest, 23 Oct. 1989, *Wong* 1582 (K, L). SARAWAK. Miri div., Niah, Gunong Subis area, Jan. 1961, *S (Mohidin)* 21689 (K); Miri div., Sg. Ulu Bakong on ridge, 100 ft, 15 March 1966, *S (Sibat ak Luang)* 24372 (BO, K, KEP, L, SAR); Miri distr., proposed Lambir Hills National Park, nearby stream, 3 May 1966, *S (Banyeng ak Nudong)* 25060 (K); Bintulu distr., Tubau, Bukit Sekiwa, mixed dipterocarp forest, 25 June 1986, *S (Abang Mohtar et al.)* 53915 (K, KEP, L, SAR). KALIMANTAN. Behind Embalah [probably: Embaluh], swamp forest, 17 Oct. 1949, Expedition E. Polak, *Main* 2195 (K, L).

4. Antidesma brevipes *Petra Hoffm.*, Kew Bull. 54: 348 (1999). Type: Sulawesi, Sulawesi Utara (Manado distr.), N of Mt Klabat, Wiau complex, Mt Tuandei, Sani, 700 m, forest, 29 June 1956, *Forman* 316 (K!, holotype; L!, US!, isotypes).

Tree, up to 15 m. *Young twigs* terete, shortly ferrugineous-hispid, soon becoming glabrous, dark brown at first, soon becoming very light brown with a red tinge. *Stipules* early-caducous, deltoid, c. 1.5 × 0.6 mm, shortly ferrugineous-hispid. *Petioles* hardly channelled adaxially, 1.5 – 2 × 1.2 – 1.5 mm, glabrous, soon becoming whitish grey and rugose. *Leaf blades* oblong, apically acuminate (all tips damaged), basally acute, mostly concave, 5 – 8 × 1.7 – 2.8 cm, 2.8 – 3.3 times as long as wide, eglandular, coriaceous, glabrous, shiny adaxially, dull abaxially, midvein flat adaxially, tertiary veins reticulate, widely spaced, hardly prominent, drying dark reddish brown adaxially, bright reddish brown abaxially, domatia absent. *Staminate inflorescences* 2 – 3 cm long, axillary, simple, axes c. 0.4 mm wide, pilose. *Bracts* ovate to lanceolate, apically acute, 0.2 – 0.3 × c. 0.2 mm, pilose. *Staminate flowers* c. 2 × 1 – 1.5 mm. *Pedicels* c. 0.1 mm long, not articulated, glabrous. *Calyx* c. 0.8 × 1 mm, conical to cupular, sepals 4, fused for c. $^{1}/_{2}$ of their length, deltoid, thick, apically acute, the abaxial lobe smaller than the other three, ferrugineous-hispid outside, nearly glabrous inside, margin entire. *Disc* cushion-shaped, fully enclosing the bases of the filaments, pilose. *Stamens* 2 (in one flower there are 3 stamens, 2 of which are inserted in the same disc excavation), antisepalous with regard to the two lateral sepals, c. 1.5 mm long, exserted c. 1 mm from the calyx, anthers c. 0.3 × 0.5 mm. *Pistillode* absent. For *pistillate plants* and *fruits* see note below. Fig. 6.

DISTRIBUTION. Sulawesi. Map 5.

HABITAT. In forest. 700 m altitude.

CONSERVATION STATUS. Data Deficient (DD): only known from the staminate type and three tentatively matched pistillate collections.

ETYMOLOGY. The epithet refers to the very short petioles (Latin, brevis = short; pes = foot).

KEY CHARACTERS. Plant almost glabrous; leaves nearly sessile, coriaceous; inflorescences short; disc hairy; stamens 2; pistillode absent.

SIMILAR TAXA. *A. contractum* has longer petioles and stipules, larger leaves with raised midveins and distinctly decurrent leaf bases, longer pedicels in staminate flowers, free sepals and a glabrous disc. — *A. coriaceum* has longer petioles and stipules, larger leaves with impressed midveins, sessile staminate flowers, free sepals, free glabrous disc lobes, 3 – 5 stamens and a pistillode. — *A. neurocarpum* var. *hosei* has a glabrous disc and 3 – 5 stamens.

NOTE. At the proof stage of this publication, five superficially very similar collections from Northern Sulawesi came to my attention (250 km W of Gorontalo, 75 km inland from Papayuto, on tributary of Sungai Papayuto, 0°45'N, 121°30'E, 150 m, 26 – 30 March 1990, *Burley et al.* 4067, 4131, 4209, 4219, 4222, all A). The two staminate specimens (Burley *et al.* 4067 and 4222) agree in every detail with *A. riparium* var. *riparium* which is known from Central and Southeast Sulawesi.

The fruits of the three pistillate specimens, however, are very different from those of *A. riparium* and agree more with *A. celebicum.* The plant differs from *A. celebicum* in the shorter, thicker petioles, the leaves drying reddish, the absence of domatia, the inflorescences fasciculate but never branched, more stigmas, shorter fruiting pedicels but slightly larger fruits. Besides the fruits, differences to *A. riparium* include the shorter petiole, shorter, more strongly fasciculate

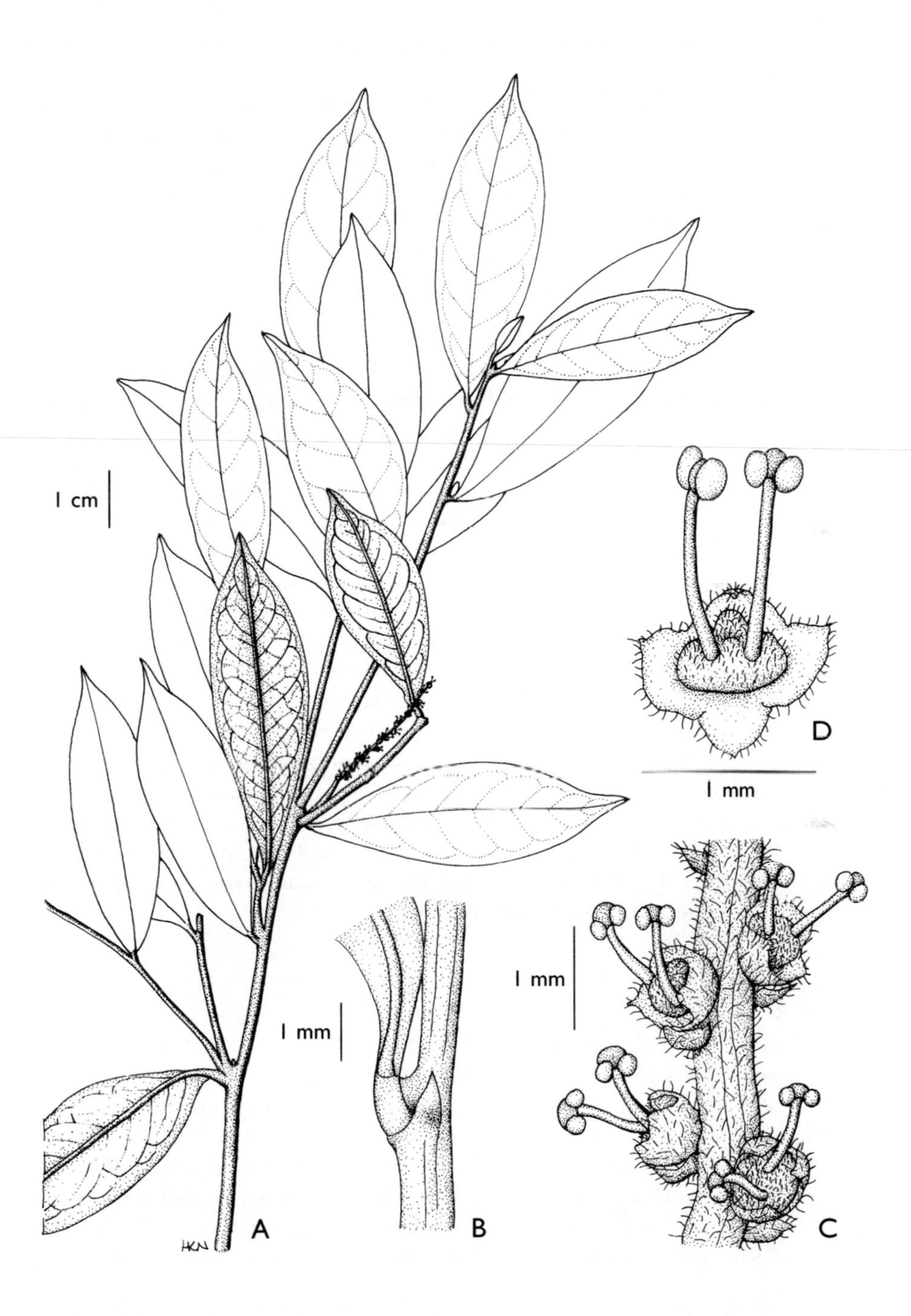

Fig. 6. *Antidesma brevipes.* **A** habit with staminate inflorescence; **B** part of branch with stipule; **C** part of staminate inflorescence; **D** staminate flower. From *Forman* 316 (L). Drawn by Holly Nixon.

inflorescences, smaller stipules, bracts and flowers, shorter pedicels and, again, higher number of stigmas.

In vegetative and inflorescence structure, the three pistillate specimens agree very well with *A. brevipes* which is only known from the staminate type collection. Although it is difficult to imagine that the staminate and pistillate specimens collected so closely together do not belong to the same taxon, it seems even more unlikely that an undescribed species should have staminate individuals identical with *A. riparium*, and completely different pistillate plants. The pistillate material (*Burley et al.* 4131, 4209 and 4219) is therefore tentatively assigned to *A. brevipes*, and a separate description is given:

Tree, up to 12 m, diameter up to 12 cm. Bark grey to greyish brown, smooth, inner bark yellowish brown or reddish brown. *Young twigs* terete, ferrugineous-pilose, glabrescent, first dark brown, light grey to brown when older. *Stipules* caducous, deltoid, 3 × 1 mm, appressed ferrugineous-pilose. *Petioles* channelled or flat adaxially, 1 – 3 × c. 1 mm, sparsely ferrugineous-pilose, becoming glabrous when old. *Leaf blades* elliptic, more rarely oblong, apically acuminate-mucronate, basally acute, 7 – 13.5 × 3 – 4.5 cm, 2.3 – 3.1 times as long as wide, eglandular, subcoriaceous, glabrous, moderately shiny on both surfaces, major veins flat adaxially, tertiary veins reticulate, drying reddish brown adaxially, slightly lighter abaxially, domatia absent. *Pistillate inflorescences* 3 – 5 cm long, axillary, simple, 3 – 7 per fascicle, axes ferrugineous-pilose. *Bracts* narrowly deltoid to linear, apically acute, 0.3 – 0.5 × 0.3 – 0.5 mm, pilose. *Pistillate flowers* 0.7 × 1.5 mm. *Pedicels* 0 – 0.2 mm long. *Calyx* c. 0.4 × 1 mm, shallowly cupular, sepals 5, fused for c. $^{3}/_{4}$ of their length, apically acute, sinuses rounded, glabrous on both sides, margin fimbriate. *Disc* shorter than the sepals, but indumentum sometimes exserted from the sepals, pubescent only at the margin. *Ovary* ellipsoid, glabrous, style terminal, stigmas 6 – 9. *Infructescences* 3 – 8 cm long. *Fruiting pedicels* 0.1 – 0.3 mm long. *Fruits* ellipsoid, laterally compressed, basally symmetrical, with a terminal style, 4 – 6 × 3 – 4 mm, glabrous, sometimes white-pustulate, aerolate when dry.

5. Antidesma bunius *(L.) Spreng.*, Syst. Veg. 1: 826 (1824); Wight, Icon. Pl. Ind. Orient. 3(1): t. 819 (1844 – 45); Pax & K. Hoffm. in Engl., Pflanzenr. 81: 112, fig. 12 E, G (1922). — *Bunius sativus* Rumph., Herb. Amboin. 3: 204, t. 131 (1743), prelinnean. — *Stilago bunius* L., Mant. Pl. 1: 122 (1767). — *A. stilago* Poir., Encycl. suppl. 1: 403 (1811), p.p., nom. illeg. — *A. bunius* (L.) Spreng. var. *genuinum* Müll. Arg. in DC., Prodr. 15(2): 262 (1866), nom. inval. Type: "Bunius sativus", Rumphius, Herb. Amboin. 3: t. 131 (1743), designated by Airy Shaw 1980b: 693. — Note: The iconotype is rather crude and provides no analysis, but captures the habit of the plant well enough. The plate *Icones Roxburghianae* 1704 at Kew, published with alterations in Wight (1844 – 45: t. 819), shows more detail.

Tree, rarely shrub, up to 30 m, clear bole up to 10 m, diameter up to 1 m, usually straight, often fluted or with buttresses (up to 3 m tall and 10 cm out), sometimes with more than one stem. Bark brown, yellow brown, greyish caramel, tan or grey, thin (1 – 2 mm thick), rough or smooth, usually more or less fissured or vertically cracked, sometimes flaking or with scattered small bumps, soft; inner bark reddish brown, red, pinkish, orange white, white or straw, 3 – 6 mm thick, fibrous; sapwood pale red, pale orange or pale brown, heartwood reddish brown, hard and dense, sometimes slight odour to cut wood. *Young twigs* terete or flattened, glabrous to densely ferrugineous-pubescent, brown. *Stipules* early-caducous, linear, 4 – 6 × 1.5 – 2 mm, pubescent. *Petioles* channelled adaxially, 3 – 10(– 17) × 1.5 – 2 mm, glabrous to densely ferrugineous-pubescent. *Leaf blades* oblong to elliptic, more rarely obovate, apically acuminate-mucronate (mucro up to 3 mm long), often with a rounded apiculum, basally acute to rounded, rarely cordate, (5 –)10 – 18(– 32) × (2 –)4 – 6(– 9) cm, (2 –)2.5 – 3(– 6) times as long as wide, eglandular, coriaceous, glabrous, or pilose to ferrugineous-pubescent only along the midvein adaxially, glabrous to ferrugineous-pubescent all over abaxially, shiny on both surfaces, midvein flat adaxially, tertiary veins reticulate, drying olive-green adaxially, lighter and more reddish abaxially, domatia absent. *Staminate inflorescences* (6 –) 15 – 25 cm long, axillary, consisting of 3 – 8(– 14) branches, axes pubescent to glabrous. *Bracts* deltoid to lanceolate, c. 0.5 × 0.5 mm, pubescent. *Staminate flowers* 3 – 4 × c. 3 mm. *Pedicel* absent. *Calyx* c. 1 × 1.5 mm, cupular, sepals 3 – 4, rarely 5 (in the same inflorescence), fused for $^{2}/_{3}$ – $^{3}/_{4}$ of their length, apically obtuse, glabrous to ferrugineous-pubescent on both sides, margin fimbriate. *Disc* variable, extrastaminal-annular to lobed alternistaminally, lobes free or enclosing the bases of the stamens and pistillode, much shorter than or up to the same length as the sepals, glabrous. *Stamens* 3 – 4(– 5), 2 – 3 mm long, exserted 1.5 – 2 mm from the calyx, anthers c. 0.5 × 0.7 mm. *Pistillode* clavate to cylindrical, c. 1 × 0.5 mm, usually exserted 0.5 – 1 mm from, rarely extending to the same length as the sepals, pubescent, sometimes glabrous apically. *Pistillate inflorescences* (4 –)8 – 17 cm long, axillary, simple or more rarely consisting of up to 4 branches, axes glabrous to pubescent. *Bracts* deltoid, 0.5 – 1.5 × 0.5 – 1 mm, pubescent to pilose. *Pistillate flowers* 2.5 – 3 × 1.5 mm. *Pedicels* 0.5 – 1(– 2) mm long, pubescent to glabrous. *Calyx* 1 – 1.5 × c. 1.5 mm, cupular, sepals 3, fused for $^{2}/_{3}$ – $^{3}/_{4}$ of their length, thick, apically obtuse to rounded, glabrous to pilose on both sides, margin usually erose. *Disc* shorter than the sepals, glabrous. *Ovary* ellipsoid, glabrous to very sparsely pilose, style terminal to subterminal, stigmas 3 – 4(– 6). *Infructescences* 10 – 17 cm long, robust. *Fruiting pedicels* 2 – 4(– 9) mm long, pubescent to glabrous. *Fruits* ellipsoid, laterally compressed, basally symmetrical, with a terminal to slightly subterminal style, 5 – 11 × 4 – 7 mm, glabrous, rarely pilose, often white-pustulate, reticulate when dry.

5a. var. **bunius**

A. sylvestre Lam., Encycl. 1: 207 (1783), "sylvestris". — *A. bunius* (L.) Spreng. var. *sylvestre* (Lam.) Müll. Arg. in DC., Prodr. 15(2): 263 (1866). Type: Rheede, Hort. Malab. 5: 51, t. 26 (1685), lectotype, here designated.

A. ciliatum C. Presl, Epimel. Bot.: 234 (1849). Type: Philippines, Luzon, prov. Batangas, *Cuming* 1446 (PRC!, holotype; CGE!, E!, FHO!, G!, K!, L!, P!, isotypes).

A. cordifolium C. Presl, Epimel. Bot.: 235 (1849). — *A. bunius* (L.) Spreng. var. *cordifolium* (C. Presl) Müll. Arg. in DC., Prodr. 15(2): 262 (1866). Type:

Philippines, Luzon, prov. Laguna, Calauang, *Cuming* 474 (PRC!, holotype; CGE!, E!, G!, FHO!, K!, L!, MEL!, P!, isotypes).

A. floribundum Tul., Ann. Sci. Nat. Bot., Sér. 3: 189 (1851). — *A. bunius* (L.) Spreng. var. *floribundum* (Tul.) Müll. Arg. in DC., Prodr. 15(2): 263 (1866). Type: Sri Lanka, circa Kandy, *Macrae* s.n. (CGE!, lectotype, here designated).

A. bunius (L.) Spreng. var. *wallichii* Müll. Arg. in DC., Prodr. 15(2): 263 (1866). Type: India orientalis, *Wallich* 7282 (G-DC [microfiche], lectotype, here designated; K-W!, OXF!, P!, isolectotypes).

A. andamanicum Hook. f., Fl. Brit. India 5: 364 (1887); Chakrabarty & Gangopadhyay, J. Econ. Taxon. Bot. 24: 14 (2000), as synon. nov. Type: South Andaman Islands, *Kurz* s.n. (CAL, lectotype, designated by Mandal & Panigrahi 1983: 256; G!, K!, isolectotypes).

Sapium crassifolium Elmer, Leafl. Philipp. Bot. 2: 485 (1908). — *A. crassifolium* (Elmer) Merr., Philipp. J. Sci., C, 7: 383 (1912). Type: Philippines, Negros, Prov. Negros Oriental, Cuernos Mts, Dumaguete, June 1908, *Elmer* 10312 (US!, lectotype, here designated; BM!, E!, G!, K!, L!, isolectotypes).

A. collettii Craib, Bull. Misc. Inform., Kew 1911(10): 461 (1911). Type: Burma, Shan Hills, 1500 m, *Collett* 636 (K!, lectotype, here designated). Original Syntypes: Thailand, Chiengmai, by streams in evergreen jungle on Doi Sootep, 660 – 9000 m, *Kerr* 1115 (K!, L!, P!, TCD!), *Kerr* 1117 (K!, P!, TCD!), *Kerr* 1253 (K!).

A. thorelianum Gagnep., Bull. Soc. Bot. France 70: 124 (1923), Airy Shaw, Kew. Bull. 26: 353 (1972), as synon. nov. Type: Laos, Vien-thian (expedition du Me-Kong), *Thorel* 684 p.p. (P!, lectotype, here designated; K!, P!, isolectotypes). — Note: There are five duplicates of the type number in P, two of which are annotated as *A. thorelianum* by Gagnepain.

Young twigs glabrous to very shortly pubescent. *Petioles* glabrous to pubescent (especially adaxially). *Leaf blades* glabrous, or pilose only along the midvein, often minutely white-pustulate. *Inflorescence* axes glabrous to pubescent. *Staminate calyx* glabrous to pubescent outside, ferrugineous-pubescent at the base inside, hairs often exceeding the calyx. *Pistillate calyx* glabrous to sparsely pilose outside, ferrugineous-pubescent at the base inside, hairs often exceeding the calyx. *Ovary* glabrous.

DISTRIBUTION. India (incl. Andaman & Nicobar Islands), Sri Lanka, southern China (Hainan and Guangdong prov.), Burma, Laos, Vietnam, Thailand, Sumatra, Singapore, Borneo, Java, Philippines, Sulawesi, Lesser Sunda Islands, Moluccas, New Guinea, Christmas Islands (Indian Ocean, Australia), Tahiti, Hawaiian Islands. As *A. bunius* is widely cultivated as a fruit tree, it is impossible to distinguish truly wild occurrences and to establish the natural geographic distribution of the species. It seems to be absent in Peninsular Malaysia (except Singapore) and nearly absent in Borneo. Exceptions are one collection from West Kalimantan without precise locality (*Teysmann*), one from East Kalimantan, one collection from cultivation in Brunei, and *Castro & Melegrito* 1519 from Banguey Island, cited by Merrill (1926: 380), which could not be located and examined for this study and is therefore not

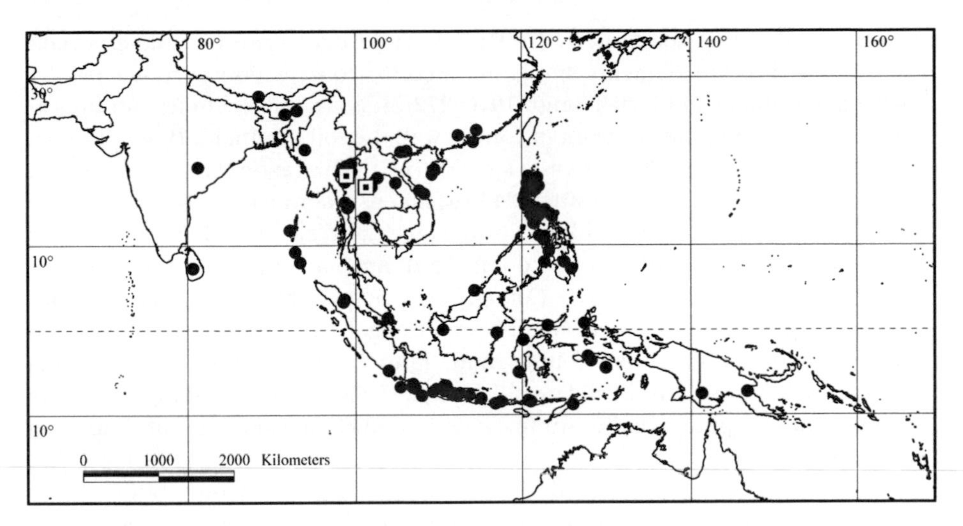

MAP 6. Distribution of *A. bunius* var. *bunius* in Asia and Malesia (●) and var. *pubescens* (▣).

included in the distribution map. In contrast, 105 specimens from Java and 70 from the Philippines were examined. Map 6.

HABITAT. In wet evergreen forest, dipterocarp forest and teak forest; on river banks, at forest edges, along roadsides; in bamboo thickets; in semi-cultivated and cultivated areas; in shady or open habitats; usually in secondary but also in primary vegetation. It thrives best in open, sunny places (A. Lamb, pers. comm.), and is said to be common in the early stages of secondary forest succession, invading marginal grassland (Gruèzo 1991: 79). On sand, loam or clay over (coral) limestone or granite bedrock. 0 – 2100 m altitude.

USES. Widely cultivated as a fruit tree, especially in Java and the Philippines. In Java, the trees are carefully tended and the fruits are highly regarded and sold. Rumphius (1743: 204) introduced the species to Ambon from Sulawesi and found them easily propagated by cuttings as well as fairly fast-growing. Hasskarl (1854: 54) reported that the trees are sometimes surrounded by bamboo scaffolding to make harvesting easier. Smith (1910: 273) claimed that *A. bunius* is to be found in every village in Java. Today, the bunches are harvested by hand with the help of a bamboo pole, preferably with a net bag to collect the detached bunches (Gruèzo 1991: 79). The fruits can be eaten raw, but as the drupes in a raceme do not ripen evenly, they are often used to produce jam or jelly. They are also made into wine in the Philippines (L. Bululacao and A. Lamb, pers. comm.) and into a liqueur in Java (Smith 1910: 273). The reason why the species is not cultivated in Peninsular Malaysia and Borneo is probably cultural rather than ecological (L. G. Saw and A. Lamb, pers. comm.). The relatively high standard of living in Malaysia explains a preference for high-yielding, more commercial fruits. In Kalimantan, bread is still largely unknown and therefore there is also no interest in jam, while alcoholic beverages are usually made from starchy crops such as rice and cassava rather than fruit. In Indonesia, a sour sauce for fish is prepared from the fruits. The leaves are eaten as vegetable, in salads and cooked with rice (Gruèzo 1991: 78). For more details on cultivation see Gruèzo (1991: 78). The wood is not durable especially when exposed to the elements, but is still used for construction purposes (Ridley 1903: 92; Rumphius 1743: 204; Smith 1910: 272). It is also subject to termite attacks and therefore unsuitable for work in contact with the soil (Vietnam, *Poilane* 10901).

CONSERVATION STATUS. Least Concern (LC). This taxon has a vast extent of occurrence, and is widely cultivated. 344 specimens examined.

VERNACULAR NAMES. China: kozut hau (Lai), zui din (Lai); Sri Lanka: karawela-kabela; Thailand: ma mao dong (Lao), ma-mao-khwai, sa phou pho-mao san; Philippines: bignai or bignay (Mang. Tag.), bognay, bugnai (Mang.), bugnay (Bulodtol, Igar., Ilk.), bugne n. iitan, bugne oongal (Ifugao); Sumatra, Java and Sulawesi: buni, boni, buneh, buni, huni, huni berak or huni gede barunei (Sundanese), woni, wuni; Sulawesi: buneh; Lesser Sunda Islands: barune (Sumbawanese), guna, hadju wune, wuler; Moluccas: kata kuti or kuti kata (Bandanese).

ETYMOLOGY. The epithet is derived from the vernacular name “buni”.

KEY CHARACTERS. Large, glabrous, shiny, coriaceous leaves; glabrous disc and fruits; large staminate flowers with a cheesy odour; perianth and androeceum mostly 3-merous.

SIMILAR SPECIES. The Australian *A. dallachyanum* Baill. (Baillon 1866: 337) has been treated as a synonym of *A. bunius* by Airy Shaw (1980b: 693). The few collections from Queensland at Kew are vegetatively very similar. Their fruits, however, are 10 – 18 × 6 – 10 mm in size, subsessile and reported to be greenish-white, white or cream, whereas the fruits in *A. bunius* are smaller and ripen from green via scarlet to black. This synonym is therefore excluded until the Australian species have been studied in more detail. — Pistillate collections of *A. erostre* F. Muell. ex Benth. (1873: 87) can be very similar to *A. bunius* but are generally more slender, have acute sepals which are fused for only half of their length as well as smaller fruits annd leaves. Staminate flowers can be easily distinguished by their deeply divided calyx and thus clearly visible, free disc lobes. Airy Shaw (1973: 279, 1980a: 212 – 213, 1981b: 637) cited two fruiting collections of this species from Papua New Guinea, *Brass* 7689 and *NGF (Ridsdale, Henty & Galore)* 31944. The former specimen is *A. bunius* and the latter *A. excavatum*, leaving *A. erostre* confined to Australia. — *A. excavatum* has a hairy disc and fruits with lateral styles. — *A. leucocladon* has whitish young twigs, thinner petioles and leaves, shorter inflorescences and subterminal to lateral styles at fruit maturity. — *A. macgregorii* has whitish twigs, less coriaceous and reddish drying leaves, shorter inflorescences and fruiting pedicels, smaller flowers, shorter stamens and lacks a pistillode. — Distinguished from *A. montanum* by its larger staminate flowers and anthers, its more highly fused sepals, its compressed, usually larger mature fruits and leaves as well as in its generally more robust habit. — *A. puncticulatum* has cauline inflorescences and a more deeply divided calyx.

SELECTED SPECIMENS. INDIA. Assam, Khasia, 1 – 3000 ft, *Hooker & Thomson* s.n. (G, L). SRI LANKA. s. loc., *Thwaites* C.P. 660 (BM, FR, G, MEL). CHINA. Hainan, Ngai distr., Fung Leng, swamp, moist, level land, sandy soil, 11 Sept. 1932, *Lau* 498 (G); Guangdong prov., Kao-Yao distr., Ting Woo Shan, near Tik village, beside stream, 22 – 29 July 1932, *Lau* 20232 (L). VIETNAM. Tonkin, Dam-ha, Sai

Wong Mo Shan (Sai Vong Mo Leng), Lomg Ngong village, 18 July – 9 Sept. 1940, *Tsang* 30196 (B, G, K, L). BURMA. W Central Burma, Mindat, 4100 ft, 28 April 1956, *Kingdon Ward* 22165 (BM). THAILAND. Northern, Chiang Mai prov., Payap prov., Pang Tawn 4 May 1931, *Put* 3897 (K, L); South-western, Kanchanaburi prov., Erawan National Park, between Khwae Noi and Mae Klong rivers, along path from guesthouse up to the hills, poor deciduous forest and bamboo jungle, in shady place along small stream, in limestone area, c. 200 – 600 m, 18 April 1968, *Van Beusekom & Phlengkhlai* 503 (K, L). SINGAPORE. Junction of Holland Road and Farrer Road, open area near a house, 5 m, 12 May 1983, *Maxwell* 83-28 (L). SUMATRA. Palembang res., c. 5 km O van Soerabaja, N zijde Ranaumeer, oerwoud op helling, 520 m, 28 Oct. 1929, *Van Steenis* 3289 (BO, L). BORNEO. Brunei, Belait distr., Seria, in compound of a house at the Panaga residential area, road G 5, 13 April 1988, *Wong* 351 (AAU, K, KEP, L); E Kalimantan, along road Lojanan to Tenggaron off km 23, Kampung Rampaya, river bank, 26 Oct. 1995, *Ambri, Arifin & Arbainsyah* AA 1413 (K, L). JAVA. C Java, Gunung Muria, Tjollo, N of Kudus, 6°36'S, 110°53'E, 800 m, 26 Nov. 1951, *Kostermans* 6296 (A, K, L); Mandalawangi, Pandeglang secondary forest, 25 April 1974, *Wiriadinata* 65 (K, L). PHILIPPINES. Mindanao, Bukidnon prov., Mt Katanglad, S slope of E peak, at So. Miarai, Katoan, dense mossy forest, 1800 m, 27 Feb. 1949, *PNH (Sulit)* 9933 (A, L); Luzon, Bataan prov., Lamao R., Mt Mariveles, Aug. 1904, *FB (Borden)* 1778 (E, K, NY, US). SULAWESI. S Sulawesi, Maros, Tompokbalang, among the bamboo grooves, c. 20 m, 27 Sept. 1975, *Soenarko* 318 (BO, K). LESSER SUNDA ISLANDS. Sumbawa, W Sumbawa, Pernek, Olat Seli, 12 km S of Sumbawa Besar, monsoon forest, c. 100 m, 20 May 1961, *Kartawinata* 251 (BO, K, L); Flores, W Flores, S part Mt Ndeki, 250 m, 11 April 1965, *Kostermans & Wirawan* 113 (BO, K, L). MOLUCCAS. Ambon, s. loc., July – Nov. 1913, *Robinson* Plantae Rumphianae Amboinenses 334 (BO, GH, K, L, NY, US). PAPUA NEW GUINEA. Lake Daviumbu, Middle Fly R., rain forest, Sept. 1936, *Brass* 7689 (A, K, L). AUSTRALIA. Christmas Island (Indian Ocean), Feb. 1984, *Powell & H'ng Kim Chey* 708 (K). TAHITI. N Tahiti, above Taharaa, semicultivated land, scattered specimens in lowland, 300 m, 29 Sept. 1971, *Van Balgooy* 1940 (L).

5b. var. **pubescens** *Petra Hoffm.*, Kew Bull. 54: 350 (1999). Type: Thailand, Lamphun prov., Me Lee, not far from stream, 2100 ft, 25 April 1915, *Winit* 295 (K!, holotype). Paratypes: Thailand, Phetchabun prov., Nam naw, 3 May 1953, *Nilphanit* 10524 (K!); Phetchabun prov., Lom Kao, 8 May 1955, *Smitinand* 11761 (K!).

Young twigs, petioles and *inflorescence axes* densely ferrugineous-pubescent. *Leaf blades* glabrous, or ferrugineous-pubescent only along the midvein adaxially, ferrugineous-pubescent all over abaxially, especially along the veins. *Staminate calyx* densely ferrugineous-pubescent on both sides. *Pistillate calyx* pilose on both sides. *Ovary* very sparsely pilose. *Infructescences* and *fruits* not seen. Fig. 7.

DISTRIBUTION. Northern Thailand: Lamphun province; Northeastern Thailand: Phetchabun province. Map 6.

HABITAT. In evergreen forest by stream. 700 – 1020 m altitude.

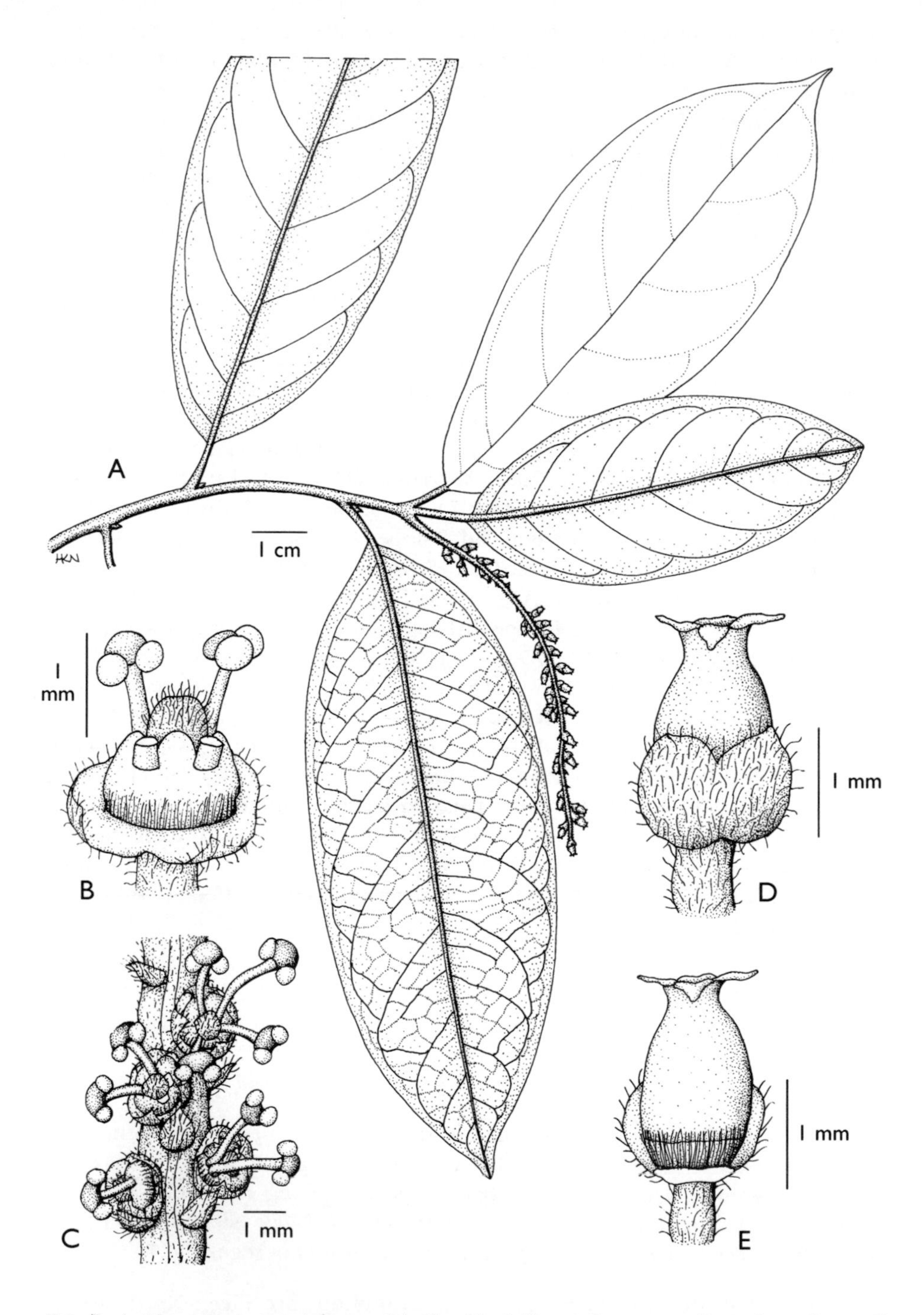

FIG. 7. *Antidesma bunius* var. *pubescens*. **A** habit with pistillate inflorescence; **B** staminate flower; **C** part of staminate inflorescence; **D** pistillate flower; **E** pistillate flower, calyx partly removed. **A**, **D** – **E** from *Winit* 295 (K); **B** – **C** from *Nilphanit* 10524 (K). Drawn by Holly Nixon.

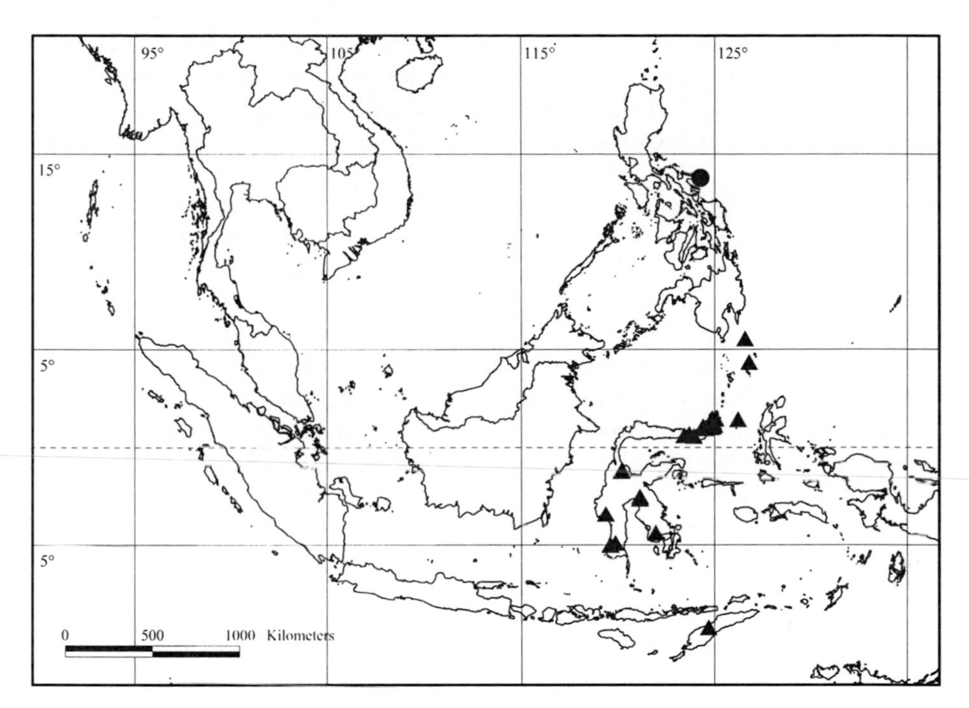

MAP 7. Distribution of *A. catanduanense* (●) and *A. celebicum* (▲).

CONSERVATION STATUS. Data Deficient (DD). Only three specimens (from three different localities) have been examined.

ETYMOLOGY. The epithet refers to the indumentum of the plant (Latin, pubescens = hairy).

NOTE. These specimens have been referred to by Airy Shaw (1972a: 359) as *A.* cf. *nienkui* Merr. & Chun (Merrill & Chun 1935: 263, pl. 54; Chun & Chang 1965: 119; Li Ping Tao 1994: 62). Apart from the considerable disjunction, *A. bunius* var. *pubescens* differs from the type and only specimen of *A. nienkui*, from Hainan [*Chun & Tso* 43995 (A!, K!, NY!, P!)], in the shorter petioles and fused sepals. Both taxa, however, have a similar dense ferrugineous indumentum in most parts, which distinguishes them from the nearly glabrous *A. bunius* var. *bunius*. *A. nienkui* has also been reported from mainland China, Guangdong prov. (Li Ping Tao 1994: 62), but I have not seen this material.

6. Antidesma catanduanense *Merr.,* Philipp. J. Sci. 16: 549 (1920). Type: Philippines, Catanduanes, Mt Mariguiodon, 13 Nov. 1917, *BS (Ramos)* 30515 (A!, lectotype, here designated; K!, US!, isolectotypes).

Young twigs terete, glabrous or nearly so, whitish grey. *Stipules* not seen. *Petioles* basally and distally pulvinate over 2 – 3 mm, 13 – 17 × 1 – 1.2 mm, shortly pilose, very light brown to whitish grey, becoming glabrous when old. *Leaf blades* elliptic, apically acuminate-mucronate, basally obtuse to rounded, 10 – 13.3 × 5.3 – 6.9 cm,

1.8 – 2 times as long as wide, eglandular, subcoriaceous, glabrous adaxially, glabrous or slightly pilose only along the major veins abaxially, shiny on both surfaces, midvein shallowly impressed to flat adaxially, tertiary veins reticulate to weakly percurrent, drying olive-green, lighter abaxially, domatia absent. *Staminate plants* unknown. *Pistillate inflorescences* axillary or terminal, branched regularly, consisting of 3 – 20 branches, sometimes in fascicles of 2 inflorescences, axes pilose. *Bracts* deltoid, apically acute, c. 0.3 × 0.3 mm, pilose. *Pistillate flowers* not known. *Calyx* in fruit 0.5 – 0.6 × 0.7 – 1 mm, cupular, sepals 4, fused for c. $^4/_5$ of their length, individual sepals inconspicuous, apically broadly obtuse, sinuses broadly rounded, glabrous outside and inside, margin entire. *Disc* extending to the same length as the sepals, sparsely whitish-pilose at the margin, hairs much shorter than the disc. *Stigmas* 3 – 4. *Infructescences* 3 – 5 cm long. *Fruiting pedicels* 1 – 2 mm long, pilose. *Fruits* ellipsoid to nearly globose, calyx not splitting, staying intact and as narrow as at anthesis, moderately laterally compressed, basally symmetrical, with a subterminal style, c. 3.5 × 2 – 2.5 mm (including the calyx), glabrous, not white-pustulate, wrinkled (neither areolate nor reticulate) when dry.

DISTRIBUTION. Philippines: Catanduanes. Map 7.

HABITAT. Along streams in forests.

CONSERVATION STATUS. Data Deficient (DD): only known from the type.

ETYMOLOGY. The epithet refers to the type locality.

KEY CHARACTERS. Long petioles; highly fused sepals; disc sparsely whitish-pilose at the margin, hairs much shorter than the disc itself.

SIMILAR SPECIES. Most collections of *A. curranii* have been distributed under the name *A. catanduanense*, but the former species has a smaller and more deeply divided calyx, a terminal style, slightly larger fruits and a longer and ferrugineous disc indumentum. — *A. microcarpum* is extremely similar but has a longer and ferrugineous disc indumentum as well as usually terete fruits with a terminal style and shorter fruiting pedicels. — *A. pleuricum* has hairy fruits, a distinctly lateral style and thinner leaves.

7. Antidesma celebicum *Miq.*, Ann. Mus. Bot. Lugduno-Batavum 1: 218 (1864), non Koord. (1898). Type: Sulawesi, Distr. Menado, *Teysmann* HB 5283 (U!, holotype). — Note: Collection number not cited in protologue.

A. celebicum Koord., Meded. Lands Plantentuin 19: 580, 625 (1898), nom. illeg. (non Miq. 1864). Type: Northern Sulawesi (Celebes bor.), in oerwoud bij bivak Pinamorongan. — Note: There are two fruiting collections (*Koorders* 16790 and 16800) from this locality in BO!, both of which represent *A. celebicum* Miq.

Tree, up to 25 m, clear bole up to 9 m, diameter up to 37 cm, sometimes with more than one stem and much-branched from the ground. Bark grey or brown, 1 mm thick, smooth, cracked or scaly, soft; inner bark tan, yellow, light pinkish or reddish brown; sapwood reddish yellow, whitish orange or reddish brown. *Young twigs* terete, shortly pilose to pubescent, brown. *Stipules* early-caducous, deltoid, apically acute, c. 3 × 1.5 – 2 mm, pubescent. *Petioles* channelled adaxially, basally and distally usually pulvinate over 2 mm, sometimes distinctly geniculate, (6 –)10 – 15(– 20) × 0.8 – 1.2 mm, pubescent (usually very short) to glabrous. *Leaf blades* oblong to elliptic or slightly ovate, apically (usually long) acuminate-mucronate, basally acute to obtuse, more rarely rounded, (7.5 –)10 – 15(– 18) × (3.5 –)4.5 – 6.5(– 8) cm, (2 –)2.4(– 3) times as long as wide, eglandular, membranaceous to subchartaceous, glabrous, or pilose only along the midvein adaxially and along the major veins abaxially, dull on both surfaces, major veins flat to shallowly impressed adaxially, tertiary veins weakly percurrent, more rarely reticulate, widely spaced, drying olive- to greyish green on both surfaces, sometimes lighter abaxially, domatia usually present. *Staminate inflorescences* 5 – 14 cm long, axillary (but aggregated at the end of the branch or some short, simple inflorescences on older twigs), usually consisting of 2 – 25 branches, slender, axes pubescent, bracts subtending individual branches conspicuously large (c. 3 × 3 mm). *Bracts* broadly lanceolate to orbicular, apically obtuse to rounded, c. 0.5 × 0.4 – 0.6 mm, pubescent. *Staminate flowers* c. 1.5 × 1 mm. *Pedicel* absent. *Calyx* c. 0.5 × 0.8 – 1 mm, bowl-shaped with a truncate base, sepals 4, fused for $^2/_3$ – $^3/_4$ of their length, apically broadly rounded, sinuses narrow, rounded, shortly pilose to glabrous outside, glabrous inside, margin slightly erose. *Disc* extrastaminal-annular, hardly lobed, sometimes almost enclosing the individual filaments, as long as or extending to the same length as the sepals, constricted at the base, glabrous at the sides, ferrugineous-pubescent apically. *Stamens* 4, 0.8 – 1.5 mm long, exserted 0.5 – 1.2 mm from the calyx, anthers 0.2 – 0.3 × 0.3 – 0.4 mm. *Pistillode* clavate, slightly crateriform apically, c. 0.5 × 0.3 – 0.4 mm, exserted from the sepals, ferrugineous-pilose to pubescent. *Pistillate inflorescences* c. 4 cm long (only one pistillate flowering specimen seen), axillary (but sometimes aggregated at the end of the branch or some short, simple inflorescences on older twigs), branched regularly, in fascicles of 2 – 3 inflorescences, each consisting of 2 – 5 branches, axes pilose to pubescent. *Bracts* deltoid to broadly lanceolate, apically acute to rounded, 0.5 – 0.7 × c. 0.5 mm, pilose to pubescent. *Pistillate flowers* c. 2 × 1 mm. *Pedicels* 1 – 1.5 mm long, pilose. *Calyx* 0.5 – 0.7 × c. 1 mm, bowl-shaped, often with a truncate base, sepals 4, fused for c. $^2/_3$ of their length, apically broadly rounded, sinuses rounded, shortly pilose to glabrous outside, glabrous inside, margin slightly erose. *Disc* slightly longer than the sepals (including the indumentum), glabrous inside, glabrous or with short ferrugineous hairs outside, always long ferrugineous hairs at the margin. *Ovary* globose to ellipsoid, laterally compressed, pubescent to glabrous, style slightly subterminal, stigmas 4 – 6. *Infructescences* 5 – 9 cm long. *Fruiting pedicels* 1 – 2 mm long, spreading-pilose. *Fruits* lenticular to ellipsoid, distinctly laterally compressed, basally symmetrical, more rarely slightly asymmetrical, with a terminal or subterminal style, 3 – 4 × c. 3 mm, glabrous to pilose, sometimes white-pustulate, areolate when dry.

DISTRIBUTION. Sulawesi incl. Sangi & Talaud Islands, Lesser Sunda Islands (Timor), Moluccas (Bacan Island and Halmahera). Map 7.

HABITAT. In primary, more rarely secondary forest, usually in understorey. On deep clayey, fertile volcanic soil over andesite, phyllite and quartzite. 0 – 1200 m altitude.

CONSERVATION STATUS. Near Threatened (NT). This species has a very large extent of occurrence but its habitat is generally threatened by deforestation. 42 specimens examined.

VERNACULAR NAMES. Halmahera: o kadateke; Talau: angusip'a (Talau Kuruhit); Sulawesi: kayu tuah; lombopale hutan; ruomo.

ETYMOLOGY. The name refers to Celebes, the old name of Sulawesi.

KEY CHARACTERS. Long petioles; leaves usually with domatia; bracts wide; sepals highly fused; disc often exserted from the calyx and with long ferrugineous hairs.

SIMILAR SPECIES. *A. cuspidatum* has larger fruits, lateral styles, longer infructescences, thicker, shiny leaves, free sepals and a cushion-shaped staminate disc. — *A. digitaliforme* has smaller, more shiny leaves and free staminate disc lobes. — *A. edule* has a dense overall indumentum and thicker, reddish brown drying leaves and lateral styles. — Differs from *A. excavatum* in its smaller fruits with (sub)terminal styles, its thinner, duller leaves and in its longer staminate inflorescences; flowering specimens are difficult to tell apart. — *A. microcarpum* has whitish young twigs and thicker leaves with rounded to obtuse bases and without domatia, an urceolate pistillate calyx and smaller, terete fruits. — In *A. pleuricum* the ovary and fruits are distinctly oblique with a lateral style. — The vegetatively similar *A. tetrandrum* and *A. venenosum* differ in their glabrous disc, consisting of free lobes in staminate plants.

SELECTED SPECIMENS. SULAWESI. N SULAWESI. 220 km W of Manado, 50 km inland from Pangi, on tributary of Sungai Ilanga, 0°41'N, 123°40'E, on slope area, 350 – 750 m, 27 Feb. 1990, *Burley, Tukirin et al.* 3550 (L); c. 220 km W of Manado, km 50 inland from Pangi on Sungai Ilanga, 0°41'N, 123°40'E, primary lowland forest, 500 m, 28 Feb. 1990, *Burley, Tukirin et al.* 3558 (E, K, L); Bolaang Mongondow, Dumoga Bone National Park, Toraut Dam, 0°34'N, 123°54'E, rather undisturbed primary forest some 35 m high, in valley near small streamlet, deep clayey volcanic soil, terrain level, 220 m, 15 March 1985, *De Vogel & Vermeulen* 6555 (K, L, NY); N slope of Mt Klabat, forest, 500 m, 27 June 1956, *Forman* 253 (BM, K, L, NY); Gorontalo distr., Dumoga Bone National Park, Momaliada'a Camp on the Sungai Olama, 0°36'N, 123°24'E, primary forest on ridge (topped with *Casuarina* trees) close to the river, steep slope, clayey soil, 475 m, 14 Aug. 1991, *Milliken* 917 (E, K, L). C SULAWESI. SW of Tongoa, 1°10'S, 120°10'E, partly felled primary forest, 650 m, 15 March 1981, *Johansson, Nybom & Riebe* 386 (K, L); Sungai Sadaunta, hillside forest, Oct. – Dec. 1974, *Musser* S-4 (K). S SULAWESI. Malili afd., nabij Kampoeng Kawata, oud bosch, 250 m, 1932 – 33, *Boschproefstation cel/V (Waturandang)* 125 (BO, L). SANGI AND TALAUD ISL. Talaud, Karakelang, S slope of G. Duata, oud bosch, 50 m, 3 May 1926, *Lam* 2785 (A, BO, L, U). LESSER SUNDA ISLANDS. Timor, Belu, Hutan Fatukaduak, 16 Feb. 1980, *Widjaja* 1202 (L).

8. Antidesma chalaranthum *Airy Shaw,* Kew Bull. 33: 424 (1979). Type: Northeastern New Guinea, Eastern Highlands Distr., Goroka subdistr., Asaro-Mairifutica divide, 0.8 km S of Daulo camp, ridge forest, 2400 m, 4 Aug. 1957, *Pullen* 456 (K!, holotype; L!, US!, isotypes).

Shrub or tree, up to 7 m. *Young twigs* terete to slightly striate, ferrugineous-pubescent, light brown. *Stipules* persistent, deltoid, apically acute, 1.5 – 2.5 × 0.5 – 1 mm, ferrugineous-pubescent, becoming glabrous when old. *Petioles* channelled adaxially, 3 – 6 × 1 – 1.5 mm, pubescent to pilose. *Leaf blades* elliptic, apically acuminate-mucronate, sometimes with a rounded or retuse apiculum, basally acute, 5.5 – 10 × 2.5 – 5 cm, 2 – 2.8 times as long as wide, eglandular, chartaceous to subcoriaceous, glabrous, or sparsely pilose only along the midvein adaxially, sparsely pilose all over abaxially, especially along the major veins, moderately shiny to dull adaxially, dull abaxially, midvein impressed adaxially, tertiary veins reticulate, widely spaced, drying reddish to greyish brown, domatia weakly developed, inconspicuous. *Staminate inflorescences* 3 – 5 cm long, axillary and cauline, consisting of 2 – 4 branches, very slender, axes 0.2 – 0.5 mm wide, sparsely and spreading-hirsute. *Bracts* deltoid, apically acute, c. 0.7 × 0.5 mm, spreading-pilose. *Staminate flowers* 2.5 – 3 × 1.5 – 2 mm. *Pedicels* 2 – 4 mm long, not articulated, spreading-pilose. *Calyx* c. 1.5 × 0.7 mm, cylindrical, sepals 4, fused for $^1/_4$ – $^1/_3$ of their length, deltoid, apically obtuse, sinuses obtuse, appressed-pilose outside, glabrous inside, margin slightly erose. *Disc* extrastaminal-annular, extending to the same length as the sepals, lobed towards the centre, or lobes ± detached from the annular part of the disc, subulate and standing between the stamens, lobes slightly longer than annular part of disc and ± highly connate to the pistillode; disc glabrous at the sides, pubescent (partly ferrugineous) apically. *Stamens* 3 – 4, 2 – 2.5 mm long, exserted 1.5 – 2 mm from the calyx, anthers c. 0.5 × 0.5 – 1 mm. *Pistillode* clavate, c. 1.5 × 0.5 – 1 mm, exserted c. 1 mm from the sepals, with a fold on one side, c. 0.5 mm long and 0.5 mm deep, ferrugineous-pubescent. *Pistillate inflorescences* axillary to cauline, simple or consisting of up to 3 branches, axes sparsely and spreading-hirsute. *Bracts* deltoid, apically acute, c. 0.5 × 0.5 mm, pilose. *Pistillate flowers* not known. *Calyx* in fruit c. 1 × 2 mm, bowl-shaped, sepals 4, fused for $^2/_3$ – $^3/_4$ of their length, deltoid, apically obtuse, sinuses obtuse, appressed-pilose outside, glabrous inside, margin slightly erose. *Disc* shorter than the sepals, pubescent at the margin. *Stigmas* 3 – 5. *Infructescences* 3 – 5 cm long, axes c. 0.5 mm wide. *Fruiting pedicels* slender, 3 – 4 × c. 0.3 mm, spreading-pilose. *Fruits* lenticular, laterally compressed, basally symmetrical, with a distinctly lateral style, 5 – 8 × 5 – 8 mm, glabrous or nearly so, white-pustulate, reticulate when dry.

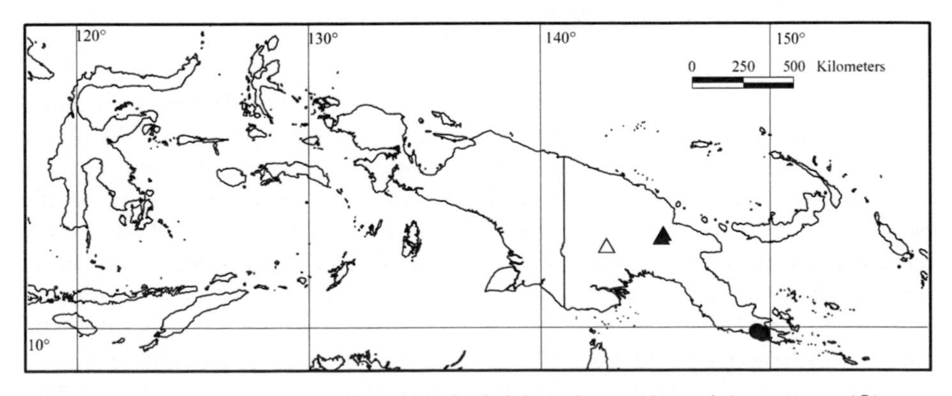

MAP 8. Distribution of *A. chalaranthum* (▲), doubtful specimens (△), and *A. concinnum* (●).

DISTRIBUTION. Papua New Guinea: Eastern Highlands and probably also Southern Highlands provinces. Map 8.

HABITAT. In mossy forest on ridge; in primary vegetation; associated with *Nothofagus*. 2200 – 2600 m altitude.

CONSERVATION STATUS. Vulnerable (VU D1). Only 2 specimens safely identified are known of this distinctive taxon.

VERNACULAR NAMES. ndiyer.

ETYMOLOGY. The epithet refers to the inflorescence (Greek, chalaros = lax; anthos = flower).

KEY CHARACTERS. Slender, short, inflorescences; long slender pedicels; highly fused sepals; hairy disc; long, clavate pistillode; fruits with lateral styles; 2200 – 2400 m altitude.

SIMILAR SPECIES. Very distinct in staminate flower; fruits similar to *A. excavatum* which has more robust inflorescence axes and pedicels.

NOTE. Originally described only from the staminate type. Airy Shaw suggested that *NGF (Streimann & Kairo)* 27636 from Aseki, Morobe province, might be a matching pistillate collection, but this specimen certainly belongs to *A. baccatum* which is not known to occur in the Eastern Highlands province, nor is it known from more than 1500 m altitude. The fruiting collection *LAE (Kerenga & Symon)* 56821 from Daulo Pass is a better match for *A. chalaranthum*; it was collected close to the type locality at 2200 m altitude, and besides a great overall similarity has the same slender inflorescence axes and pedicels. Two very similar fruiting collections, though not as close in locality and altitude, are *Jacobs* s.n. (6 Oct. 1973) and *LAE (Damas)* 58888 from Mt Bosavi, Southern Highlands province.

FURTHER SPECIMEN. PAPUA NEW GUINEA. Eastern Highlands distr., Watabung subdistr., Daulo Pass, 6°05'S, 145°13'E, regrowth in moss forest, 2200 m, 17 June 1984, *LAE (Kerenga & Symon)* 56821 (K, L).

9. Antidesma concinnum *Airy Shaw*, Kew Bull. 33: 425 (1979). Type: Papua New Guinea, Central Distr., near Nunumai, 12 km N of Amazon Bay, 10°11'S, 149°23'E, forest on stony hills, 52 m, 11 June 1969, *Pullen* 7541 (K!, holotype; L!, isotype). Paratype: Milne Bay distr. & subdistr., Origuina R., 10°17'S, 149°38'E, riverine forest, 60 m, 29 March 1970, *NGF (Henty & Katik)* 42936 (K!, L!).

Shrub or treelet, up to 1.5 m. *Young twigs* terete, very slender, pubescent, brown. *Stipules* persistent, subulate, 1 – 2 × 0.2 – 0.3 mm, pilose. *Petioles* nearly terete to channelled adaxially, 1.5 – 3 × c. 0.5 mm, pilose. *Leaf blades* ovate, apically acuminate to caudate-mucronate, with a rounded to retuse apiculum, basally acute, 2.5 – 5.5 × 1 – 2 cm, 2.3 – 3.4 times as long as wide, eglandular, thinly chartaceous, glabrous, or pilose only along the midvein adaxially, sparsely pilose all over abaxially, more densely so along the veins, shiny on both surfaces, midvein flat to impressed adaxially, tertiary veins reticulate, drying olive-green, domatia absent. *Staminate inflorescences* 4 – 6 cm long, appearing terminal, simple, axes c. 0.2 mm wide, pilose. *Bracts* ovate to deltoid, apically acute to obtuse, c. 0.3 × 0.3 mm, glabrous. *Staminate flowers* 1 – 2 × 1 – 1.5 mm. *Pedicel* absent. *Calyx* c. 0.7 × 1 mm, bowl-shaped, sepals 3, fused for c. $^{3}/_{4}$ of their length, apically acute to obtuse, sinuses rounded, wide, shallow, glabrous outside and inside, margin sparsely fimbriate. *Disc* consisting of 3 free alternistaminal lobes, lobes ± obconical, well-separated from each other, c. 0.3 – 0.4 × 0.5 mm, glabrous at the sides, ferrugineous-pubescent apically. *Stamens* 3, immature c. 1 mm long, exserted c. 0.5 mm from the calyx, anthers 0.3 – 0.4 × c. 0.5 mm. *Pistillode* subulate, 0.3 – 0.5 × 0.1 – 0.2 mm, extending to the same length as the disc, slightly exserted from the sepals, ferrugineous-pubescent. *Young infructescences* c. 5 cm long, axillary or terminal, simple, very slender, axes sparsely puberulent. *Bracts* nearly orbicular to oblong, 0.5 – 0.6 × 0.4 – 0.5 mm, glabrous, margin fimbriate. *Young fruits* c. 2 × 1.5 mm. *Pedicels* 1 – 1.5 mm long, pilose. *Calyx* in young fruit c. 0.5 × 1 mm, bowl-shaped with a truncate base, sepals 4, fused for c. $^{4}/_{5}$ of their length, apically acute, sinuses wide, rounded, glabrous outside and inside, margin slightly erose. *Disc* exserted from or extending to the same length as the sepals, ochraceous- to ferrugineous-tomentose at the margin, indumentum usually exserted from the sepals. *Ovary* lenticular, pilose, style lateral, stigmas 4 – 5. *Infructescences* 2 – 5 cm long, axes up to 0.5 mm wide. *Fruiting pedicels* 1 – 2 mm long, sparsely puberulous. *Fruits* lenticular, distinctly laterally compressed, basally symmetrical, with a distinctly lateral style, c. 5 × 5 mm, sparsely appressed-pilose, not white-pustulate, areolate when dry.

DISTRIBUTION. Papua New Guinea: Central and Milne Bay provinces. Map 8.

HABITAT. In primary forest. 50 – 100 m altitude.

CONSERVATION STATUS. Vulnerable (VU D1). Only 3 specimens examined; this could be a very local endemic.

ETYMOLOGY. The epithet refers to the general appearance of the plant (Latin, concinnus = neat, pretty).

KEY CHARACTERS. Very slender habit; small, ovate leaves; slender inflorescences; small staminate flowers with free disc lobes.

SIMILAR SPECIES. Differs from *A. excavatum* in its free staminate disc lobes. — *A. jucundum* has a long spreading indumentum and pedicellate staminate flowers with a glabrous disc.

NOTE. The slender habit and small, ovate leaves make this taxon rather distinctive. However, the pistillate flowers and fruits of the type are identical with those of *A. excavatum* which is vegetatively so variable that it would be hard to justify recognition of a species on the basis of leaf shape and size alone. Several fruiting collections from different parts of Papua New Guinea more or less bridge the gap between the types of *A. concinnum* and typical *A. excavatum* in leaf shape and size (*Kanehira & Hatusima* 12057 from Nabire; *Carr* 12511 from Koitaki; *Brass* 3809 from Dieni). The staminate paratype of *A. concinnum*, on the other hand, has free disc lobes which are never observed in *A. excavatum*.

SELECTED SPECIMENS. PAPUA NEW GUINEA. Central distr., Abau subdistr., c. 12 km N of Amazon Bay, N of Nunumai village across Ulumanok R., 10°11'S, 149°23'E, low ridge in primary forest, c. 100 m, 16 June 1969, *Kanis* 1030 (L).

10. Antidesma contractum *J. J. Sm.* in Lorentz, Nova Guinea 8(1): 229, t. LVI (1910). Type: Dutch New Guinea, Etna Bai, Dec. 1904, *Koch* 78 (L!, lectotype, here designated). Original syntypes: auf den Hügeln südlich des Geluks-Hügels in c. 50 m Meereshöhe im Urwalde, Sept. 1907, *Versteeg* 1730 (L!, U!), 1734 (K!, L!, P!); an dem Noord-Fluß am Fuße des Nepenthes-Hügels, Sept. 1907, *Versteeg* 1770 (L!).

A. ledermannianum Pax & K. Hoffm. in Engl., Pflanzenr. 81: 133 (1922), **synon. nov.** (e descr.) Type: Neu Guinea, Kaiser Wilhelmsland, am Sepik, Station Gubugai, Sumpfwald, *Ledermann* 7759. — Note: Type probably destroyed, no isotypes located. The original description, however, leaves little doubt as to the identity of this plant: the combination of large, narrow, glabrous leaves with a slightly decurrent base, short petioles, short, fascicled, axillary to cauline inflorescences, pedicellate staminate flowers with a deeply divided calyx, a glabrous, cushion-shaped disc, 3 stamens and no pistillode is found only in *A. contractum* Pax & Hoffmann, who compared the floral structure of their new species to *A. contractum*, described it probably because of its only slightly decurrent leaf base ("limbus ... basi obtusus, vix vel brevissime contractus"), while in *A. contractum* the leaf blade is narrowed into a distinct "pseudo-petiole" (Fig. 2G, see their key on page 123 – 124). This character is, however, variable in *A. contractum.*

A. impressivenum Kaneh. & Hatus., "impressivena", nom. nud. — In sched.: Dutch New Guinea, Nabire, in the rain-forest, 20 m, 27 Feb. 1940, *Kanehira & Hatusima* 11699 (BO!).

Shrub or tree, up to 8 m, diameter up to 8 cm, often poorly branched, spindly and straggling with overhanging branches. Twigs pale green and knobby. Bark pale grey or fawn, thin, rather smooth, with scattered lenticels; wood light pink. *Young twigs* terete, glabrous, light brown to whitish grey. *Stipules* usually caducous, narrowly deltoid to linear, apically acute, 3 – 5(– 10) × 0.5 – 1 mm, glabrous to sparsely pilose. *Petioles* channelled to flat adaxially, often rugose and whitish grey, 3 – 10 mm long (but apparently up to 25 mm long including the decurrent leaf base), 1.5 – 3 mm wide, pilose to glabrous. *Leaf blades* ovate to oblong, more rarely elliptic or slightly obovate, apically acuminate- to caudate-mucronate, basally acute to rounded, 2 – 17 mm long more or less sharply decurrent, (6 –)15 – 20(– 30) × (2 –)4 – 7(– 10) cm, (2 –)3 – 3.5(– 5.2) times as long as wide, eglandular, chartaceous to coriaceous, glabrous, or sparsely pilose only along the midvein adaxially and along the major veins abaxially, moderately shiny on both surfaces, sometimes bullate, all veins distinctly raised adaxially, rarely almost flat, tertiary veins reticulate to weakly percurrent, usually strong perpendicular intersecondary veins, drying dark reddish brown on both surfaces, or lighter abaxially, domatia absent. *Staminate inflorescences* 4 – 7 cm long, axillary to almost cauline, simple, solitary or up to 7 per fascicle, very slender, axes ferrugineous-hispid to nearly glabrous. *Bracts* narrowly deltoid, apically acute, 0.5 – 0.7 × c. 0.3 mm, pilose. *Staminate flowers* 1.5 – 2.5 × 1.5 – 2.5 mm. *Pedicels* 0.5 – 1 mm long, not articulated, glabrous to very sparsely pilose. *Calyx* c. 0.8 × 1 mm, sepals (3 –)4(– 5), free or nearly so, c. 0.6 mm wide, deltoid to ovate, apically acute to shortly acuminate, glabrous outside, pilose inside, margin entire, long fimbriate. *Disc* cushion-shaped, fully enclosing the bases of the filaments and, if present, the pistillode, glabrous. *Stamens* 2 – 3, 1.5 – 2 mm long, exserted 1 – 1.5 mm from the calyx, anthers c. 0.5 × 0.6 mm. *Pistillode* absent or subulate, c. 0.2 × 0.1 mm, shorter than the sepals, glabrous. *Pistillate inflorescences* 3 – 10 cm long, axillary to almost cauline, simple, solitary or up to 4 per fascicle, very slender, axes ferrugineous-hispid to nearly glabrous. *Bracts* narrowly lanceolate, 0.5 – 1 × c. 0.3 mm, pilose. *Pistillate flowers* 1 – 2 × 1 – 2 mm. *Pedicels* 0 – 1.5 mm long, glabrous to sparsely pilose. *Calyx* c. 1 × 1 – 1.5 mm, sepals (3 –)4, free or nearly so, 0.5 – 1 mm wide, deltoid, ovate or orbicular, apically acute, acuminate or rounded, glabrous outside, pilose inside, margin slightly erose, fimbriate. *Disc* shorter than the sepals, glabrous. *Ovary* ellipsoid, glabrous, style terminal, stigmas 3 – 6, rather long and thick. *Infructescences* 5 – 15(– 28) cm long, axes 0.5 – 0.8(– 1) mm thick. *Fruiting pedicels* 0 – 4 mm long, glabrous to pilose. *Fruits* ellipsoid to ovoid, laterally compressed, basally symmetrical, with a terminal style, 5 – 8(– 10) × 4 – 5(– 6) mm, glabrous, sometimes finely white-pustulate, reticulate when dry. Fig. 8.

DISTRIBUTION. New Guinea. Map 9.

HABITAT. In undergrowth of primary rain forest (rarely old secondary forest), sometimes by rivers and seasonally flooded; *Metroxylon* swamp; near the beach. On old, well-drained volcanic soil, over sandstone. 1 – 1200 m altitude.

USES. The wood is used in house construction.

CONSERVATION STATUS. Least Concern (LC). The extent of occurrence spans over 250,000 km^2 and the species is locally common (specimen labels and T. Utteridge, pers. comm.). 36 specimens examined.

VERNACULAR NAMES. Indonesia: Papua: ara kra mkek, ara mkek, tfan kek; PNG: fe-ce (Tokples).

ETYMOLOGY. The epithet refers to the contracted leaf-base.

KEY CHARACTERS. Veins raised adaxially; large, narrow, glabrous, decurrent leaves; short petioles; short, fascicled inflorescences; sepals ± free; glabrous disc; stamens 2 – 3.

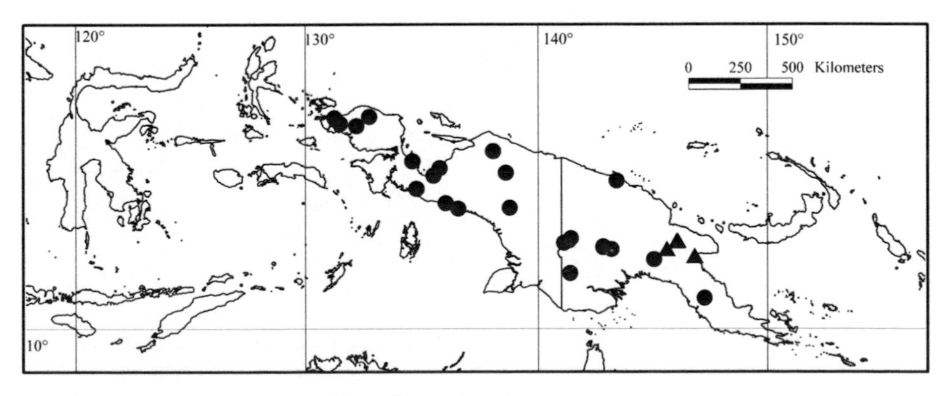

MAP 9. Distribution of *A. contractum* (●) and *A. ferrugineum* (▲).

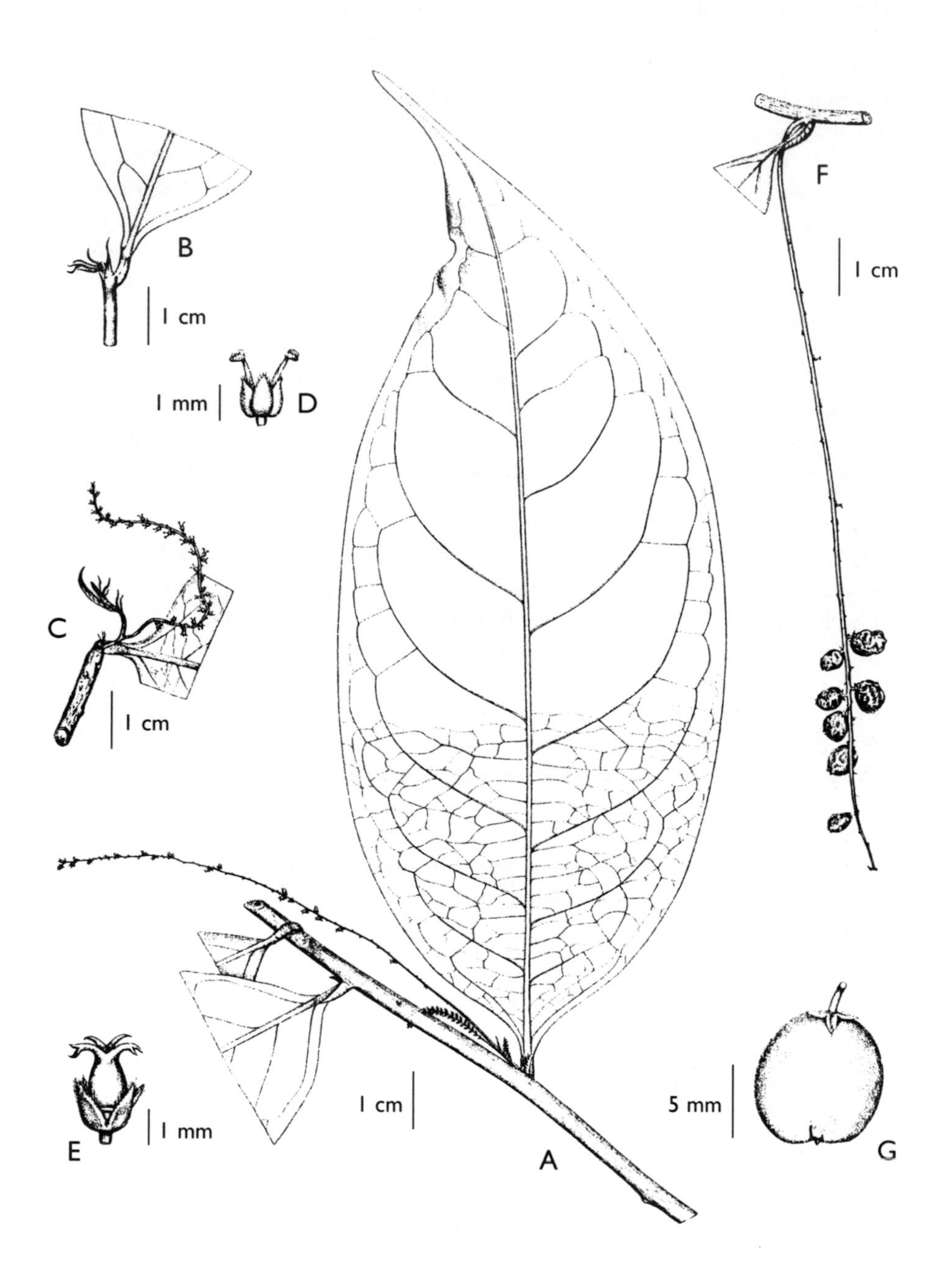

FIG. 8. *Antidesma contractum.* **A** pistillate branch; **B** leaf bases; **C** part of staminate inflorescence; **D** staminate flower; **E** pistillate flower; **F** fruiting inflorescence; **G** fruit. Reproduced from Lorentz 1910: t. LVI.

SIMILAR SPECIES. Differs from *A. brevipes* in its longer petioles and stipules, its larger leaves with a raised midvein and its distinctly decurrent leaf base, its longer pedicels in staminate flowers, its free sepals and in its glabrous disc.

SELECTED SPECIMENS. INDONESIA: PAPUA. S New Guinea, Sg. Aendoea near Oeta, 3 m, 7 July 1941, Expedition Lundquist, *Aet* 449 (K, L); near Prauwen Bivouac, primary forest, 10 m, 23 July 1920, *Lam* 707 (K, L); NW Guinea, Sorong, near Klamono, open terrain, near riverside, 20 m, 19 Aug. 1948, *Pleyte* 619 (K, L); Vogelkop Peninsula, Aifat river valley, N slope of the E part of the Tohkiri Range, path Sururem–Timtum, in *Castanopsis acuminatissima* undergrowth, 950 – 1050 m, 21 Oct. 1961, *Van Royen & Sleumer* 6913 (K, L). PAPUA NEW GUINEA. Western distr., Kiunga subdistr., Elevala river, near Gusure village, side of river in seasonally flooded forest, 90 ft, 14 Aug. 1971, *LAE (Streimann et al.)* 51901 (K, L); Western distr., near Erekta, c. 8 miles S of D'Albertis Junction, upper Fly R., rainforest on frequently flooded valley floor, c. 60 ft, 20 Sept. 1967, *Pullen* 7344 (K, L, US).

11. Antidesma coriaceum *Tul.*, Ann. Sci. Nat. Bot., Sér. 3: 204 (1851). Type: Peninsular Malaysia, Penang, *Hb. Wallich (G. Porter coll.)* 8584 (P!, lectotype, here designated; CGE!, E!, G!, G-DC [microfiche], K!, L!, NY!, OXF!, TCD!, isolectotypes).

A. fallax Müll. Arg., Linnaea 34: 68 (1865 – 66). Type: Malaysia, in collibus ins. Penang, *Hb. Wallich (G. Porter coll.)* 9101 (G-DC [microfiche], lectotype, here designated; K!, isolectotype).

Aporosa griffithii Hook. f., Fl. Brit. India 5: 353 (1887). Type: Burma and Malay Peninsula, 24 Jan. 1845, *Griffith*, Kew Distr. No. 4955 (K!, holotype).

Antidesma pachyphyllum Merr., Philipp. J. Sci., C, 11: 58 (1916). Type: Sarawak, Baram distr., Miri R., Jan. 1895, *Hose* 69 (K!, lectotype, here designated; A! [photo of destroyed holotype ex PNH]; E!, L!, P!, isolectotypes).

A. nitens Pax & K. Hoffm. in Engl., Pflanzenr. 81: 136 (1922). Type: Nord-Borneo, *Beccari* PB 1404 (K!, lectotype, here designated; P!, isolectotype).

A. cordatum Airy Shaw, Kew Bull. 26: 465 (1972), **synon. nov.** Type: Malay Peninsula, Selangor, Ulu Gombak Forest Reserve, primary forest on hillside, alt. 6000 m, 10 Nov. 1959, *KEP (Kochummen)* 94034 (K!, holotype; A!, KEP!, L!, SING, isotypes). — Note: Leaf shape in *A. coriaceum* is very variable and ranges continuously from acute to a cordate leaf bases. The latter corresponds with a higher number of secondary veins because it occurs in specimens with larger leaves.

Tree, rarely shrub, up to 29 m, clear bole up to 15 m, diameter up to 75 cm, usually straight. Twigs light greyish brown. Bark whitish, grey or brown, sometimes white-pink, red-greenish or reddish brown, thin (c. 3 mm thick), smooth, more or less fissured longitudinally, soft, fibrous, inner bark whitish, cream, yellowish brown, brown, reddish brown, red, pink or green, thin (2 – 5 mm thick), fibrous, laminated, soft; cambium brownish or light yellow; wood very hard, sapwood orange, yellow, yellowish brown, brown or whitish grey. *Young twigs* terete or striate, ferrugineous-hispid, brown. *Stipules* early-caducous, subulate to linear, 5 – 7(– 15) × 0.5 – 1(– 2) mm, ferrugineous-hispid. *Petioles* channelled adaxially, basally and distally sometimes slightly pulvinate and geniculate, (4 –)8 – 20(– 30) × 1 – 2 mm, shortly ferrugineous-hispid, becoming glabrous when old. *Leaf blades* oblong to ovate or obovate, apically long acute to acuminate-mucronate (mucro up to 4 mm long), basally acute to cordate, (6 –)10 – 15(– 21) × (2 –)4 – 6(– 9) cm, (2.1 –)2.6(– 3.3) times as long as wide, eglandular, coriaceous, glabrous, or ferrugineous-hispid only along the major veins abaxially, shiny on both surfaces, midvein impressed adaxially, tertiary veins coarsely reticulate, quaternary veins hardly visible, drying grey to reddish brown adaxially, reddish brown abaxially, domatia absent. *Staminate inflorescences* 2 – 7 cm long, cauline or axillary, simple or branched mostly near the base, consisting of up to 12 branches, axes ferrugineous-hispid. *Bracts* lanceolate to deltoid, apically acute, 0.5 – 1 × 0.5 mm, hispid. *Staminate flowers* c. 1.5 × 1 mm. *Pedicel* absent. *Calyx* 0.5 – 1 mm long, sepals 4 – 5, free or nearly so, 0.3 – 0.5 mm wide, deltoid, apically acute, hispid outside, hispid to glabrous inside, long hairs at least at the base, margin entire. *Disc* consisting of 3 – 5 free alternistaminal lobes, lobes ± obconical, sometimes with two shallow imprints apically, c. 0.5 × 0.5 – 1 mm, glabrous. *Stamens* 3 – 5, 1 – 1.5 mm long, exserted 1 – 1.5 mm from the calyx, anthers 0.3 – 0.4 × 0.4 – 0.5 mm. *Pistillode* cylindrical to obconical, 0.3 – 0.5 × c. 0.2 mm, extending to the same length or exserted from the sepals, hispid. *Pistillate inflorescences* 2 – 3 cm long, cauline or axillary, simple or consisting of up to 10 branches, axes ferrugineous-hispid. *Bracts* lanceolate to deltoid, apically acute, c. 1 × 0.5 mm, ferrugineous-hispid. *Pistillate flowers* c. 2 × 1.5 mm. *Calyx* 0.7 – 1.2 mm long, sepals 4 – 5, free or nearly so, c. 0.5 mm wide, deltoid, apically acute, hispid outside, hispid to glabrous inside, margin entire. *Disc* shorter than the sepals, glabrous. *Ovary* ellipsoid, glabrous, style slightly subterminal, usually thick, stigmas 4 – 10, often conspicuously short and thick. *Infructescences* 2 – 5(– 9) cm long, weak and often bent. *Fruiting pedicels* 0.3 – 1.5 mm long, ferrugineous-hispid. *Fruits* lenticular to ellipsoid, sometimes wider than long, laterally compressed, very rarely terete, basally symmetrical to slightly asymmetrical, with a subterminal to nearly terminal style, (3 –)4 – 8(– 13) × 4 – 6 mm, glabrous, often white-pustulate, areolate or fleshy when dry. Fig. 9A – B.

DISTRIBUTION. Peninsular Malaysia, Singapore, Sumatra, Borneo. Map 10.

HABITAT. Mostly in peat swamp forest, associated with *Casuarina*; in mangroves; also in "kerangas" (dry shrubby heath forest on poor sand), associated with *Agathis* and *Dacrydium*; more rarely in lowland dipterocarp forest and sugar plantations; often in much disturbed but also in primary vegetation; open to shaded habitats. On sandy soil, (giant) podsol, black soil, over white sandstone/granitic sand. 0 – 1200 m altitude.

CONSERVATION STATUS. Least Concern (LC). This is a common species with a large extent of occurrence. 199 specimens examined.

VERNACULAR NAMES. Sumatra: buruk butu (Indonesian), kawa kawa (Minangkabau), mangas (Indonesian), mempadu padang, palempangmerawat, sepatjet, sungkai alas; Malay Peninsula: sebasa (name for *Aporosa*); Sabah:

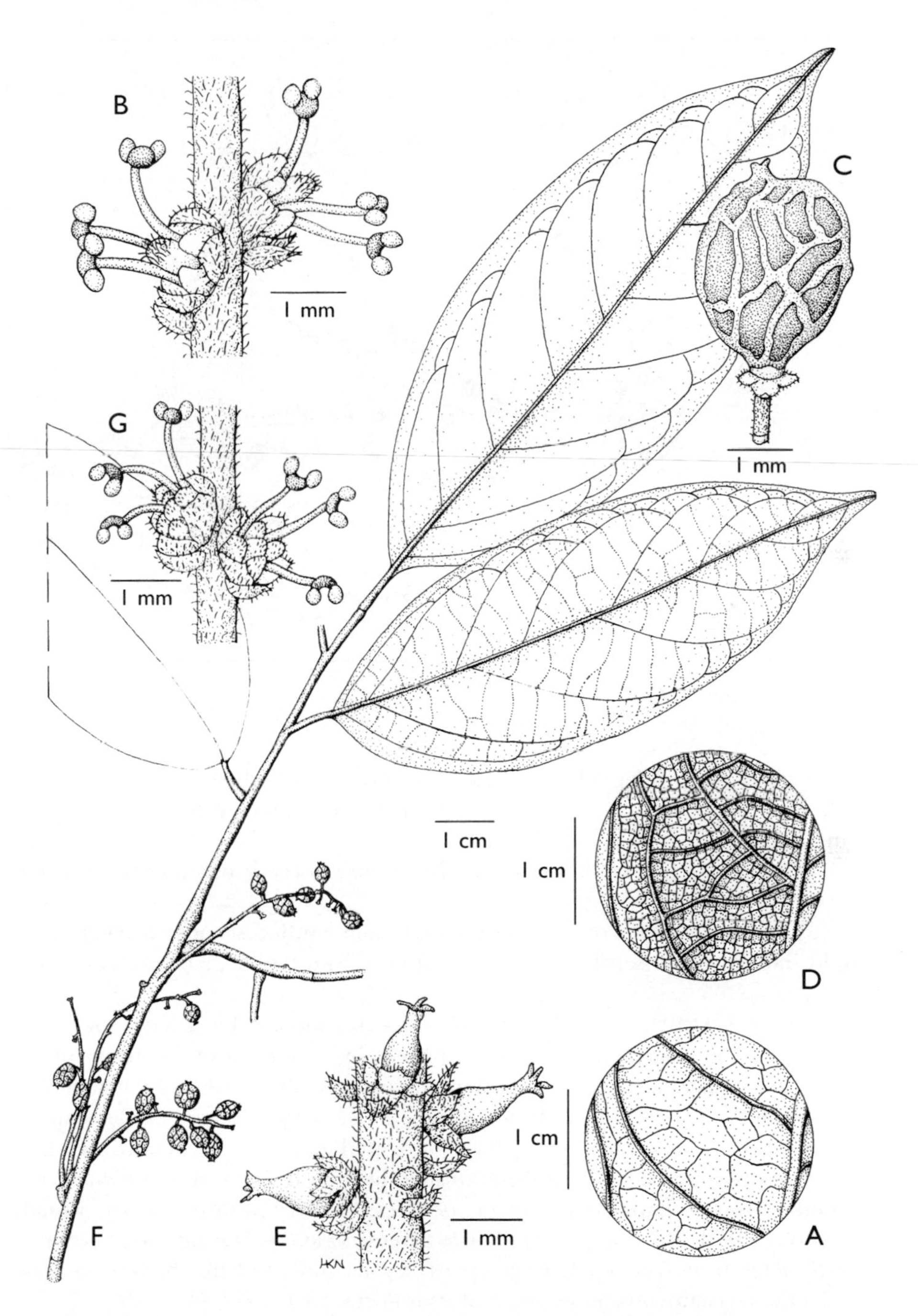

FIG. 9. *Antidesma coriaceum* and *A. puncticulatum*. **A – B** *A. coriaceum*. **A** detail of leaf venation; **B** part of staminate inflorescence. **C – G** *A. puncticulatum*. **C** fruit; **D** detail of finely tessellated ultimate leaf venation; **E** part of pistillate inflorescence; **F** twig with leaves and infructescence; **G** part of staminate inflorescence; **A – B** from *Alston* 13171 (L); **C – D, F** from *SAN* 25755 (L); **E** from *SAN* 36431 (L); **G** from Bogor Bot. Gard. VIII.B.62a (L). Drawn by Holly Nixon.

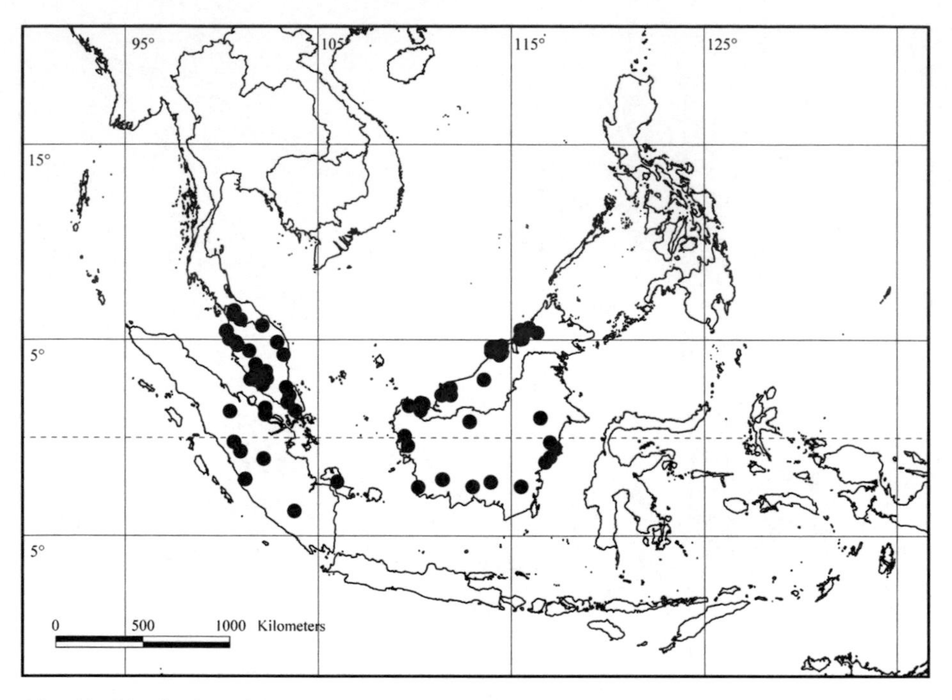

MAP 10. Distribution of *A. coriaceum.*

cheremai (Kedayan), lagas lagas (Dusun), rayan; Sarawak: buah jerawai, dolkuyat (Milanau), nepis kalit; Brunei: ubis (Dusun); Kalimantan: cabi-cabi, mempene item, pomjut.

ETYMOLOGY. The epithet refers to the texture of the leaves (Latin, coriaceus = leathery).

KEY CHARACTERS. Long petioles; short, cauline inflorescences; ferrugineous-hispid indumentum; sessile to subsessile flowers and fruits; glabrous fruits; free sepals.

SIMILAR SPECIES. *A. brevipes*, see there. — *A. cruciforme* has a hairy lower leaf surface. — *A. cuspidatum* is distinguished by length and position of its inflorescences as well as its cushion-shaped staminate disc. — *A. helferi* has shorter petioles and a hairy disc. — Pistillate specimens with shorter petioles can be confused with *A. montanum* if they have elongated and almost symmetrical fruits; staminate specimens are clearly distinguished by their discs. — *A. neurocarpum* has a similar overall habit but shorter petioles, axillary inflorescences, usually foliaceous stipules and a subterminal to lateral style. — Distinguished from *A. puncticulatum* by its free sepals, its glabrous disc, its non-tessellated higher venation and by the ferrugineous pubescence of its inflorescences (Fig. 9).

SELECTED SPECIMENS. PENINSULAR MALAYSIA. Pahang/Selangor, new road to Genting Highlands, forested hill side, 2500 ft, 25 Nov. 1966, *FRI (Whitmore)* 916 (K, KEP, L); Terengganu, Kemaman, Rasau Kerteh Forest Reserve, undulating logged forest, 11 May 1976, *FRI (Chan)* 25012 (K, L, SAR); Negeri Sembilan, Jelebu, primary forest, habitat undulating, ridge top, 1500 ft, 10 Oct. 1969, *KEP (Everett)* 104938 (K, KEP, L). SINGAPORE. Bukit Timah Nature Reserve, along the trail in Fern Valley, shaded evergreen forest, 125 m, 11 Nov. 1982, *Maxwell* 82-284 (L). SUMATRA. Sumatera Selatan, Bangka, G. Mangkol, granitic sand, c. 50 m, 14 Sept. 1949, *Kostermans & Anta* 689 (A, BO, K, L). BORNEO. Brunei, Belait distr., Badas Forest Reserve, 14 July 1988, *Haslani Abdullah* 50 (K, KEP, L); Sabah, Pedalaman, Sipitang distr., Pantai, sandy flat soil, 20 ft, 29 Nov. 1961, *SAN (Thaufeck)* 27958 (BO, K, KEP, L); Sarawak, Kuching div., near Kuching, 10 Jan. 1895, *Haviland & Hose* 3671 (BM, K, L); Sarawak, Sarikei div., Binatang (= Bintangor) distr., Lassa P. F., Tj. Sumong, peat swamp forest (mixed swamp forest), c. 12 ft, 9 March 1960, *S (Anderson)* 12416 (K, L, SAR); C Kalimantan, 8 km E of Sampit, Sampit R., near sea level, 21 Jan. 1954, *Alston* 13171 (K, L, US); W Kalimantan, Pontianak, Sei Poetat, 16 March 1931, *Mondi* 62 (K, L, NY); E Kalimantan, Samarinda, Lon Lempong, Blajan R., primary forest, 18 March 1955, *Nedi* 727 (B, K, L).

12. Antidesma cruciforme *Gage,* Rec. Bot. Surv. India 9: 226 (1922). Type: Perak, Gunong Batu Puteh, *Wray* 912 (SING!, lectotype, here designated, Kew Negative No. 11316). Original syntype: *Wray* 1183.

Shrub or tree, up to 6 m, diameter 8 cm. *Young twigs* terete to striate, spreading-ferrugineous-pubescent, brown. *Stipules* persistent or caducous, subulate, apically acute, 7 – 14 × 0.5 – 1.5 mm, ferrugineous-pubescent. *Petioles* channelled adaxially, basally and distally slightly pulvinate and geniculate, (12 –)17 – 35 mm long, 1.2 – 2 mm wide in the middle, ferrugineous-pubescent, becoming glabrous when old. *Leaf blades* oblong to elliptic, apically acuminate-mucronate, basally obtuse to rounded, (11 –)16 – 21(– 25.5) × (4 –)6 – 8(– 10) cm, (2 –)2.6(– 3.2) times as long as wide, eglandular, chartaceous to subcoriaceous, glabrous adaxially, pilose all over abaxially, shiny adaxially, shiny or dull abaxially, midvein impressed adaxially, tertiary veins percurrent to almost reticulate, close together, drying reddish brown to olive-green, domatia absent. *Staminate plants* unknown. *Pistillate inflorescences* 3 – 5 cm long, axillary, simple, axes densely ferrugineous-pubescent. *Bracts* lanceolate to subulate, 0.7 – 1.5 × 0.3 – 0.5 mm, ferrugineous-pubescent. *Pistillate flowers* 1 – 1.5 × 1 – 1.5 mm. *Calyx* c. 1 mm long, sepals 4, free, 0.3 – 0.5 mm wide, narrowly deltoid, apically acute, pubescent outside, glabrous inside except for long hairs at the base, margin entire. *Disc* much shorter than the sepals, glabrous. *Ovary* ellipsoid, glabrous or with few hairs, style subterminal to terminal, stigmas 4 – 6, rather regular, flatly spread out. *Infructescences* 12 – 17 cm long. *Fruiting pedicels* 0 – 1(– 3) mm long, pilose to pubescent. *Fruits* ellipsoid, laterally compressed, basally symmetrical, more rarely asymmetrical, with a terminal, more rarely subterminal style, 8 – 10 × 6 – 8 mm, glabrous, sometimes white-pustulate, fleshy to reticulate when dry. Fig. 10.

DISTRIBUTION. Peninsular Malaysia: Kedah, Pahang, Perak, Selangor. Map 11.
HABITAT. In forest. 130 – 1450 m altitude.
CONSERVATION STATUS. Near Threatened (NT). Extent of occurrence is c. 22,000 km^2, but continuing deforestation could make it necessary to move this species to a higher category in the future. Only 10 specimens examined.

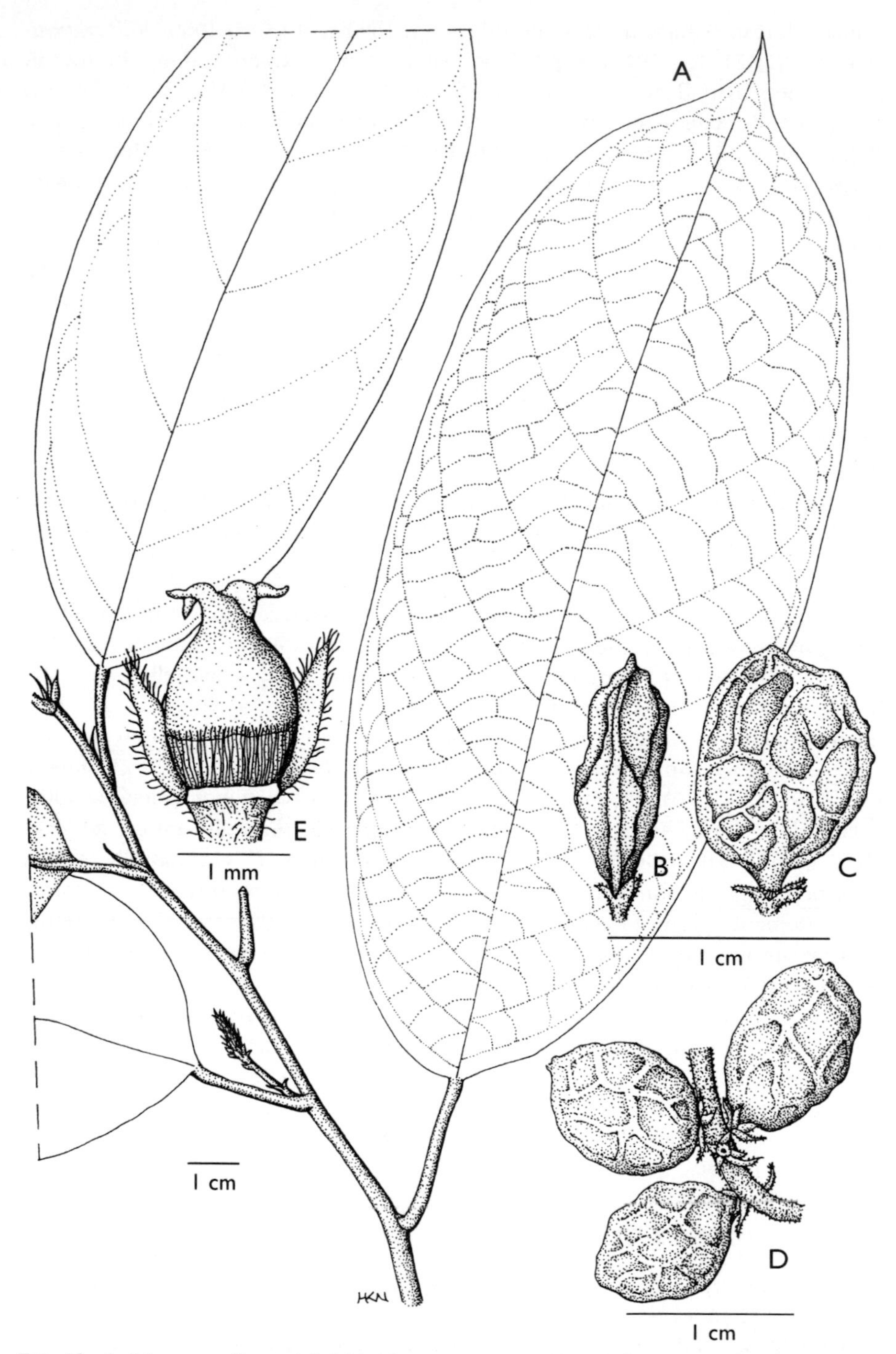

FIG. 10. *Antidesma cruciforme.* **A** habit with very young pistillate inflorescence; **B** fruit in dorsal view; **C** fruit in lateral view; **D** part of infructescence with bracts and sepals; **E** pistillate flower, sepals partly removed. **A – D** from *FRI* 5989 (L); **E** from *KEP* 78634 (L). Drawn by Holly Nixon.

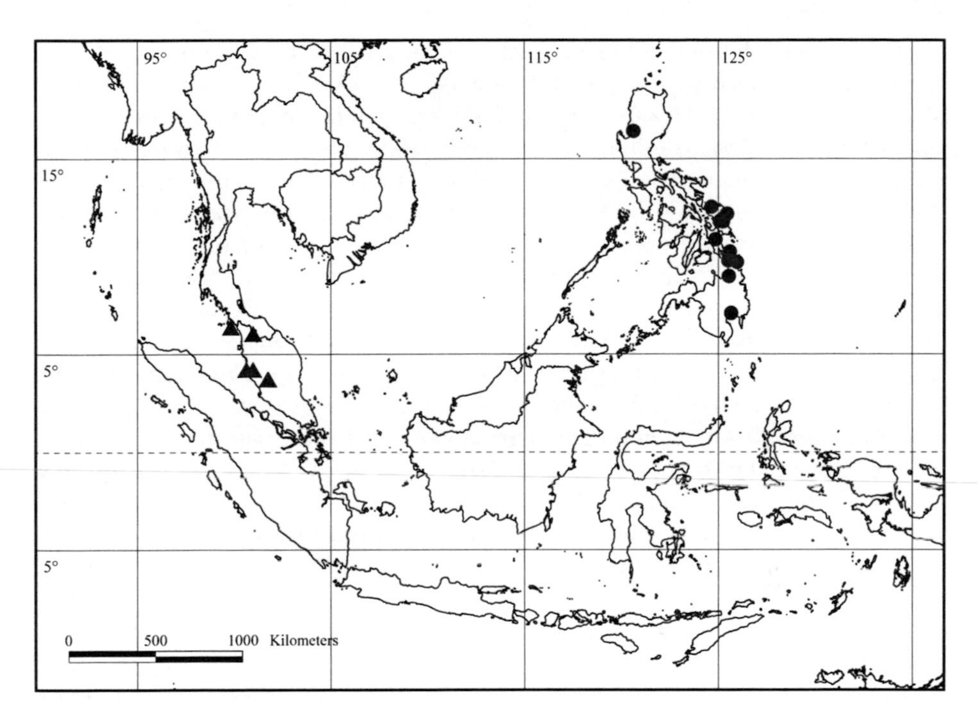

MAP 11. Distribution of *A. cruciforme* (▲) and *A. curranii* (●).

ETYMOLOGY. The epithet refers to the cross-like appearance of the flowers (Latin, crux, crucis = cross; forma = shape).

KEY CHARACTERS. Petioles long; sepals free; disc glabrous; style (sub)terminal; fruits glabrous, sessile; indumentum ferrugineous-tomentose.

SIMILAR SPECIES. *A. coriaceum*, see there. — *A. orthogyne* has shorter petioles, a hairy disc and ovary as well as dorsiventrally compressed fruits.

SELECTED SPECIMENS. PENINSULAR MALAYSIA. Kedah, Pulau Langkawi, Jan. 1897, *Ridley* 8344 (SING); Pahang, Fraser's Hill, upon the Selangor border, 4000 – 4370 ft, 16 – 30 Sept. 1922, *SF (Burkill & Holttum)* 8515 (SING); Pahang, Pine Tree Hill, 4800 ft, 5 Sept. 1923, *SF (Nur)* 11276 (SING); Perak, Melidang Saving Forest Reserve, 7 Feb. 1934, *For. Dept. F.M.S.* 33635 (SING); Kledang Sajong Forest Reserve, Batu Gajah, newly worked forest, hillside, 500 ft, 3 April 1968, *FRI (Ng)* 5989 (K, L); Kampung Gajah Forest Reserve, ridge top, 400 ft, 12 Nov. 1958, *KEP (Kochummen)* 78634 (K, L); Selangor, Gunong Ulu Semangkoh, 28 Sept. 1922, *SF (Burkill)* 8878 (SING).

13. Antidesma curranii *Merr.*, Philipp. J. Sci., C, 9: 466 (1914). Type: Philippines, Luzon, Subprov. Benguet, Baguio, in stream depressions, Aug. 1906, *FB (Curran)* 5087 (US!, lectotype, here designated).

Shrub or tree, up to 10 m, diameter up to 20 cm. Wood reddish. *Young twigs* terete to angular, sparsely spreading-pilose, very light to medium brown. *Stipules* early-caducous, deltoid to linear, apically acute, 3 – 5 × 1 – 1.5 mm, appressed-pilose. *Petioles* narrowly channelled adaxially, basally and distally sometimes slightly pulvinate, 3 – 15 × 0.7 – 1.2 mm, pilose, soon becoming glabrous, rugose and whitish. *Leaf blades* elliptic, more rarely ovate, oblong or obovate, apically acuminate-mucronate, basally acute to obtuse, more rarely rounded, (4.5 –)6 – 10(– 15.5) × (2 –)3 – 5(– 6.5) cm, (1.5 –)2 – 2.3(– 2.5) times as long as wide, eglandular, chartaceous to subcoriaceous, glabrous except for some hairs along the major veins, dull, more rarely shiny on both surfaces, midvein impressed adaxially, tertiary veins reticulate to weakly percurrent, drying olive-green, usually lighter abaxially, domatia absent. *Staminate inflorescences* 4 – 7 cm long, axillary to terminal, consisting of 3 – 6 branches, axes pilose. *Bracts* lanceolate to ovate, 0.7 – 1 × 0.5 – 0.7 mm, glabrous, margin erose. *Staminate flowers* 1.5 – 2 × 1 – 2 mm. *Pedicels* 0 – 0.2 mm long, not articulated, glabrous. *Calyx* 0.8 – 1 × 1 – 1.5 mm, cupular to bowl-shaped, sepals 4, fused for $^1/_2$ – $^2/_3$ of their length, apically acute to rounded, glabrous on both surfaces, margin erose. *Disc* consisting of 4 free alternistaminal lobes, lobes ± obconical, with two shallow imprints apically, c. 0.6 × 0.6 – 0.8 mm, exserted from the sepals, ferrugineous-pilose apically. *Stamens* 4, c. 1.5 mm long, exserted c. 1 mm from the disc, exserted c. 1.5 mm from the calyx, anthers 0.3 – 0.4 × c. 0.7 mm. *Pistillode* obconical to cylindrical, 0.4 – 0.5 × 0.3 – 0.4 mm, slightly exserted from the disc, ferrugineous-pubescent, especially apically. *Pistillate inflorescences* axillary, sometimes terminal, simple or once-branched, more rarely consisting of up to 6 branches, sometimes in fascicles of 2 inflorescences, axes glabrous to spreading-pilose. *Bracts* lanceolate to ovate, apically acute, 0.5 – 1 × 0.5 – 0.6 mm, glabrous, margin erose. *Pistillate flowers* not known. *Calyx* in fruit 1 – 1.5 × 1.5 – 2 mm, sepals 4, at anthesis probably fused to halfway, in fruit irregularly laciniate, apically obtuse to rounded, glabrous outside and inside, margin erose, fimbriate. *Disc* much shorter than the sepals, ferrugineous-pubescent, especially at the margin. *Stigmas* 4 – 6. *Infructescences* 3 – 6 cm long. *Fruting pedicels* 0.5 – 2 mm long, glabrous to pilose. *Fruits* ellipsoid to lenticular, laterally compressed, basally symmetrical, with a terminal to slightly subterminal style, 4 – 5 × 3 – 4 mm, glabrous, sometimes white-pustulate, pale yellow to light brown, more rarely brick-red when dry, areolate when dry.

DISTRIBUTION. Philippines: Bucas Grande, Dinagat, Leyte, Mindanao, Samar. Map 11.

HABITAT. In (dry) dipterocarp forest; in primary or secondary vegetation; in shady or half-shady habitats. On clay soil, also reported from ultrabasic soil. 50 – 700 m altitude.

CONSERVATION STATUS. Near Threatened (NT). The extent of occurrence is c. 150,000 km^2, however, the Philippines have a particularly high rate of deforestation. 18 specimens examined.

VERNACULAR NAMES. Samar: aalimara (Bis).

ETYMOLOGY. The epithet refers to the collector of the type specimen.

KEY CHARACTERS. Branches light brown; petioles fairly long; leaves glabrous; calyx lacerate in fruit; disc with ferrugineous indumentum; free staminate disc lobes.

SIMILAR SPECIES. Most collections of *A. curranii* have been distributed under the name *A. catanduanense*, see there. — *A. digitaliforme* has shorter and more highly fused sepals which are not laciniate at fruit maturity, a pistillate disc indumentum which is exserted from the sepals at fruit maturity, and smaller stipules. — *A. microcarpum* has an extrastaminally fused staminate disc, smaller fruits, more highly fused sepals and thinner, usually shorter fruiting pedicels. — *A. pleuricum* has more highly fused sepals, pilose ovaries, smaller fruits with a lateral style and an extrastaminally fused staminate disc.

SELECTED SPECIMENS. PHILIPPINES. Bucas Grande Isl., s. loc., June 1919, *BS (Ramos & Pascasio)* 35108 (A); Leyte, s. loc. 15 June 1913, *Wenzel* 115 (A, G, US); 30 June 1913, *Wenzel* 319 (A, G, L, US); Mindanao, Davao prov., Santa Cruz, 3 June 1905, *Williams* 2912 (NY); Mindanao, Surigao del Norte prov., Dinagat Island, in forest, somewhat dry, low altitude, 25 May 1931, *BS (Ramos & Convocar)* 83820 (NY); Mindanao, Surigao prov., Mabuhay Mining Camp, virgin forest, 25 May 1950, *PNH (Anonuevo)* 13457 (A, K, L); Mindanao, Surigao prov., Surigao, Sukailang, in forest ridge, dipterocarp type, 330 m, Feb. – March 1949, *PNH (Mendoza & Convocar)* 10273 (A, K, L); Samar, Bagacay, Concord, dipterocarp forest ridge (flat), under shade, 300 m, April – May 1948, *Sulit* 6257 (A, K, L); Samar, Mt Cansayao, Lope de Vega, Catarman, logged-over area, dipterocarp forest under partial shade, 200 m, 14 April 1951, *PNH (Sulit)* 14420 (A, K, L).

14. Antidesma cuspidatum *Müll. Arg.*, Linnaea 34: 67 (1865 – 1866). Type: In Malacca Indiae or., *Leman [Lemann]* s.n. (G-DC [microfiche], lectotype, here designated; G!, K!, isolectotypes).

Tree, rarely shrub, up to 17 m, clear bole up to 4 m, diameter up to 20 cm, straight, branching, twisted or crooked, sometimes fluted. Twigs grey-green or bright green. Bark whitish grey, grey, brown, greenish grey or greenish brown, thin, smooth, flaky or slightly cracking, brittle, soft; inner bark brown, pink, red, yellow or white, thin, fibrous; cambium white, hard; sapwood white, red, reddish brown or pale yellow, heartwood red. *Young twigs* terete to striate, shortly pubescent, brown. *Stipules* persistent or caducous, lanceolate to ovate, apically acute to acuminate, 5 – 10 × 1 – 3 mm, pubescent. *Petioles* channelled adaxially, basally and distally sometimes slightly pulvinate and geniculate, (5 –)7 – 20(– 25) × 1 – 1.5 mm, pubescent. *Leaf blades* oblong to ovate, apically long-acute to acuminate-mucronate, basally obtuse to truncate, (7 –)11 – 16(– 23) × (2.5 –)4 – 6(– 9) cm, (2 –)2.5(– 3) times as long as wide, eglandular, chartaceous, glabrous except for some hairs along the major veins on both surfaces or only abaxially, shiny on both surfaces, midvein impressed adaxially, tertiary veins reticulate, drying dark olive-green to reddish brown, lighter abaxially, domatia absent. *Staminate inflorescences* 5 – 12 cm long, terminal and axillary, simple or consisting of up to 10(– 20) branches, axes shortly pubescent. *Bracts* linear to deltoid, apically acute, 0.4 – 1.5 × 0.2 – 0.3 mm, pubescent. *Staminate flowers* 1.5 – 3 × c. 2 mm. *Pedicels* 0.3 – 1 mm long, not articulated, pubescent. *Calyx* 0.5 – 0.8 mm long, sepals 3 – 5, free, sometimes unequal, 0.5 – 1 mm wide, spreading, nearly orbicular, apically truncate to obtuse, pubescent outside, pilose to glabrous inside, long hairs at least at the base, margin erose, glandular-fimbriate to lacerate. *Disc* cushion-shaped, fully enclosing the bases of the filaments and pistillode, constricted at the base, shortly pubescent to glabrous. *Stamens* 3 – 5, c. 2 mm long, exserted 1.5 – 2 mm from the calyx, anthers 0.3 – 0.4 × 0.3 – 0.4 mm. *Pistillode* clavate and apically crateriform to subulate, 0.3 – 0.5 × 0.2 – 0.4 mm, extending to the same length or slightly exserted from the sepals, shortly pubescent to glabrous. *Pistillate inflorescences* 6 – 12(– 18) cm long, terminally or axillary, simple or branched regularly, mostly near the base, consisting of up to 5 branches, axes shortly pubescent. *Bracts* linear to deltoid, apically acute, 0.4 – 1.5 × c. 0.4 mm, pubescent. *Pistillate flowers* 1.5 – 2 × 0.8 – 1.5 mm. *Pedicels* 0.5 – 1.5 mm long, pubescent. *Calyx* 0.5 – 1 mm long, sepals 3 – 6, free or fused for c. $^{2}/_{3}$ of their length, pilose to pubescent outside, pilose to glabrous inside with but long hairs at least at the base, margin erose, glandular-fimbriate to lacerate. *Disc* shorter than the sepals, shortly hirsute to glabrous. *Ovary* lenticular, glabrous to pilose, style subterminal, stigmas 3 – 6. *Infructescences* 12 – 18 cm long, straight and robust. *Fruiting pedicels* 1 – 4 mm long, pubescent. *Fruits* lenticular, distinctly laterally compressed, basally symmetrical to slightly asymmetrical, with a lateral style, 5 – 7 × 4 – 6 mm, glabrous to pilose, sometimes white-pustulate, often conspicuously black and shiny when dry, areolate when dry.

14a. var. **cuspidatum**

A. rotatum Müll. Arg. in DC., Prodr. 15(2): 256 (1866); Hook. f., Fl. Brit. India 5: 360 (1887), pro synon.; Ridl., Fl. Malay Penins. 3: 232 (1924), pro synon. Type: Malacca Indiae or., *Jagor "Jaeger"* 303 (NY! ex B, lectotype, here designated). — Note: Holotype in B was destroyed in WW II; no material in G-DC on microfiche. The collector was Jagor (van Steenis 1950b: 259), but the handwriting on his labels is hard to decipher. For Müller it was natural to read it as "Jaeger", a common German surname. All other data on the label as well as the fact that it is a staminate specimen (only the staminate flowers are described in the protologue), identify the lectotype as a duplicate of the destroyed holotype.

A. cuspidatum Müll. Arg. var. *leiodiscus* Ridl., Fl. Malay Penins. 3: 232 (1924), **synon. nov.** Type: Malaysia, Perak, Larut, recd from Dr King 1887, *Scortechini* s.n. (K!, lectotype, here designated). — Note: This variety differs from the type mainly by its glabrous disc. The indumentum of the calyx, disc, pistillode and ovary in *A. cuspidatum* is, however, very variable. Interestingly, in most specimens a glabrous disc is correlated with bracts shorter than 1 mm while most collections with a hairy disc have bracts 1 – 2 mm long. Several specimens, including that cited by Ridley for var. *leiodiscus*, combine a glabrous disc with short bracts. Both forms occur in the same locality (e.g., *FRI* 11326 & 11350) in what seems to be the same population.

A. cuspidatum Müll. Arg. var. *borneense* Airy Shaw, Kew Bull., Addit. Ser. 4: 210 (1975), **synon. nov.** Type: Sarawak, Kuching Distr., 12th mile, Penrissen Road, hill slope, near tree no. 4547, 1 Sept. 1966, *S (Banyeng & Sibat)* 24915 (K!, holotype; A!, SAR!, isotypes).

Stipules usually persistent, lanceolate to ovate, apically acute to acuminate, 5 – 10 × 1 – 3 mm, pubescent. *Pistillate flowers* 1 – 1.5 mm wide. *Calyx* 0.5 – 0.8 mm long, sepals 3 – 6, free, c. 0.5 mm wide, spreading, deltoid to nearly orbicular, sometimes unequal, apically acute to truncate. *Disc* shortly hirsute to glabrous. *Ovary* glabrous to pilose.

DISTRIBUTION. Peninsular Malaysia, Singapore, Sumatra, Borneo (Kalimantan, western Sarawak). Chakrabarty & Gangopadhyay (1997: 479) reported *A. cuspidatum* also from Java on the basis of *Forbes* 895 (CAL), but the duplicates of this collection in BM and L are *A. tetrandrum.* Map 12.

HABITAT. In dipterocarp forest; sometimes on marshy or swampy ground; primary to disturbed, sometimes partly open vegetation. On yellow sand, loam or peat. 0 – 1600 m altitude.

USES. The wood is hard and fine-grained but usually only available in small quantities so that it is mainly used for small pieces of work (Ridley 1903: 92); it is also suitable as firewood.

CONSERVATION STATUS. Least Concern (LC). This is a widely distributed and well-collected species. 177 specimens examined.

VERNACULAR NAMES. Malay Peninsula: buni, kayu buloh (Temuan), okor kuching, selumar; Sumatra: kayu bata, kayu dakka tolu, kayu manuk-manuk, kayu simburo; Kalimantan: uhai.

ETYMOLOGY. The epithet refers to the leaf-apex (Latin, cuspidatus = pointed).

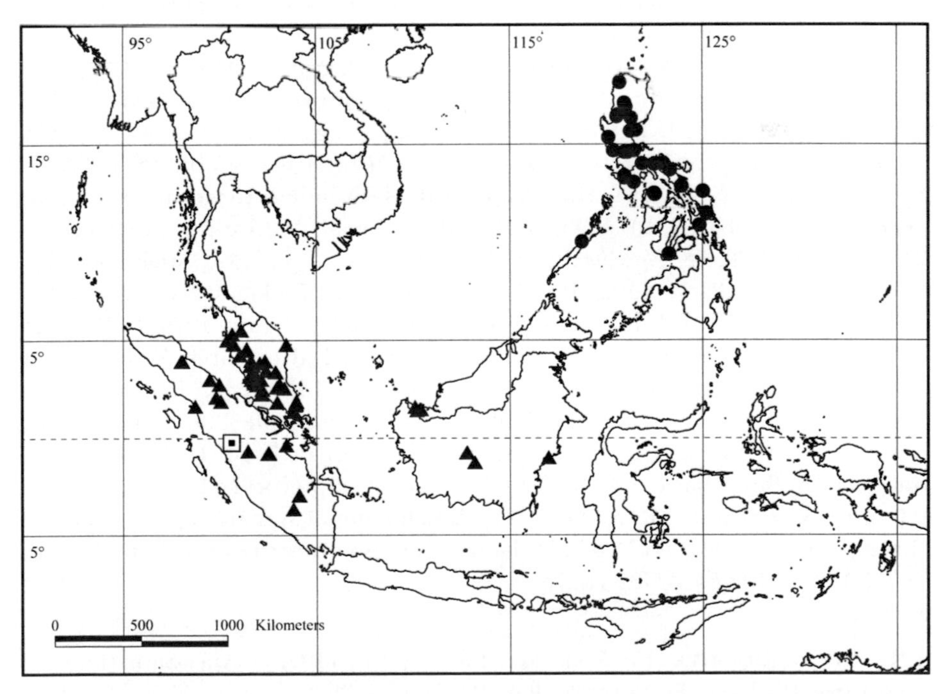

MAP 12. Distribution of *A. cuspidatum* var. *cuspidatum* (▲), *A. cuspidatum* var. *orthocalyx* (⊡) and *A. digitaliforme* (●).

KEY CHARACTERS. Long petioles; strongly compressed, lenticular ovaries and fruits; usually rounded, glandular-fimbriate sepals.

SIMILAR SPECIES. *A. celebicum* and *A. coriaceum*, see there. — *A. forbesii* has smaller, caducuous stipules, thinner, duller leaves, a narrower leaf base, shorter, axillary inflorescences, shorter bracts, smaller flowers with shorter stamens, partly fused sepals and free staminate disc lobes. — *A. cuspidatum* is vegetatively similar to *A. montanum* but has longer petioles and thicker leaves. The compressed, asymmetrical ovaries cannot be mistaken for *A. montanum*. Staminate inflorescences of *A. cuspidatum* are usually longer and branched at the base only, disc and pistillode often hairy. — Pistillate specimens with glabrous discs and ovaries and missing stipules resemble *A. tetrandrum* which has equally long petioles but much smaller, fused, eglandular sepals.

NOTES. In Sumatra, the boundary between *A. cuspidatum* and *A. montanum* becomes blurred. There are groups of collections with different combinations of characters from both species: (1) The staminate collection *de Wilde & de Wilde-Duyfjes* 18518 from Gunung Leuser National Park has short petioles (4 mm long) and thin leaves like *A. montanum*, but the leaf shape and flowers are typical of *A. cuspidatum*, with a hairy disc and completely free, fairly wide, rounded sepals. (2) The pistillate and fruiting collections *de Wilde & de Wilde-Duyfjes* 12099, 12318, 12372, 12834, 18150, 18219, also from Gunung Leuser National Park, have longer petioles (7 – 12 mm long), but also thin leaves with sometimes acute bases, which is atypical of *A. cuspidatum*, but normal for *A. montanum*. The inflorescences, flowers and fruits are, again, typical of *A. cuspidatum*. These specimens are here identified as *A.* cf. *cuspidatum*.

SELECTED SPECIMENS. PENINSULAR MALAYSIA. Pahang, Cameron Highlands, Mt Berembun, 4°29'N, 101°24'E, on peaty soil, 5000 ft, 3 Oct. 1963, *Chew* 763 (A, G, K, L); Kelantan, SE Kelantan, Sungai Lebir, 2 miles E Kuala Aring, ridge top, primary forest, 13 Sept. 1967, *FRI (Cockburn)* 7109 (K, L, SAR); Selangor, road to Fraser's Hill, Gap road, 47th mile, 15 April 1980, *FRI (Kochummen)* 29088 (K, L). SINGAPORE. Bukit Timah Nature Reserve, tree no. 214, 14 Oct. 1938, *SF (Ngadiman)* 35922 (BO, E, K). SUMATRA. Atjeh, Gunung Leuser National Park, Sekundur Forest Reserve, c. 75 km WNW of Medan, 3°55'N, 98°05'E, recently logged over forest, 50 – 100 m, 7 Aug. 1979, *De Wilde & De Wilde-Duyfjes* 19515 (K, L, US); Palembang distr., Tjaban Forest Reserve near Muara Enim, low altitude, Feb. 195, *Kostermans* S 142 (A, K, L). BORNEO. Sarawak, Kuching div., Sampadi Forest Reserve, 26th mile Bau/Lundu road, hillside, yellow loam, 650 ft, 19 June 1968, *S (Ilias Paie)* 24949 (E, K); C/W Kalimantan, Bukit Raya, 0°45'S, 112°47'E, primary dipterocarp forest, c. 130 m, 9 Jan. 1983, *Nooteboom* 4482 (BO, K, NY); E Kalimantan, Wanariset research area, Rintis Soejawo, 1°00'S, 117°00'E, lowland mixed dipterocarp forest, 50 m, 13 May 1991, *Ambri & Arifin* W 734 (K, L).

14b. var. **orthocalyx** *Airy Shaw*, Kew Bull. 28: 274 (1973). Type: Sumatra, Res. West Coast (Padangsche Bovenlanden), Mt Merapi, 20 Nov. 1956, *Meijer* 7657 (K!, holotype; L!, SING!, isotypes).

Stipules not seen. *Staminate plants* not known. *Pistillate flowers* c. 0.8 mm wide. *Calyx* c. 1 × 0.7 mm, urceolate, sepals 4, fused for c. $^2/_3$ of their length, erect, sinuses rounded. *Disc* glabrous. *Ovary* pilose. *Infructescences* and *fruits* not known.

DISTRIBUTION. West Sumatra. Map 12.
HABITAT. 1400 m altitude.
CONSERVATION STATUS. Data Deficient (DD): only known from the type.
ETYMOLOGY. The epithet refers to the shape of the calyx (Greek, orthos = straight).
KEY CHARACTERS. Sepals fused; otherwise similar to the type variety.
NOTE. This specimen might well represent a different species, as pointed out in the protologue, but the material is too incomplete to be certain. It is also similar to *A. forbesii* but differs in the large, coriaceous leaves with stout petioles, as well as in the stiff and much-branched inflorescences.

15. Antidesma digitaliforme *Tul.*, Ann. Sci. Nat. Bot., Sér. 3: 191 (1851). Type: Philippines, Luzon, apud Igolotas, in sylvis montium, prope Manillam, *Callery* 42 (P!, lectotype, here designated; A!, G!, isolectotypes). — Note: The lectotype bears only a few heavily disturbed flowers, whereas the isolectotype in G has more and normally developed flowers.

A. lucidum Merr., Philipp. J. Sci. 1, suppl.: 78 (1906). Type: Philippines, Lamao Forest Reserve, March, *Whitford* 1135 (US!, lectotype, here designated; G!, K!, NY!, isolectotypes). Original syntypes: Feb., *Meyer* 2642 (K!, NY!, US!); *Meyer* 2775 (K!, NY!, US!); Nov., *Elmer* 7005 (E!; G!, K!, NY!).

A. luzonicum Merr., Philipp. J. Sci., C, 9: 464 (1914). Type: Philippines, Luzon, Prov. Camarines, Mount Isarog, in forests, 24 Nov. 1913, *Ramos* (Phil. Pl.) 1555 (L!, lectotype, here designated; A!, BO!, G!, NY!, SING!, isolectotypes).

Shrub or tree, up to 12 m, diameter up to 27 cm. *Young twigs* terete, shortly pubescent to glabrous, brown. *Stipules* early-caducous, deltoid, apically acute, c. 0.5 × 0.5 mm, pubescent. *Petioles* channelled adaxially, often becoming rugose when old, 3 – 8(– 10) × 0.7 – 1.3(– 1.5) mm, pilose to glabrous, sometimes very light brown. *Leaf blades* elliptic (margins often apically and distally concave), more rarely ovate, apically acuminate-mucronate, often with an obtuse apiculum, basally acute, more rarely obtuse or decurrent, (4 –)5 – 9(– 11.5) × (1 –)2 – 3.5(– 5.5) cm, (1.7 –)2.5(– 3) times as long as wide, eglandular, chartaceous to subcoriaceous, glabrous, or with some short hairs only along the major veins abaxially, shiny on both surfaces, midvein impressed adaxially, tertiary veins reticulate, drying dark olive-green to greyish brown adaxially, lighter olive-green to reddish brown abaxially, distinctly discolorous to almost concolorous, domatia usually present. *Staminate inflorescences* 2 – 8(– 12) cm long, axillary, solitary or 2 per fascicle, simple or consisting of up to 5(– 8) branches, axes glabrous to pilose. *Bracts* deltoid to ovate, apically acute to rounded, 0.5 – 1 × 0.3 – 1 mm, pilose to glabrous, margin fimbriate. *Staminate flowers* 1.5 – 2 × 1.5 – 2 mm. *Pedicel* absent. *Calyx* 0.5 – 0.8 × 1 – 1.5 mm, bowl-shaped to cupular, sepals 4 – 6, fused for $^2/_3$ or more of their length, broadly deltoid, apically acute, glabrous to slightly pilose outside, glabrous inside, margin entire. *Disc* consisting of 4 – 6 free alternistaminal lobes, lobes ± obconical, with two shallow imprints apically, 0.3 – 0.4 × 0.5 – 0.6 mm, or partly or completely extrastaminally fused, usually exserted from the calyx, shortly ferrugineous-pubescent to pilose. *Stamens* (1 –)4 – 6, 1 – 1.5 mm long, exserted c. 1 mm from the calyx, anthers c. 0.3 × 0.5 – 0.6 mm. *Pistillode* cylindrical to globose, 0.5 – 0.8 × 0.3 – 0.5 mm, extending to the same length or exserted from the sepals, slightly ferrugineous-pilose. *Pistillate inflorescences* 2 – 3 cm long, axillary, simple or consisting of 1 – 4 branches, sometimes in fascicles of 2 inflorescences, axes glabrous to pilose. *Bracts* deltoid to ovate, apically acute to obtuse, 0.4 – 0.5 × 0.3 – 0.5 mm, glabrous to pilose, margin entire, sometimes fimbriate. *Pistillate flowers* c. 1.5 × 0.5 – 1 mm. *Pedicels* 0 – 0.5 mm long, glabrous. *Calyx* 0.8 – 1 × 0.5 – 1 mm, urceolate, sepals 4(– 5), fused for c. $^3/_4$ of their length, broadly deltoid, apically acute to rounded, glabrous on both sides, margin sometimes fimbriate or erose. *Disc* shorter than the sepals, but indumentum usually exserted from the sepals, ferrugineous to ochraceous hairs at the margin, otherwise glabrous. *Ovary* ellipsoid, laterally compressed, glabrous, more rarely sparsely pilose, style subterminal, stigmas 3 – 5. *Infructescences* 2 – 5 cm long. *Fruiting pedicels* 0.5 – 1.2 mm long, glabrous or pilose. *Fruits* ellipsoid to lenticular, laterally compressed, basally symmetrical or asymmetrical, the narrow calyx usually conspicuous, with a subterminal style, (2.5 –)3 – 6 × 2 – 4 mm, glabrous, sometimes white-pustulate, areolate when dry.

DISTRIBUTION. Philippines: Leyte, Luzon, Mindoro, Negros, Palawan, Samar, Sibuyan. Map 12.

HABITAT. In the understorey of dipterocarp and elfin/montane forest; in primary or secondary vegetation; sometimes on exposed ridges. On clay-loam or loose rocky soil. 10 – 1600 m altitude.

USES. Firewood.

CONSERVATION STATUS. Near threatened (NT). The species is widespread in the Philippines (extent of occurrence almost 400,000 km^2), but continuing habitat destruction can be expected to threaten it in the future. 51 specimens examined.

ETYMOLOGY. The epithet refers to the shape of the calyx, resembling the flowers of *Digitalis* L. (Scrophulariaceae).

KEY CHARACTERS. Sepals highly fused; calyx urceolate; disc ferrugineous- to ochraceous-pubescent; staminate disc lobes free; bracts wide; dry leaves often discolorous.

SIMILAR SPECIES. *A. celebicum* and *A. curranii*, see there. — Vegetatively similar to *A. japonicum* which has a glabrous, cushion-shaped (in staminate flowers) disc, and less highly fused sepals. — *A. pleuricum* has longer petioles, obtuse to truncate leaf bases, fused staminate disc lobes and fruits with lateral styles.

NOTES. Some specimens (including the type) have deformed staminate flowers: the calyx is almost globose with only a very small apical opening; the disc is often enlarged and the number of stamens reduced, sometimes to one stamen only. Similar aberrations occur in *A. japonicum*. These flowers seem to be affected by some genetic or pathological disorder.

The collection *Soejarto et al.* 7790 from Luzon differs by its staminate flowers on pedicels 0.5 – 1 mm long, and the lack of a pistillode. In *PNH (Mendoza & Convocar)* 10273 from Mindanao the stipules are exceptionally large (10 × 2 mm). *Elmer* 7005 from Luzon has fruits up to 8 × 6 mm which are also less oblique and more fleshy than in other collections. The calyx is missing on the mature fruit and the disc is nearly glabrous. Although an original syntype of *A. lucidum*, this specimen is therefore included here only doubtfully.

SELECTED SPECIMENS. PHILIPPINES. Leyte, s. loc., 15 June 1913, *Wenzel* 116 (A, G, US); Luzon, Benguet prov., Baguio, 21 June 1904, *Williams* 1141 (K, NY, US); Luzon, Benguet subprov., s. loc., May 1911, *Merrill* 893 (A, FR, G, U, US, WRSL); Mindoro, Pinamalayan, June 1922, *BS (Ramos)* 40931 (A, K, L, US); Negros, Negros Oriental prov., Sibulan, Kabalinan, L. Balinsasayao, primary forest, lakeside, loose, rocky soil, 23 May 1991, *PPI (Reynoso, Fuentes & Garcia)* 1092 (L); Palawan, Bahile, *Vidal* 1762 (A); Samar, Catubig river, Feb. – March 1916, *BS (Ramos)* 24436 (K, NY, US); Sibuyan Isl., Capiz prov., Mt Giting-Giting, Magallanes, wooded ridge, in moist red soil and with some rocks, 4750 ft or higher, May 1910, *Elmer* 12519 (BM, E, G, K, L, NY, US, WRSL).

16. Antidesma edule *Merr.*, Bur. Govt. Lab. Bull. (Philippines) 17: 26 (1904); Sp. Blancoan.: 219 (1918). Type: Philippines, Luzon, Bataan prov., Lamao R., Jan. 1904, *FB (Barnes)* 167 (US!, lectotype, here designated; BM!, K!, NY!, isolectotypes). Original syntypes: Oct. 1903, *Merrill* 3148 (K!, P!, NY!, US!); Jan. 1904; *Merrill* 3784 (K!, NY!, P!, US!); March 1904, *FB (Barnes)* 574 (in BM!, K!, NY!, P! and US! there are nos. 547 in which plants and labels match perfectly. This might be a printing error either on the sheets or in the protologue). — Note: Merrill (1918: 219 – 220) reduced this species to *A. spicatum* Blanco, but did not cite any original material. He selected two "illustrative specimens" of *A. edule* for his *Species Blancoanae* (nos. 718, K! & 915, K!) to represent *A. spicatum*. As this term is not equivalent to the term "type" in the sense of Art. 7.11. (Greuter *et al.* 2000), this did not effect neotypification. There are several inconsistencies between Blanco's description and the types of *A. edule*. The name may well represent *A. ghaesembilla*, as Pax & Hoffmann (1922: 155) suggested with reference to Index Kewensis. As Merrill gave no satisfactory proof that Blanco's species is identical with *A. edule*, the latter name is used here, and *A. spicatum* treated as incompletely known until original material is found.

Tree, up to 20 m, diameter up to 30 cm. *Young twigs* terete to striate, densely pubescent, brown. *Stipules* early-caducous, deltoid, apically acute, 3 – 6 × 1 – 2 mm, pubescent. *Petioles* channelled adaxially, basally and distally sometimes slightly pulvinate, (4 –)10 – 20(– 30) × 1.2 – 2(– 3) mm, pilose to pubescent. *Leaf blades* elliptic to ovate or obovate, apically acuminate-mucronate, basally rounded to acute, more rarely slightly cordate, (7.5 –)13 – 17(– 23) × (2.3 –)5 – 8(– 12) cm, (1.4 –)2.4(– 3.4) times as long as wide, eglandular, thinly chartaceous to thickly coriaceous, glabrous to sparsely pilose all over and pubescent along the major veins adaxially, pubescent all over or only along the midvein abaxially, dull on both surfaces, midvein impressed, rarely flat adaxially, tertiary veins reticulate to weakly percurrent, widely spaced, drying grey, more rarely reddish brown adaxially, reddish brown, more rarely olive-green abaxially, domatia absent. *Staminate inflorescences* up to 10 cm long, axillary, consisting of up to 25 branches, axes pubescent. *Bracts* lanceolate to elliptic or orbicular, apically acute, 0.5 – 1 × 0.5 – 1 mm, pubescent to sparsely pilose. *Staminate flowers* 1.5 – 2 × 1 – 1.5 mm. *Pedicels* 0 – 1 mm long, not articulated, pubescent. *Calyx* c. 1 × 1 – 1.5 mm, cupular, sepals usually 6, fused for c. $^1/_2$ of their length, unequal, pilose to pubescent outside, glabrous inside. *Disc* consisting of 5 – 8 ± free lobes placed between and in front of the stamens; lobes of unequal size, ± obconical, tightly adhering to one another, c. 0.5 × up to 0.5 mm, initially appearing cushion-shaped, pubescent. *Stamens* 4 – 5(– 7), sometimes fused, 1.5 – 2 mm long, exserted 1 – 1.5 mm from the calyx, anthers c. 0.5 × 0.5 – 0.6 mm. *Pistillode* subulate, 0.2 – 0.5 × 0.1 – 0.2 mm, much shorter to rarely exserted from the disc, pubescent. *Pistillate inflorescences* 1 – 9 cm long, axillary, but often aggregated at the end of the branch, simple or consisting of 1 – 25 branches, axes pubescent. *Bracts* lanceolate, 1 – 1.5 × 0.5 – 1 mm, pubescent. *Pistillate flowers* c. 1.5 × 1 mm. *Pedicels* c. 0.5 mm long, pilose. *Calyx* 0.5 – 1 × c. 1 mm, urceolate, sepals 4 – 5, fused for c. $^2/_3$ of their length, often unequal, apically acute, pilose, rarely glabrous outside, glabrous inside. *Disc* shorter than the sepals, but indumentum extending to the same length as or exserted from the sepals, long (0.3 – 0.5 mm) hairs at the margin extending to the same length as the sepals, otherwise glabrous. *Ovary* ellipsoid, pilose, style lateral, distinct, stigmas 3 – 6. *Infructescences* 2 – 10 cm long. *Fruiting pedicels* c. 1 mm long, pilose. *Fruits* bean-shaped to lenticular, laterally compressed, basally asymmetrical, with a distinctly lateral style, 3 – 4(– 8) × 2.5 – 3(– 9) mm, glabrous to sparsely pilose, white-pustulate, areolate or reticulate when dry.

16a. var. **edule**

Tree, up to 13 m, diameter up to 30 cm. *Petioles* basally and distally sometimes slightly pulvinate, (5 –)10 – 20(– 30) × 1.5 – 2(– 3) mm, pilose to pubescent. *Leaf blades* elliptic to ovate, thinly chartaceous to coriaceous, glabrous to sparsely pilose all over and pubescent along the major veins adaxially, pubescent all over abaxially. *Staminate inflorescences* 3 – 10 cm long, consisting of 7 – 25 branches. *Pistillate inflorescences* consisting of 2 – 25 branches. *Fruits* bean-shaped to obliquely lenticular, 3 – 4(– 6) × 2.5 – 3(– 4) mm, glabrous to sparsely pilose.

DISTRIBUTION. Philippines: Luzon, Mindanao, Mindoro, Polillo, Samar. Map 13.

HABITAT. In mixed lowland to hill forest; sometimes in damp forests by streams. On rocky volcanic/ultrabasic soil. 70 – 400 m altitude.

USES. The fruits are edible.

CONSERVATION STATUS. Near Threatened (NT). Like *A. digitaliforme*, this species seems to be collected fairly frequently but is classified here in view of the extent of deforestation in the Philippines. 69 specimens examined.

VERNACULAR NAMES. apanaug (Waray waray); tanigi (T., Pamp.).

ETYMOLOGY. The epithet refers to the edible fruits (Latin, edulis = edible).

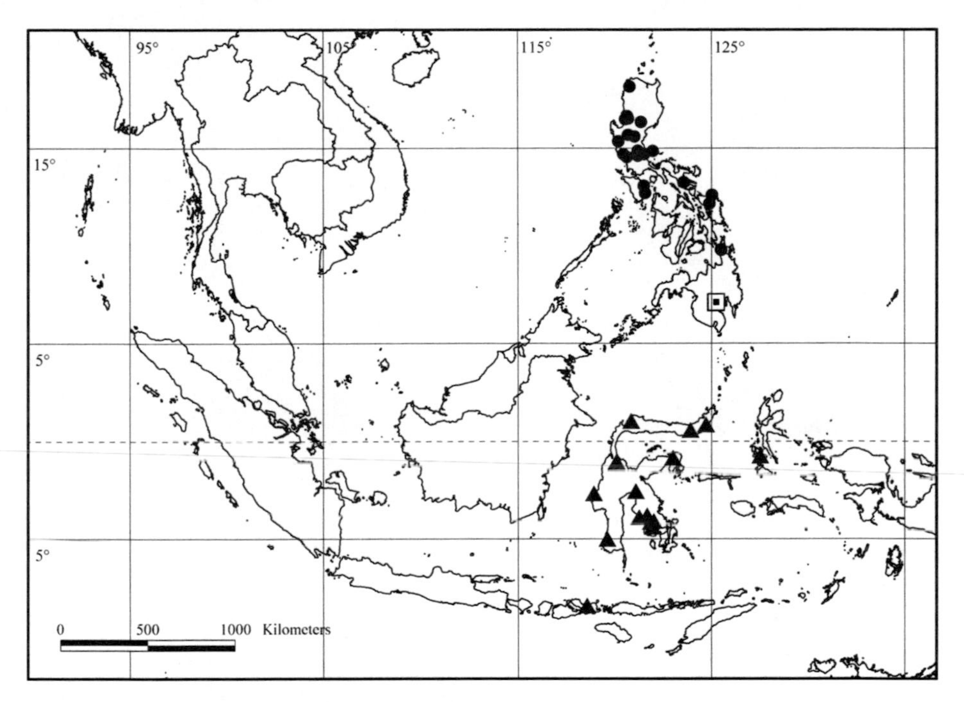

MAP 13. Distribution of *A. edule* var. *edule* (●), *A. edule* var. *apoense* (⊡) and *A. elbertii* (▲).

KEY CHARACTERS. Dense indumentum; long petioles; large leaves; highly fused sepals.

SIMILAR SPECIES. *A. celebicum*, see there. — *A. ghaesembilla* has free sepals, fruits with terminal styles, smaller, thinner, olive-green drying leaves and thinner petioles. — Often misidentified as *A. montanum* which has much shorter petioles, glabrous discs and glabrous, terete fruits with a terminal style. — *A. subcordatum* has ellipsoid fruits with terminal styles, thinner petioles and leaves, which do not dry reddish brown, and a more cordate leaf base.

SELECTED SPECIMENS. PHILIPPINES. Luzon, Bataan prov., Lamao Forest Reserve, Oct. 1906, *BS (Foxworthy)* 1625 (BO, NY, US); Luzon, Zambales prov., Mt Marangay, Nov. – Dec. 1924, *BS (Ramos & Edaño)* 44532 (BM, G, MEL, NY); Luzon, Albay prov., 15 Feb. 1921, *Cuming* 1348 (CGE, FHO, G, K, L); Luzon, Benguet prov., Sablan, 18 Nov. 1904, *Williams* 1372 (GH, K, NY, US); Mindanao, Surigao, 22 March 1928, *Wenzel* 3194 (A, G, K, NY); Mindoro, Pinamalayan, June 1922, *BS (Ramos)* 40923 (A, US); Polillo Isl., Oct. – Nov. 1909, *McGregor* 10279 (L, P); Samar, Catubig R., Feb. – March 1916, *BS (Ramos)* 24176 (A, BM, K, L, NY, US).

16b. var. **apoense** *Petra Hoffm.*, Kew Bull. 54: 350 (1999). Type: Philippines, Mindanao, North Cotabato prov., Kidapawan municipality, Mt Apo Geothermal Project Site B, 7°05'N, 125°14'E, midmontane forest with average canopy height of c. 20 m, alt. c. 1390 m, 24 Oct. 1990, *Co* 3142 (A!, holotype). Paratype: same locality, *Co* 3139 (A, L!, PNH, PUH).

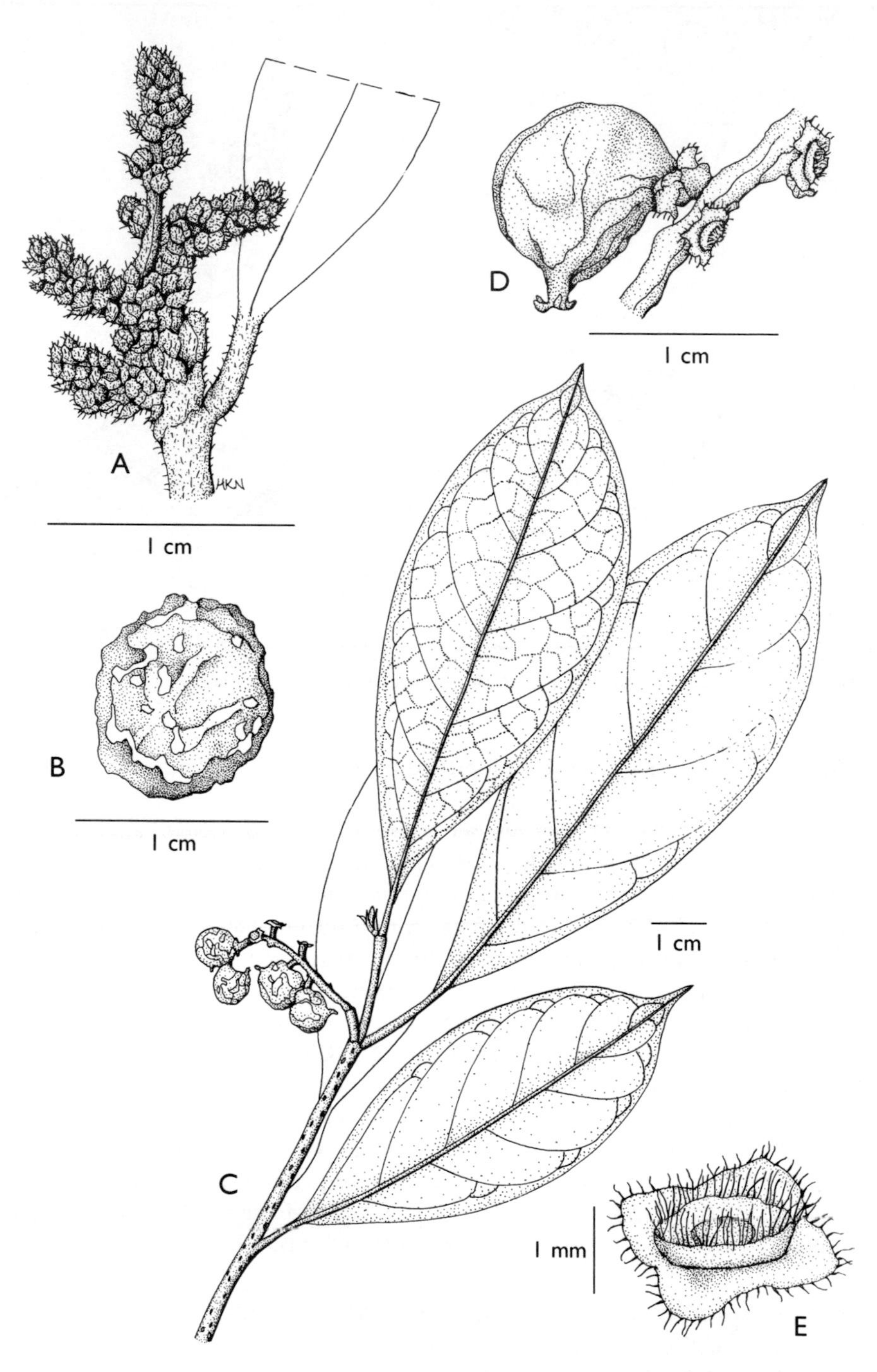

FIG. 11. *Antidesma edule* var. *apoense*. **A** young staminate inflorescence; **B** sculptured endocarp; **C** habit with infructescence; **D** part of infructescence; **E** pistillate calyx after fruit abscission. **A** from *Co* 3139 (L); **B** – **E** from *Co* 3142 (A). Drawn by Holly Nixon.

Tree, up to 20 m. *Petioles* not pulvinate, 4 – 12 × 1.2 – 1.8 mm, pilose, soon becoming glabrous and rugose. *Leaf blades* elliptic to obovate, thickly coriaceous, glabrous, or sparsely pilose only at the base of the midvein adaxially and along the midvein abaxially. *Staminate inflorescences* (only young inflorescences seen) up to 3 cm long, consisting of up to 7 branches, sometimes in fascicles of 2 – 3 inflorescences. *Pistillate inflorescences* simple. *Flowers* not known. *Fruits* lenticular, slightly wider than long, 6 – 8 × 7 – 9 mm, glabrous. Fig. 11.

DISTRIBUTION. Philippines, Mindanao. Map 13.

HABITAT. In montane forest with average canopy height of c. 20 m. 1400 m altitude.

CONSERVATION STATUS. Data Deficient (DD). This taxon is known only from two specimens from a single location but there is not enough information about the locality to classify it in a high threat category.

ETYMOLOGY. The epithet refers to the type locality.

KEY CHARACTERS. Glabrous fruits; simple infructescences; relatively short petioles; glabrous, strongly coriaceous leaves.

17. Antidesma elbertii *Petra Hoffm.*, Kew Bull. 54: 352 (1999). Type: South-eastern Sulawesi, Landschaft Rumbia, Lawankudu Fluss, tropischer Regenwald, wasserreich, Phyllite, 150 – 497 m, 21 Sept. 1909, *Elbert* 3139 (L!, holotype).

Shrub or treelet, up to 9 m, diameter up to 7 cm. Bark dark purplish, thin; wood very hard, sapwood white, heartwood very dark red. *Young twigs* terete, densely ochraceous spreading-hirsute, brown. *Stipules* persistent, linear to narrowly deltoid, 3 – 7 × 0.5 – 1.2 mm, densely hirsute. *Petioles* channelled adaxially, 3 – 8 × 1 – 2 mm, densely spreading-hirsute. *Leaf blades* oblong, oblong-ovate or elliptic, apically acuminate-mucronate, basally acute to rounded, (5 –)9 – 13(– 20) × (2.5 –)3.5 – 5(– 7) cm, (1.8 –)2.5 – 3(– 3.8) times as long as wide, eglandular, membranaceous to chartaceous, glabrous except along the major veins, more rarely pilose all over adaxially, spreading-hirsute all over abaxially, especially along the veins, shiny to dull adaxially, dull to moderately shiny abaxially, midvein impressed adaxially, tertiary veins reticulate to weakly percurrent, usually widely spaced, drying olive-green, greyish or dark reddish green, lighter abaxially. *Staminate inflorescences* 4 – 6 cm long, axillary, simple or once-branched, axes spreading-hirsute. *Bracts* lanceolate, 0.5 – 0.7 × 0.2 – 0.3 mm, hirsute, margin sometimes glandular. *Staminate flowers* 2 – 3 × c. 2 mm. *Pedicels* 0.2 – 1 mm long, not articulated, glabrous to pilose. *Calyx* 0.8 – 1 × 0.8 – 1 mm, conical, sepals 4, fused for $^1/_3$ – $^2/_3$ of their length, deltoid, apically acute, hirsute outside, glabrous inside, margin fimbriate to almost lacerate, sometimes glandular. *Disc* cushion-shaped, enclosing the bases of the filaments (deeply inserted) and pistillode (shallowly inserted), 0.7 – 0.8 × 0.7 – 0.8 mm, constricted at the base, glabrous at the sides, whitish-pubescent to sparsely pilose apically, hairs 0.2 – 0.3 mm long, extending to the length of the pistillode. *Stamens* 4, 2 – 2.5 mm long, exserted 1.5 – 2 mm from the calyx, anthers 0.3 – 0.5 × 0.3 – 0.5 mm. *Pistillode* obconical to cylindrical, slightly crateriform apically, c. 0.3 × 0.2 – 0.3 mm, exserted from the sepals, pilose. *Pistillate inflorescences* (4 –)7 – 8 cm long, cauline to terminal, simple, more rarely once-branched (consisting of up to 6 branches in *s. coll.* CEL/III-32), axes c. 1.2 mm wide, spreading-hirsute. *Bracts* narrowly deltoid, 0.7 – 1 × 0.2 – 0.5 mm, spreading-hirsute. *Pistillate flowers* 2 – 3 × 1 – 1.5 mm. *Pedicels* 0.2 – 0.5 mm long, spreading-hirsute. *Calyx* c. 1 × 1.5 – 2 mm, sepals 4 – 5, free to fused for up to half of their length (then calyx conical), c. 0.5 mm wide, deltoid to semiorbicular, apically acute to rounded, hirsute outside, glabrous inside but with long hairs at the base, margin entire to erose, sometimes glandular-fimbriate. *Disc* shorter than the sepals, glabrous or sparsely hirsute. *Ovary* globose to lenticular, spreading-hirsute, style lateral to subterminal, thin, very distinct, stigmas 3 – 6, c. 0.5 mm long, not particularly thin. *Infructescences* (4 –)8 – 23 cm long (in *s. coll.* CEL/III-32: 2 – 3 cm long), axes 1 – 2 mm wide. *Fruiting pedicels* 1 – 3 mm long, spreading-hirsute. *Fruits* lenticular to ellipsoid, often distinctly (c. 2 mm long) beaked, laterally compressed, basally symmetrical to slightly asymmetrical, with a lateral style, 6 – 8 × 4 – 6 mm, spreading-pilose to nearly glabrous, usually conspicuously and coarsely white-pustulate, reticulate when dry. Fig. 12.

DISTRIBUTION. Sulawesi, Lesser Sunda Islands (Sumbawa) and Moluccas (Bacan Island). Map 13.

HABITAT. In forests up to 60 m tall in river valleys; often in damp places; in primary to disturbed vegetation. On deep clayey soil derived from granite, coralline limestone and volcanic sand. 20 – 1150 m altitude.

CONSERVATION STATUS. Near Threatened (NT). The extent of occurrence is more than 500,000 km^2 but the species is not well collected. 19 specimens examined.

VERNACULAR NAMES. Sulawesi: maitam bokas.

ETYMOLOGY. The epithet refers to the collector of the type specimen.

KEY CHARACTERS. Indumentum spreading-hirsute; staminate calyx conical; staminate disc hairy on top, extending to the same length as the sepals; fruits with lateral styles.

SIMILAR SPECIES. *A. montanum* has cupular to bowl-shaped staminate calyces with the sepals free, to fused to halfway, a glabrous (rarely shortly hairy), flat to hemispherical staminate disc which is shorter than the calyx and never basally constricted, glabrous, terete fruits with terminal styles, and a less strongly developed abaxial leaf indumentum. — *A. velutinosum* has longer bracts, free sepals, a glabrous disc, more sepals and stamens as well as percurrent secondary leaf venation.

SELECTED SPECIMENS. SULAWESI. S Sulawesi, Mamuju Kab., Kaluku Kec., Desa Sondoang, Dusun Rea, c. 100 m, disturbed forest, open place, common, on slope, 2 Feb. 1993, *Afriastini* 2001 (K, L); C Sulawesi, Luwuk area, NE of Luwuk, on road from Kayutanyu to Siuna, 0°51'S, 123°00'E, c. 60 m, disturbed forest in river valley, raised coralline limestone, 7 Oct. 1989, *Coode* 5838 (AAU, K, L); same locality, 8 Oct. 1989, *Coode* 5844 (K, L); SE Sulawesi, Kolaka area, Gunung Watuwila foothills, above Sanggona, 'Mokuwu camp', valley of Mokuwu R., 3°48'S, 121°39'E, c. 200 m, disturbed forest in river valley, 30 Oct. 1989, *Coode* 6068 (K, L); SE Sulawesi, Rumbia region, Wambako?roe, 4°23'S, 121°55'E, 40 – 130 m, monsoon forest, humid, 9 Sept. 1909, *Elbert* 3079 (L); Manado prov., camp Totok close to Ratatotok, 200 m, primary forest, fertile volcanic sand, 18 March 1895, *Koorders* 16797 b (BO); S Sulawesi, SW peninsula, NE of Makassar within 54 – 60 km on the

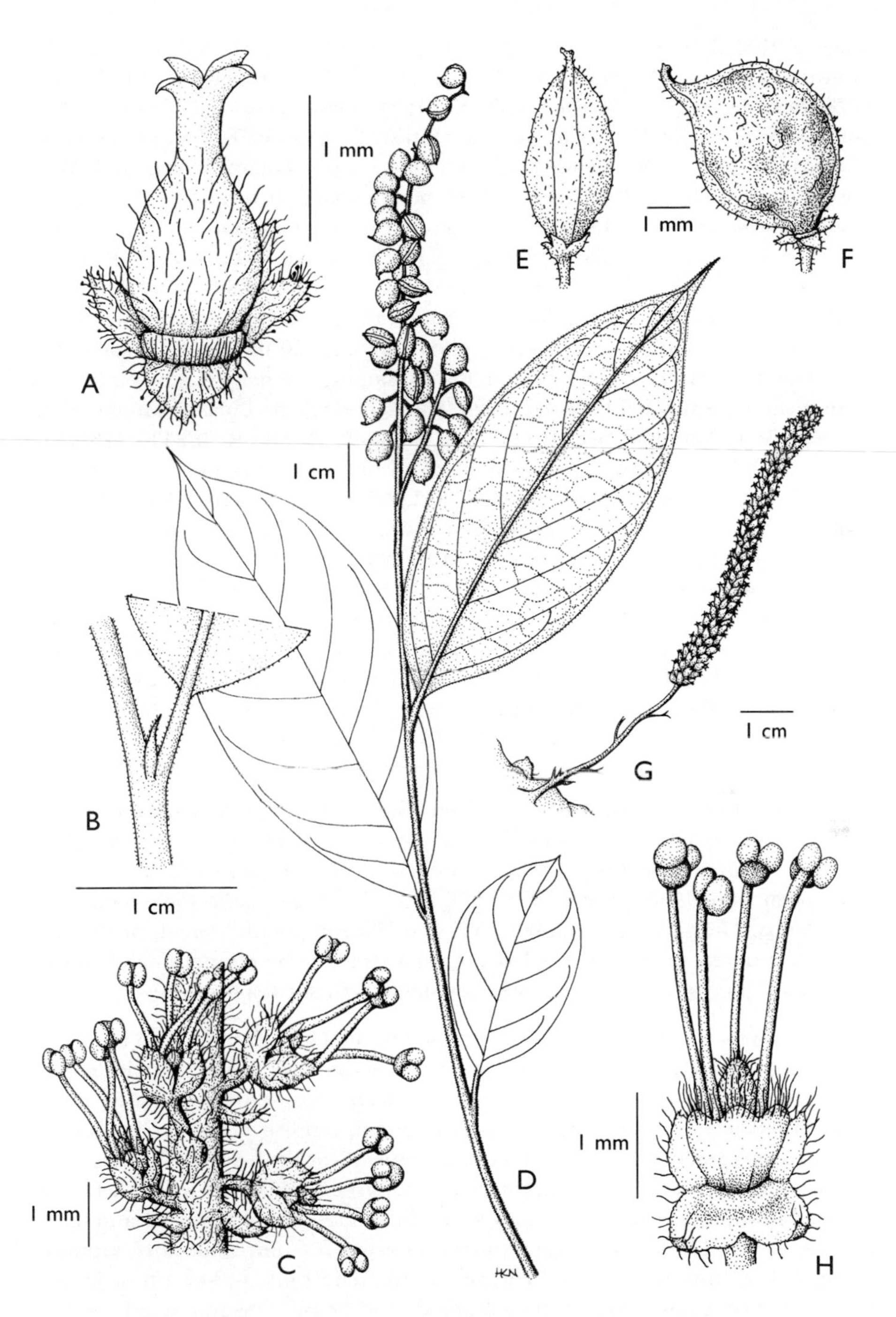

FIG. 12. *Antidesma elbertii.* **A** pistillate flower; **B** part of branch with stipule; **C** part of staminate inflorescence; **D** habit with infructescence; **E** fruit in dorsal view; **F** fruit in lateral view; **G** cauline pistillate inflorescence; **H** staminate flower, calyx partly removed. **A – B**, **G** from *Coode* 5838 (L); **C**, **H** from *Elbert* 3139 (L); **D – F** from *Ramlanto & Zainal Fanani* 675 (L). Drawn by Holly Nixon.

road, 5°01'S, 119°35'E, 4 July 1976, *Meijer* 10811 (L); SE Sulawesi, Kolaka, Tirawuta, Ladongi Mountain, 3°54'S, 121°14'E, 150 m, primary forest, 17 Oct. 1978, *Prawiroatmodjo & Maskuri* 1233 (K, L); SE Sulawesi, Kolaka, Tirawuta, Polipolia, 4°08'S, 121°50'E, 100 m, primary forest, 21 Oct. 1978, *Prawiroatmodjo & Maskuri* 1417 (K, L); SE Sulawesi, around Opa [Aopa] swamp, W side, hills W of Polipolia, 4°05'S, 122°00'E, 20 – 250 m, primary forest, 22 Nov. 1978, *Prawiroatmodjo & Soewoko* 1887 (L); C Sulawesi, Gunung Gindopo, Basidondo, Toli-Toli, Tungkuan river, 1°02'N, 120°49'E, 250 m, river bank, 8 March 1985, *Ramlanto & Zainal Fanani* 675 (K, L); S Sulawesi, Malili, Oesoe, 2°34'S, 121°04'E, 28 Oct. 1931, *s. coll.* CEL/III 32 (L); C Sulawesi, Sopu valley, c. 60 km SSE of Palu, margin of concession PT Kebun Sari near Berdikari, 1°03'S, 120°03'E, 650 m, somewhat disturbed primary forest c. 40 m high, terrain sloping, on deep clayey soil derived from granite, 30 May 1979, *De Vogel* 5625 (L); Manado distr., Bolaang Mongondow, Dumoga Bone National Park, Toraut Dam, along the Toraut R., 0°34'N, 123°54'E, 220 m, slightly disturbed primary forest 35 m high, on bank near the water, 13 March 1985, *De Vogel & Vermeulen* 6508 (K, L, NY); same locality, 14 March 1985, *De Vogel & Vermeulen* 6517 (L); Manado distr., Lama, Dumoga Bone proposed National Park, Doloduo, Tumokang, 0°34'N, 123°54'E, 300 m, ridge top, 20 Sept. 1984, *Whitmore & Sidiyasa* 3465 (K, L). LESSER SUNDA ISLANDS. Sumbawa, Bima, NTB, Rora, Padende, Mt Lahio, 1150 m, mountain slope, 21 Sept. 1982, *Sarkat Danimihardja* 2119 (L). MOLUCCAS. Bacan Isl., Gunung Sibela near Waiaua, 0°45'S, 127°32'E, 250 m, very open disturbed forest near river bank, to 60 m and much regeneration of trees to 5 m high, sloping hill side, deep clayey soil mixed with stones, bedrock grey schists, 27 Oct. 1974, *De Vogel* 3699 (K, L).

18. Antidesma excavatum *Miq.*, Ann. Mus. Bot. Lugduno-Batavum 1: 218 (1864). Type: Sulawesi, Menado distr., *Teysmann* HB 5589 (U!, holotype; BO!, isotype). —Note: Collection number not cited in protologue, so the 5589 might have been added afterwards. The BO sheet bears no collector's name but "Antidesma, Menado", the number 5589 and "Palang poetih", which might be a vernacular name, on an original label in what seems to be Miquel's handwriting. The appearance of the specimen strongly suggests that it is an isotype.

Tree, rarely shrub, up to 25 m, clear bole up to 11 m, diameter up to 50 cm, bole crooked or straight, sometimes twisted, sometimes with more than one stem; thick, equal buttresses of 7.5 cm reported from the Solomon Islands (*BSIP (Mauriasi)* 8807). Bark brown, grey, greyish green, reddish brown, yellow-brown, fawn or white, smooth or rough, flaky or ± deeply vertically fissured, fibrous, often pustular or warty, soft; inner bark green-white, green, pink, red, reddish brown, orange brown, brown, dark red, straw or cream, 6 – 12 mm thick, weakly and short-fibrous; clear, white or yellowish exudate reported by some collectors; sapwood white, yellow, straw or reddish; heartwood pink, dark pink, red-brown or brown, hard (some collectors report soft slash wood) and dense, fine-grained, dense fine medullary rays, minute pores in short radial chains. *Young twigs* terete, pilose to tomentose, medium to light brown. *Stipules* early-caducous, linear to narrowly deltoid, apically acute, 1.5 – 6 × 0.4 – 1 mm, pubescent. *Petioles* flat to channelled adaxially, basally and distally sometimes pulvinate and geniculate for up to 10 mm, (2 –)6 – 20(– 30) × 1 – 2(– 3) mm, pubescent to glabrous. *Leaf blades* elliptic, more rarely oblong, ovate or slightly obovate, apically acuminate-mucronate, more rarely with a slightly retuse apiculum, basally acute to rounded, often decurrent ("pseudo-petiole", Fig. 2G), very rarely truncate or subcordate, (5 –)10 – 15(– 30) × (2 –)4 – 6(– 13.5) cm, (1.5 –)2.5 – 3(– 5) times as long as wide, eglandular, chartaceous to coriaceous, glabrous or pubescent only along the major veins adaxially, glabrous to pubescent abaxially, shiny adaxially, shiny or dull abaxially, major veins flat to impressed adaxially, tertiary veins reticulate to percurrent, usually widely spaced, drying olive-green, sometimes more reddish brown abaxially, more rarely greyish-green, domatia present. *Staminate inflorescences* 2 – 12 cm long, axillary, consisting of 2 – 5(– 10) branches, more rarely simple, robust, axes pilose to pubescent. *Bracts* deltoid to orbicular, apically acute to rounded, 0.5 – 1.2 × 0.3 – 0.6 mm, glabrous to pubescent, margin slightly erose. *Staminate flowers* 1.5 – 2 × 1 – 2 mm. *Pedicel* absent. *Calyx* 0.5 – 0.8 × 0.8 – 1.2 mm, cupular to globose or bowl-shaped with a truncate base, sepals usually 4, fused for $^{2}/_{3}$ – $^{4}/_{5}$ of their length, usually deltoid, apically obtuse to acute, sinuses usually wide, rounded, glabrous to very shortly pilose on both sides, glabrous inside, margin fimbriate to entire. *Disc* extrastaminal-annular, more or less lobed, lobes pointing inwards, sometimes appearing cushion-shaped, exserted from or extending to the same length as the sepals, constricted at the base, glabrous at the sides, ferrugineous- to ochraceous-pubescent apically. *Stamens* 3 – 4, 1 – 1.7 mm long, exserted 0.5 – 1 mm from the calyx, anthers 0.2 – 0.3(– 0.5) × 0.3 – 0.4(– 0.7) mm. *Pistillode* subulate, more rarely clavate, 0.4 – 1 × 0.1 – 0.5 mm, extending to the same length or exserted from the sepals, ferrugineous- to ochraceous-pubescent to ochraceous-pilose. *Pistillate inflorescences* 2 – 7(– 17) cm long, axillary, but often aggregated at the end of the branch, branched regularly, consisting of 2 – 7(– 11) branches, more rarely simple, sometimes in fascicles of 2 – 3 inflorescences, axes glabrous to pubescent. *Bracts* deltoid to orbicular, apically acute to rounded, 0.4 – 0.8 × 0.3 – 0.6 mm, pilose, margin erose. *Pistillate flowers* 1.5 – 2.5 × 1 – 2 mm. *Pedicels* 0.2 – 1 mm long, glabrous to pubescent. *Calyx* 0.5 – 1 × (0.5 –) 1 – 2 mm, bowl-shaped, more rarely cupular, sepals 3 – 5, fused for $^{2}/_{3}$ – $^{4}/_{5}$ of their length, erect, usually deltoid, often unequal, apically obtuse to acute, sinuses wide, rounded, glabrous to pubescent outside, glabrous inside, margin erose or entire, slightly fimbriate. *Disc* exserted from or extending to the same length as, in fruit sometimes slightly shorter than the sepals, ochraceous- to ferrugineous-tomentose at the margin, indumentum usually exserted from the sepals. *Ovary* lenticular, glabrous to tomentose, style lateral, stigmas (3 –)4 – 6, sometimes long, thick, irregularly fused. *Infructescences* 4 – 7(– 17) cm long. *Fruiting pedicels* 0.5 – 3(– 5) mm long, glabrous to pubescent. *Fruits* lenticular, more rarely bean-shaped, laterally compressed, basally symmetrical or asymmetrical, with a distinctly lateral style, 3 – 8(– 15) × 3 – 6(– 10) mm, pilose to glabrous, sometimes white-pustulate, areolate or fleshy when dry.

NOTES. The name *A. excavatum* was treated as a synonym of *A. celebicum* by Pax & Hoffmann (1922: 129), who followed the observations of Müller (1866: 256).

The type of *A. excavatum* in the Miquel Herbarium in Utrecht was probably not consulted by Airy Shaw, and the name *A. excavatum* is not included in his account on Central Malesia (1982: 5 – 7) or on New Guinea (1980a: 208 – 219).

A very variable species, especially in size and indumentum of all parts. This includes the abaxial leaf indumentum which Airy Shaw used (1980a: 209 – 210) to distinguish between four species, all of which are here treated under *A. excavatum*. There are transitions from completely glabrous leaves to a dense indumentum all over the abaxial leaf surfaces. However, as the latter occurs only in New Guinea, these plants are here recognised as var. *indutum*.

Leaf shape is also extremely variable and although some forms are rather conspicuous, they intergrade too much with the typical forms to be recognised as distinct taxa. This includes the large- and narrow-leaved forms described as *A. tagulae* and *A. katikii* as well as the forms with very small, coriaceous leaves with only 3 – 6 pairs of secondary veins from Rossel Island, Louisiade Archipelago (e.g., *Brass* 28430, *LAE (Katik)* 70937).

Another name to be subsumed here is *A. sphaerocarpum*. The only differential character given by Airy Shaw (1980a: 209 – 210, 218) is the length of the petioles: under 1.5 cm for *A. moluccanum*, *A. olivaceum* and *A. polyanthum* (here all synonyms of *A. excavatum*), and 1.5 – 3 cm for *A. sphaerocarpum* in the key, but 1 – 3 cm for the latter species in the description on p. 218. The petioles of the type of *A. excavatum* are 10 – 13 mm long and the range of the petiole length on Samoa is (7 –)13 – 23(– 30) mm.

The leaf domatia in *A. excavatum* and *A. celebicum* from Sulawesi are glabrous, whereas those in *A. excavatum* from New Guinea are usually covered with hairs. The infructescences sometimes bear characteristic clavate galls 0.5 – 2 × 0.2 – 0.4 cm in size (e.g., *Robinson* 1799 (A, BM, K, L), *Kostermans* 1658 (A, K, L)).

18a. var. **excavatum**

A. sphaerocarpum Müll. Arg. in DC., Prodr. 15(2): 255 (1866); Airy Shaw, Kew Bull. 28: 276 (1973), **synon. nov.** Type: In insulis Archipelagi Samoa juxta Fidji Ins., Herb. U.S. Explor. Expedition, *Capt. Wilkes* (G-DC! [microfiche], lectotype, here designated; K!, isolectotype) — Note: The lectotype is in fruit and has Müller's handwriting on the label.

A. polyanthum K. Schum. & Lauterb., Fl. Schutzgeb. Südsee: 392 (1900); Pax & Hoffmann in Engl., Pflanzenr. 81: 112, fig. 12 B, F (1922), **synon. nov.** Type: Bismarck-Archipel, Neu-Lauenburg-Gruppe, Insel Kerawara, im lichten Wald, 14 Mai 1890, *Lauterbach* 102 (WRSL!, lectotype, here designated). Original syntypes: New Guinea, Kaiser Wilhelmsland, Finschhafen, *Warburg* 20518; Sattelberg, 12 Dez. 1896, *Bamler* 10 (WRSL!); Augustafluss, 2. Station, *Hollrung* 734 (BO, K!, MEL!); Ramufluss, 27. Juli 1898, *Tappenbeck* 140 (WRSL!). — Note: Tappenbeck 140 is also an original syntype of *A. warburgii* K. Schum.

A. oligonervium Lauterb. in K. Schum. & Lauterb., Nachtr. Fl. Schutzgeb. Südsee: 294 (1905), **synon. nov.** Type: New Guinea, Kaiser Wilhelmsland, Torricelli-Gebirge, 600 m, April 1902, *Schlechter* 14311 (WRSL!, lectotype, here designated; BM!, BO!, K!, P!, isolectotypes). — Notes: Airy Shaw (1980a: 216) spelled this name "*A. oligoneurum* [sphalm. "-nervium"]", probably because he disliked the combination of Greek and Latin. Such a change, however, is not permitted under Art. 60.1 (Greuter *et al.* 2000: 92). Known only from the type, which is in very young pistillate flower. The leaves have only 3 or 4 pairs of secondary veins, but strong intersecondary veins. This is unusual but not unusual enough to deserve taxonomic recognition. Airy Shaw stated that the leaves are "... contracted at base into a distinct short pseudo-petiole (which is pallid when dry), ...". The swelling of the leaf base is caused by arthropods (probably mites, remains of which are still present in the type but cannot be examined further without danger to the material) in the domatia.

A. warburgii K. Schum. in K. Schum. & Lauterb., Nachtr. Fl. Schutzgeb. Südsee: 293 (1905). Type: Neu Guinea, Kaiser Wilhelmsland, bei Finschhafen, 6 Jan. 1891, *Lauterbach* 1434 (WRSL!, lectotype, here designated). Original syntypes: Augustafluß, am Ufer bei der 2. Station, *Hollrung* 882 (K!); Ramufluß, 27 Juli 1898, *Tappenbeck* 140 (WRSL!); Ramuflußgebiet, Hochwald, 90 m, 28 Okt. 1899, *Lauterbach* 3132 (WRSL!). — Note: *Tappenbeck* 140 is also a syntype of *A. polyanthum* K. Schum. & Lauterb.

A. novoguineense Pax & K. Hoffm. in Engl., Pflanzenr. 81: 153 (1922), **synon. nov.** Type: Neu Guinea, Kaiser Wilhelmsland, Etappenberg, *Ledermann* 9415 (B!, lectotype, here designated; L!, isolectotype). Original syntypes: *Ledermann* 9612; Etappenberg, *Ledermann* 9143 (B!), 9280 (B!, L!), 9547 (B!); Lordberg, *Ledermann* 9891 (B!), 10022 (B!).

A. kusaiense Kaneh., Bot. Mag. (Tokyo) 46: 456 (1932), **synon. nov.** Type: Caroline Islands, Kusai, in primary forests, alt. 300 – 400 m, July 1931, *Kanehira* 1325 (NY!, lectotype, here designated; US!, isolectotype). Original syntype: *Kanehira* 1433 (NY!).

A. moluccanum Airy Shaw, Kew Bull. 23: 284 (1969), Kew Bull. 28: 276 (1973), **synon. nov.** Type: Moluccas, Morotai, Totodoku, 5 May 1949, *Kostermans* 624 (K!, holotype; A!, L!, isotypes).

A. sarcocarpum Airy Shaw, Kew Bull. 23: 288 (1969), **synon. nov.** Type: Papua New Guinea, Central Division, Rona, Laloki R., rain-forest, alt. 450 m, April 1933, *Brass* 3677 (K!, lectotype, here designated; A!, NY!, US!, isolectotype).

A. tagulae Airy Shaw, Kew Bull. 23: 289 (1969), **synon. nov.** Type: Louisiade Arch., Sudest Island (Tagula), Mt Riu, west slopes, bank of a stream in rain-forest, alt. 250 m, 31 Aug. 1956, *Brass* 27947 (K!, lectotype, here designated; A!, L!, US!, isolectotype). — Note: Besides the type, there is one very similar collection from Tagula (*LAE (Katik et al.)* 70851) and one from Maipa, Central province (*Darbyshire* 933). These collections have long, narrow leaves (19 – 22 × 3.5 – 6.5 cm, 4 – 5 times as long as wide) and long, unbranched to once-branched inflorescences. Leaf shape and habitat information indicates a mildly rheophytic ecology, but more material is be needed to establish this. The leaf shape in *A. excavatum* is extremely variable, and all intermediates from the typical, elliptic leaves to ovate and lanceolate-elongate leaves can be observed without correlation to other morphological characters, ecology or geography. See also under the synonym *A. katikii* below.

A. katikii Airy Shaw, Kew Bull. 28: 278 (1973), **synon. nov.** Type: New Guinea, *NGF (Coode & Katik) 32762* (K!, holotype; A!, CANB!, L!, isotypes). — Note: Known only from the type collection which has long, narrow leaves (13 – 23 × 5 – 6 cm, 3 – 5 times as long as wide), but is *A. excavatum* in all other respects. The specimen bridges the gap between *A. tagulae* and typical *A. excavatum* with respect to leaf shape and size, but the infructescence is robust and branched as in typical *A. excavatum.*

A. boridiense Airy Shaw, Kew Bull. 33: 16 (1978), **synon. nov.** Type: Papua New Guinea, Boridi, forest by a stream, 1200 m, 14 Nov. 1935, *Carr* 14942 (K!, lectotype, here designated; A!, CANB!, BM!, K!, L!, NY!, isolectotypes). Paratypes: *Carr* 14691 (BM!, K!, L!, NY!), 14832 (BM!, K!, L!). — Note: The type and two paratypes bear immature fruits, very young pistillate buds and even younger staminate buds, respectively. No further collections have been identified as *A. boridiense.* The only differential character is the size of the fruits which, as mentioned even in the original description, are immature. In the protologue, Airy Shaw compared his new species with *A. venenosum* from Borneo, with which it has little in common apart from the long petioles and the lateral styles. The hairy disc of *A. boridiense* versus the always completely glabrous disc of *A. venenosum* is enough to distinguish the two species. The large leaves and long petioles of the type material of *A. boridiense* are very similar to *A. petiolatum,* but the styles at the immature fruits are lateral while they are terminal to subterminal in the latter species. Unfortunately, the types of *A. boridiense* have lost all their stipules, the only other good diagnostic character.

A. orarium Airy Shaw, Kew Bull. 33: 17 (1978), **synon. nov.** Type: North-eastern New Guinea, Morobe Distr., south-east of Lae, on the coast, opposite Lasanga Island, 147°10'E, 07°25'S, tall primary forest on rocky slope facing the sea, 200 – 300 m, 12 Nov. 1973, *Jacobs* 9587 (L!, holotype; US!, isotype). — Note: This taxon is known only from the type. Airy Shaw stated: "It appears to stand midway between the four species mentioned in the diagnosis, agreeing with *A. polyanthum* and *A. olivaceum* in its puberulous fruits, with *A. moluccanum* (apparently) in its small fruits and completely glabrous leaves, and with forms of *A. sphaerocarpum* in its large leaves and long petioles." All four names mentioned are here subsumed under the oldest name *A. excavatum.*

A. pseudopetiolatum Airy Shaw, Kew Bull. 33: 423 (1979), **synon. nov.** Type: Indonesia: Papua, subdistr. Hollandia, "Dok II", flat country, scattered in secondary forest, stony clayey soil, 100 m, 29 May 1956, *BW (Versteegh)* 3824 (K!, holotype; L!, isotype).

A. hylandii Airy Shaw, Kew Bull. 36: 636 (1981), **synon. nov.** Type: Australia, Queensland, Cook distr., Cape York Peninsula, between Lockerbie and Somerset, 10°47'S, 142°31'E, rain forest, 80 m, 3 Feb. 1980, *Hyland* 10265 (K!, holotype; CANB!, KEP!, L!, MEL!, isotypes).

Tree, rarely shrub, up to 25 m, clear bole up to 11 m, reported to be a climber on some specimen labels. Bark 1 – 8 mm thick, outer bark 0.2 mm thick. *Young twigs* shortly pilose. *Stipules* 1.5 – 3 × 0.4 – 0.7 mm. *Petioles* shortly pubescent to glabrous. *Leaf blades* glabrous, sometimes some short bristly hairs at the base of the midvein and in the axils between midvein and secondary veins abaxially, shiny on both surfaces. *Bracts* broadly lanceolate to orbicular, apically obtuse to rounded, c. 0.5 × 0.5 – 0.6 mm, glabrous to pilose. *Staminate sepals* glabrous or rarely pilose outside. *Ovary* pilose to glabrous, rarely tomentose. *Fruiting pedicels* glabrous to spreading-pilose. *Fruits* sparsely pilose to glabrous. Fig. 5D – F (p. 68).

DISTRIBUTION. This is the most widespread Eastern Malesian species: Sulawesi, Moluccas, New Guinea incl. Bismarck and Louisiade Archipelagos, Solomon Islands, Australia (Queensland, Cape York Peninsula), Caroline Islands (Kusaie), Fiji (Rotuma Islands), Wallis & Futuna Islands, Samoa. Its occurrence in Borneo and the Philippines is somewhat doubtful. Only one collection is known from Borneo: East Kalimantan, Berau, 1°30'N, 117°20'E, *Kato, Okamoto & Ueda* 11736 (L!). There are three collections from the Philippines: *BS (Ramos & Edaño)* 39026 from Mindanao is the only typical specimen; *BS (Barbos)* 24836 *and BS (Ramos)* 33057 from Luzon are doubtful. New Caledonia has a distinct, endemic species, *A. messianianum*. The distribution area given in Baker *et al.* (1998: 252, fig. 6B) has been substantially extended by additional material. Map 14.

HABITAT. In (rain) forest, associated with *Anisoptera, Araucaria, Calophyllum, Castanopsis*, Cunoniaceae, *Eucalyptopsis, Eugenia, Hopea, Melaleuca, Quercus, Semecarpus* and *Vatica*; *Agathis* forest; low shrubby forest; coastal plain forest; gallery forest; sometimes on alluvial, swampy or seasonally flooded ground; on grassland; in mangrove forest with *Bruguiera, Rhizophora, Sonneratia, Xylocarpus*; at

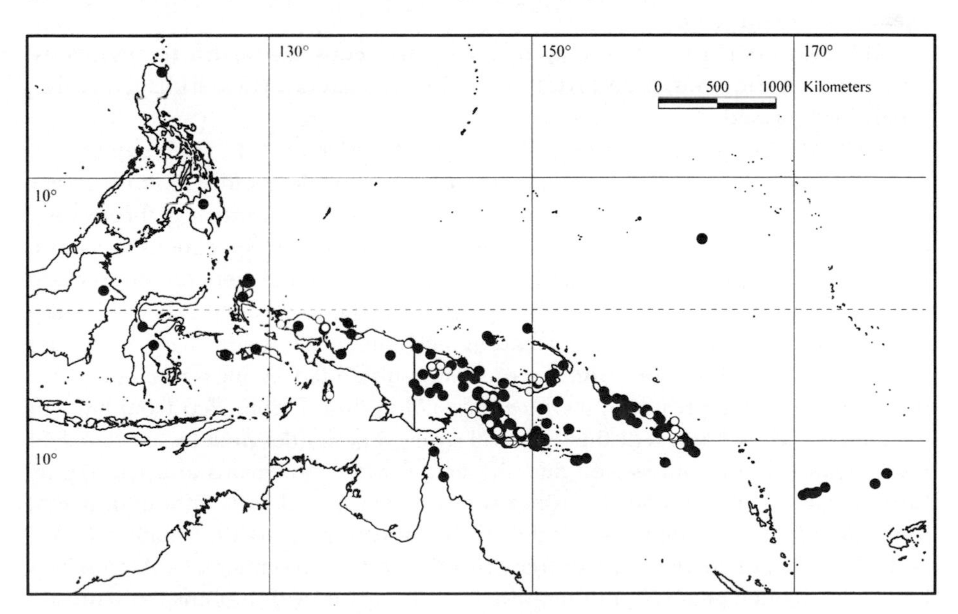

MAP 14. Distribution of *A. excavatum* var. *excavatum* (●) and var. *indutum* (○).

back of beaches, associated with *Pandanus* and *Cocos*; on coral shores; on an extinct volcano; on semi-cultivated and cultivated land; in lowland to mid-montane, primary or secondary vegetation. On white sand, sandy loam soil, alluvial soil derived from granite, volcanic and ultrabasic soil, limestone rock. 0 – 3600 m altitude.

USES. The fruits are eaten; they are said to make a good jam and yield a purple dye. *A. excavatum* seems to be cultivated as a minor fruit tree in eastern Malesia (similar to *A. bunius* in western Malesia) which means that there may have been a certain amount of dispersal by humans. The wood is used for construction, posts, taro planting and digging sticks (does not blunt easily), bird hide sites and firewood. Wasku people in New Guinea use it to extract sago from the palm trunk prior to washing.

CONSERVATION STATUS. Least Concern (LC). This is a very widespread and common taxon. 351 specimens examined.

VERNACULAR NAMES. Sulawesi: hikulu (Besoa); Indonesia: Papua: adap (Papua); kawian kek, tefan kek; PNG: agor (Wanigela, Onjob, Koreaf), agoro (Onjob, Naukwate, Minufia), as imendal (Mianmin), babe (Kabuli) or babi (Lessau), domu (R), eta-kanki'pe, gira, gitapo (Wario), hinuha, ipeipa (Mekeo), katechomi (Mbuke), kupuki, miel (Wagu), pipen (Kurte), sigoreh (Orokaiva), sisila (Baruga), sisira (Mambare and Minufia), tue tue, uluwinik (Waskuk); Solomon Islands: aenia'a (Kwara'ae), babaram, boborama or poporama (Kwara'ae), gaki or ngaki (Rennell), lasi (Kwara'ae), ngoape (Santa Cruz), ngoungou (Kwara'ae), oa (Kwara'ae), oaniara (Kwara'ae), lasoi-guy; lassau, malamala oa (Kwara'ae), maru-pul-pu, saola or saula (Kwara'ae); Caroline Islands: seelel; Fiji: komo.

ETYMOLOGY. The epithet refers to the domatia in the leaf-axils (Latin, excavatus = hollowed out).

KEY CHARACTERS. Often long petioles; coriaceous leaves; disc ferrugineous-pubescent on top; ovaries and fruits with distinctly lateral styles; fruits lenticular; sepals highly fused.

SIMILAR TAXA. *A. baccatum, A. bunius, A. celebicum, A. chalaranthum, A. concinnum*, see there. — *A. jucundum* has a long, spreading indumentum, a very slender habit, small leaves and pedicellate staminate flowers with a glabrous disc. — *A. myriocarpum* has fruits with a terminal style; staminate specimens are more difficult to distinguish, but the medium-brown twigs, thinner petioles, often ovate leaves, rounded to cordate leaf bases, inconspicuous tertiary relative to the conspicuous higher order leaf venation and the often pedicellate staminate flowers with free disc lobes which are usually subtended by narrower bracts are useful distinguishing features. — *A. pacificum* Müll. Arg. (1866: 254) from Fiji and the New Hebrides has less distinctly lateral styles, very slender petioles, cordate leaf bases, leaves drying olive-green and the acute, more numerous and less highly fused sepals. — *A. petiolatum* has wider stipules, very large leaves without domatia, long petioles and glabrous ovaries with subterminal styles, smaller fruits, pedicellate staminate flowers with free disc lobes and 2 – 3 stamens. — *A. pleuricum* has more slender petioles, thinner, usually ovate leaves with a obtuse to truncate base, no domatia, shorter, laxer and more branched staminate inflorescences and more hairy fruits. — *A. rhynchophyllum* has delicate, pedicellate, 2 – 3-staminate flowers and lacks a pistillode; pistillate material differs mainly in its smaller, obovate, membranaceous to chartaceous leaves with an abruptly acuminate apex and in its more conspicuous hairtuft-domatia.

SELECTED SPECIMENS. BORNEO. E Kalimantan, Berau, c. 70 km S of Tanjung Redeb, 1°30'N, 117°20'E, in a moist sunken place near a peak in limestone mountains, on limestone rock, 150 – 750 m, 28 Aug. – 1 Sept. 1981, *Kato, Okamoto & Ueda* B 11736 (L). PHILIPPINES. Mindanao, Bukidnon subprov., Mt Dumalucpihan, June – July 1920, *BS (Ramos & Edaño)* 39026 (A, K, L, US). SULAWESI. S Sulawesi, Laron[d]a proposed Hydro-electric Dam Reserve, W of Towuti Lake, E of Malili, 2°40'S, 121°10'E, ultrabasic region, 19 July 1976, *Meijer* 11289 (L). MOLUCCAS. Morotai, Gunung Sangowo, 800 m, 25 May 1949, *Kostermans* 973 (A, K, L). INDONESIA: PAPUA. Yapen, Serui, 15 Aug. 1939, *Aet & Idjan* 454 (A, K, L); Sukarnapura (Djajapura), valley, 100 m, 29 July 1966, *Kostermans & Soegang* 77 (BO, K, L). PAPUA NEW GUINEA. Territory of New Guinea, Sepik R. (Augustafluss), 2. Augusta-Station, Sept. 1887, *Hollrung* 766 (K, L, MEL, WRSL); Milne Bay distr., D'Entrecasteaux Islands, Bolu Bolu subprov., Goodenough Isl., Mt Oiatawa'a, 9°16'S, 150°18'E, undisturbed primary forest, 920 m, 21 Dec. 1977, *LAE (Benjamin)* 67927 (K, L, US); Morobe distr., on main Buang Track, along ridge above Gabensis, 1900 ft, 18 May 1955, *NGF (Floyd)* 7260 (A, BM, K, L, MEL, US); Central div., Brown River Forest Reserve, 20 miles E of Port Moresby, 9°15'S, 147°20'E, 200 ft, Jan. 1960, *NGF (McDonald)* 8221 (K, L, MEL); Bismarck Archipelago, Admiralty Islands, Mbuke Island, ridge top above Mbuke village, 2°22'S, 146°50'E, lowland rainforest, in shade, 150 m, 2 Dec. 1975, *Sands* 2974 (K, L, US). SOLOMON ISLANDS. Santa Ysabel (Bughotu), NW Santa Ysabel, Binusa, primary forest, ridge top, 17 Jan. 1966, *BSIP (Beer's collectors)* 6759 (K, L, US). CAROLINE ISLANDS. Kusaie, Mt Matante, 13 Aug. 1938, *Hosokawa* 9470 (L, US). WESTERN SAMOA. Upolu Island, Lefaga, S of Mata'utu, forest, sea level, 1 April 1968, *Bristol* 1992 (K, L, US). FIJI. Rotuma Isl., Itulia distr., Kilinga, 12°30'S, 177°05'E, woods, on rocky (tuff) headland, 200 ft, 4 July 1938, *St John* 19036 (US). AUSTRALIA. Queensland, Claudie R., 12°44'S, 143°15'E, rainforest, 40 m, 21 Jan. 1982, *Hyland* 11538 (K).

18b. var. **indutum** *(Airy Shaw) Petra Hoffm.*, Kew Bull. 54: 355 (1999). — *A. moluccanum* Airy Shaw var. *indutum* Airy Shaw, Kew Bull. 33: 16 (1978). Type: North-eastern New Guinea, Morobe Distr., Huon Peninsula, between Masba Creek and Pependango, 3 km south of Pindiu, in tall secondary forest on gentle slope, 840 m, 17 May 1964, *Hoogland* 8977 (K!, holotype; A!, CANB!, L!, isotypes).

A. olivaceum K. Schum. in K. Schum. & Hollrung, Fl. Kais. Wilh. Land: 76 (1889). Type: New Guinea, Kaiser Wilhelmsland, Lagerberg der 2. Augusta-Station, *Hollrung* 757 (WRSL!, lectotype, designated in Hoffmann 1999a: 357; BO, K!, L!, MEL!, P!, isolectotypes).

Tree, rarely shrub (*fide NGF* 35125: woody vine), up to 17 m, clear bole up to 8 m, diameter up to 38 cm, crooked or straight, sometimes twisted. Bark 5 – 8 mm thick, outer bark c. 1 mm thick. *Young twigs* pale yellow to ferrugineous-tomentose.

Stipules 4 – 6 × 0.7 – 1 mm. *Petioles* densely pubescent. *Leaf blades* glabrous except for the pubescent major veins, more rarely sparsely pilose all over adaxially, ochraceous- to ferrugineous-pubescent all over abaxially, dull or shiny adaxially, dull abaxially. *Bracts* deltoid to lanceolate, apically acute, 0.8 – 1.2 × 0.3 – 0.5 mm, pubescent. *Staminate sepals* sparsely pilose to pubescent outside. *Ovary* tomentose to pilose. *Fruiting pedicels* ochraceous- to ferrugineous-pubescent. *Fruits* pilose.

DISTRIBUTION. Moluccas (Aru Islands only), New Guinea, Solomon Islands. Map 14.

HABITAT. In (rain) forest, associated with *Ficus*, *Hopea*, *Pometia* and *Syzygium*; on river banks; in beach rain forest; in gardens; sometimes on swampy ground; in primary and secondary vegetation. On sand and clay over limestone. 0 – 1700 m altitude.

USES. See under var. *excavatum*.

CONSERVATION STATUS. Least Concern (LC). This taxon is not very distinct from the type variety, and is often collected. 53 specimens examined.

VERNACULAR NAMES. Moluccas: kwada atam; Indonesia: Papua: kabiu (Amberbaken), pukepga (Manikiong), wohhoika; PNG: banamih'van (Asengseng), kamia (Kukukuku, Tauri), la malas tititili (W. Nakanai), samanga (Waskuk); Solomon Islands: aidori (Kwara'a), boborama, boborma, bobrama or poporama (Kwara'ae), la-so; New Britain: range biara (Garumaia).

ETYMOLOGY. The epithet refers to the indumentum (Latin, indutus = clothed).

KEY CHARACTERS. Leaves pubescent abaxially; otherwise similar to the type variety.

SIMILAR SPECIES. *A. ferrugineum* has pedicellate staminate flowers, a more deeply divided, bell-shaped pistillate calyx, which is longer relative to the disc, and smaller fruits.

SELECTED SPECIMENS. MOLUCCAS. Aru Islands, Baun Island, 6°30'S, 134°35'E, rainforest dominated by *Ficus*, *Hopea*, *Pometia* and *Syzygium*, 5 m, 1 Nov. 1994, *Van Balgooy* 6787 (K). INDONESIA: PAPUA. Kaloal, Salawati Isl., primair bos, vlak terrein in regentijd onder water staand, zandige kleigrond, 0 m, 30 Oct. 1956, *BW (Versteegh)* 4672 (L). PAPUA NEW GUINEA. New Britain, W Nakanai, Galilo Village near Cape Hoskins, 28 Feb. 1954, *Floyd* 3471 (BM, K, L, MEL, US); Northern distr., Tufi subdistr., near Koreaf village, dense low secondary forest, c. 25 m, 23 Sept. 1954, *Hoogland* 4783 (K, L); Central distr., Abau subdistr., Mori river, 10°10'S, 148°35'E, lowland rainforest bordering recently constructed roadway close to river, 50 ft, 23 April 1964, *NGF (Sayers)* 19672 (K, L, US); Gulf distr., Lohiki village area, junction of Lohoki and Vailala rivers, primary forest, c. 70 ft, 24 Jan. 1966, *Schodde & Craven* 4286 (K, L, US). SOLOMON ISLANDS. Santa Ysabel (Bughotu), N Santa Ysabel, Kwamaebusia R., primary forest, valley bottom, 1 Sept. 1966, *BSIP (Beer's collectors)* 7694 (K, L, US).

19. Antidesma ferrugineum *Airy Shaw*, Kew Bull. 26: 462 (1972). Type: New Guinea, Morobe distr., Wagau, 6°50'S, 146°50'E, rain forest, alt. 1760 m, 10 June 1964, *NGF (Millar)* 23458 (K!, lectotype, here designated; A!, L!, isolectotypes).

Paratypes: Eastern Highlands distr., Kainantu subdistr., Kassam Pass, 6°12'S, 146°02'E, forest, alt. 1200 m, 9 Jan. 1968, *NGF (Henty & Coode)* 29180 (K!, L!); Kassam Pass, 6°10'S, 146°05'E, forest, alt. 1260 m, 15 Jan. 1968, *NGF (Coode)* 32686 (K!, L!).

Shrub or tree, up to 6 m. *Young twigs* terete, ferrugineous-pubescent, brown. *Stipules* early-caducous, linear to deltoid, apically acute, 2 – 4 × 0.5 – 1.5 mm, ferrugineous-tomentose to glabrous. *Petioles* channelled adaxially, basally and distally sometimes slightly pulvinate and geniculate, 7 – 18 × 1 – 2 mm, ferrugineous-tomentose to glabrous. *Leaf blades* elliptic to ovate, apically acuminate-mucronate, basally acute to rounded, 7 – 13.5 × 3 – 7.7 cm, 1.6 – 3.1 times as long as wide, eglandular, chartaceous, glabrous except for the pubescent major veins adaxially, ferrugineous-pubescent all over abaxially (*NGF* 29932 glabrous except for the sparsely puberulent midvein on both surfaces), dull on both surfaces, major veins impressed adaxially, tertiary veins percurrent to reticulate, drying olive-green to reddish brown, lighter abaxially. *Staminate inflorescences* 2 – 7 cm long, axillary, sometimes cauline, consisting of 2 – 11 branches, sometimes in fascicles of 2 inflorescences, axes densely ferrugineous-tomentose. *Bracts* ovate to lanceolate, apically acute to rounded, 1 – 1.2 × 0.5 – 1 mm, pubescent. *Staminate flowers* c. 2 × 2 mm. *Pedicels* (0.5 –)1 – 2 mm long, not articulated, glabrous to sparsely pilose. *Calyx* 0.5 – 1 × 1 – 1.5 mm, cupular with a truncate base, sepals usually 5, fused for $^2/_3$ – $^4/_5$ of their length, sometimes hard to make out individually, apically usually acute, sinuses rounded to acute, glabrous to pilose outside, glabrous inside, margin erose. *Disc* extrastaminal-annular, more or less lobed, extending to the same length as the sepals, glabrous at the sides, pubescent apically. *Stamens* 3 – 4(– 5), 1.5 – 1.8 mm long, exserted 1 – 1.2 mm from the calyx, anthers 0.2 – 0.4 × 0.4 – 0.7 mm. *Pistillode* subulate to cylindrical, 0.3 – 0.6 × 0.1 – 0.3 mm, exserted from the sepals, pubescent. *Pistillate inflorescences* axillary or terminal, branched regularly, consisting of 2 – 20 branches, sometimes in fascicles of 2 inflorescences, axes densely ferrugineous-tomentose. *Bracts* deltoid to linear, 0.7 – 1 × c. 0.5 mm, ferrugineous-pubescent. *Pistillate flowers* not known. *Calyx* in fruit c. 1 × 1 – 1.5 mm, campanulate with a truncate base, sepals 4 – 6, fused for $^1/_2$ – $^2/_3$ of their length, deltoid, apically acute, pilose to glabrous outside, glabrous inside, margin erose or entire, fimbriate. *Disc* shorter than the sepals, ferrugineous-tomentose at the margin, indumentum not exserted from the sepals. *Stigmas* 4 – 5. *Infructescences* 2 – 8 cm long. *Fruiting pedicels* 0.2 – 1.5 × c. 0.3 mm, ferrugineous-pilose. *Fruits* lenticular to almost bean-shaped, laterally compressed, basally asymmetrical, with a distinctly lateral style, 2.5 – 3.5 × 2 – 3 mm, pilose, white-pustulate, areolate when dry.

DISTRIBUTION. Papua New Guinea: Eastern Highlands and Morobe provinces. Map 9.

HABITAT. In (disturbed) rain forest. 1300 – 2000 m altitude.

CONSERVATION STATUS. Endangered (EN B1ab(i,ii)). Only 4 specimens are known of this species, from an area of less than 4000 km^2.

VERNACULAR NAMES. abuk - possum food; ikabin (Fore).

ETYMOLOGY. The epithet refers to the colour of the indumentum (Latin, ferrugineus = rusty, light red-brown).

KEY CHARACTERS. Abaxial leaf surfaces ferrugineous-pubescent all over; disc pubescent; fruits small, pilose, with lateral styles; staminate flowers pedicellate; above 1300 m altitude.

SIMILAR TAXA. *A. excavatum* var. *indutum*, see there. — *A. myriocarpum* var. *puberulum* has a cupular to urceolate calyx, glabrous ovaries and fruits with terminal styles, fine tertiary and more conspicuous higher order leaf venation, leaf domatia, narrower bracts, a longer calyx, a disc which is pubescent all over, and wider, less hairy pistillodes. — *A. petiolatum* has wider stipules, longer petioles and larger leaves that dry olive-green, as well as usually free disc lobes.

FURTHER SPECIMEN. PAPUA NEW GUINEA. Eastern Highlands distr., Okapa subdistr., Anumba to Ilafo near Okasa, 6°32'S, 145°37'E, damp deep gully in overgrown disturbed area, 6100 ft, 24 May 1967, *NGF (Coode & Lelean)* 29932 (K, L, US).

20. Antidesma forbesii *Pax & K.Hoffm.* in Engl., Pflanzenr. 81: 153 (1922). Type: Sumatra, *Forbes* 1972 (BM!, lectotype, here designated; L!, isolectotype). Original syntypes: Sumatra, *Forbes* 2451 (BM!, CAL, L!), 2519 (BM!, L!).

A. plagiorrhynchum Airy Shaw, Kew Bull. 28: 272 (1973); Kew Bull. 36: 356, fig. C1-2, 363 (1981), **synon. nov.** Type: Sumatra, West Coast Res., Padangsche Bovenlanden, Mt Singalan, June – July 1878, *Beccari* PS 55 (K!, holotype; BM!, L!, MEL!, isotypes). Paratype: *Beccari* PS 245 (K!, L!).

A. pradoshii Chakrab. & Gang., J. Econ. Taxon. Bot. 21: 479 – 480, fig. 1 (1997), **synon. nov.** Type: Sumatra, W. Paue, 4300 ft alt., 1881, *Forbes* 2451 (CAL (herb. acc. no. 407806), holotype; BM!, CAL (herb. acc. no. 407807), L!, isotypes). Paratype: Sumatra, Priaman, *Diepenhorst* HB 2352 (CAL, K!, U!). — Note: The type of this name is a syntype of *A. forbesii,* and the paratype is the type of *A. salicifolium* Miq. (non C. Presl) [= *A. neurocarpum* var. *neurocarpum*].

Treelet, up to 8(– 12.5) m, diameter 8(– 15.6) cm. *Young twigs* terete, glabrous to puberulent, first dark brown with conspicuous lenticels, then whitish to light grey. *Stipules* early-caducous, deltoid, apically acute, c. 1 × 0.5 mm, glabrous. *Petioles* channelled adaxially, basally and, more so, distally often pulvinate and geniculate, (4 –)6 – 17 × 0.5 – 1 mm, pilose to puberulent. *Leaf blades* ovate to elliptic, apically acuminate-mucronate, with a rounded apiculum, basally acute to obtuse, (4.5 –)6 – 11(– 16) × (1.5 –)2 – 4(– 5.5) cm, 2.3 – 4.1 times as long as wide, eglandular, chartaceous to submembranaceous, glabrous, or pilose only along the midvein adaxially and/or abaxially, dull on both surfaces, midvein impressed adaxially, tertiary veins reticulate, drying olive-green, lighter abaxially. *Staminate inflorescences* 2 – 4 cm long, axillary, simple or consisting of up to 4 branches, slender, axes pubescent. *Bracts* elliptic, apically acute, 0.4 – 0.5 × c. 0.4 mm, pilose. *Staminate flowers* 1 – 1.5 × 1 – 2 mm. *Pedicel* absent. *Calyx* c. 0.5 × 0.8 – 1 mm, bowl-shaped, sepals 5, fused for 2/3 of their

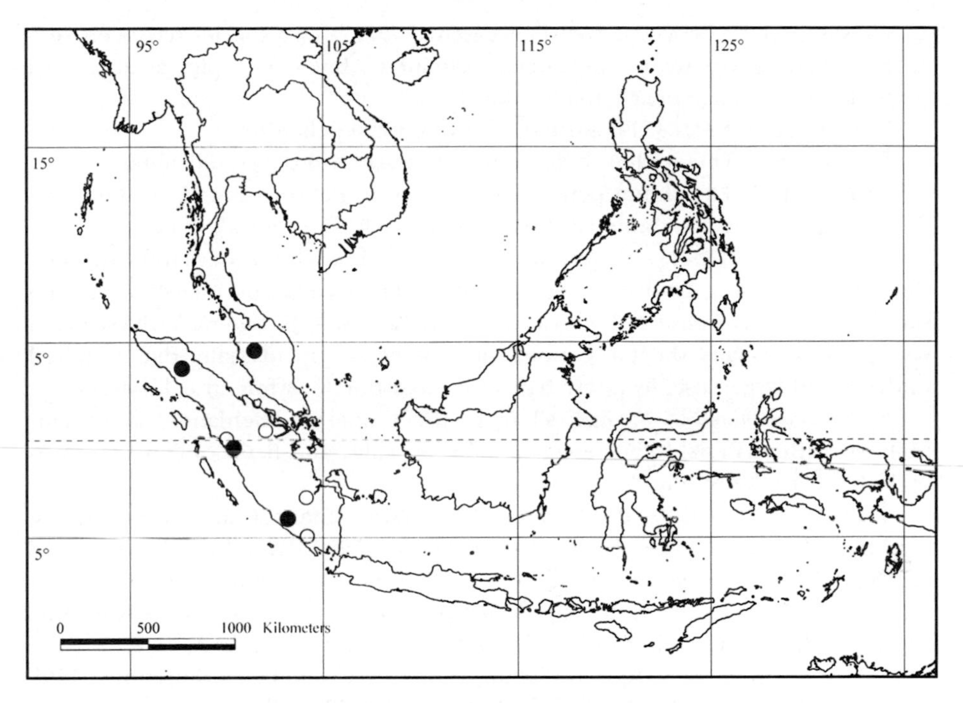

MAP 15. Distribution of *A. forbesii* (●), doubtful specimens (○).

length, apically rounded, shortly pilose outside, glabrous inside but with long hairs at the base, margin hyaline, ciliate. *Disc* consisting of 4 – 5 free alternistaminal lobes, lobes ± obconical, with two shallow imprints apically, c. 0.2 × 0.3 mm, extending to the same length as the sepals, glabrous. *Stamens* 4 – 5, 1 – 1.5 mm long, exserted c. 0.5 mm from the calyx, anthers c. 0.2 × 0.3 mm. *Pistillode* cylindrical, c. 0.8 × 0.3 mm, exserted from the sepals, pubescent. *Pistillate inflorescences* 2 – 5 cm long, axillary, simple, solitary or sometimes 2 per fascicle, axes shortly pubescent. *Bracts* elliptic, c. 0.5 × 0.5 mm, pilose. *Pistillate flowers* c. 1.5 × 1.5 mm. *Pedicels* 0.7 – 1.2 mm long, pilose. *Calyx* c. 0.7 × 1 mm, cupular, sepals 4 – 5, fused for $^1/_2$ or a little more of their length, apically rounded, pilose outside, glabrous inside but with long hairs at the base, margin fimbriate. *Disc* shorter than the sepals, glabrous. *Ovary* nearly globose, laterally compressed, glabrous to sparsely pilose, style lateral, thick, stigmas 3 – 6. *Infructescences* 5 – 11 cm long. *Fruiting pedicels* 2 – 5 mm long, puberulent. *Fruits* roughly bean-shaped, with a ventral bulge, laterally compressed, basally asymmetrical, with a lateral style, 5 – 8 × 4 – 6 mm, glabrous to very sparsely pilose, usually white-pustulate, areolate or fleshy when dry.

DISTRIBUTION. Peninsular Thailand (one collection from Pangnga province), Peninsular Malaysia (only Cameron Highlands, Pahang), Sumatra. Map 15.

HABITAT. In montane rain and mossy forest. On limestone (but very little geological data available). 1000 – 1900 m altitude.

CONSERVATION STATUS. Near Threatened (NT). This species has been very rarely collected to date (8 specimens examined), but it occupies a large area (extent of occurrence nearly 150,000 km^2).

VERNACULAR NAMES. Peninsular Malaysia: mata pelanduk.

ETYMOLOGY. The epithet refers to the collector of the type specimen.

KEY CHARACTERS. Long petioles; fused sepals; glabrous disc; free staminate disc lobes; fruits with asymmetrical base, lateral style and ventral bulge.

SIMILAR SPECIES. *A. cuspidatum*, see there. — *A. japonicum* has larger stipules, thicker, shiny leaves, pedicellate staminate flowers, free sepals, a cushion-shaped staminate disc, glabrous pistillodes and terminal styles. — *A. leucocladon* has whitish young twigs, usually shorter petioles and pedicels, an adaxially flat to raised midvein and larger, usually pilose fruits with a distinctly asymmetrical base.

NOTE. *Soepadmo & Suhaimi* s9 from the Cameron Highlands, Peninsular Malaysia with terminal infructescences, shorter calyx and fruits only 6 × 4 mm is included here doubtfully.

SELECTED SPECIMENS. PENINSULAR MALAYSIA. Pahang, Cameron Highlands, 4°34'N, 101°25'E, montane forest, 970 m, 11 Nov. 1989, *Soepadmo & Suhaimi* s46 (L, NY). SUMATRA. Atjeh, Gunung Leuser National Park, Gunung Ketambe and vicinity, 8 – 15 km SW from the mouth of Lau Ketambe, c. 40 km NW of Kutajane, camp 4 to 3, montane rain forest and mossy forest, 1750 m, 17 Aug. 1972, *De Wilde & De Wilde-Duyfjes* 14341 (K, L, US); camp 3, montane rain forest, limestone, 1700 – 1900 m, 19 July 1972, *De Wilde & De Wilde-Duyfjes* 13770 (K, L, US).

21. Antidesma ghaesembilla *Gaertn.*, Fruct. Sem. Pl. 1: 189, t. 39 (1788); Wight, Icon. Pl. Ind. Orient. 3(1): t. 820 (1844 – 45); Gagnep. in Lecomte, Fl. Indo-Chine 5(5/6): 505, fig. 64.11 – 19 (1926). — *A. ghaesembilla* Gaertn. var. *genuinum* Müll. Arg. in DC., Prodr. 15(2): 251 (1866), nom. inval. Type: Gaertner, Fruct. Sem. Pl. 1: t. 39 (designated by Philcox 1997: 276). — Notes: Gaertner cited a specimen "e collect. sem. hort. lugdb.", but there is no such specimen in the van Royen collection at Leiden today (cf. also Stafleu & Cowan 1976: 902).

The spelling of the epithet in the original publication of the name is twice "ghaesembilla" and twice "ghesaembilla". As the name is derived from the vernacular name "Ghaesembilla", it seemed best to follow current usage and Art. 60.1., Ex. 2 (Greuter *et al.* 2000: 92) and to adopt the former spelling.

Almeida & Almeida (1987: 492 – 493) argued that the conservation of the generic name *Embelia* Burm. (Myristicaceae) against *Ghesaembilla* Adans. (Greuter *et al.* 2000: 300) made the name *A. ghaesembilla* Gaertn. illegitimate and that therefore the name *A. pubescens* Roxb. had to be adopted. Conserved names, however, are conserved only against names in the same rank (Greuter *et al.* 2000: 27, Art. 14.4.). Nicolson *et al.* (1988: 193) also disagreed with the Almeidas on that point.

A. pubescens Roxb., Pl. Coromandel 2: 35, pl. 167 (1802). Type: Icones Roxburghianae No. 108 at Kew (lectotype, here designated). — Note: The actual specimen after which the plate was drawn in India is unlikely to have been preserved. There is no material of this species in BR (*fide* Merrill 1952). The plate in the protologue is a copy of the iconotype.

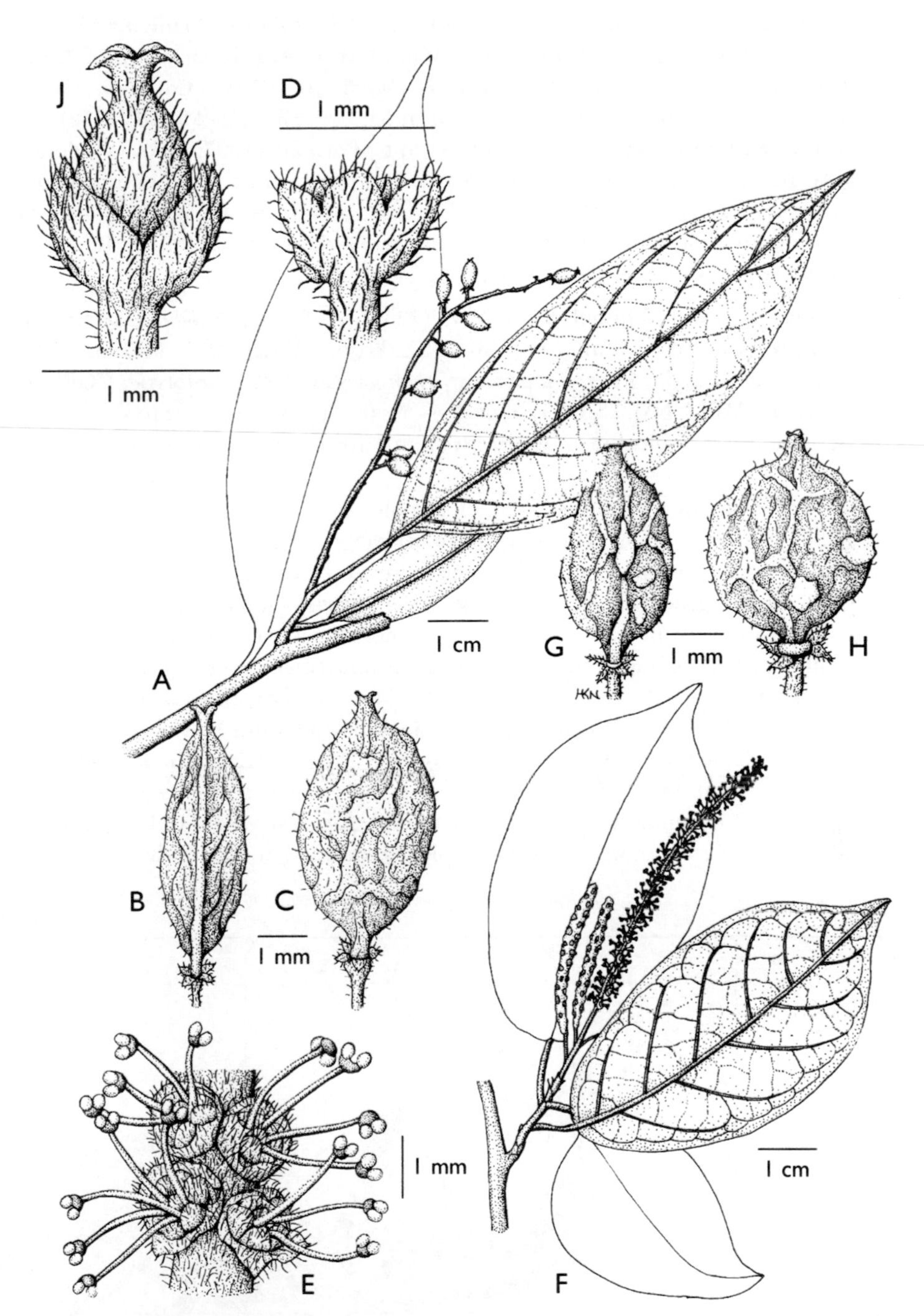

FIG. 13. *Antidesma ghaesembilla* and *A. subcordatum*. **A – D** *A. subcordatum*. **A** habit with infructescence; **B** fruit in dorsal view; **C** fruit in lateral view; **D** pistillate calyx. **E – J** *A. ghaesembilla*. **E** part of staminate inflorescence; **F** habit with staminate inflorescence; **G** fruit in dorsal view; **H** fruit in lateral view; **J** pistillate flower. **A – D** from *Elmer* 18083 (L); **E – F** from *PNH (Sulit)* 15719 (L); **G – J** from *Suvarnakoses* 2833 (L). Drawn by Holly Nixon.

A. paniculatum Willd., Sp. Pl., ed. 4, vol. 4.2.: 764 (1806), "paniculata". — *A. ghaesembilla* Gaertn. var. *paniculatum* (Willd.) Müll. Arg. in DC., Prodr. 15(2): 251 (1866). Type: India orientalis, *Roxburgh* s.n. (B-W!, Cat. No. 18350, holotype). — Note: Authentic material in K!, US!, BR (*fide* Merrill 1952). No further authentic material was discovered by Forman (1997: 523). Willdenow gave the synonym *A. paniculatum* Roxb. without further reference. No such name appears in any of Roxburgh's works, and thus the name has to be credited to Willdenow. The *Icones Roxburghianae* drawing no. 1297 at Kew most probably also shows *A. ghaesembilla*.

A. vestitum C. Presl, Epimel. Bot.: 232 (1849). — *A. ghaesembilla* Gaertn. var. *vestitum* (C. Presl) Müll. Arg. in DC., Prodr. 15(2): 251 (1866). Type: Philippines, Luzon, Prov. Pangasanar, *Cuming* 986 (PRC!, holotype; CGE!, E!, G!, K!, L!, MEL!, NY!, TCD!, isotypes). — Note: The type has five to six stamens per flower as is typical of *A. ghaesembilla*, not three as stated in the protologue.

Shrub, treelet or tree, up to 20 m, clear bole up to 8 m, diameter up to 32 cm (*SAN* 30313: height 32 m, clear bole 24 m, diameter 33 cm), mostly crooked and gnarled, usually low-branched with dense crown (sometimes flat or umbrella-shaped). Twigs greenish, brown, grey, young twigs sometimes pinkish. Bark whitish, green, grey, brown, red-brown or almost black, smooth or rough, often vertically cracked or flaky, fibrous, soft, 2 – 6 mm thick; inner bark pink, pale green, yellow, brown or reddish brown, 10 – 12 mm thick, fibrous; purplish sap reported on *Hartley* 9877; cambium whitish; sapwood whitish, yellow or green-yellow, soft. *Young twigs* terete, pubescent, brown. *Stipules* early-caducous, subulate, 3 – 6 × 0.5 – 1 mm, pubescent. *Petioles* terete or narrowly channelled adaxially, 4 – 10(– 17) × 0.7 – 1 mm, pubescent, becoming glabrous when old. *Leaf blades* oblong, rarely slightly ovate or obovate, apically rounded, more rarely obtuse, very

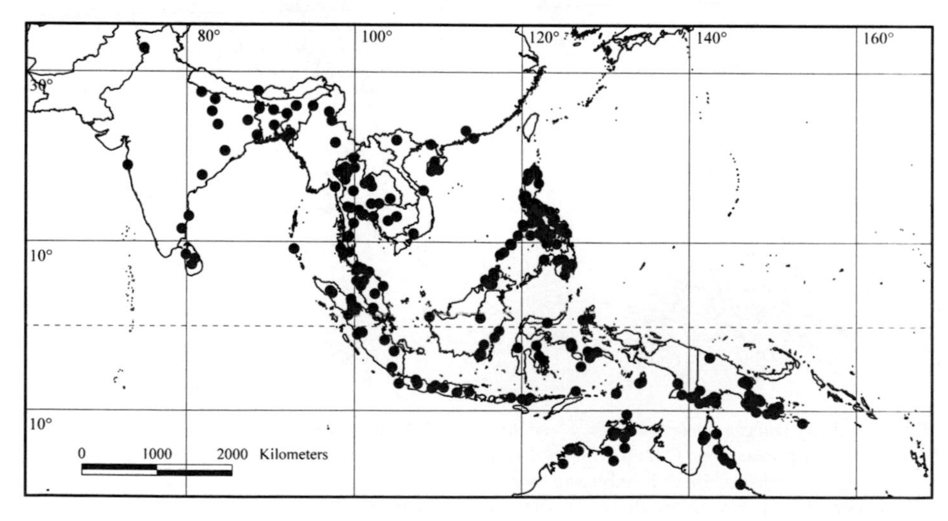

MAP 15. Distribution of *A. ghaesembilla*

shortly acuminate, truncate or retuse, usually shortly mucronate, basally rounded to cordate, very rarely acute, (2 –)4 – 7(– 16) × (2 –)3 – 4.5(– 9) cm, (1 –)1.3 – 1.7(– 2.25) times as long as wide, eglandular, chartaceous to coriaceous, pubescent to glabrous, often only the major veins of both surfaces and the margin pubescent, dull or shiny adaxially, dull abaxially, midvein flat adaxially, tertiary veins reticulate to weakly percurrent, drying olive-green. *Staminate inflorescences* 4 – 8 cm long, axillary, consisting of 10 – 20 branches, axes ferrugineous-pubescent. *Bracts* lanceolate, c. 0.5 – 1 × 0.5 mm, pubescent. *Staminate flowers* 2 – 3 × 2 – 3 mm. *Pedicel* absent. *Calyx* 0.5 – 1 × c. 1.5 mm, sepals 4 – 6, free, deltoid to oblong, apically acute to obtuse, pubescent outside, more or less glabrous inside, margin entire. *Disc* consisting of 4 – 6 free alternistaminal lobes, lobes ± obconical, well-separated from each other, c. 0.5 × 0.5 mm, pubescent. *Stamens* 4 – 6, c. 2 mm long, exserted 1 – 1.5 mm from the calyx, anthers c. 0.5 × 0.5 mm. *Pistillode* obconical, c. 0.5 × 0.2 – 0.3 mm, extending to the same length or slightly exserted from the sepals, pubescent. *Pistillate inflorescences* 2 – 3 cm long, axillary, consisting of (1 –) 10 – 20 branches, axes ferrugineous-pubescent. *Bracts* lanceolate, 0.8 × 0.4 mm, pubescent. *Pistillate flowers* 1.5 – 2 × 1 – 1.5 mm. *Pedicels* 0 – 1 mm long, pubescent. *Calyx* c. 0.7 × 1 – 1.5 mm, sepals 5 – 6, free, deltoid, apically acute, pubescent outside, glabrous inside, margin entire. *Disc* shorter than the sepals, glabrous to pubescent especially at the margin. *Ovary* ovoid to globose, pubescent, style usually subterminal, glabrous, stigmas (2 –)3(– 5). *Infructescences* 4 – 7 cm long. *Fruiting pedicels* 0 – 1 mm long, pubescent. *Fruits* lenticular, more rarely ellipsoid, laterally compressed, basally symmetrical, with a terminal, more rarely subterminal style, 3 – 4(– 5) × 2.5 – 3(– 3.5) mm, pilose, areolate when dry. Fig. 13E – J.

DISTRIBUTION. India incl. Nicobar Islands, Sri Lanka, southern China (Hainan, Guangdong prov.), Bangladesh, Burma, Laos, Vietnam, Cambodia, Thailand, Peninsular Malaysia, Sumatra, Borneo (Kalimantan, Sabah), Java, Philippines, Sulawesi, Lesser Sunda Islands, Moluccas, New Guinea, northern Australia. Map 16.

HABITAT. In savannah, grassland, open forest, dense scrubby forest and vine thickets; in fresh-water swamps, at the edge of mangrove, in coastal fringes; in secondary vegetation around human habitation (Malay: "belukar"); also in mixed evergreen or deciduous forest, associated with *Bischofia*, Dipterocarpaceae, *Eucalyptus*, *Ficus*, *Grevillea*, *Macaranga*, *Quercus*, *Terminalia*; in teak-forest in Java; along roadsides and river banks; on dry to swampy ground; usually in secondary vegetation. On clay, sand, lateritic soil, black peat, limestone; sometimes on ultrabasic soil; over shale and granite bedrock. 0 – 1250 m altitude.

The plant occurs often in regularly burnt habitats and is said to be fire resistant (Gruèzo 1991: 78; label of *NGF* 9809, see also under *A. rumphii* in Doubtful Species, p. 242). In the 1920's, the species escaped from the Botanic Gardens in Georgetown, British Guiana (now Guyana) and became a troublesome weed (for details see p. 12).

USES. The fruits are eaten locally. The wood is cheap but soft and splits when dried. It is nevertheless used for roof construction (Ridley 1903: 92).

CONSERVATION STATUS. Least Concern (LC). This is the species with the widest distribution in the entire genus, and is very common. 672 specimens examined.

VERNACULAR NAMES. India: creya, nunkuli; Sri Lanka: ambilla cingal, ghaesembilla; Bangladesh: takauliya; China: mai sui tsz, tso wo muk; Burma: byisen; Cambodia: daem dankkiep khdam; Thailand: ma mao sai, ma mao, ma-mao-khao-bao, mak mao, mamao kai; Sumatra: banoton, bohneh, bohnei, bungorak, kucir, mekremie, monton, tingiran puni; Malay Peninsula: guncak or cuncak; Sabah: andarupis, anjarubi, anjuripes, indarupis, ondurupis, tandurupis or tendrupis (Dusun), borotindik (Dusun), dempul (Dusun), guchek, gunchin, gunipot (Dusun), kakapal (Bisaya), obah; Java: andi, ki valot, onjam, sepat, wuni dedek, wuni jaran dawuk; Philippines: anyam, arosep (Il.), bananyo (Tagbanua), barunasi, baso-baso, bignay-pugo, binayuyo, bignai-pogo, bignayoyo (Tagbanua), dangol, grumun, holat-baguis, imian, inang or iniam, kabogbog (Tagbanua); Lesser Sunda Islands: kunfunu (Mollo), luna, piras, wuler ku or wuler satar; Moluccas: babine igo, bidara, kutikata-gunung, o lalade (Tobelorese); Indonesia: Papua: kwaik (Bian); PNG: fair (Wanigela), sigoreh (Orokaiva), sila, tagi (Onjob), Australia: murrungun or native plum.

ETYMOLOGY. The epithet is derived from a vernacular name.

KEY CHARACTERS. Leaf apex obtuse or rounded; leaves and petioles slender; inflorescences much-branched; flowers 5-merous; staminate disc lobes free, hairy.

SIMILAR SPECIES. *A. edule*, see there. — *A. myriocarpum* has acuminate leaves with a higher length/width-ratio, domatia, fused sepals and smaller staminate flowers with a glabrous to sparsely pilose pistillode. — *A. schultzii* Benth. (Bentham 1873: 86) from Australia has been treated as a synonym of *A. ghaesembilla* by Airy Shaw (1980b: 694), but may well be a separate taxon. The leaves are elliptical and not oblong, the apex and base are acute instead of mostly truncate. The fruits (6 – 8 mm long) are larger than in *A. ghaesembilla*. The staminate flowers have mostly only four stamens and the staminate disc is peculiar. The disc lobes are close to the pistillode and fused intrastaminally. One flower out of the seven collections of *A. schultzii* examined (*Clarkson* 4117 (K)), however, had two free and two fused disc lobes. The ovary of *A. schultzii* is glabrous whereas it is pubescent in *A. ghaesembilla*, the pistillate disc is wide and thin only in the type, but like typical *A. ghaesembilla* in similar specimens with elliptical leaves. Until further data become available, *A. schultzii* is not treated as a synonym of *A. ghaesembilla*. The typical *A. ghaesembilla* is, however, also present in Australia (e.g., *Keneally* 8635 (K)). — *A. subcordatum* has fused pistillate sepals, elongate-ellipsoid ovaries and, to a lesser extent, fruits, less-branched, more densely pubescent inflorescences and acuminate leaves, which are usually more than twice as long as wide (Fig. 13); staminate specimens, however, can hardly be distinguished.

NOTE. The label of the Kew specimen *Hyland* 7819 from Australia was annotated by Reid: "Rust is *Crossopsora antidesmae-dioicae*".

SELECTED SPECIMENS. INDIA. Uttar Pradesh, Varanasi distr., Silhat, forest, near a nala, 27 July 1966, *Panigrahi* 11188 (L). SRI LANKA. 18th mile Amaparai-Kandy road, dry forest, low altitude, 26 May 1973, *Kostermans* 24853 (G, K, L). CHINA. Hainan, Taam-chau distr., Sha Po Shan (Shui Mei Hang) and vicinity, Ts'o Wo Kwo

Shue, 24 Aug. 1927, *Tsang* LU 16085 (G, K). BANGLADESH. Chittagong, Chural Forest Reserve, 16/18 May 1966, *Majumber & Islam* 123 (K, L). CAMBODIA. Cantonnement du Tonle-Sap, poste du Kompong-Thom, Dang Kiep Kdam, recd Jan. 1912, *Lecomte & Finet* s.n. (G). VIETNAM. Tonkin, entre Hanoi et Bac-Ninh, bois de Co-Phah, 3 May 1891, *Balansa* 4809 (G, K). BURMA. Tenasserim prov., Moulmein, *Falconer* 115 (G, L, MEL). THAILAND. Northern, Lamphun prov., 23 km N of Li and 80 km S of Lampoon, open forest with trees to 12 m tall, scattered Cycads, soil a tan, pebbly sand, c. 420 m, 5 June 1963, *Merrill King et al.* 5448 (K, L, MEL). PENINSULAR MALAYSIA. Kelantan, Ulu Sat Forest Reserve primary forest, hillside, 16 June 1968, *KEP (Suppiah)* 104564 (K, KEP, L). SUMATRA. E coast, Batoe Bara, along the Sg. Moeka between Tanah Datar and Tandjoeng Tiram, 10 April 1927, *Bartlett* 7152 (L, NY, US). BORNEO. Sabah, Pantai Barat, Kota Belud distr., Bukit Kibalui, 6 April 1950, *SF (Henderson)* 38956 (E, K, L); E Kalimantan, W Koetai, no. 7, near M. Moentai, bank of Mahakam R., 1°00'N, 115°00'E, secondary forest, open spot on bank of river, 10 m, 9 July 1925, *Endert* 1985 (A, K, L). JAVA. W Java, Ujungkulon, coral limestone, 5 m, 5 Aug. 1955, *Kostermans* 9987 (A, K, L). PHILIPPINES. Luzon, Tayabas prov., Lucban, near Butano Falls, in shrubbery, May 1907, *Elmer* 9183 (A, E, G, K, L, NY, US, WRSL); Palawan, Brooks Point (Addison Peak), cogon formation, compact blackish soil, 50 ft, Feb. 1911, *Elmer* 12703 (A, E, G, K, L, NY, U, US, WRSL). SULAWESI. Sangi and Talaud Isl., Talaud, Nanusa, Merampi, G. Maranggi, 4°45'S, 127°07'E, open secondary forest, 150 m, 13 June 1936, *Lam* 3424 (A, K, L). LESSER SUNDA ISLANDS. Flores, W Flores, S part, near Mborong, 11 April 1965, *Kostermans* 22217 (K, L). MOLUCCAS. Ambon, s. loc., July – Nov. 1913, *Robinson* 1710 (K, L, NY, US). INDONESIA: PAPUA, SE West Irian, Bagam Pa, Maio R., 20 m, 30 April 1967, *Soegeng Reksodihardjo* 230 (L, US); PAPUA NEW GUINEA, Northern div., Oro bay, 1 km S of Beamu village, on badly eroded rocky soil, c. 60 m, 14 Sept. 1953, *Hoogland* 3889 (K, L, MEL, US). AUSTRALIA. Queensland, S of Ingham, 18°50'S, 146°07'E, beside a dry waterhole, open forest, heavy soil, 60 m, 14 Oct. 1976, *Hyland* 9143 (K, KEP, L).

22. Antidesma helferi *Hook. f.*, Fl. Brit. India 5: 357 (1887). Type: Tenasserim (or Andaman Isl.), *Helfer* Kew Distr. No. 4942 (K!, holotype).

A. pachystemon Airy Shaw, Kew Bull. 23: 279 (1969), Kew Bull. 26: 354 (1972), as synon. nov. Type: Thailand, peninsular region, Puket Circle, Kao Pawta Luang Kêo, Ranawng, evergreen forest, alt. 900 m, 1 Feb. 1929, *Kerr* 16928 (K!, holotype; L!, isotype).

Shrub or tree, up to 20 cm, diameter up to 15 cm. Twigs grey-white. *Young twigs* terete to striate, glabrous or nearly so, light grey. *Stipules* caducous or persistent, rarely foliaceous (*Geesink & Santisuk* 5166 from Thailand), linear to oblanceolate, apically acute, 3 – 8(– 20) × 0.5 – 1(– 5) mm, glabrous. *Petioles* channelled adaxially, becoming rugose when old, sometimes sharply keeled abaxially, the keel not extending into the leaf blades, 3 – 10 × (0.7 –)1 – 2 mm, glabrous. *Leaf blades* elliptic, oblong or ovate, apically acute- or acuminate-mucronate, basally acute, sometimes decurrent, (5 –)9 – 14(– 19) × (2 –)3.5 – 5.5(– 7.5) cm, (1.9 –)2.6(– 4.3) times as

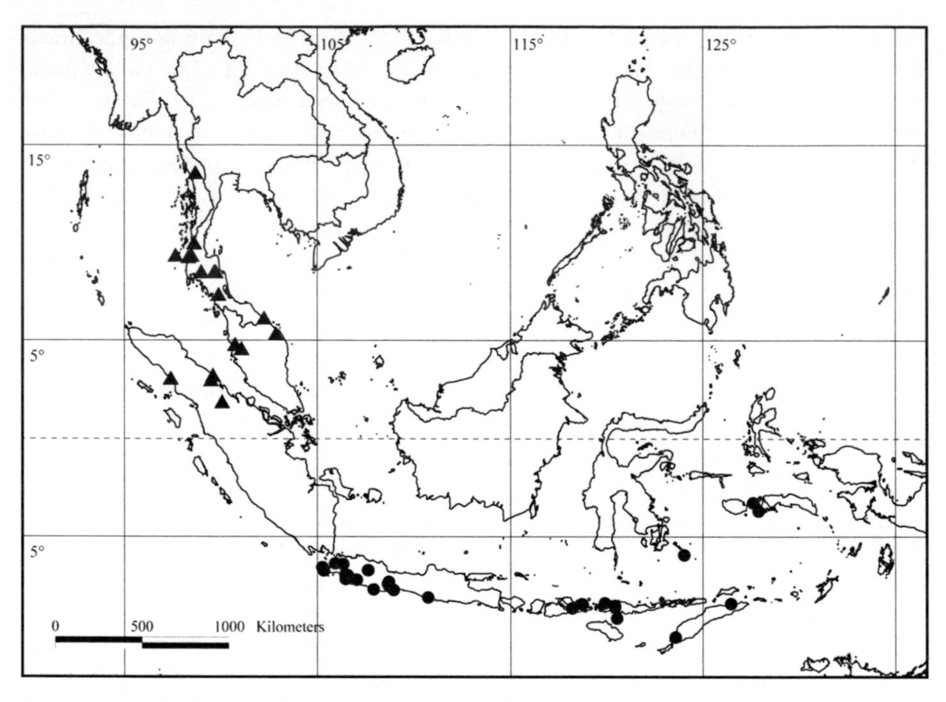

MAP 17. Distribution of *A. helferi* (▲) and *A. heterophyllum* (●).

long as wide, eglandular, chartaceous to coriaceous, glabrous, shiny or dull on both surfaces, midvein flat to shallowly impressed adaxially, tertiary veins reticulate, widely spaced (4 – 7 between every two secondary veins), drying greyish green to reddish brown, domatia absent. *Staminate inflorescences* 3 – 8 cm long, axillary to almost cauline, simple or once-branched at the base, axes pilose to glabrous. *Bracts* deltoid to linear, 0.7 – 1.2 × 0.3 – 0.5 mm, pilose or glabrous. *Staminate flowers* 2 – 3 × 1.5 – 2 mm. *Pedicels* 0 – 1 mm long, not articulated, pilose to glabrous. *Calyx* 0.7 – 1 × 1.2 – 1.5 mm, sepals 4, fused for c. $^1/_3$ of their length, 0.7 – 1 mm wide, apically rounded, pilose to glabrous outside, glabrous inside. *Disc* cushion-shaped, enclosing the bases of the filaments and pistillode, pilose. *Stamens* 4, 2 – 2.5 mm long, exserted 1.5 – 2 mm from the calyx, anthers 0.5 – 0.7 × 0.5 – 1 mm. *Pistillode* conical or globose, 0.1 – 0.3 × 0.1 – 0.3 mm, shorter than the sepals, glabrous. *Pistillate inflorescences* 4 – 10 cm long, terminal, axillary or almost cauline, simple, rarely branched once or twice at the base, solitary or sometimes 2 per fascicle, axes spreading-pubescent to glabrous. *Bracts* narrowly deltoid to linear, 0.5 – 1(– 1.5) × 0.3 – 0.4 mm, pubescent to glabrous. *Pistillate flowers* 1.5 – 3 × 1.5 – 2 mm. *Pedicels* 0 – 1 mm long, setulous. *Calyx* 0.7 – 1.2 × 1.2 – 1.5 mm, sepals 4 – 5(– 6), free or nearly so, 0.5 – 0.8 mm wide, orbicular to almost linear, apically rounded to acuminate, pilose to glabrous outside, glabrous inside, margin entire to erose. *Disc* shorter than or extending to the same length as the sepals, setulous outside and at the margin. *Ovary* ellipsoid to globose, sparsely to densely setulous, style subterminal to lateral, stigmas (2 –)3 – 4, very (up to 1.5 mm) long. *Infructescences* 4 – 15 cm long. *Fruiting pedicels* 0.5 – 2.5 mm long, glabrous to setulous. *Fruits* ellipsoid to lenticular, laterally compressed, basally symmetrical or asymmetrical, with a subterminal to lateral, rarely terminal style, 6 – 8 × 4 – 6 mm, glabrous to setulous, sometimes finely white-pustulate, often brick-red when dry, areolate or reticulate when dry.

DISTRIBUTION. Peninsular Burma, Peninsular Thailand, Peninsular Malaysia, Northern Sumatra. Map 17.

HABITAT. In primary evergreen or mossy forest; often near water or on marshy ground. On alluvial and greyish clay-mud soil. 100 – 1300 m altitude.

CONSERVATION STATUS. Near Threatened (NT). The species is relatively widely distributed but only 28 specimens are known.

ETYMOLOGY. The epithet refers to the collector of the type specimen.

KEY CHARACTERS. Nearly free, rounded sepals with often parallel sides; hairy disc; pubescent ovaries and fruits; long, slender inflorescences; thin petioles.

SIMILAR SPECIES. *A. brachybotrys* and *A. coriaceum*, see there. — *A. laurifolium* has longer, narrower leaves with a distinctly raised midvein adaxially, shorter bracts, a glabrous disc and larger fruits. — *A. leucocladon* has more highly fused sepals, longer pistillodes and larger, ovoid fruits. — *A. helferi* is distinguished from *A. leucopodum* by its widely spaced tertiary veins, subterminal styles, shorter inflorescences, smaller leaves and stipules. — *A. macgregorii* is vegetatively almost identical but has more highly fused sepals, a glabrous disc, larger fruits with terminal styles, and lacks a pistillode. — *A. montis-silam* has thicker petioles, longer leaves and more highly fused sepals. — *A. neurocarpum* has a glabrous disc, smaller flowers and acute sepals.

NOTES. Airy Shaw (1972a: 354) synonymised *A. helferi* p.p. with *A. cambodianum* Gagnep. This cannot be confirmed here. The type material of *A. cambodianum* examined (Tonking, environs d'Ouonbi, *Balansa* 1496 (P!); Laos, Massie, plateau d'Attopeu, *Harmand in hb. Pierre* s.n.; Cambodge, monts de Knang-krepeuh, prov. de Thepong, *Pierre* s.n. (K!, NY!, P!); monts Camchay, *Pierre* s.n. (P!)) differs consistently from *A. helferi*, e.g., by its glabrous discs and ovaries and resembles *A. montanum*.

The specimens from Peninsular Malaysia and from Sumatra have larger, more elliptic, thinner and duller leaves than those from Burma and Thailand, whereas the floral and fruiting characters are uniform throughout the distribution area. The fruiting specimens *Kerr* 18703 from Krabi and *Larsen et al.* 45988 from Nakhon Si Thammarat province, Thailand, differ from the remainder of the collections in their glabrous discs. In this and all vegetative characters, the latter specimen resembles the unmatched collection of *van Beusekom & Phengkhlai* 967 (p. 241) from the same locality and altitude.

SELECTED SPECIMENS. BURMA. Mergui distr., Myinmoletkat, 1125 m, 20 Jan. 1930, *Parker* 3131 (K). THAILAND. Peninsular, Phangnga prov., Khao Pawta Luang Keow, 9°15'N, 98°20'E, hill evergreen forest, 800 – 1000 m, 2 May 1973, *Geesink & Santisuk* 5166 (L); Ranong prov., Muang Len, 13 Jan. 1966, *Hansen & Smitinand* 37260 (K, L); Nakhon Si Thammarat prov., Kao Luang, evergreen forest, 900 – 1100 m, 29 April 1925, *Kerr* 15467 (K, L); Trang prov., Kao Soi Dao, evergreen forest, c. 400 m, 27 April 1930, *Kerr* 19147 (K). PENINSULAR MALAYSIA. Perak,

Bota Kiri Forest Reserve, near Ipoh, forest, 50 – 100 ft, 10 March 1958, *Shah* 301 (K, L, SING). SUMATRA. E coast, Asahan, Silo Maradja, April 1927, *Bartlett* 7237 (L, NY, US); E coast, Laboehan Batoe subdiv., Kota Pinang distr., Goenoeng Si Papan (in Concession Kaloebi: Topographic Sheet 41, SE quarter), 7 – 14 April 1933, *Rahmat Si Toroes* 3907 (A, L, NY, US); E coast, s. loc., acc. 14 Jan. 1930, *Yates* 1329 (B, NY).

23. Antidesma heterophyllum *Blume*, Bijdr. Fl. Ned. Ind.: 1123 (1826 – 27). Type: North Java, *Hb. Blume* s.n., (L! herb. no. 910222-1244, lectotype, here designated). — Note: The lectotype is the only eligible specimen in L, as the voucher for the variety, although annotated by Blume, cannot be used as a type (Blume explicitly stated that it is not typical). Another sheet was collected by *Kuhl & van Hasselt*, who went to Java only after the name was published. The designated lectotype bears no data from the protologue, but the species name in Blume's handwriting.

A. zippelii Airy Shaw, Kew Bull. 37: 6 (1982), **synon. nov.** Type: Timor, absque loc. exact. vel notulis, 1841, "Antidesma undulatum Zp.", *Zippelius* 32/6 (L! Herb. No. 910222-1193, lectotype, here designated; L!, isolectotype).

Shrub or tree, up to 8 m, diameter up to 6 cm. *Young twigs* terete, shortly spreading-pilose to pubescent, brown. *Stipules* persistent, linear, 3 – 5(– 10) × 0.5 – 1 mm, glabrous to pilose. *Petioles* channelled adaxially, 2 – 5(– 10) × 1 – 1.5(– 2) mm, pilose to pubescent. *Leaf blades* oblong, more rarely elliptic or obovate, apically acuminate-mucronate to acute-mucronate, often with a rounded apiculum, basally acute, more rarely obtuse to rounded, (4 –)10 – 15(– 25) × (1.5 –)3 – 5(– 8.5) cm, (2.1 –)3.1(– 4.6) times as long as wide, eglandular, chartaceous to membranaceous, glabrous, sometimes pilose only along the midvein, dull or shiny on both surfaces, midvein impressed adaxially, tertiary veins reticulate to weakly percurrent, widely spaced, rather oblique, quarternary and finer veins hardly prominent on either leaf surface in dry material, intersecondary veins not conspicuous, drying olive-green. *Staminate inflorescences* 3 – 5 cm long, axillary, simple or consisting of up to 10 branches, axes pilose to pubescent. *Bracts* deltoid to linear, 0.3 – 1 × 0.3 – 0.5 mm, glabrous to pilose. *Staminate flowers* c. 2 × 1 – 2 mm. *Pedicels* 0 – 2 mm long, not articulated, glabrous to pilose. *Calyx* 0.5 – 0.7 mm long, sepals 3 – 4, free to fused for $^1/_2$ of their length, deltoid to nearly orbicular, apically acuminate to rounded, glabrous outside, glabrous inside but sometimes with long hairs at the base, margin erose, sometimes glandular-fimbriate. *Disc* cushion-shaped, hemispherical, fully enclosing the bases of the filaments and pistillode, often 3-lobed, glabrous. *Stamens* 3 – 4, 1 – 1.5 mm long, exserted 1 – 1.5 mm from the calyx, anthers 0.3 – 0.4 × 0.3 – 0.4 mm. *Pistillode* subulate to flat, 0.2 – 0.5 × 0.1 – 0.2 mm, shorter than to exserted from the sepals, glabrous. *Pistillate inflorescences* 3 – 6 cm long, axillary, may be aggregated at the end of the branch, simple, more rarely consisting of up to 3 branches, axes pilose to pubescent. *Bracts* deltoid to linear, 0.5 – 1.5 × 0.3 – 0.4 mm, glabrous to pilose, margin sometimes glandular-fimbriate. *Pistillate flowers* c. 2 × 1 mm. *Pedicels* 0.5 – 1 mm long, spreading-pilose. *Calyx* 0.5 – 0.8 mm long, sepals 3 – 4(– 5), free to fused for up to

$^1/_2$ of their length, deltoid to almost orbicular, apically acute to rounded, glabrous outside, glabrous inside but sometimes with long hairs at the base, margin glandular-fimbriate, erose or entire. *Disc* shorter than the sepals, glabrous. *Ovary* pyriform to globose, laterally compressed, glabrous, style subterminal to almost lateral, stigmas 3 – 5, rather long, sometimes irregularly fused. *Infructescences* 5 – 9 cm long. *Fruiting pedicels* 1 – 3 mm long, glabrous to pilose. *Fruits* ellipsoid to lenticular, laterally compressed, basally symmetrical to slightly asymmetrical, with a terminal to slightly subterminal style, 5 – 7 × 4 – 6 mm, glabrous, rarely finely white-pustulate, areolate when dry.

DISTRIBUTION. Java, south-eastern Sulawesi (Binongko Island), Lesser Sunda Islands, Moluccas (Ambon, Manipa Island). Map 17.

HABITAT. In primary or secondary forest, thickets or dry shrubby heath forest ("kerangas"). On coral sand, sandy loam and limestone; andesite breccia. 1 – 750 m altitude.

USES. Used as a remedy against toothache (Hasskarl 1845: 71).

CONSERVATION STATUS. Near Threatened (NT). This species occupies a large area but there are few recent collections. 41 specimens examined.

VERNACULAR NAMES. hachunian or huhunian (Sundanese), seueur (Sundanese), tamandilang. Hasskarl (1845: 71): kibangbara, kitakketan, kitjalikket.

ETYMOLOGY. The epithet refers to the leaves (Greek, heteros = different; phyllon = leaf).

KEY CHARACTERS. Fruits ellipsoid, laterally compressed; leaves oblong; bracts usually long and narrow, persistent.

SIMILAR SPECIES. *A. minus* has larger, ovoid fruits, more elliptic, reddish drying leaves, and whitish twigs. — *A. heterophyllum* and *A. montanum* are indistinguishable if not in fruit.

NOTE. Very close to *A. montanum*. Staminate specimens can only tentatively be identified. *A. heterophyllum* is confined to a coherent geographic area characterised by a drier and more seasonal climate than ever-wet western Malesia. Whether it is a good species or a mere ecological variant of *A. montanum* cannot be answered by morphological analysis alone.

Blume's (1826 – 27: 1123) variety "Var. foliis angustioribus, spicis subsolitariis" is not validly published according to Art. 23.6(a) (Greuter *et al.* 2000: 43).

SELECTED SPECIMENS. JAVA. W Java, Ujungkulon, Oct. 1913, *Amdjah* 39 (BO, L); Banjumas res., Tjilatjap afd. & distr., Nusa Kambangan, Lebak paetjoeng, boschterrein, 12 June 1898, *Koorders* 30293 (BO, L); Banjumas prov., Nusa Kambangan Isl., SW part beach W of K. Babakan, 20 Nov. 1938, *Kostermans & Van Woerden* 32 (K, L); W Java, Pulau Panaitan (Prinseneiland), Mt Parat, forest, 50 m, 2 Oct. 1951, *Van Borssum Waalkes* 793 (BO, K, L); Priangan res., Karang House, Wynkoopslaai, 10 m, *Van Leeuwen-Reijnvaan* 14191 (L). SULAWESI. SE Sulawesi, Binongko (Benongko), trocken, Korallensand, 23 July 1909, *Elbert* 2553 (BO, L). LESSER SUNDA ISLANDS. Sumbawa, Sultanat Bima, Landschaft Donggo (Donggo Mts), Oo, Parklandschaft, trocken, Andesitbreccie, 250 – 500 m, 3 Dec. 1909, *Elbert* 3515 (A, L); Sumbawa, Sultanat Dompu, Kowanko v. d. Sahleh-Bai (= Kawangko = Bangkulua), Monsun-Schlucht, Wala, Flusstal, Andesitbreccien, 20 – 150 m, 26

Dec. 1909, *Elbert* 4090 (A, K, L); Timor, E Portuguese Timor, between Baucau (Baukau) and Vemassi (Vermasse), forest patch along ravine, limestone plateau escarpment, 17 Dec. 1953, *Van Steenis* 18078 (BM, L).

24. Antidesma japonicum *Siebold & Zucc.*, Abh. Math.-Phys. Cl. Königl. Bayer. Akad. Wiss. 4(3): 212 (1846), repr. as Fl. Jap. Fam. Nat. 2: 88 (1846). Type: Japan, *Siebold & Zuccarini* s.n. (L! Herb. No. 903154-234, lectotype, here designated; L!, LE?, isolectotypes) — Note: No type material in MAK *fide* J. Murata, in litt.

Shrub, treelet or tree, up to 5(– 8) m, diameter up to 10 cm. *Young twigs* terete, pubescent to pilose, brown. *Stipules* usually caducous, narrowly deltoid to subulate, apically acute, 2 – 5 × 0.7 – 1 mm, puberulent to pilose. *Petioles* channelled adaxially, basally and distally sometimes slightly pulvinate, 2 – 8 × 0.5 – 1.5 mm, pubescent to almost glabrous. *Leaf blades* narrowly elliptic-oblong to ovate or obovate, apically acuminate-mucronate to acute-mucronate, sometimes caudate, basally acute to rounded, (3.5 –)6 – 10(– 13) × (1 –)2 – 3.5(– 4.7) cm, (1.2 –)3(– 4.4) times as long as wide, eglandular, subcoriaceous to chartaceous, glabrous, sometimes some hairs only along the midvein, shiny on both surfaces, midvein impressed to flat or distinctly raised adaxially, tertiary veins weakly percurrent in wider, reticulate in narrower leaves, widely spaced, tertiary veins and finer venation thin, inconspicuous, drying light olive- to greyish green. *Staminate inflorescences* 2 – 6 cm long, axillary, sometimes aggregated at the end of the branch, simple or consisting of up to 5 branches, slender, axes pilose. *Bracts* deltoid to lanceolate, 0.5 – 0.7(– 1) × 0.3 – 0.5 mm, glabrous to pilose, margin often glandular-fimbriate. *Staminate flowers* 1 – 1.5 × 1 – 1.5 mm. *Pedicels* (0.5 –)1 – 1.5 mm long, not articulated, pilose. *Calyx* 0.4 – 0.6 mm long, sepals 3 – 5, free, spreading, deltoid to lanceolate, apically acute to rounded, glabrous outside and inside, margin erose. *Disc* cushion-shaped, fully or partially enclosing the bases of the filaments and pistillode, glabrous. *Stamens* 3 – 5, 1 – 2 mm long, fully exserted from the spreading sepals, anthers 0.3 – 0.5 × 0.3 – 0.5 mm. *Pistillode* absent, flat, clavate, cylindrical or 3-fid, 0.1 – 0.5 × c. 0.2 mm, extending to the same length and blending into the disc to distinctly exserted from the disc, glabrous. *Pistillate inflorescences* 2 – 5 cm long, axillary to terminal, simple or branched once, rarely twice, axes nearly glabrous to puberulent. *Bracts* deltoid to linear, apically acute, 0.3 – 1 × 0.2 – 0.5 mm, pubescent to almost glabrous, margin sometimes glandular-fimbriate. *Pistillate flowers* 1 – 1.5 × 1 mm. *Pedicels* (0.5 –)1 – 1.5 mm long, pilose. *Calyx* 0.5 – 0.7 × c. 0.7 mm, sepals 3 – 5(– 7), free, 0.3 – 0.5 mm wide, deltoid, apically acute to rounded, glabrous outside, glabrous inside but sometimes with long hairs at the base, margin entire to erose. *Disc* much shorter than or extending to the same length as the sepals, glabrous. *Ovary* fusiform to ellipsoid, glabrous, style terminal, stigmas 3 – 5. *Infructescences* 4 – 9 cm long. *Fruiting pedicels* (0 –)3 – 6 mm long, pilose to glabrous. *Fruits* ellipsoid to lenticular, laterally compressed, basally asymmetrical to slightly symmetrical, with a subterminal, rarely nearly terminal style, 5 – 6(– 8) × 4 – 5(– 6) mm, glabrous, rarely white-pustulate, areolate or fleshy when dry.

24a. var. japonicum

?*A. gracillimum* Gage, Rec. Bot. Surv. India 9: 227 (1922). Type: Malaysia, Perak, Gunong Inas, *Wray* 4064? (CAL, herb. acc. no. 409068, lectotype, designated by Chakrabarty & Gangopadhyay 1997: 479; CAL, isolectotype). — Note: The question mark after the number 4064 appears in the original description. Original syntypes: Perak, Maxwell's Hill, *Wray* 2946 (CAL, SING!); 3237 (CAL, SING!). — Notes: This name was subsumed under *A. japonicum* by Airy Shaw (1972a: 354) and Whitmore (1973: 57). The two syntypes *Wray* 3237 (in staminate bud) and *Wray* 2946 (in pistillate flower and young fruit) were examined for this revision and support this conclusion. Chakrabarty & Gangopadhyay, however, maintained *A. gracillimum* as a separate species on account of its sessile flowers as well as the subglobose, somewhat fleshy fruits, 4 – 5 mm in diameter with terminal or almost terminal styles. The lectotype chosen by Chakrabarty & Gangopadhyay could not be examined for this revision. It may, unlike the two original syntypes, not belong to *A. japonicum*. From the description given by Chakrabarty & Gangopadhyay this is most likely to be the very similar *A. montanum* Blume. In any event, the differential characters do not justify specific rank for *A. gracillimum*.

A. acuminatissimum Quisumb. & Merr., Philipp. J. Sci. 37: 159 (1928), **synon. nov.** Type: Philippines, Luzon, Tayabas Prov., Casiguran, 9 June 1925, in the open and on river flats, alt. c. 200 m, *BS (Ramos & Edaño)* 45189 (A!, lectotype, here designated; B!, K!, P!, US!, isolectotypes). — Note: Known only from the type in young fruit. These are already compressed with an oblique style. Petioles, pedicels and inflorescences are slender as in typical *A. japonicum*.

Petioles basally and distally sometimes pulvinate, 0.5 – 1 mm wide in the middle. *Leaf blades* (3.5 –)6 – 10(– 13) × (1 –)2 – 3.5(– 4.7) cm, midvein impressed to flat adaxially. *Pistillate sepals* apically acute. *Infructescences* slender. *Fruiting pedicels* very thin, spreading, (2 –)3 – 6 mm long.

DISTRIBUTION. Japan (Kyushu and Ryukyu Islands), Taiwan, southern China (Guangdong, Guangxi, Guizhou, Hainan, Hunan, Guangxi, Sichuan, Zhejiang provinces, not in Yunnan province), Burma (Tenasserim & Thaton distr.), Vietnam, Thailand, Peninsular Malaysia (Kedah, Perak, Pahang), Philippines (Luzon, Mindanao). Map 18.

HABITAT. In (evergreen) forest; in virgin forest, dipterocarp forest and dry thickets; on limestone cliffs; along roadsides; in primary or secondary vegetation; on wet or dry ground. On sandy soil, clay, limestone, over granite bedrock. 0 – 1700 m altitude.

USES. Firewood.

CONSERVATION STATUS. Least Concern (LC). This taxon is widespread and frequently collected. 134 specimens examined.

VERNACULAR NAMES. China: ngo ch'un tsz.

ETYMOLOGY. The epithet refers to the type locality.

KEY CHARACTERS. Plants ± glabrous; inflorescences, pedicels and petioles short, slender; disc often fleshy and longer than the calyx; fruits distinctly laterally compressed.

SIMILAR SPECIES. *A. digitaliforme* and *A. forbesii*, see there. — *A. leucocladon* has whitish young twigs, adaxially raised midveins, a hairy disc and larger fruits. — *A. japonicum* differs from *A. montanum* mainly in the distinctly laterally compressed fruits with subterminal style and fewer, smaller white pustules, firmer, glossier leaves and the generally more slender appearance of the plant (especially petioles, inflorescences and pedicels). The two species are, however, extremely similar. The concept of *A. pentandrum* (Blanco) Merr. (here treated as a synonym of *A. montanum*) in some respects bridges the gap between *A. japonicum* and *A. montanum*. Further studies in the northern geographic range of the genus might show the need to subsume *A. japonicum* under *A. montanum*.

NOTE. Staminate flowers are often deformed throughout the distribution area, with irregularly fused filaments and aberrant anthers, similar to *A. digitaliforme*.

SELECTED SPECIMENS. JAPAN. Ryukyu Islands (Nansei-Shoto), Iriomote Isl., low jungle, 15 m, Aug. 1934, *Linsley Gressitt* 538 (G, K, L); Kyushu, Nagasaki, 1862, *Oldham* 744 (G, K, L, NY). TAIWAN. L. Candidius, hill side, 750 m, May – Aug. 1934, *Linsley Gressitt* 212 (G, K, L). CHINA. Hong Kong, Victoria Island, peak to Pokfulam Reservoir, 2 May 1969, *Shiu Ying Hu* 7220 (K); Guangdong prov., Sin-Fung distr., Lo-Lo-ha village, Sha Lo Shan, forest, 6 – 25 July 1938, *Taam* 948 (G, K). VIETNAM. Tonkin, Taai Wong Mo Shan, near Chuk-phai, Ha-coi, in thicket, on dry clayey soil, 3 May – 22 June 1939, *Tsang* 29038 (B, G, K). BURMA. Thaton distr., Dawna Range, Paingkyu to Tale, 4 – 5000 ft, 22 Feb. 1909, *Lace* 4617 (K). THAILAND. Northern, Phitsanulok prov., Tung Salaeng Luang, evergreen forest, 550 m, 19 July 1966, *Larsen, Smitinand & Warncke* 485 (L); South-western, Prachuap Khiri Khan prov., Kao Luang, by stream in evergreen forest, c. 300 m, 3 July 1926, *Kerr* 10791 (K, L). PENINSULAR MALAYSIA. Perak, Kanthan, 13 miles N of Ipoh, limestone cliffs at P.W.D. Quarry, 24 Oct. 1958, *Sinclair* 9865 (E, K,

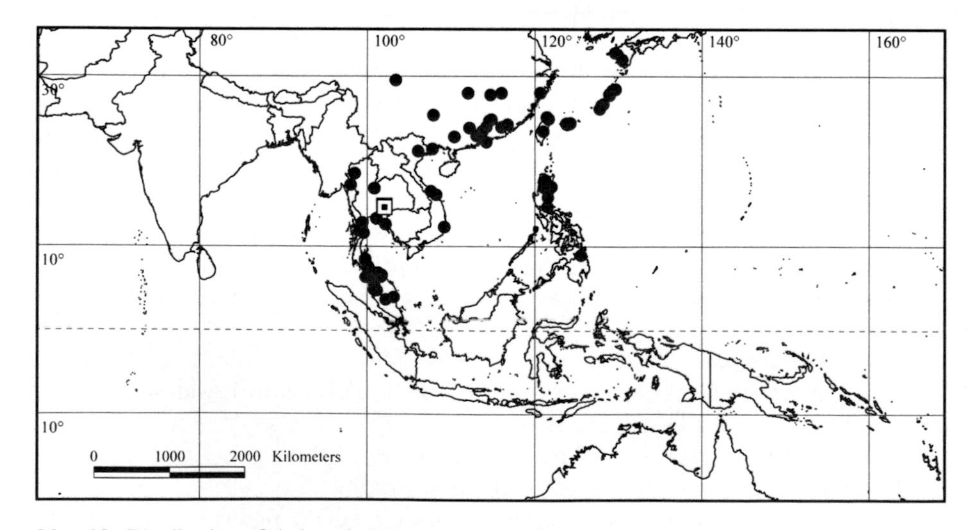

MAP 18. Distribution of *A. japonicum* var. *japonicum* (●) and var. *robustius* (⊡).

SING). PHILIPPINES. Luzon, Isabela, Sierra Madre Mts, San Mariano, Bo. Disulap, along Disusuan creek, dipterocarp forest, 400 ft, 3 May 1961, *PNH (Gutierrez)* 78159 (A, K, L); Mindanao, Butuan subprov., s. loc., Dec. 1911, *Weber* 1492 (A, E, G, K, MEL, US).

24b. var. **robustius** *Airy Shaw,* Kew Bull. 26: 355 (1972). Type: Thailand, Pak Thon Chai ("TREND Camp"), eastern part of Khao Yai National Park, evergreen forest, alt. 500 m, 8 Aug. 1968, *Larsen, Santisuk & Warncke* 3137 (K!, holotype; AAU, L!, isotypes).

Petioles basally and distally hardly pulvinate, 0.5 – 1.5 mm wide. *Leaf blades* 6 – 7.5 × 2 – 4.5 cm, midvein distinctly raised adaxially. *Pistillate sepals* apically rounded. *Infructescences* not particularly slender. *Fruiting pedicels* not thin as in type variety, 0 – 1 mm long.

DISTRIBUTION. Eastern Thailand: Nakhon Ratchasima province only. The variety seems to be restricted to the Pak Thong Chai/Sakaerat area. Map 18.

HABITAT. In dry evergreen forest. 350 – 500 m altitude.

CONSERVATION STATUS. Vulnerable (VU D1 + 2). This variety is only known 6 specimens from an area of about 15 km^2 in Sakaerat Forest Reserve/Khao Yai National Park. According to the information on the herbarium labels, it is locally not uncommon, but there are many common species that can easily be confused with this taxon.

ETYMOLOGY. The epithet refers to the habit of the plant (Latin, robustius = more robust).

KEY CHARACTERS. Midvein raised adaxially; less delicate than type variety.

SIMILAR TAXA. *A. japonicum* var. *robustius* shares the adaxially raised midvein only with *A. laurifolium* from which it differs in its smaller leaves and fruits, its glabrous ovaries, its shorter infructescences and in its leaves drying green.

NOTES. Airy Shaw (1972a: 355) synonymised *A. japonicum* var. *robustius* p.p. with *A. cambodianum* Gagnep. This cannot be confirmed here. The type material of *A. cambodianum* examined (Tonking, environs d'Ouonbi, *Balansa* 1496 (P!); Laos, Massie, plateau d'Attopeu, Harmand in hb. Pierre s.n.; Cambodge, monts de Knang-krepeuh, prov. de Thepong, *Pierre* s.n. (K!, NY!, P!); monts Camchay, *Pierre* s.n. (P!)) differs consistently from *A. japonicum* var. *robustius* in, for example, the impressed midveins, and resembles *A. montanum.*

SELECTED SPECIMENS. THAILAND, EASTERN, NAKHON RATCHASIMA PROV. Pak Thong Chai ("TREND Camp"), E part of Khao Yai National Park, evergreen forest, 500 m, 8 Aug. 1968, *Larsen, Santisuk & Warncke* 3118 (K); Pak Thon Chai distr., Sakaerat Exp. Stat., shaded, dry evergreen forest, 350 m, 24 Aug. 1975, *Maxwell* 75-921 (L); Pak Thon Chai, Wang Nam Khieo, in dry evergreen forest, 495 m, 16 June 1968, *Phengnaren* 36332 (K); Pak Thong Chai, Sakaerat Forest Reserve, 14°45'N, 102°00'E, dry evergreen forest, c. 400 m, 28 Oct. 1969, *Van Beusekom & Charoenphol* 1877 (L); Sakaerat, 14°40'N, 102°02'E, dry evergreen forest, 400 m, 23 Oct. 1971, *Van Beusekom & Geesink* 3322 (K, L).

25. Antidesma jucundum *Airy Shaw,* Kew Bull. 33: 426 (1979). Type: Papua New Guinea, Central province, Abau subdistrict, Cape Rodney, 10°04'S, 148°32'E, disturbed rain-forest on riverine alluvium, 30 m, 29 Aug. 1969, *Pullen* 8163 (K!, holotype; B!, CANB!, L!, isotypes).

Shrub or tree, up to 4 m, with drooping branchlets. *Young twigs* terete, very slender, long spreading-pubescent, brown. *Stipules* persistent, subulate, 2 – 6 × 0.1 – 0.5 mm, pilose. *Petioles* nearly terete to narrowly channelled adaxially, 1 – 3 × 0.5 – 1 mm, long spreading-pubescent. *Leaf blades* ovate, rarely elliptic, apically acuminate- to caudate-mucronate, with a rounded to retuse apiculum, basally rounded to obtuse, more rarely acute, (2.2 –)3 – 5(– 6.5) × 1 – 3 cm, 1.7 – 3 times as long as wide, eglandular, chartaceous, glabrous except along the midvein, rarely pilose all over adaxially, long spreading-pilose all over abaxially, more densely so along the veins, shiny adaxially, dull abaxially, midvein impressed adaxially, tertiary veins reticulate, drying olive-green, domatia absent. *Staminate plants* — see below under *A.* cf. *jucundum. Pistillate inflorescences* c. 2.5 cm long, axillary, simple or once-branched, axes long spreading-pilose. *Bracts* ovate, almost laciniate, c. 0.5 × 0.5 mm, long spreading-pilose. *Pistillate flowers* 1.5 – 2 × 1 – 1.5 mm. *Pedicels* c. 0.5 mm long, long spreading-pilose. *Calyx* c. 0.5 × 1 mm, bowl-shaped, with a truncate base, sepals 4, fused for $^{2}/_{3}$ – $^{4}/_{5}$ of their length, apically acute, sinuses wide, rounded, long spreading-pilose outside, glabrous inside, margin fimbriate to laciniate. *Disc* extending to the same length as the sepals, long spreading-pilose at the margin, indumentum exserted from the sepals. *Ovary* lenticular, long spreading-pilose, style lateral, stigmas 4 – 7, thin. *Infructescences* 1.5 – 4 cm long, axes c. 0.5 mm wide. *Fruiting pedicels* 1 – 2 mm long, long spreading-pilose. *Fruits* lenticular, distinctly laterally compressed, basally symmetrical or asymmetrical, with a distinctly lateral style, 4 – 6 × 4 – 6 mm, long spreading-pilose, not white-pustulate, areolate when dry.

DISTRIBUTION. Papua New Guinea, Central and Milne Bay provinces; staminate paratype of doubtful affinity from Northern province, see below under *A.* cf. *jucundum.* Map 19.

HABITAT. In mixed rainforest; in primary or disturbed vegetation. On grey-brown heavy soil over red clay, lateritic soil or riverine alluvium. 10 – 200 m altitude.

CONSERVATION STATUS. Vulnerable (VU B1ab(i,ii)). The species deserves this category because its extent of occurrence is only about c. 15000 km^2. 8 specimens examined.

ETYMOLOGY. The epithet refers to the general appearance of the plant (Latin, jucundum = pleasant, lovely).

KEY CHARACTERS. Long, spreading indumentum of all parts; slender habit; small, mainly ovate leaves; pedicellate staminate flowers with a glabrous disc.

SIMILAR SPECIES. *A. concinnum* and *A. excavatum,* see there.

NOTE. The long, spreading indumentum, slender habit and small, mainly ovate leaves make the collections cited by Airy Shaw in the protologue rather conspicuous. However, the pistillate flowers and fruits of the type are identical with those of *A. excavatum* which is vegetatively so variable that it would hardly be

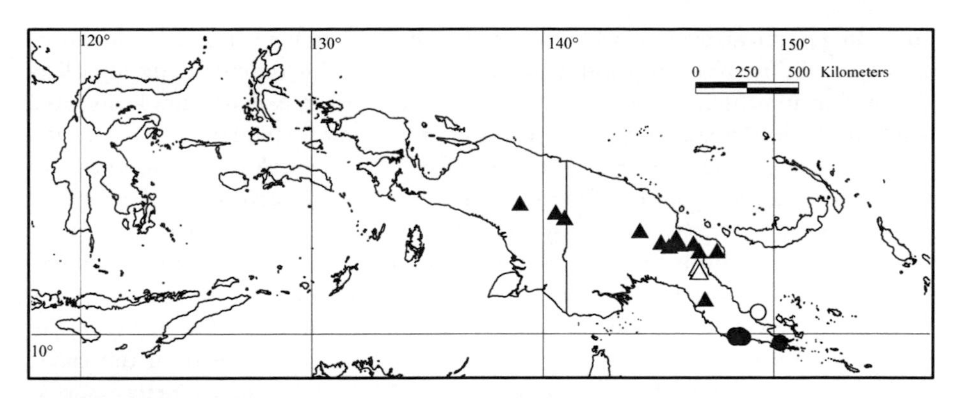

MAP 19. Distribution of *A. jucundum* (●), staminate paratype (○) and *A. myriocarpum* var. *myriocarpum* (▲) and var. *puberulum* (△).

justified to recognise a species on the basis of leaf shape and size alone. The only staminate paratype, on the other hand, has pedicellate flowers and a glabrous disc, both characters that are not observed in *A. excavatum*. Airy Shaw included it only doubtfully because of its larger leaves and distant locality. For these reasons, *A. jucundum* is maintained here until more material can be examined, but separate descriptions are given for the type and matching pistillate paratypes from the Central province and the possibly mis-matched staminate paratype from the Northern province.

SELECTED SPECIMENS. PAPUA NEW GUINEA. Central distr., Abau subdistr., Minari logging area, 10°10'S, 148°40'E, lowland rainforest, 10 m, 9 July 1974, *LAE (Katik)* 62148 (K, L, US); Abau subdistr., Mori R., 10°10'S, 148°35'E, lowland forest bordering recently cleared roadway close to river, 50 ft, 23 April 1964, *NGF (Sayers)* 19675 (K, L, US); Abau subdistr., Cape Rodney, near P.I.T., 10°05'S, 148°18'E, mixed forest on undulating country, grey-brown heavy soil over red clay, 200 ft, 18 June 1968, *NGF (Henty)* 38521 (K, L); Abau subdistr., Mori R., 10°10'S, 148°20'E, forest on flat land, lateritic soil, 200 ft, 16 Feb. 1969, *NGF (Henty & Lelean)* 41887 (K, L); Milne Bay distr., Alotau subdistr., Sagarai, S slopes of Pini Range, 10°23'S, 150°15'E, in primary rainforest near campsite, 200 m, 2 March 1984, *LAE (Gideon)* 76946 (K, L); Eastern div., Mori R., 23 May 1926, *Brass* 1531 (K, P).

A. cf. **jucundum**, staminate paratype: Papua New Guinea, Northern province, Tufi subdistrict, near Budi Barracks, in dense tall rainforest, c. 75 m, 30 Aug. 1954, *Hoogland* 4621 (K!, L!, MEL!, US!).

Bark grey-brown, 1 mm thick, thin-flaky, wood pinkish-straw. *Young twigs* terete, very slender, spreading-pubescent, brown. *Stipules* persistent, subulate, 3 – 5 × 0.2 – 0.5 mm, pilose. *Petioles* nearly terete, 2 – 3 × c. 0.5 mm, spreading-pubescent. *Leaf blades* ovate, oblong or obovate, apically acuminate- to caudate-mucronate, with an acute apiculum, basally acute to obtuse, 6 – 8.5 × 1.5 – 3 cm, 2.3 – 3.5 times as long as wide, eglandular, chartaceous, glabrous except along the midvein adaxially, spreading-pilose all over abaxially, more densely so along the veins, shiny adaxially, dull abaxially, midvein impressed adaxially, tertiary veins reticulate, drying olive-green, domatia absent. *Staminate inflorescences* c. 2 cm long, axillary, simple, axes 0.2 – 0.3 mm wide, spreading-pubescent. *Bracts* ovate, almost laciniate, c. 1 × 0.2 – 0.3 mm, spreading-pilose. *Staminate flowers* c. 2 × 1.5 mm. *Pedicels* c. 1.5 mm long, not articulated, spreading-pilose. *Calyx* c. 1 × 1.5 mm, bowl-shaped, sepals 4, fused for $^1/_2$ – $^3/_4$ of their length, irregularly shaped, apically usually rounded, spreading-pilose outside, glabrous inside, margin fimbriate to laciniate. *Disc* extrastaminal-annular, much shorter than the sepals, lobed towards the centre or lobes ± detached from the annular part of the disc, subulate and standing between the stamens, glabrous. *Stamens* 4, c. 2 mm long, exserted c. 1 mm from the calyx, anthers c. 0.3 × 0.7 mm. *Pistillode* subulate, c. 0.5 × 0.1 – 0.2 mm, distinctly exserted from the disc, extending to the same length as the sepals, sparsely pilose at the tip.

26. Antidesma kunstleri *Gage*, Rec. Bot. Surv. India 9: 225 (1922). Type: Malaysia, Perak, Maxwell's Hill, *Ridley* 2975 (SING!, lectotype, here designated). Original syntypes: Perak, *Scortechini* 629; Larut, *King's collector* 2645, 2927, 3219, 5012; Gunong Inas, *Wray* 4077. — Note: The name was considered a synonym of *A. stipulare* by Whitmore (1973: 56). Airy Shaw (1975: 216) followed Whitmore, but had obviously seen only the photographs of the lectotype kept at Kew.

Shrub, 1.2 m. *Young twigs* terete, spreading-ferrugineous-pilose, brown. *Stipules* persistent, foliaceous, ovate, apically caudate, basally asymmetrical, reticulately veined, 10 – 30 × 5 – 12 mm, pilose or glabrous. *Petioles* terete or channelled adaxially, basally and distally slightly pulvinate and geniculate, 10 – 20 × 1 – 1.5 mm, spreading-pilose or glabrous. *Leaf blades* oblong to slightly ovate or obovate, apically caudate-mucronate, basally obtuse to rounded, 13 – 21 × 3.5 – 6.5 cm, 3 – 4 times as long as wide, eglandular, subcoriaceous to chartaceous, glabrous adaxially, glabrous or pilose only along the midvein abaxially, shiny on both surfaces, midvein impressed to flat adaxially, tertiary veins reticulate to percurrent, drying olive-green. *Staminate inflorescences* (only young inflorescences seen) 2 – 3.5 cm long, axillary, simple, axes ochraceous-pubescent. *Bracts* deltoid, c. 0.8 × 0.5 mm, ochraceous-pubescent. *Buds* sessile just before anthesis. *Staminate flowers* not known at anthesis. *Calyx* in bud c. 0.8 mm long, sepals 4, fused only at the base, deltoid, apically acute, densely pilose outside, glabrous inside, margin entire. *Stamens* 4. *Pistillode* apparently present, probably hairy. *Pistillate inflorescences* 4 – 12 cm long, axillary, simple or once-branched, axes densely ferrugineous-pubescent. *Bracts* lanceolate, 0.7 – 1 × c. 0.5 mm, ferrugineous-pubescent. *Pistillate flowers* c. 2 × 1.5 mm. *Pedicels* c. 0.5 mm long, pubescent. *Calyx* c. 1.2 × 1.5 mm, sepals 4, fused only at the base, deltoid, apically acute, pubescent outside, glabrous inside but with long hairs at the base, margin entire. *Disc* much shorter than the sepals, glabrous. *Ovary* ellipsoid, glabrous or with few hairs, style subterminal, stigmas 6, rather regular. *Fruits* not known.

DISTRIBUTION. Peninsular Malaysia: Perak, only known from Maxwell's Hill. Map 20.

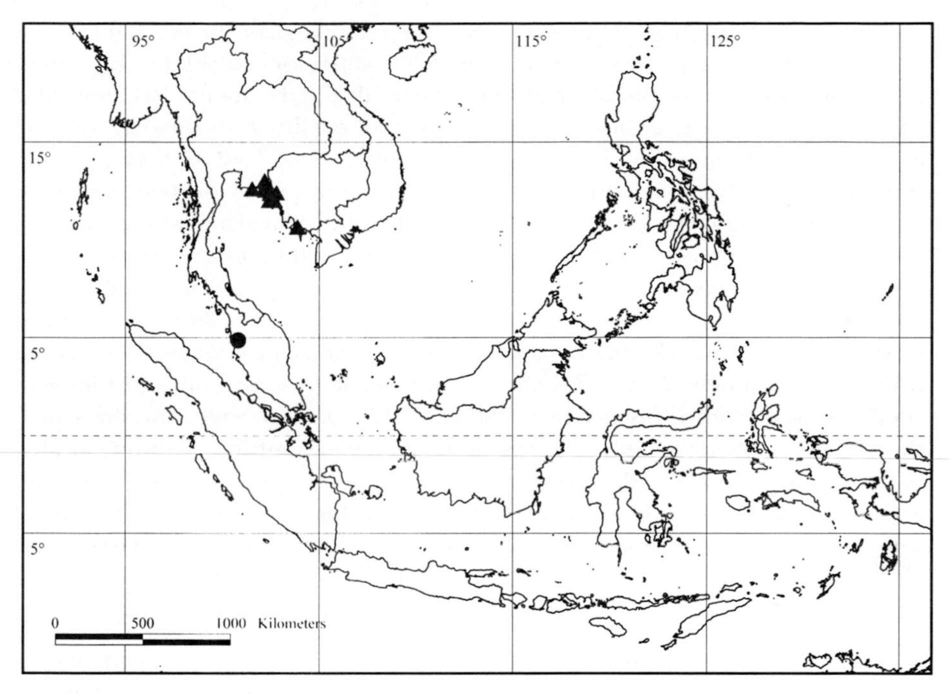

MAP 20. Distribution of *A. kunstleri* (●) and *A. laurifolium* (▲).

HABITAT. In forest. 1300 m altitude.

CONSERVATION STATUS. Critically Endangered (CR B1ab(i,ii,iii,iv,v)). This distinctive species is known only from a single locality. The general collection density in Peninsular Malaysia is fairly high. The fact that *A. kunstleri* has only ever been found three times, and always in the same accessible and well-collected locality, provides sufficient evidence to classify it as a highly endangered species.

ETYMOLOGY. The epithet pays homage to the German explorer and plant collector H. H. Kunstler who collected for G. King in Malaysia (also known as "King's collector").

KEY CHARACTERS. Foliaceous stipules; long petioles; inflorescence axes, pedicels and calyx densely ferrugineous-pubescent.

SIMILAR SPECIES. *A. neurocarpum* has shorter petioles and reddish or greyish drying leaves. — *A. kunstleri* differs from *A. stipulare* in the free sepals, more slender petioles and a ferrugineous-pubescent inflorescence axes, pedicels and calyx.

FURTHER SPECIMENS. PENINSULAR MALAYSIA, PERAK. Gunong Hijau (= Maxwell's Hill), forest, 4000 ft, 4 March 1980, *Stone* 14392 (L); Maxwell's Hill, Aug. 1888, *Wray* 2831 (SING).

27. Antidesma laurifolium *Airy Shaw*, Kew Bull. 26: 356 (1972). Type: Thailand, Kao Sabap, Chantaburi ("Chantabun"), evergreen forest, alt. 400 m, 8 Jan. 1930, *Kerr* 18012 (K!, holotype; L!, isotype).

Shrub or slender tree, up to 4 m. *Young twigs* terete, glabrous, whitish grey or brown. *Stipules* caducous, linear, apically acute, 4 – 10 × 0.5 – 1.5 mm, ferrugineous-pilose, becoming glabrous. *Petioles* distinctly channelled adaxially, (4 –)6 – 9 × 1 – 3 mm, glabrous. *Leaf blades* narrowly elliptic to oblong, apically acuminate-mucronate, basally acute, rarely obtuse, 13 – 27 × 2.5 – 8 cm, 4 – 5.5 times as long as wide, eglandular, coriaceous, glabrous, shiny on both surfaces, major veins distinctly raised adaxially, tertiary veins reticulate, widely spaced, drying greyish green to reddish brown adaxially, reddish to greyish abaxially. *Staminate plants* unknown. *Pistillate inflorescences* c. 3 cm long, axillary, simple, axes ferrugineous-pubescent. *Bracts* ovate to deltoid, apically acute, shortly acuminate or obtuse, 0.3 – 0.7 × 0.3 – 0.5 mm, ferrugineous-pubescent. *Pistillate flowers* c. 1.5 × 1 mm. *Pedicel* absent. *Calyx* 0.7 – 0.8 × c. 1 mm, sepals 4 (sometimes an additional smaller sepal present), fused only at the base, 0.5 – 0.8 mm wide, apically acute, shortly acuminate or rounded, ferrugineous-pubescent outside, glabrous inside, margin fimbriate. *Disc* shorter than the sepals, glabrous. *Ovary* ellipsoid, appressed-pubescent, style terminal, stigmas 3 – 6. *Infructescences* 7 – 15 cm long. *Fruiting pedicels* 0 – 1 mm long, glabrous to sparsely pilose. *Fruits* ellipsoid to slightly ovoid, laterally compressed, basally symmetrical, with a terminal style, 8 – 11 × 5 – 8 mm, pilose, not white-pustulate, reticulate when dry.

DISTRIBUTION. Cambodia (Kampot province), South-East Thailand (Chon Buri, Chanthaburi and Trat provinces) and Southern Peninsular Malaysia (Johore). The disjunction between the everwet Malay Peninsula and the slopes and foothills of the Cardamon mountains extending eastwards into Cambodia as well as to the islands off-shore has been described by Whitmore (1975: 164) for *Diospyros hermaphroditica* (Ebenaceae). This indicates the existence of small, isolated patches of true evergreen rain forest in the "Chantaburi pocket", set in more seasonal forest. Map 20.

HABITAT. In evergreen forest and secondary growth. 0 – 600 m altitude.

CONSERVATION STATUS. Vulnerable (VU B1ab(i,ii)). The extent of occurrence in Thailand and Cambodia is 16000 km^2, and this disjunct species is known from two localities in southern Peninsular Malaysia. 13 specimens examined.

VERNACULAR NAMES. Cambodia: kracate audaek.

ETYMOLOGY. The epithet refers to the leaf-shape (Latin, *Laurus* = bay-tree, laurel; folium = leaf) which resembles *Laurus* L. (Lauraceae).

KEY CHARACTER. Midvein in dry material sharply raised adaxially (Fig. 2E), distinctly perceptible to the touch.

SIMILAR TAXA. *A. helferi*, *A. japonicum* var. *robustius*, see there.

NOTE. Erroneously published as "sp. nov." a second time (Airy Shaw 1972b: 458).

SELECTED SPECIMENS. CAMBODIA. Kampot prov., reserve Veal Renh, forêt dense, 6 Dec. 1974, *Marvis (or Maroris?)* 1172 (K). THAILAND, SOUTH-EASTERN. Chanthaburi prov., Kao Soi Dao, evergreen forest, c. 600 m, 14 Dec. 1924, *Kerr* 9673 (K, L); Chanthaburi prov., Kao Sabab, E of Makam, 1 – 300 m, 18 Jan. 1958, *Sorensen, Larsen & Hansen* 479 (L); Trat prov., Ko Kut Island, SE part, 0 – 70 m, 20 Nov. 1970, *Charoenphol, Larsen & Warncke* 5064 (K); Trat prov., Kaw Chang (Ko Chang) Island, Klawng Mayom Salak Kawk, evergreen forest, c. 10 m, 5 April 1923,

Kerr 6969 (K); Trat prov., Kao Kuap, evergreen forest, c. 200 m, 23 Dec. 1919, *Kerr* 18076 (K, L); Trat prov., Kao Saming, 24 Jan. 1927, *Put* 555 (K); Trat prov., Kao Kuap, *Put* 2949 (K); Trat prov., Koh Chang Island, 12°00'N, 102°40'E, dry evergreen forest and secondary growth, low altitude, 11 March 1970, *Van Beusekom & Santisuk* 3163 (L, K). PENINSULAR MALAYSIA. Johore, Gunong Blumut, upper camp, rocky stream bed, 360 m, May 1968, *FRI (Whitmore)* 8786 (KEP), 8853 (KEP); Johore, Endau-Rompin National Park, trail to the summit of Gunong Jampin Barat, 300 m, 8 Oct. 1994, *Periera et al.* 49 (KEP).

28. Antidesma leucocladon *Hook. f.*, Fl. Brit. India 5: 358 (1887). Type: Malaysia, Perak, at Sunga Ryah, *King's collector* [759] (K!, lectotype, here designated; BM!, isolectotype). Original syntypes: Malaysia, Penang, *Wallich* [7283] (K!); Perak, *Scortechini* s.n. — Note: Only the collectors but not the collection numbers [here in square brackets] are cited in the protologue. The localities, however, are the same as those in the protologue, and all sheets are annotated in Hooker's handwriting.

Shrub or tree (*fide Wray* 2042: climber), up to 13 m, diameter up to 10 cm. *Young twigs* terete to angular, glabrous or pilose and soon becoming glabrous, whitish grey. *Stipules* usually caducous, narrowly deltoid, apically acute, 2 – 5(– 10) × 0.5 – 1 mm, pilose to glabrous. *Petioles* channelled adaxially, 2 – 8(– 11) × 0.7 – 1.5 mm, pilose to glabrous. *Leaf blades* elliptic-oblong to ovate-oblong, apically acuminate- to caudate-mucronate, basally acute to rounded, (5 –)8 – 15(– 20) × (2 –)3.1(– 4.7) cm, (2 –) 3.1(– 4.7) times as long as wide, eglandular, membranaceous to chartaceous, glabrous, or pilose only along the midvein on both surfaces, shiny on both surfaces, midvein slightly raised to flat adaxially, tertiary veins reticulate, widely spaced, drying olive-green to brownish. *Staminate inflorescences* 6 – 12 cm long, axillary, sometimes aggregated at the end of the branch, simple or once-branched, slender, axes shortly pubescent. *Bracts* lanceolate to linear, 0.5 – 1.2 × 0.2 – 0.3 mm, pilose to glabrous. *Buds* elongate-ellipsoid, widely spaced on inflorescence axes. *Staminate flowers* 2 – 3 × c. 2 mm. *Pedicel* absent. *Calyx* 0.6 – 0.8 × c. 1 mm, cupular to cylindrical, sepals 4, fused for $^2/_3$ – $^3/_4$ of their length, apically rounded, glabrous to sparsely pilose outside, glabrous inside, margin erose. *Disc* consisting of 4 free alternistaminal lobes, lobes ± obconical, c. 0.3 × 0.5 mm, sometimes extrastaminally fused, hirsute. *Stamens* 4, 2 – 2.5 mm long, exserted 1.5 – 2 mm from the calyx, anthers 0.4 – 0.5 × 0.4 – 0.5 mm. *Pistillode* clavate to cylindrical, c. 0.8 × 0.3 mm, exserted 0.5 mm from the sepals, pilose, especially at the base. *Pistillate inflorescences* 4 – 10 cm long, axillary or terminal, simple or consisting of 1 – 2(– 4) branches, axes shortly pubescent. *Bracts* lanceolate, apically acute, 0.5 – 1 × c. 0.3 mm, pilose. *Pistillate flowers* c. 2 × 1 mm. *Pedicels* 1 – 1.5 mm long, relatively thick, shortly hirsute. *Calyx* 0.5 – 0.8 × c. 1 mm, cylindrical to urceolate, sepals 4, fused for $^2/_3$ – $^3/_4$ of their length, apically broadly rounded to broadly acute, glabrous to shortly pilose outside, glabrous inside, sometimes long hairs at the base, margin erose. *Disc* shorter than the sepals, shortly hirsute at the margin. *Ovary* ovoid, appressed-hirsute, style terminal to subterminal, stigmas 3 – 5. *Infructescences* 7 – 10 cm long. *Fruiting pedicels* 1 – 2.5 mm long, pubescent to pilose. *Fruits* ovoid to almost lens- or slightly bean-shaped, up to 2 mm long beaked, laterally compressed, basally symmetrical, with a subterminal to lateral style, 10 – 11 × 6 – 7 mm, pilose to almost glabrous, sometimes white-pustulate, areolate to reticulate when dry.

DISTRIBUTION. Thailand (Peninsula and one collection from Southeast), Peninsular Malaysia, East Sumatra. Map 21.

HABITAT. In primary evergreen forest, sometimes along streams or in fresh water swamps. 0 – 800 m altitude.

CONSERVATION STATUS. Near Threatened (NT). This is a relatively well-collected and fairly widespread species, however, habitat destruction in Peninsular Malaysia is too intense to classify it under Least Concern. 30 specimens examined.

VERNACULAR NAMES. Thailand: phak wan lang kao; Malay Peninsula: dada ruan; gurusek puteh.

ETYMOLOGY. The epithet refers to the colour of the bark (Greek, leucon = white colour; clados = branch).

KEY CHARACTERS. Plant nearly glabrous; young twigs whitish; midrib slightly raised adaxially; sepals fused; disc shortly hirsute; fruits beaked.

SIMILAR SPECIES. *A. bunius*, *A. forbesii*, *A. helferi*, *A. japonicum*, see there. — *A. montanum* has brown young twigs, adaxially impressed midveins, glabrous discs, more or less free sepals and terete fruits.

SELECTED SPECIMENS. THAILAND. South-eastern, Chanthaburi prov., Pliew Falls, shaded dense forest slopes, 4 April 1971, *Maxwell* 71-248 (L); Peninsular,

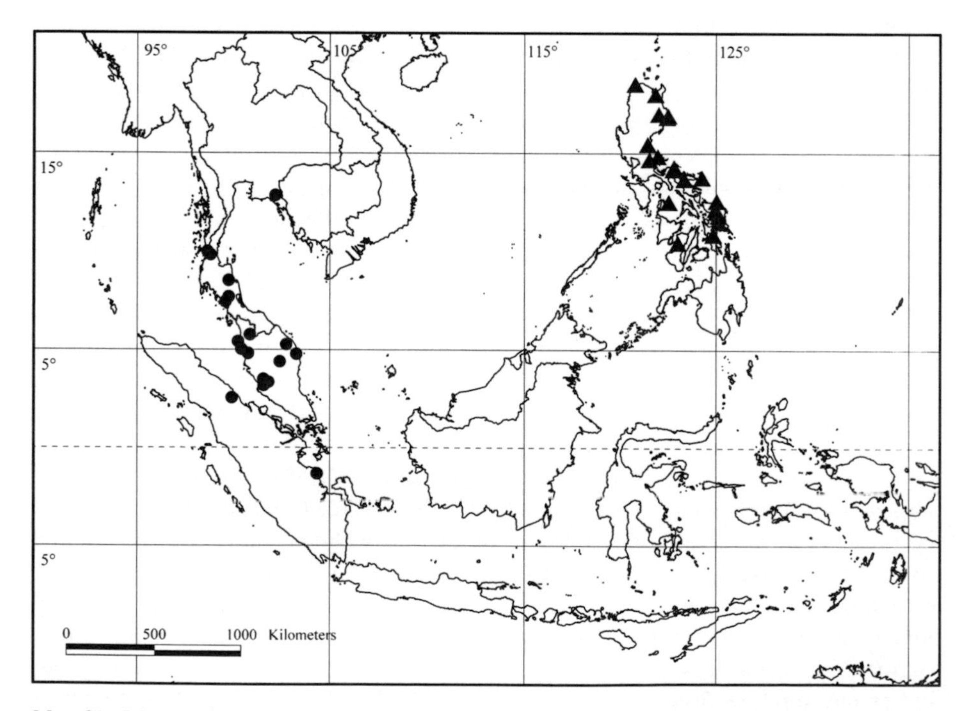

MAP 21. Distribution of *A. leucocladon* (●) and *A. macgregorii* (▲).

Chumphon prov., Pato (Phato), Langsuan, evergreen forest, c. 200 m, 3 March 1927, *Kerr* 12190 (K); Peninsular, Surat Thani prov., Kao Saming, edge of evergreen forest, under 50 m, 31 Dec. 1929, *Kerr* 17891 (K). PENINSULAR MALAYSIA. Malacca, Bukit Shigei, 30 March 1886, *Alvins* s.n. (SING); Kelantan, SE Kelantan, Sungai Lebir near Kuala Ternya, low undulating country, 6 Sept. 1967, *FRI (Whitmore)* 4308 (K, L); Pahang, Taman Negara, Bukit Tersik at Kuala Tahan, steep slopes, 20 April 1971, *FRI (Whitmore)* 15992 (K, L); Kedah, C Kedah, Jerai Forest Reserve catchment area, 24 Feb. 1960, *KEP (Kochummen)* 80933 (K, L, SING); Perak, Larut, dense jungle, low hilly locality, 300 – 500 ft, Feb. 1883, *King's collector* 3845 (K); Perak, near Ulu Kerling, dense old jungle, 500 – 800 ft, March 1886, *King's collector* 8626 (BM, BO, SING); Selangor, Sungai Buloh Forest Reserve, flat land, 31 March 1967, *Kochummen* 2271 (K, L, SING). SUMATRA. Jambi prov., Berbak reserve, Air Hitam Laut (river), study plot III, 1°20'S, 104°20'E, fresh water swamp forest, just above sea level, 1980, *Franken & Roos* TFB 1683 (L); Sumatera Utara, Aek Kanopan, Loendoet Concession, Koelaoe, old jungle, 18 April 1927, *Bartlett* 7336 (K, L, NY, US).

29. Antidesma leucopodum *Miq.*, Fl. Ned. Ind., Eerste Bijv.: 465 (1861). Type: Sumatra or., prov. Palembang, prope Muara-enim, *Teysmann* HB 3816 (U!, lectotype, here designated; BO!, isolectotype); s. loc., ex Herbario Miquel, s.n. (K!, isolectotype?). Original syntype: *Teysmann* HB 3661 (BO!, U!) — Note: Collection number not cited in protologue.

A. clementis Merr., Philipp. J. Sci., C, 9: 465 (1914), non Merr. 1917. Type: Philippines, Mindanao, Distr. Lanao, Camp Keithley, in thickets along streams, Jan. 1907, *M. S. Clemens* 884 (G!, lectotype, here designated). Paratype: same locality, March 1906, *M. S. Clemens* 339 (G!).

A. cauliflorum W. W. Sm., Notes Roy. Bot. Gard. Edinburgh 8: 316 (1915), non Merr. (1917). Type: Borneo, Sarawak, near Kuching, *Haviland* 730 [or 981/730] (K!, lectotype, here designated); BM!, CGE!, K!, L!, SAR!, isolectotypes). Original syntypes: Kuching, April 1914, *Native Collector* C 133 (E!, P!); Borneo, Sarawak, Limbang, *Haviland* 726 (K!); Baram, May 1894, *Haviland* 3263 (K!); July 1894, *Haviland & Hose* 3264 (K!, SAR!).

A. trunciflorum Merr., Bibl. Enum. Born. Pl.: 333 (1921), nom. nov. — *A. cauliflorum* Merr., J. Straits Branch Roy. Asiat. Soc. 76: 89 (1917), nom. illeg. (non W. W. Sm. 1915). Type: Sabah, Mt Kinabalu, Gurulau Spur and Kiau, Nov. 1915, *Clemens* 10790 (A! [photo of destroyed holotype ex PNH], lectotype, here designated). Paratype: same locality, Nov. 1915, *Clemens* 9944 (A!).

A. hirtellum Ridl., Bull Misc. Inform., Kew 1923 (10): 366 (1923). Type: Malaysia, Perak, Bujong Malacca, in woods on the hills, *Ridley* 9581 (K!, lectotype, here designated; SING, isolectotype).

A. caudatum Pax & K. Hoffm., Mitt. Inst. Allg. Bot. Hamburg 7: 223 (1931). Type: West-Borneo, am Unterlauf des Serawei, um 70 m, 19 Nov. 1924, *Winkler* 210 (HBG, holotype; BO!, isotype).

A. leucopodum Miq. var. *kinabaluense* Airy Shaw, Kew Bull. 28: 273 (1973), **synon. nov.** Type: Borneo, Sabah, *Chew & Corner* RSNB 4067 (K!, holotype; A!, L!, NY!, SAR!, US!, isotypes). — Note: Airy Shaw distinguished the new variety as follows: "In the glabrous ovary, glabrous tumid disk and large fruits this differs strikingly from the common form of this species...". In the type specimens of *A. leucopodum* var. *leucopodum* in U (which he may not have seen), however, the discs are glabrous and the fruits up to 6 mm long. Specimens with a densely setulous disc (tumidity of the disc and large fruits are not good diagnostic characters) may be more common than those with glabrous discs, but they do not correspond with the type of the species. Specimens with glabrous discs were collected in Sumatra, Borneo and the Malay Peninsula. There are also some intermediate collections with sparsely pilose discs (*Clemens & Clemens* 40878 (A, K, L) and 26530 (A, B, K, L, NY) from Sabah; *Ridley* 9581 (K, SING) from Peninsular Malaysia).

A. leucopodum Miq. var. *platyphyllum* Airy Shaw, Kew Bull. 28: 273 (1973), **synon. nov.** Type: Borneo, Sarawak, 4th Division, Baram, May 1894, *Haviland & Hose* 3263 (K!, holotype; SAR!, isotype).

Tree or treelet, rarely shrub, up to 30 m, clear bole up to 8 m, diameter up to 22 cm, usually straight, sometimes with more than one stem. Young twigs medium to whitish grey. Bark white, greenish, brown, pale yellowish, c. 0.5 mm thick, smooth, sometimes scaly, papery or brittle, easily detaching; inner bark grey-white, greenish, yellow, brown, pink or red, 0.5 – 3 mm thick, fibrous, soft; cambium white, brownish or red; wood hard and dense; sapwood grey, white, cream, yellow, reddish or brown; heartwood reddish. *Young twigs* terete to angular, pubescent to glabrous, brown. *Stipules* usually persistent, sometimes almost foliaceous, linear to lanceolate, slightly falcate, apically acute, 8 – 17 × 1 – 3(– 4) mm, pubescent. *Petioles* 4-angular to slightly channelled adaxially, 2 – 7(– 15) × 1.5 – 2.5(– 4) mm, pubescent, becoming glabrous when old. *Leaf blades* oblong to narrowly elliptic, apically caudate-mucronate, basally obtuse to rounded, more rarely acute or subcordate, (7 –)13 – 24(– 50) × (2.5 –)4 – 8(– 15) cm, (2.1 –)3.3(– 6.6) times as long as wide, eglandular, chartaceous to coriaceous, glabrous, or pilose only along the midvein adaxially, pilose all over abaxially, pubescent along the major veins, moderately shiny, usually with minute, regularly spaced white pustules adaxially, moderately shiny to dull abaxially, midvein impressed, secondary veins often raised adaxially, tertiary veins percurrent, close together (10 – 20 between every two secondary veins), prominent on both surfaces, drying olive-green to dark reddish brown adaxially, lighter abaxially. *Staminate inflorescences* (4 –)15 – 24 cm long, cauline, very rarely axillary, simple or branched once or twice at the base, solitary or up to 10 per fascicle, axes pubescent. *Bracts* deltoid, apically acute, 0.5 – 0.7 × 0.4 – 0.5 mm, pubescent. *Staminate flowers* 1.5 – 2 × 1.5 – 2 mm. *Pedicels* 0.2 – 1 mm long, not articulated, pubescent to pilose. *Calyx* c. 0.5 × 1 mm, sepals 3 – 5, almost free to fused for $^1/_2$ of their length, c. 0.7 mm wide, usually spreading, broadly deltoid to orbicular, apically acute to rounded, shortly pubescent to glabrous outside, pubescent or glabrous inside but with long hairs at least at the base, margin often fimbriate. *Disc* consisting of 3 – 4 alternistaminal lobes (lobes ± obconical, 0.2 – 0.3

× 0.2 – 0.3 mm), free or fused intrastaminally and partially enclosing the stamens, or completely fused and fully enclosing the stamens, constricted at the base, glabrous to shortly pubescent. *Stamens* 3 – 5, 1 – 1.5 mm long, exserted 1 – 1.5 mm from the calyx, anthers 0.3 – 0.4 × 0.4 – 0.5 mm. *Pistillode* absent or subulate, up to 0.2 × c. 0.1 mm, hardly exserted from the disc, pubescent. *Pistillate inflorescences* 5 – 7 cm long, cauline, very rarely axillary, simple or branched once or twice at the base, in fascicles of up to 13 inflorescences, axes pubescent. *Bracts* deltoid, 0.5 – 1 × 0.4 – 0.5 mm, pubescent. *Pistillate flowers* c. 1.5 × 1 mm. *Pedicels* 0 – 1 mm long, pubescent to glabrous. *Calyx* 0.3 – 0.5 × 0.7 – 1 mm, sepals 3 – 5, almost free to fused for up to $^{1}/_{2}$ of their length, c. 0.5 mm wide, broadly deltoid to orbicular, apically acute to rounded, shortly pubescent to glabrous outside, pubescent or glabrous inside with long hairs at the base, margin often fimbriate. *Disc* extending to about the same length as the (usually spreading) sepals, glabrous to densely pubescent. *Ovary* ellipsoid, pubescent to nearly glabrous, style terminal, often large and deeply 2- to 4-fid, stigmas 4 – 6. *Infructescences* 10 – 25(– 69) cm long. *Fruiting pedicels* 0.5 – 4(– 9) mm long, pilose. *Fruits* lenticular to ellipsoid, rarely ovoid and beaked, distinctly laterally compressed to terete, basally symmetrical or slightly asymmetrical, with a terminal to subterminal style, 4 – 8(– 10) × 3 – 6(– 7) mm, pilose to glabrous, white-pustulate, often pale yellow to light brown, areolate with a finely wrinkled epidermis when dry. Fig. 18A – C (p. 199).

DISTRIBUTION. Peninsular Thailand (Narathiwat and Yala provinces), Peninsular Malaysia, Sumatra, Borneo, Philippines (Basilan, Mindanao). Map 22.

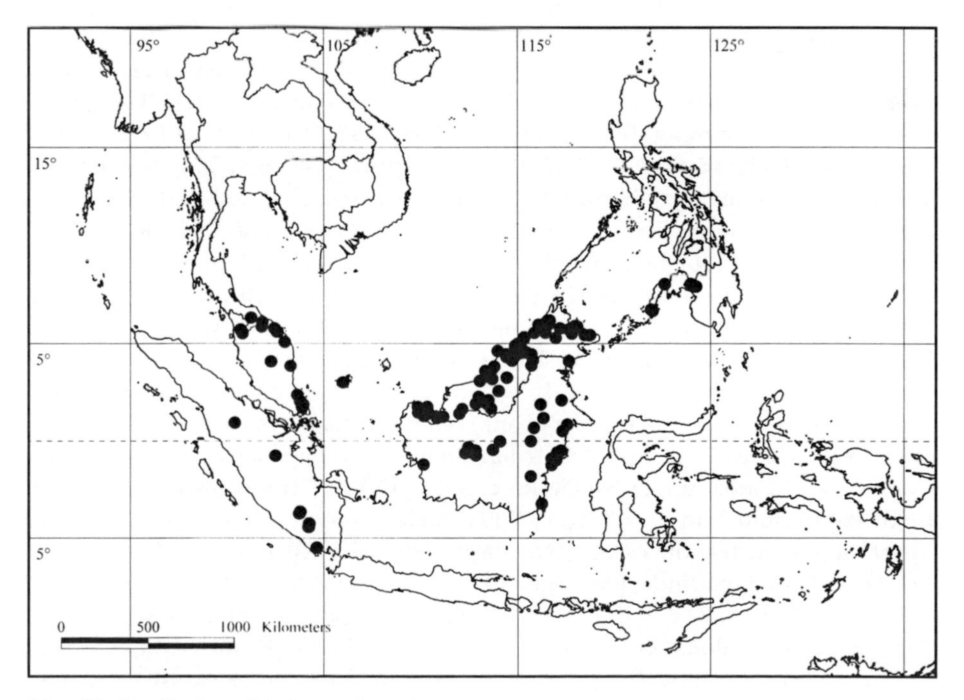

MAP 22. Distribution of *A. leucopodum*.

HABITAT. In lowland to lower montane (mixed) dipterocarp forest up to 50 m tall; in mossy forest; in riverine forest; usually in damp understorey, associated with *Castanopsis, Dipterocarpus, Durio, Shorea*; also along roadsides; in primary or secondary vegetation. On loam and clay over sandstone and shale, alluvial soil and limestone rocks. 0 – 2200 m altitude.

USES. Firewood.

CONSERVATION STATUS. Least Concern (LC). This is a widespread and common species. 212 specimens examined.

VERNACULAR NAMES. Thailand: mow polo; Sumatra: bernai paddie, bernai talang; Sabah: erayu gabok, jarupis (Malay), kilas primpiam, rain laki (Kedayan), rainbini (Kedayan), rayan (Kedayan); Sarawak: bajan (Iban), lembai temai (Kenyah); Brunei: engkuni (Iban), rambai tikus; Kalimantan: kayu mawar, ubah tatau, uhai puruk; Philippines: teksanbagia (Sub.).

ETYMOLOGY. The epithet refers to the colour of the bark (Greek, leucon = white colour; pous, podos = foot).

KEY CHARACTERS. Cauline, often fasciculate inflorescences; tertiary leaf veins parallel, close together; abundant lenticular fruits with (sub)terminal styles.

SIMILAR SPECIES. *A. helferi*, see there. — The vegetatively extremely similar *A. polystylum* (Fig. 18) has a denser indumentum of the inflorescences and abaxial leaf surfaces, longer and more elongate fruits, shorter, more robust infructescences and more numerous stigmas. — *A. tomentosum* has axillary inflorescences, mango-shaped fruits, stamens shorter exserted from the calyx, and leaves drying reddish.

SELECTED SPECIMENS. THAILAND. Peninsular, Narathiwat prov., Chatvarin Falls at Sungai Padi, 6°04'N, 101°52'E, evergreen forest along stream, 18 – 19 Oct. 1970, *Charoenphol, Larsen & Warncke* 3982 (K, L). PENINSULAR MALAYSIA. Johore, Kuala Sedili New road [same collection as *Sinclair* 10678], primary forest, 0 – 250 m, 28 Nov. 1961, *Chew* 288 (G, K, L, NY, SING); Terengganu, Jerteh, Trengganu Hutam Simpan, Gunong Tebu compt. 90, 350 ft, 13 Oct. 1971, *FRI (Zainuddin Sohadi)* 17974 (K, L, SING). SUMATRA. Riau prov., Tigapulu Mts, 5 km W of Talanglakat on Rengat-Jambi road, Bukit Karampal area, near Sesirih R. source, on ridge near river source, 100 m, 5 Dec. 1988, *Burley, Tukirin et al.* 1887 (K, KEP, L). ANAMBAS & NATUNA ISL. Anambas Isl., Gunung Adong, N Temaja, c. 800 ft, 13 April 1928, *SF (Henderson)* 20373 (K, L, NY). BORNEO. Brunei, Belait distr., between Mendaram and Teraja, ridgetop forest, 0 m, 1 May 1988, *Wong* 79 (A, AAU, K, L, SAR); Sabah, Mt Kinabalu, Dallas, forest trails, 3000 ft, 4 – 6 Sept. 1931, *Clemens & Clemens* 26530 (A, B, K, L, NY); Sarawak, Kuching div., Gunong Undan, 17th mile Bau/Lundu road, stream bank, soil sandy loam, 50 m, 30 April 1983, *S (Yii)* 45941 (K, KEP, L, SAR); C Kalimantan, Bukit Raya and upper Katingan (Mendawai) R. area, upper Samba R., c. 5 km NNW of Tumbang Tosah descent Bukit Raya Camp 3 to 1, 0°30'S, 112°50'E, well-drained primary lower montane forest, SE side, c. 600 m, 7 Dec. 1982, *Mogea* 4005 (K, KEP, L, US); E Kalimantan, Loa Haur, W of Samarinda, low ridges, loam soil on sandstone, 40 m, 15 May 1952, *Kostermans* 6952 (A, B, BO, K, L, NY); W Kalimantan, Sintang, Bukit Baka National Park, S of camp along Bukit Asing ridge, 0°38'S, 112°17'E, mixed dipterocarp forest, associates include *Dipterocarpus*, Myrtaceae, Meliaceae, Anacardiaceae, red clay soil, 525 m, 24 Oct. 1993, *Church* 342 (L). PHILIPPINES. Basilan Island, s. loc.,

Aug. 1912, *BS (Reillo)* 15439 (A, L, US); Mindanao, Zamboanga distr., Malangas, Oct. – Nov. 1919, *BS (Ramos & Edaño)* 37337 (A, BM, BO, K, US).

30. Antidesma macgregorii *C. B. Rob.*, Philipp. J. Sci., C, 6: 207 (1911). Type: Philippines, Polillo, *BS (McGregor)* 10280 (US!, lectotype, here designated; K!, isolectotype). Paratypes: Luzon, Province of Cagayan, Claveria, *BS (Ramos)* 7397 (L!, NY!, US!); East coast, locality with no known name, *BS (McGregor)* 10583 (US!). — Note: Airy Shaw first synonymised this species with *A. helferi* (1972a: 354), but resurrected it later (1983: 6).

Shrub or tree, up to 7 m, diameter up to 13 cm. *Young twigs* terete, glabrous, whitish grey, more rarely dark brown and puberulent to begin with. *Stipules* early caducous, linear, apically acute, 2 – 5(– 10) × 0.5 – 1(– 1.5) mm, pilose. *Petioles* strongly channelled to flat adaxially, (2 –)4 – 11 × 1.2 – 2 mm, puberulent, soon becoming glabrous, rugose and whitish grey. *Leaf blades* elliptic, more rarely slightly obovate or ovate, apically acuminate-mucronate, basally acute, more rarely obtuse, margin usually concave, (5 –)9 – 13(– 18.5) × (2 –)3.5 – 5.5(– 9.5) cm, (1.5 –)2.2 – 2.7(– 3.2) times as long as wide, eglandular, coriaceous to chartaceous, glabrous, dull, more rarely shiny on both surfaces, midvein flat adaxially, tertiary veins reticulate, widely spaced (3 – 5 between every two secondary veins), drying dark reddish brown to purplish grey adaxially, lighter reddish brown abaxially, more rarely light greyish brown. *Staminate inflorescences* 5 – 9 cm long, axillary to almost cauline, simple or consisting of up to 3 branches, axes ferrugineous-pilose. *Bracts* lanceolate to spathulate, 0.3 – 0.7 × 0.2 – 0.4 mm, ferrugineous-pilose. *Staminate flowers* 1.5 – 2 × 1.5 – 2 mm. *Pedicels* 0 – 0.2 mm long, not articulated, glabrous. *Calyx* 0.5 – 1 × 1 – 1.2 mm, cupular, sepals 3 – 4, fused for $^3/_4$ or more of their length, apically obtuse to rounded, glabrous to sparsely pilose outside, glabrous inside, margin entire, slightly fimbriate. *Disc* cushion-shaped, enclosing the bases of the filaments, very small, much shorter than the sepals, glabrous. *Stamens* 3 – 4, 1.2 – 1.5 mm long, exserted 1 – 1.2 mm from the calyx, anthers 0.3 – 0.4 × 0.4 – 0.6 mm. *Pistillode* absent. *Pistillate inflorescences* 2 – 4 cm long, axillary, simple, more rarely branched once or twice at the base or in fascicles of 2 – 3 inflorescences, axes ferrugineous-puberulent. *Bracts* broadly ovate to lanceolate, 0.2 – 0.7 × 0.3 – 0.5 mm, ferrugineous-pubescent. *Pistillate flowers* 1.5 – 2 × c. 1 mm. *Pedicels* 0 – 0.5 mm long, glabrous. *Calyx* 0.5 – 1 × c. 1 mm, cupular to urceolate (bowl-shaped in fruit), sepals 3 – 5, fused for $^2/_3$ – $^3/_4$ of their length, apically acute to rounded, pilose to glabrous outside, glabrous inside but sometimes with long hairs at the base, margin entire, sparsely ciliate. *Disc* shorter than the sepals, glabrous. *Ovary* ellipsoid, glabrous to pubescent, style terminal, not distinct, thick, stigmas 5 – 8(– 10). *Infructescences* 3 – 14 cm long, robust. *Fruiting pedicels* 0 – 1.5 mm long, glabrous. *Fruits* ellipsoid, laterally compressed, basally symmetrical, with a terminal to slightly subterminal style, 7 – 10 × 5 – 7 mm, glabrous to puberulent, usually white-pustulate, areolate or fleshy when dry.

DISTRIBUTION. Philippines: Catanduanes, Leyte, Luzon, Negros, Polillo, Samar, Sibuyan. The specimen *Curran* 3477(A) from Masamba, southern Sulawesi, is the only collection of *A. macgregorii* from outside the Philippines. It matches the Philippine specimens but bears only very young fruits and is therefore here regarded as doubtful. Map 21.

HABITAT. In dipterocarp forest; along forest edges; primary or disturbed vegetation. On brown clay and ultrabasic soil. 50 – 900 m altitude.

USES. Firewood.

CONSERVATION STATUS. Near Threatened (NT). The species is widespread in the Philippines but could become a concern as vegetation cover continues to decline. 37 specimens examined.

VERNACULAR NAMES. Luzon: buhlong.

ETYMOLOGY. The epithet refers to the collector of the type specimen.

KEY CHARACTERS. Plant ± glabrous; twigs and petioles whitish; leaves drying reddish; sepals fused; staminate disc much shorter than the calyx; pistillodes absent; stigmas terminal.

SIMILAR SPECIES. *A. brachybotrys, A. bunius, A. helferi*, see there. — *A. minus* has longer staminate pedicels, free sepals, a pistillode and consistently glabrous ovaries and fruits.

SELECTED SPECIMENS. PHILIPPINES. Catanduanes, s. loc., 14 Nov. – 11 Dec. 1917, *BS (Ramos)* 30542 (A, NY, US); Leyte, Mt Abucayan, Feb. 1923, *BS (Edaño)* 41841 (A, K, US); Luzon, Ilocos Norte prov., Bangui to Claveria, Aug. 1918, *BS (Ramos)* 33023 (GH, K, US); Luzon, Camarines Norte prov., Paracale, Nov. – Dec. 1918, *BS (Ramos & Edaño)* 33638 (A, K, US); Luzon, Cagayan prov., s. loc., Jan. 1912, *FB (Curran)* 19575 (BM, K, L, US); Luzon, Isabela prov., Palanan, Diguyo, 16°55'N, 122°30'E, low regrowth, canopy completely logged, ultrabasic, *Ridsdale, Dejan & Baquiran* ISU 270 (K, L); Luzon, Cagayan prov., Abulug R., Jan. 1912, *Weber* 1577 (A, E, G, K, US); Samar, Catubig R., Feb. – March 1916, *BS (Ramos)* 24375 (K, L, US); Samar, Mt Sarawag, along forest edge, 150 m, 3 Dec. 1951, *PNH (Edaño)* 15250 (A, K, L); Sibuyan Isl., Romblon prov., Magdiwang, Mt Guiting-Guiting, along the trail, May 1972, *PNH (Reynoso & Espiritu)* 118752 (K).

31. Antidesma microcarpum *Elmer*, Leafl. Philipp. Bot. 2: 487 (1908). Type: Philippines, Negros, Prov. Negros Oriental, Cuernos Mts, Dumaguete, *Elmer* 9668 (E!, lectotype, here designated; G!, isolectotype).

A. santosii Merr., Philipp. J. Sci. 16: 550 (1920). Type: Philippines, Luzon, Laguna Prov., Mount Banahao, 28 Aug. 1916, on slopes along small streams, alt. ab. 600 m, *FB (Santos)* 26300 (US!, lectotype, here designated; K!, isolectotype). — Note: The type of this name and four other fruiting collections from Luzon and Panay (*PNH (Banlugan et al.)* 72730 (A, K, L), *Elmer* 18105 (A, BM, G, K, L, NY, U, US), *BS (Edaño & Martelino)* 35715 (A, BM, K, L, US), *PNH (Sulit)* 6911 (K, L, US)) correspond in all vegetative and calyx characters to *A. microcarpum*, whereas the fruits are 3 – 4 mm long, laterally compressed, the pistillate pedicels up to 0.8 mm and fruiting pedicels 1 – 1.5 mm long, as in *A. celebicum*. The name is included here under *A. microcarpum* because of the tightly urceolate calyx, the leaf shape and texture and the light colour of the young twigs.

A. maesoides Pax & K. Hoffm. in Engl., Pflanzenr. 81: 164 (1922). – Type: Philippinen, Leyte, Dagami, *Ramos* 15251 (A!, lectotype, here designated). Original syntypes: Mindanao, Davao, Waldrand, *Warburg* 14355; Lake Lanao, *M. S. Clemens* s.n. (G!).

A. frutiferum Elmer, Leafl. Philipp. Bot. 10: 3731 (1939), nom. inval. as diagnosis not in Latin. Based on: Philippines, Luzon, Prov. Sorsogon, Irosin (Mt Bulusan), damp rocky ground of humid woods, alt. 2500 ft, Oct. 1915, *Elmer* 14508 (BM!, BO!, G!, GH!, K!, L!, NY!, U!, US!).

Tree, rarely shrub, up to 11 m, diameter up to 20 cm. Bark grey, smooth; wood moderately soft to hard. *Young twigs* terete, spreading-pilose but soon becoming glabrous, first brown, whitish grey when older. *Stipules* early-caducous, linear, 1.5 – 3 × 0.5 – 1 mm, pubescent. *Petioles* narrowly channelled adaxially, basally and distally pulvinate for 2 – 4 mm, 6 – 18(– 27) × 1 – 1.5 mm, densely pilose to glabrous. *Leaf blades* ovate to oblong or elliptic, apically acuminate- to caudate-mucronate, sometimes with a rounded apiculum, basally rounded to obtuse, often slightly folded when dry, (5 –)10 – 15(– 20) × (2.5 –)4 – 7(– 10) cm, (1.5 –)2 – 2.4(– 3.3) times as long as wide, eglandular, chartaceous to subcoriaceous, glabrous, or slightly pilose only along the major veins on both surfaces, shiny or dull on both surfaces, midvein impressed to flat adaxially, tertiary veins percurrent to weakly percurrent, drying olive-green, often lighter abaxially. *Staminate inflorescences* 4 – 14 cm long, axillary, lax, consisting of 2 – 10 branches, sometimes fascicles of 2 inflorescences, axes pilose. *Bracts* lanceolate to ovate, 0.6 – 0.8 × 0.4 – 0.5 mm, glabrous to pilose, margin ciliate. *Staminate flowers* 1.5 – 2 × 1 – 1.5 mm. *Pedicel* absent. *Calyx* 0.5 – 1 × c. 1 mm, cupular to bowl-shaped, sepals 4, fused for c. $^3/_4$ of their length, pointing inwards, apically rounded, glabrous on both sides, margin slightly erose, ciliate. *Disc* extrastaminal-annular (but appearing cushion-shaped), extending to the same length as the sepals, constricted at the base, (3 –)4-lobed, the lobes filling the space between the filaments and the pistillode, ferrugineous-tomentose. *Stamens* (3 –)4, c. 1.5 mm long, exserted c. 1 mm from the calyx, anthers c. 0.2 × 0.3 mm. *Pistillode* cylindrical, 0.4 – 0.5 × 0.2 – 0.3 mm, exserted from the disc, white pilose. *Pistillate inflorescences* 2 – 10 cm long, axillary, branched regularly, consisting of 2 – 14 branches, sometimes in fascicles of 2 inflorescences, axes pilose to pubescent. *Bracts* lanceolate to ovate, apically acute to rounded, 0.3 – 0.7 × 0.3 – 0.4 mm, pilose. *Pistillate flowers* 1 – 2 × 0.7 – 1 mm. *Pedicels* 0.2 – 0.5(– 0.8) mm long, pilose. *Calyx* 0.7 – 1 × 0.7 – 1 mm, urceolate, sepals 4, fused for ($^1/_2$ –)$^3/_4$ – $^4/_5$ of their length, deltoid to oblong, apically acute to rounded, glabrous, more rarely pilose outside, glabrous inside, margin erose, fimbriate. *Disc* shorter than the sepals but indumentum usually exserted from the sepals, ferrugineous-pubescent at the margin, otherwise glabrous, hairs as long as or longer than the disc. *Ovary* ellipsoid, glabrous to sparsely pilose, style terminal, stigmas 4 – 5. *Infructescences* 2 – 10 cm long. *Fruiting pedicels* 0.2 – 0.5(– 1.5) mm long, glabrous. *Fruits* ellipsoid to nearly globose, terete (but see note under the synonym *A. santosii*), basally symmetrical, with a terminal style, 2 – 3(– 4) × 1 – 1.5(– 2) mm including the calyx, glabrous to very sparsely pilose, not white-pustulate, usually wrinkled (neither areolate nor reticulate), rarely areolate when dry, calyx not splitting, staying intact and usually as narrow as at anthesis.

DISTRIBUTION. Philippines: Biliran, Bohol, Leyte, Luzon, Mindanao, Negros, Panay, Samar. Map 23.

HABITAT. Mainly in secondary forest. 60 – 1050 m altitude.

USES. Used as firewood, and also for fish poisoning (*PNH (Anonuevo)* 13568).

CONSERVATION STATUS. Near threatened (NT). Like *A. macgregorii*, this species is not threatened at present but may become so with continuing deforestation in the Philippines. 26 pistillate specimens examined.

VERNACULAR NAMES. baruruan, bignai-pugo (Tag.), buro-bignai (Bic.), putukan (If.); tagobinlod.

ETYMOLOGY. The epithet refers to the size of the fruits (Greek, micro- = small; carpos = fruit).

KEY CHARACTERS. Very small, glabrous, terete fruits with a terminal style; fused sepals; hairy disc; branched inflorescences; long petioles; whitish twigs.

SIMILAR SPECIES. *A. catanduanense, A. celebicum, A. curranii*, see there. — Sterile and staminate specimens are virtually indistinguishable from the sympatric *A. pleuricum* in which the twigs are a slightly lighter colour; pistillate specimens are distinguished by terminal styles and smaller, terete fruits. Because of the similarity of the staminate flowers, all staminate collections are here listed either as *A.* cf. *microcarpum* or *A.* cf. *pleuricum.*

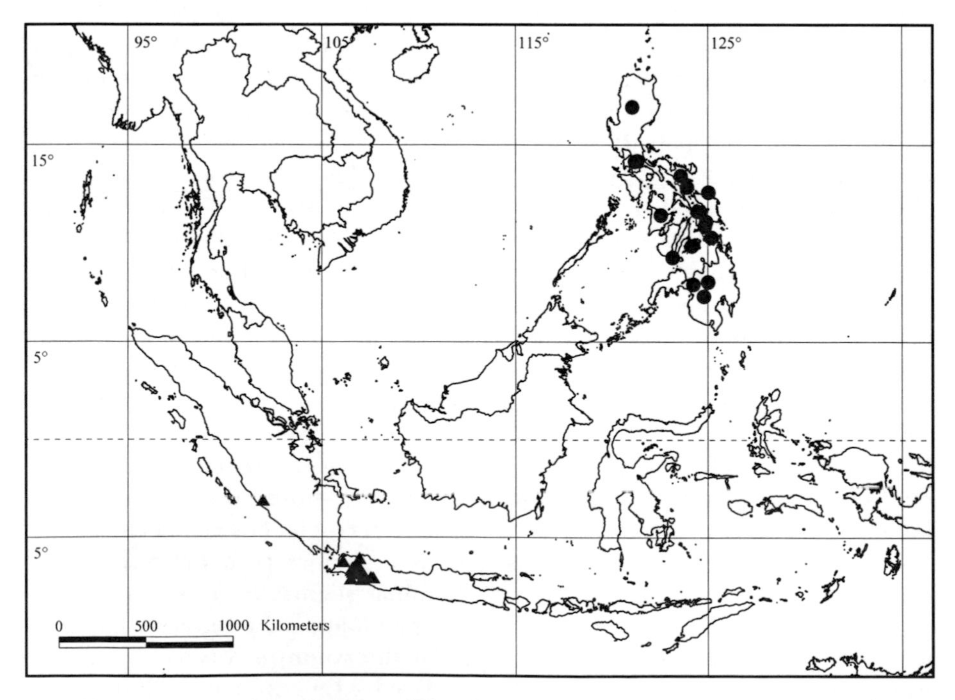

MAP 23. Distribution of *A. microcarpum* (●) and *A. minus* (▲).

SELECTED SPECIMENS. PHILIPPINES. Biliran Island, Mt Suiro (N slope), very steep slope, wet soil, 680 m, April – May 1954, *PNH (Sulit)* 21591 (K, L); Bohol, s. loc., Aug. – Oct. 1923, *BS (Ramos)* 43059 (A, K); Leyte, Cabalian, Dec. 1922, *BS (Ramos)* 41519 (A, K, L, US); Luzon, Laguna prov., Los Banos (Mt Maquiling), June – July 1917, *Elmer* 18105 (A, BM, G, K, L, NY, U, US); Luzon, Albay prov., Mt Malinao, forested slopes, 700 m, 4 Feb. 1956, *PNH (Edaño)* 34472 (A, BM, K, L); Luzon, Mountain prov., Banaue, Lugu, secondary forest, 20 Aug. 1961, *PNH (Banlugan et al.)* 72730 (A, K, L); Luzon, Sorsogon prov., s. loc., June – Aug. 1915, *Ramos* 23524 (A, BM, K, L, NY, US); Panay, Capiz prov., Mt Bulilao, June 1919, *BS (Edaño & Martelino)* 35715 (A, BM, K, L, US); Samar, Catubig R./Dumaguete, Feb. – March 1916, *BS (Ramos)* 24220 (A, BM, K, NY, US).

32. Antidesma minus *Blume*, Bijdr. Fl. Ned. Ind.: 1123 (1826 – 27). Type: Java, *Blume* s.n. (L! herb. no. 903154-263, lectotype, here designated).

A. zollingeri Müll. Arg., in sched., nom. nud., *fide* Pax & K. Hoffm. in Engl., Pflanzenr. 81: 132 (1922).

Shrub or tree, up to 10 m, diameter up to 16 cm. *Young twigs* terete, long spreading-pubescent but soon becoming glabrous, whitish grey. *Stipules* caducous, subulate, 2 – 8 × 0.5 – 1 mm, pilose to pubescent. *Petioles* channelled adaxially, 2 – 5(– 10) × c. 1 mm, pilose to pubescent (especially adaxially), becoming glabrous, rugose and white when older. *Leaf blades* narrowly elliptic, apically long acuminate-mucronate, basally acute, (6.5 –)8 – 13(– 15.5) × (1.5 –)2 – 4(– 5) cm, (2.7 –)3.4(– 4.2) times as long as wide, eglandular, chartaceous, glabrous, or pilose only along the midvein, moderately shiny on both surfaces, midvein impressed adaxially, tertiary veins reticulate, finer veins distinctly prominent on both surfaces in dry material, dense intersecondary and perpendicular tertiary veins conspicuous near the midvein, drying dark reddish brown adaxially, lighter abaxially. *Staminate inflorescences* 4 – 7 cm long, axillary, simple, slender, axes glabrous to pubescent. *Bracts* narrowly deltoid, apically acute, 0.4 – 0.7 × 0.2 – 0.3 mm, pilose, margin fimbriate. *Staminate flowers* c. 1.5 × 1.5 – 2 mm. *Pedicels* 0.2 – 1 mm long, not articulated, glabrous to pilose. *Calyx* 0.5 – 1 mm long, sepals 4, free, 0.3 – 0.5 mm wide, ± reflexed, deltoid, apically acute, glabrous outside, glabrous inside but with dense hairs at the base, margin entire, fimbriate. *Disc* cushion-shaped, hemispherical to almost mushroom-shaped, fully enclosing the bases of the filaments and pistillode, glabrous. *Stamens* 4, 0.8 – 1.2 mm long, exserted c. 1.2 mm from the calyx, anthers 0.3 – 0.5 × 0.4 – 0.5 mm. *Pistillode* absent or subulate, up to 0.2 × c. 0.1 mm, hardly exserted from the disc, glabrous. *Pistillate inflorescences* 3 – 4 cm long, axillary, simple, axes glabrous to pilose. *Bracts* deltoid, apically acute, 0.5 – 1 × 0.2 – 0.3 mm, pubescent. *Pistillate flowers* c. 1.5 × 1 mm. *Pedicels* 0 – 1 mm long, glabrous to pilose. *Calyx* c. 0.5 mm long, sepals 4(– 5), free, 0.3 – 0.5 mm wide, deltoid, apically acute, glabrous outside, glabrous inside with dense hairs at the base, margin entire, fimbriate. *Disc* extending to the same length as the sepals, conspicuous and fleshy, glabrous. *Ovary* ellipsoid to cylindrical, glabrous, style terminal, not distinct, thick, stigmas 3. *Infructescences* 7 – 9 cm long. *Fruiting pedicels* 0 – 1 mm long, glabrous to pilose. *Fruits* ellipsoid to ovoid or obovoid, sometimes slightly beaked, distinctly laterally compressed, basally asymmetrical or symmetrical, with a subterminal to terminal style, 7 – 13 × 5 – 8 mm, glabrous, sometimes slightly white-pustulate, reticulate when dry.

DISTRIBUTION. West Java, and one specimen from south-western Sumatra, Benkulu province. Map 23.

HABITAT. In (primary) rain forests. On red volcanic sand. 300 – 1350 m altitude.

CONSERVATION STATUS. Near Threatened (NT). 63 specimens of this species have been examined, but nearly all are from western Java where human population density is very high and habitat destruction intense.

VERNACULAR NAMES. huhunian, huni monjet, ki mimis, ki seueur (seueur = many, refers to the many edible fruits), kipeen, onjam, walin kecil (all Sundanese).

ETYMOLOGY. Latin, minus = smaller.

KEY CHARACTERS. Young twigs whitish; leaves drying reddish; prefers higher altitudes.

SIMILAR SPECIES. *A. heterophyllum* and *A. macgregorii*, see there. — *A. montanum* has smaller and terete fruits, brown twigs and petioles, and sepals fused to about halfway.

SELECTED SPECIMENS. SUMATRA. Bengkulu, Benkoelen res., Mt Sago, N slope, montane forest, 900 – 1200 m, 28 July 1955, *Meijer* 3787 (BO, L). JAVA. Batavia res., Buitenzorg afd., Gunung Salak bij Kamp Bobodjong, oerwoud, 700 m, 22 Nov. 1896, *Koorders* 24197 (A, BO, L); 12 Sept. 1896, *Koorders* 24220 (BO, L); Bantam prov., Gunung Karang bij ?Galoesoer, 30 May 1912, *Koorders* 41566 (BO, L); Preanger, *Ploem* s.n. (L); W Java, Kiara Dua, S of Sukabumi, remnant forest, 800 m, 6 July 1986, *Van Balgooy & Van Setten* 5680 (L, U); Batavia prov., Gunung Liiang, oerwoud, c. 600 m, 30 Oct. 1928, *Van Steenis* 2414 (BO, L, U); Preanger res., Gunung Beser, Tjidadap, Tjibeber, bosch, 1000 m, 9 Feb. 1918, *Winckel* 61 (BO, L, U); Preanger res., Gunung Toegoe (Tjimonteh), Tjidadap, Tjibeber, bosch, 1000 m, 27 May 1918, *Winckel* 175 (BO, L, U).

33. Antidesma montanum *Blume,* Bijdr. Fl. Ned. Ind.: 1124 (1826 – 27), non Thwaites 1861. Type: Java, in montosis, *Blume* s.n. (L! herb. no. 903154-278, lectotype, here designated; L!, isolectotype). — Note: Authentic material also in BO!, NY!, P! and US!

Shrub or tree (reported to be a climber on several herbarium labels), up to 20 m (*fide Noermawati* 16: 45 m), clear bole up to 7 m, diameter up to 40 cm, often branching from base or below the middle, sometimes with more than one stem, crown round, branches spreading, ultimate branches often drooping. Twigs brown, grey or green. Bark brown, grey, green or white, 0.5 – 3 mm thick, smooth, more rarely rough, sometimes vertically fissured or scaly, fibrous, soft; inner bark white, greenish-white, yellow, pink, orange, red or brown, 1 – 5 mm thick, fibrous, soft; cambium white, yellow, brown, red or pink; wood hard, somewhat brittle, finely grained, odourless and tasteless, brown, red, orange, yellow or white. *Young twigs* terete, very slightly pilose to densely pubescent, brown. *Stipules* often caducous, foliaceous and ovate only in var. *wallichii*, otherwise linear to subulate, 2 – 7(– 15) × 0.5 – 1.5(– 7) mm, pilose to pubescent. *Petioles* channelled adaxially, (0 –)2 – 6(– 15) × (0.7 –)1 – 2 mm, pilose to pubescent, rarely glabrous. *Leaf blades* elliptic to oblong, more rarely slightly ovate or obovate, or lanceolate to linear, apically acuminate-mucronate, basally acute to rounded, rarely truncate, (1.5 –)9 – 15(– 30) × (0.4 –)3 – 6(– 12) cm, (2 –)2.5 – 3.5(– 10) times as long as wide, eglandular, membranaceous to chartaceous, more rarely coriaceous, glabrous except along the major veins, rarely pilose abaxially all over, moderately shiny on both surfaces, major veins impressed, rarely flat adaxially, tertiary veins weakly percurrent (reticulate in rheophytes), widely spaced, rather oblique, quarternary and finer veins hardly prominent on either leaf surface when dry, intersecondary veins not conspicuous, drying olive-green, sometimes lighter abaxially. *Staminate inflorescences* (1 –)3 – 13 cm long, axillary, sometimes aggregated at the end of the branch, simple to consisting of up to 10(– 40) branches, axes pilose to pubescent. *Bracts* deltoid or linear, apically acute, 0.3 – 1.2 × 0.2 – 0.7 mm, pilose to pubescent, margin sometimes glandular-fimbriate. *Staminate flowers* 1 – 2.5 × 1 – 1.5 mm. *Pedicels* 0 – 1.5(– 2) mm long, not articulated, glabrous to pilose. *Calyx* 0.5 – 0.8 × 1 – 1.5 mm, cupular to bowl-shaped, sepals 3 – 4(– 5), free to fused for c. $^1/_2$ of their length, apically rounded to obtuse, more rarely acute, pilose to glabrous outside, glabrous inside but often with some long hairs at the base, margin erose to entire, glandular-fimbriate to lacerate. *Disc* cushion-shaped, fully or partially enclosing the bases of the filaments and pistillode, shorter than the sepals, glabrous, very rarely with some short hairs. *Stamens* 3 – 6, 1 – 2 mm long, exserted (0.5 –)1 – 1.5 mm from the calyx, anthers 0.2 – 0.5 × 0.2 – 0.5 mm. *Pistillode* variable, hemispherical, globose, clavate and crateriform apically, or subulate, 0.1 – 1 × 0.1 – 0.3 mm, shorter than to exserted from the sepals, glabrous, more rarely slightly pilose. *Pistillate inflorescences* (2 –)4 – 10 cm long, axillary, sometimes aggregated at the end of the branch, simple to consisting of up to 13 branches, axes glabrous to pubescent. *Bracts* deltoid to linear, apically acute, 0.3 – 1.5 × 0.2 – 0.7 mm, glabrous to pubescent, margin sometimes glandular-fimbriate. *Pistillate flowers* 1 – 2 × 1 – 1.5 mm. *Pedicels* 0.2 – 1.5 mm long, glabrous to pubescent. *Calyx* 0.4 – 1 × 0.5 – 1.5 mm, cupular, sepals 3 – 5(– 6), free to fused for up to $^1/_2$ of their length, thin, apically rounded to obtuse, more rarely acute to acuminate, pilose or glabrous, rarely pubescent outside, glabrous inside but often with some long hairs at the base, rarely pilose, margin glandular-fimbriate, lacerate, erose or entire. *Disc* much shorter than the sepals (especially in fruit), glabrous, very rarely pilose. *Ovary* ovoid, glabrous, style terminal or slightly subterminal, stigmas 3 – 6. *Infructescences* (2.5 –)6 – 20 cm long. *Fruiting pedicels* 1 – 4 mm long, glabrous to pilose, rarely pubescent. *Fruits* ellipsoid to globose or ovoid, terete to slightly laterally or dorsiventrally compressed, basally symmetrical, with a terminal, rarely slightly subterminal style, (3 –)4 – 6(– 8) × 2.5 – 4(– 6) mm, glabrous, rarely pilose, mostly white-pustulate, areolate when dry.

NOTES. This is the most common and most variable species of the genus. It is treated here in a broad sense, i.e. including forms such as the slender, almost glabrous *A. leptocladum* as well as the robust, big-leaved *A. phanerophlebium*.

Another extreme is the small-leaved, pubescent form *A. pentandrum*, retained as a separate species by Airy Shaw (1983: 6). It is found in the Ryukyu Islands (Japan), Taiwan, Hainan and in the northern part of the Philippines. South of Mindoro it intergrades imperceptibly into the typical form of *A. montanum*. The characters used to distinguish both forms in the literature (leaves smaller, acuminate with rounded-mucronate apiculum, drying yellowish green, bracts long and linear, twigs, petioles and axes of inflorescences densely pubescent) are neither consistent nor strong enough to justify specific recognition. The leaves and, in some specimens, the fruits can resemble *A. japonicum*.

A. japonicum and, at the southern border of the distribution area, *A. heterophyllum*, are accepted here as separate species although there are several transitional specimens.

33a. var. montanum

A. oblongifolium Blume, Bijdr. Fl. Ned. Ind.: 1125 (1826 – 27) non Boerl. & Koord. 1910. — *A. oblongifolium* Blume var. *genuinum* Müll. Arg. in DC., Prodr. 15(2): 264 (1866), nom. inval. Type: Java, in montanis, *Blume* s.n. (L! herb. no. 903154-312, lectotype, here designated).

Cansjera pentandra Blanco, Fl. Filip.: 73 (1837), "Cansiera"; Fl. Filip., ed. 2: 53 (1845), "Cansiera". — *A. pentandrum* (Blanco) Merr., Philipp. J. Sci., C, 9: 462 (1914), Sp. Blancoan.: 219 (1918), **synon. nov.** — *A. pentandrum* (Blanco) Merr. var. *genuinum* (Müll. Arg.) Pax & K. Hoffm. in Engl., Pflanzenr. 81: 125 (1922), nom. inval. Type: Philippines, Luzon, Rizal prov., Pasay, Sept. 1914 [but "Bulacan prov., Angat, Sept. 1913" in Merrill, Sp. Blancoan.: 219 (1918)], *Merrill* Species Blancoanae 31 (A!, neotype, here designated; GH!, K!, L!, NY!, P!, US!, isoneotypes). — Note: All original material is likely to be lost (nothing located in G-DC, K, K-W, MA, P). Merrill published his new combination without citing any material that could have been seen by Blanco. Later (1918: 219) he selected two "illustrative specimens" to represent Blanco's name. As "illustrative specimen" is not equivalent to the term "type" (Greuter, pers. comm.), this was not effective neotypification according to Art. 7.11. (Greuter *et al.* 2000: 9). Merrill's "illustrative specimens" are not accepted as neotypes here for any of Blanco's names (see list of incompletely known species) apart from *A. pentandrum*. The reason for this exception is that, unlike the other names, the name *A. pentandrum* has been widely applied in many herbaria, and therefore a neotypification is necessary.

A. barbatum C. Presl, Epimel. Bot.: 233 (1849). — *A. rostratum* Tul. var. *barbatum* (C. Presl) Müll. Arg. in DC., Prodr. 15(2): 257 (1866). — *A. pentandrum* (Blanco) Merr. var. *barbatum* (C. Presl) Merr., Philipp. J. Sci., C, 9: 463 (1914). Type: Philippines, Luzon, prov. Ilocos borealis, *Cuming* 1246 (PRC!, lectotype, here designated; CGE!, E!, G!, K!, L!, MEL!, P!, PRC!, isolectotypes).

A. salicifolium C. Presl, Epimel. Bot.: 233 (1849) non Miq. 1861. Type: Philippines, Luzon, prov. Cayagan, *Cuming* 1316 (PRC!, holotype; CGE!, E!, G!, FHO!, K!, L!, MEL!, P!, isotypes).

A. leptocladum Tul., Ann. Sci. Nat. Bot., Sér. 3: 199 (1851); Airy Shaw, Kew Bull. 37: 6 (1982). — *A. leptocladum* Tul. var. *genuinum* Müll. Arg. in DC., Prodr. 15(2): 253 (1866), nom. inval. Type: Philippines, Luzon, Prov. Batangas, *Cuming* 1513 (CGE!, lectotype, here designated; A!, CGE!, E!, G!, G-DC [microfiche], L!, K!, NY!, P!, PRC!, isolectotypes). — Note: *A. leptocladum* was retained as a species by Merrill (1923: 414) and Airy Shaw (1982: 5), but reduced to *A. montanum* by Pax & Hoffmann (1922: 158).

A. nitidum Tul., Ann. Sci. Nat. Bot., Sér. 3: 193 (1851). — *A. leptocladum* Tul. var. *nitidum* (Tul.) Müll. Arg. in DC., Prodr. 15(2): 253 (1866). Type: Philippines, Luzon, circa Manillam, *Cuming* 1511 (P!, lectotype, here designated; A!, CGE!, E!, G!, G-DC [microfiche], K!, L!, NY!, P!, PRC!, isolectotypes).

A. pubescens Roxb. var. *menasu* Tul., Ann. Sci. Nat. Bot., Sér. 3: 215 (1851). — *A. menasu* (Tul.) Müll. Arg. in DC., Prodr. 15(2): 257 (1866), not Kurz, Forest Fl. Burma 2: 360 (1877), as cited in Index Kewensis. Type: India, prope urb. Mangalor., *Hohenacker* Pl. Ind. Or. 104 (P!, lectotype, here designated; A!, FHO!, G!, G-DC [microfiche], K!, L!, MEL!, TCD!, isolectotypes). Original syntype: India, prope urb. Mangalor., *Hohenacker* Pl. Ind. Or. 459a (as *A. bunius*) (G!, K!, MEL!, P!).

A. pubescens Roxb. var. *moritzii* Tul., Ann. Sci. Nat. Bot., Sér. 3: 215 (1851). — *A. pubescens* auct. non Roxb. 1802: Moritzi, Syst. Verz.: 73 (1846), nom. illeg. according to Art. 53.1. (Greuter *et al.* 2000). — *A. moritzii* (Tul.) Müll. Arg., Linnaea 34: 67 (1865 – 66). Type: Java, *Zollinger* 485 (P!, lectotype, here designated; BM!, G!, G-DC [microfiche], K!, L!, MEL!, P!, U!, isolectotypes).

A. rostratum Tul., Ann. Sci. Nat. Bot., Sér. 3: 218 (1851). — *A. rostratum* Tul. var. *genuinum* Müll. Arg. in DC., Prodr. 15(2): 257 (1866), nom. inval. Type: Philippines, Luzon, circa Manillam, 1819, *Perrottet* s.n. (P!, lectotype, here designated; G!, L!, isolectotypes). Original syntypes: Pangasinan, 1840, *Callery* 58bis (A!, P!); nec non secus flumen manillense prope pagos Macati & S. Nicolas, *Baume* s.n. (P!).

A. rostratum Tul. var. *lobbianum* Tul., Ann. Sci. Nat. Bot., Sér. 3: 219 (1851), in adnot., "lobbiana". — *A. lobbianum* (Tul.) Müll. Arg. in DC., Prodr. 15(2): 254 (1866). — *A. pentandrum* (Blanco) Merr. var. *lobbianum* (Tul.) Merr., Philipp. J. Sci., C, 9: 463 (1914). Type: Java or Luzon, "hb. proprio", *Lobb* 460 (CGE! [Hb. Lindl.], lectotype; BM!, G!, G-DC [microfiche], K!, L!, OXF!, isolectotypes). — Note: This number could not be located in Paris. Merrill (1914: 463 – 464) wrote: "I have examined Lobb's specimen in the Kew Herbarium which is indicated as from Luzon, and which is exactly matched by *Loher* 4656. It is a well known fact that Lobb's labels were badly mixed, and although Tulasne's specimen of this same number was labeled as from Java, this is no indication that the specimen came from Java, and the same number will be doubtless found in other herbaria labeled as from Singapore or from Borneo." Backer & Bakhuizen van den Brink f. (1964: 457) stated "*A. lobbianum* M. A., a Philippine species, was erroneously mentioned for Java". In any event, there is no doubt that it belongs to *A. montanum.*

A. acuminatum Wight, Icon. Pl. Ind. Or. 6: 12, t. 1991 (1853); Chakrabarty & Gangopadhyay, J. Econ. Taxon. Bot. 24: 28 (2000) as synon. nov. Type: Hort. Bot. Calcutt., Herb. Wight s.n. (K!, lectotype [designated here]; K!, isolectotype). The lectotype has apparently hermaphroditic flowers and includes the handwritten description as in the protologue.

A. diversifolium Miq., Fl. Ned. Ind., Eerste Bijv.: 468 (1861). Type: Sumatra austr., in prov. Lampong, secus fl. Tarabangi, 18 Dec., *Teysmann* HB 4431 (U!, holotype; BO!, isotype). — Notes: Known only from the type collection which is sterile although the fruits are described in the protologue. Airy Shaw (1981a: 363), who could not trace any fertile material either, included it tentatively in *A. montanum*. Collection number not cited in protologue.

A. palembanicum Miq., Fl. Ned. Ind., Eerste Bijv.: 465 (1861). Type: Sumatra or., in prov. Palembang, in Ogan-ulu, *Teysmann* HB 3733 (U!, holotype; BO!, isotype). — Note: This species is known only from the type collection which bears young pistillate inflorescences. Airy Shaw (1981a: 363) included it tentatively in *A. montanum*. Collection number not cited in protologue.

A. simile Müll. Arg., Linnaea 34: 67 (1865 – 66). Type: India or., Silhet, *Wallich* 7282 B (G!, holotype). — Note: There is a specimen in G-DC (microfiche) annotated by Müller Argoviensis as *A. simile*, but it has no suffix.

?*A. erythrocarpum* Müll. Arg. in DC., Prodr. 15(2): 258 (1866). Type: Java, *Teysmann* (B ex hb. Miq.), not located. — Note: The type in Berlin was probably destroyed in WW II. No material was found in G or G-DC (microfiche). Judging from Müller's description, however, this is *A. montanum*. Smith (1910: 259) came to the same conclusion, while Backer & Bakhuizen van den Brink f. (1964: 457 – 460) ignored the name.

A. leptocladum Tul. var. *glabrum* Müll. Arg. in DC., Prodr. 15(2): 254 (1866). Type: Philippines, *Cuming* 1820 (G-DC [microfiche], lectotype, here designated; CGE!, FHO!, G!, K!, P!, PRC!, TCD!, isolectotypes).

A. refractum Müll. Arg. in DC., Prodr. 15(2): 257 (1866). Type: India or., Khasia, 0 – 3000 ped, *J. D. Hooker & T. Thomson* s.n. (G!, epitype, here designated); India orient., prov. Sikkim, alt. 2000 ft, *J. D. Hooker* s.n. (G!, lectotype, here designated). — Note: The specimen cited in the protologue: "In Indiae orient. prov. Sikkim alt. 2000 ped. s. m. (Hook. f.! in hb. reg. berol.)", was destroyed in WW II. The lectotype from G, with the same information on the label, bears no original handwriting of Müller but a note in a different handwriting: "Antidesma refractum Müll. Arg., det. Müll. Arg.". It possesses only fruiting pedicels whereas in the protologue the fruits are also described. The lectotype, however, shows well the "pedicellis foem. refractis", on which the specific epithet is based. As this choice of lectotype might appear ambiguous, the only other sheet with the name *A. refractum* on it is here designated as an epitype according to Art. 9.7 (Greuter *et al.* 2000: 13). This specimen has a label in Müller's handwriting: "32. A. refractum Müll. Arg." (32 is the species number of *A. refractum* in the De Candolle's "Prodromus"), which means that it was definitely examined by Müller prior to the publication of the name. There is no material annotated as *A. refractum* in G-DC on microfiche.

A. menasu (Tul.) Müll. Arg. var. *linearifolium* Hook. f., Fl. Brit. India 5: 364 (1887), "linearifolia". Type: Bombay Herbarium, *Dalzell* s.n. (K!, lectotype, here designated). Original syntype: Karwar, *Talbot* 244 (K!). — Note: Both specimens cited above bear Hooker's handwriting, and the lectotype also has a drawing by his hand. The locality cited in the protologue, "Canara", does not appear on any of the labels, but Karwar is in this district, and North Kanara belonged to the old Bombay Presidency.

A. henryi Hemsl. in H. O. Forbes & Hemsl., J. Linn. Soc., Bot., 26: 431 (1894); Pax & K. Hoffm. in Engl., Pflanzenr. 81: 159 (1922), as synon. nov.; Airy Shaw, Kew Bull. 26: 358 (1972); Li Ping Tao, Fl. Reip. Pop. Sin. 44(1): 60 (1994), non Pax & K. Hoffm. 1922: 132. Type: China, Hainan, *B. C. Henry* 8562 (K!, lectotype, here designated). Original syntypes: Hainan, *B. C. Henry* 57 (K!), 35 (K!).

A. mucronatum Boerl. & Koord. in Koord.-Schum., Syst. Verz. 2: 27 (1910), "mucronata". Type: Sumatra, bei Langgam in überschwemmtem Hochwald von *Gluta renghas* L., um 20 m, sehr gemein und dort das Hauptunterholz bildend, 20 März 1891, *Koorders* 10302 b (BO!, holotype).

A. oblongifolium Boerl. & Koord. in Koord.-Schum., Syst. Verz. 2: 27 (1910), nom. illeg. (non Blume 1826 – 27), "oblongifolia". Type: Sumatra, bei Biwak im Regenwald, auf trocknem Boden, um 20 m, 23 März 1891, *Koorders* 21753 (BO!, lectotype, here designated). Original syntype: Sumatra, bei Biwak im Regenwald, 21 März 1891, *Koorders* 21806 (BO!). — Note: Doubtful species in Pax & Hoffmann (1922: 166) and Airy Shaw (1981a: 365); the latter, however, tentatively synonymised it with *A. montanum.*

A. kerrii Craib, Bull. Misc. Inform., Kew 1911(10): 462 (1911), **synon. nov.** Type: Thailand, Chiengmai, in thick evergreen jungle on Doi Sootep, 1200 m, *Kerr* 618a (K!, lectotype, here designated; TCD!, isolectotype). Original syntype: Thailand, Chiengmai, in thick evergreen jungle on Doi Sootep, 1200 m, *Kerr* 618 (K!). — Note: The hairy ovary is the only distinguishing character of *A. kerrii,* which is only known from the type material. Although the ovary of *A. montanum* is glabrous in the vast majority of specimens, there are several collections from different localities with a more or less pilose ovary. *A. kerrii* is therefore regarded as representing an aberrant population of *A. montanum.*

A. mindanaense Merr., Philipp. J. Sci., C, 7: 383 (1912), **synon. nov.** Type: Philippines, Mindanao, Distr. of Zamboanga, Sax R., San Ramon, 5 Feb. 1905, *Williams* 2117 (NY!, lectotype, here designated; K!, NY!, isolectotypes). Paratypes: Feb. 1904, *Hallier* s.n.; near Zamboanga, Dec. 1905, *Merrill* 8274 (A!, K!, US!).

A. obliquinervium Merr., Philipp. J. Sci., C, 9: 466 (1914), **synon. nov.** Type: Philippines, Palawan, in forests, alt. ab. 150 m, Taytay-Lake Manguao trail, *Merrill* 9294 (A!, lectotype, here designated; BM!, L!, NY!, P!, US!, isolectotypes). Paratypes: *Merrill* 9295 (A!, L!, NY!), 9336 (K!, US!); *FB (Fernandez)* 21491 (K!, US!). — Note: The two staminate *Merrill* paratypes have free disc lobes which is unusual for *A. montanum,* but no other distinguishing features. The remaining paratype which is also staminate, has a normal, cushion-shaped disc.

A. palawanense Merr., Philipp. J. Sci., C, 9: 467 (1914); Pax & Hoffmann in Engl., Pflanzenr. 81: 131 (1922); Merr., Enum. Philipp. Fl. Pl. 2: 416 (1923); Airy Shaw, Euphorb. Philippines: 6 (1983). Type: Philippines, Palawan, Palawan prov., Puerto Princesa (Mt Pulgar), March 1911, *Elmer* 12808 (K!, neotype, here designated; A!, E!, G!, L!, NY!, U!, WRSL!, isoneotypes). — Note: No duplicate of the type cited in the protologue (Philippines, Palawan, Mt Victoria, along streams, 25 March 1906, *BS (Foxworthy)* 749) could be located; the material was destroyed in Manila during WW II. The description, however, is clearly of *A. montanum*, and Merrill also wrote that the new species is: "... probably most closely allied to *Antidesma pentandrum* (Blanco) Merr. ..." which is here treated as a synonym of *A. montanum*. Pax & Hoffmann (1922) cited a staminate collection from Sarawak, Borneo (*Haviland & Hose* 1762 (BM, CGE, GH, K, L)) under this name, which also represents *A. montanum*. Merrill (1923) cited two more collections beside the type for this species: *BS (Foxworthy)* 808 and *Elmer* 12808. Several duplicates of both collections have been examined, all of which represent *A. montanum*. *BS (Foxworthy)* 808 (GH, NY, US) has only very young staminate flowers, while in the protologue of *A. palawanense* only pistillate flowers are described. The pistillate collection *Elmer* 12808 was therefore chosen as the neotype.

A. pentandrum (Blanco) Merr. var. *angustifolium* Merr., Philipp. J. Sci., C, 9: 464 (1914), **synon. nov.** — *A. angustifolium* (Merr.) Pax & K. Hoffm. in Engl., Pflanzenr. 81: 165 (1922). Type: Philippines, Luzon, Benguet subprov., Twin Peaks, 23 May, *Elmer* 6327 (US!, lectotype, here designated; G!, K!, NY!, P!, isolectotypes).

A. ramosii Merr., Philipp. J. Sci., C, 9: 468 (1914), **synon. nov.** — *A. pentandrum* Merr. var. *ramosii* (Merr.) Pax & K. Hoffm. in Engl., Pflanzenr. 81: 126 (1922). Type: Philippines, Luzon, Prov. Rizal, Bosoboso, June 1906, *BS (Ramos)* 1002 (US!, lectotype, here designated; NY!, P!, isolectotypes). — Note: A perfectly typical *A. montanum*, also with regard to leaf indumentum and number of lateral veins. The staminate pedicels are rather long but within the variability of the species.

A. agusanense Elmer, Leafl. Philipp. Bot. 7: 2632 (1915), **synon. nov.** Type: Philippines, Cabadbaran (Mt Urdaneta), Prov. Agusan, Mindanao, Aug. 1912, rich ground of a densely forested flat at 750 ft, *Elmer* 13549 (GH!, lectotype, here designated; E!, G!, K!, L!, NY!, P!, U!, US!, isolectotypes). — Note: The fruits of the type are slightly laterally compressed, with a slightly asymmetric fruit base and stigma. This can be found in several collections from the Philippines and blurs the boundaries between *A. montanum* and *A. japonicum*. In the case of *A. agusanense*, however, the habit is much more indicative of *A. montanum*.

A. phanerophlebium Merr., Philipp. J. Sci., C, 11: 59 (1916). Type: Borneo, Sarawak, *BS (Native collector)* 1384 (US!, lectotype, here designated; A! [photo of destroyed holotype ex PNH], K!, isolectotypes). — Note: Synon. nov. in Airy Shaw (1975: 214), but apparently resurrected later (Airy Shaw 1982: 6).

?*A. aruanum* Pax & K. Hoffm. in Engl., Pflanzenr. 81: 149 (1922). Type: Aru Inseln, *Warburg* 20640 (A!, lectotype, here designated). — Note: This species is only known from the type in pistillate flower. It is distinguished by glabrous leaves and calyces as well as the minutely hairy disc, found only rarely in *A. montanum*. The slightly compressed ovary with a subterminal style might develop into a fruit that looks quite different from those of typical *A. montanum*. Unfortunately, this is impossible to judge from the type specimen. As it stands, this specimen is not distinctive enough to be granted specific status. *A. montanum*, on the other hand, is the most variable and widespread species of the genus, and all measurements of the type of *A. aruanum* are well within its variability. Airy Shaw (1980a: 210) did not include the species in his key and commented on it: "Until further material can be collected in the Aru Islands this plant is probably unidentifiable". A similar specimen from West Papua, Waigeo Island, *P. van Royen* 5258 (K, L), has both pistillate flowers and near-mature fruits. The fruits are laterally compressed with a terminal style as in *A. heterophyllum*, but the disc is rather densely hairy. It is possible that more and better material from the Moluccas and West Papua will show that *A. heterophyllum* intergrades into *A. montanum* and does not deserve specific rank, or that *A. aruanum* is a distinct entity after all.

A. pseudomontanum Pax & K. Hoffm. in Engl., Pflanzenr. 81: 163 (1922); Airy Shaw, Kew Bull., Addit. Ser. 4: 214 (1975), as synon. nov. Type: Borneo, Sarawak, Kuching, *Haviland & Hose* 3104 (P!, lectotype, here designated; BM!, CGE!, K! ("*Haviland* 3104"), SAR! ("coll. = *Kalong*"), SING! ("*Haviland* 3104"), isolectotypes).

A. teysmannianum Pax & K. Hoffm. in Engl., Pflanzenr. 81: 144 (1922), **synon. nov.** Type: Java, *Warburg* 1325 (B!, lectotype, here designated). Original syntype: Java, "A. lanceolatum", *Teysmann* s.n. (B!, possible duplicate: P!). — Note: The lectotype has pistillate flowers and fruits, whereas the original syntype is sterile. Pax & Hoffmann, not having seen the staminate flowers, considered the new species to be close to *A. diandrum* Roth. (syn. *A. acidum*) and placed it as the only other species into their sect. *Diandra*, distinguished from the latter only by the number of secondary leaf-veins.

A. salicinum Ridl. var. *latius* Ridl., Fl. Malay Penins. 3: 229 (1924), "latior", **synon. nov.** Type: Malaysia, Johore, Gunong Janeng, *Kelsall* 4020 (K!, holotype; SING "Lake & Kelsall", isotype).

A. paxii F. P. Metcalf, Lingnan Sci. J. 10: 485 (1931), as nom. nov. — *A. henryi* Pax & K. Hoffm. in Engl., Pflanzenr. 81: 132 (1922), nom. illeg. (non Hemsl. 1894). Type: China, Yünnan, Wälder im Süden des Roten Flusses, *Henry* 13667 (K!, lectotype, here designated; US!, isolectotype). Original syntype: same locality, *Henry* 13688 (K!, NY!). — Note: Metcalf had not seen the types but published a nomen novum only because the name *A. henryi* had been used before by Hemsley. Airy Shaw (1972a: 359) kept *A. paxii* apart from *A. montanum* on the grounds of leaf size and puberulence of the branches, both of which are variable in *A. montanum*. Li Ping Tao (1994: 62) incorrectly included this name in the synonymy of *A. acidum*.

A. discolor Airy Shaw, Kew Bull., Addit. Ser. 8: 212 (1980), as nom. nov., **synon. nov.** — *A. bicolor* Pax & K. Hoffm. in Engl., Pflanzenr. 81: 126 (1922), nom. illeg. (non Hassk. 1844). Type: Niederländisch Neu Guinea (Indonesia: Papua), *Warburg* 20650 (A!, neotype, here designated). — Note: Pax & Hoffmann obviously had only a staminate duplicate of the type collection at their disposal in B, which presumably is now destroyed. Another duplicate of the same collection bearing pistillate flowers and young fruits was, however, found unmarked in the general collection of the Arnold Arboretum. As Pax & Hoffmann explicitly stated: "Flores feminei ignoti", this duplicate must be designated as a neotype rather than a lectotype.

A. montanum Blume var. *microcarpum* Airy Shaw, Kew Bull. 36: 363 (1981), **synon. nov.** Type: Sumatra, vicinity of Aek Salaat, Asahan (north-east of Tomoean Dolok), 450 m, 15 – 26 July 1936, *Rahmat Si Boea* 9612 (L!, holotype; A!, GH!, K!, NY!, US!, isotypes). — Note: The only character distinguishing var. *microcarpum* from the type variety is the size of the fruits (given as 2 – 3 mm long in the protologue). The fruits of the type specimen are actually 3 – 4 mm long and immature (the endocarp is soft in boiled fruits).

A. leptocladum Tul. var. *schmutzii* Airy Shaw, Kew Bull. 37: 5 (1982), **synon. nov.** Type: Lesser Sunda Isl., Flores, Manggarai, 1040 m (Gipfelgebiet), 26 Dez. 1979, *Schmutz* 4513 (L!, holotype).

Stipules not foliaceous, linear to subulate, 2 – 7(– 13) × 0.5 – 1.5(– 3) mm. *Petioles* 2 – 6(– 15) × 1 – 2 mm. *Leaf blades* elliptic to oblong, more rarely slightly ovate or obovate, basally acute to obtuse, rarely rounded, (6 –)9 – 15(– 30) × (2 –)3 – 6(– 12) cm, (2 –)2.5 – 3.5(– 6.6) times as long as wide, membranaceous to chartaceous, more rarely coriaceous, glabrous except along the major veins, on one or both surfaces, rarely slightly pilose abaxially. *Staminate inflorescences* 3 – 13 cm long, axillary, simple to consisting of up to 10(– 40) branches. *Stamens* exserted 1 – 1.5 mm from the calyx. *Pistillate inflorescences* 4 – 10 cm long, axillary, sometimes aggregated at the end of the branch, simple to consisting of up to 5 branches. *Infructescences* 6 – 20 cm long.

DISTRIBUTION. India incl. Andaman & Nicobar Islands, Bhutan, Bangladesh, southern China (Hainan, Guangxi, Yunnan provinces), Japan (only Ryukyu Islands), Taiwan, Laos, Vietnam, Cambodia, Thailand, Peninsular Malaysia, Sumatra, Borneo, Java, Philippines, Sulawesi, Lesser Sunda Islands, Moluccas, Indonesia: Papua, Australia (only Prince of Wales Island, northern Queensland). Map 24.

HABITAT. In primary and secondary evergreen to deciduous vegetation; in mixed dipterocarp forest; in riparian and littoral forest; in coastal forest and on beaches; in monsoon forest and teak-forest ("djatibosch"), in *Agathis* forest; in bamboo groves; in peat swamp forest; in heath forest, thickets, grasslands and along roadsides; around human habitation (Malay: "belukar"); also in mossy montane forest; usually in understorey; from deep shade to total exposure; on dry to wet or seasonally flooded ground. Common in most places, with extremely high

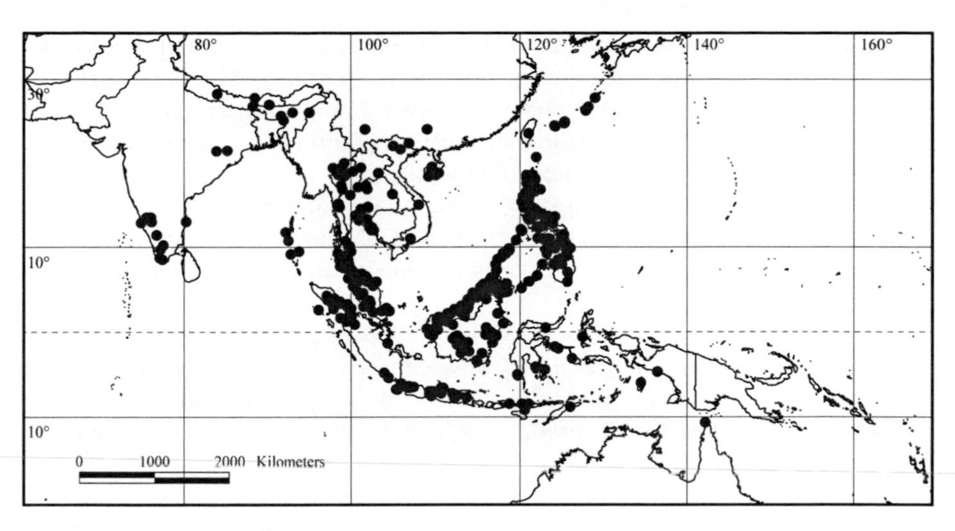

MAP 24. Distribution of *A. montanum* var. *montanum*.

ecological amplitude; it is the dominant species in inland Krakatau (Whittaker, pers. comm.). On sand, clay, loam, peat, volcanic and ultrabasic soil, over limestone, sandstone or granite. 0 – 2000 m altitude.

USES. The fruits are eaten locally. They are also used to adulterate black pepper. The Malay name "Madang lada" refers to the yellowish wood similar to "Madang" = *Litsea* (Lauraceae) and to the fruits ("lada" = pepper), thus meaning "Litsea with pepper fruits" (Yii, SAR, pers. comm.). The wood is used in house construction, for rice pestles and firewood. The leaves are applied to ulcers and lumber pains, the roots for stomach ache, and the fruits or tea from the leaves "as tonic after childbirth" (Philippines, *fide PNH* 37987, 38059, 38372, *Zwickey* 699).

CONSERVATION STATUS. Least Concern (LC). This is the most common taxon in the genus. 1387 specimens examined.

VERNACULAR NAMES. India: kaduvaivilangu (Kan.); Thailand: deeng da tee, korhier, naa dao, seh dah klee (Karen), som mao pru; Sumatra: burunai pajo, kayu aritan, kayu simburo, kayu djuhut tasu, kayu laman, kayu manuk-manuk, kayu motton, kayu si balik hangin, kayu si basa, kayu si kala, kayu si loppur, kayu si losu, kukunaw (Bisaya), lagas-lagas (Kedayan), palse palse, tutun burunai silai; Malay Peninsula: re'mool; Sabah: antatanud (Murut Bokan), apid-apid timba, bengu (Dusun), bilin sagit (Dusun), dampirut (Sungai), gagarit (Murut Tenom), gibih (Dusun), hunron (Orang Sungai), ipo-ipo (Dusun), lulunib (Murut Bu.), manggis tara rata (Dusun), menempuru (Dusun), mentegiras (Dusun), pachar ambok (Malay), ranting paya (Banjar), sinpaladuk (Sungei), tanduripis (Dusun), tanggir-angir (Kwijau), tendurusuh (Dusun), tenggilang (Dubu), totopis (Dusun); Sarawak: bruin (Iban), madang lada (Malay), Kalimantan: cabi cabi, kosaumpo; Java: honi pasir, ki jebak (Sundanese), ki seueur or seueur bener sud (Sundanese, seueur = many, refers to the many edible fruit), wuni dedek; Philippines: agosep, agosip or agusit (Tagbanua), aihip (Neg.), banuang (Cebuano), bignay pogo or binayoyo (Tagbanua) [pogo = small, chicken-like ground bird, bignay = *A. bunius*], bulinai (Pint. Sbl) or bulinay (Tagbanua), malabignay (Tagbanua) [looks like bignay (= *A. bunius*)], mataindo (Manobo), matelok (Lan.), misalagon (Tagbanua), mongay (Buk.), pagakpatulangan (Sub), pagpas (Bilaan), tegas (Sub.), tuba; Lesser Sunda Islands: ai burumi (Tetum), ai panah, eikahodok, lingko-sosor-rona (Ladju), rawamuga (Waijewa), toro.

ETYMOLOGY. Latin, montanus = growing on mountains.

KEY CHARACTERS. Leaves olive-green, glabrous except for veins; disc glabrous, much shorter than the calyx; staminate disc cushion-shaped; fruits glabrous with terminal style.

SIMILAR SPECIES. *A. acidum, A. bunius, A. coriaceum, A. cuspidatum, A. edule, A. elbertii, A. heterophyllum, A. japonicum, A. leucocladon, A. minus*, see there. — *A. orthogyne* always has a hairy disc and ovary. — *A. sootepense* has finely tessellated higher venation, more highly fused, non-glandular sepals and an extrastaminally fused staminate disc.

NOTES. The staminate disc is sometimes not fully closed around the filaments which may be due to flower size. Three collections from the Philippines, *Merrill* 9295 (A, L, NY), 9336 (K, US) from Palawan and *BS (Fenix)* 15665 (E, US) from Calusa Island, have staminate discs with free lobes but are like *A. montanum* in all other respects.

For galls see Pax & Hoffmann (1922: 159 – 160).

SELECTED SPECIMENS. JAPAN. Ryukyu Islands (Nansei-Shoto), Miyako Isl., 0.5 km NW of Karimata, dense scrub forest, limestone ridge, 20 – 30 m, 24 Aug. 1956, *Fosberg* 38387 (K, L). INDIA. Kerala, S Kerala, road to Poonmudi, wet evergreen forest, 800 m, 8 June 1976, *Kostermans* 26037 (G, K, KEP). NEPAL. Pokhara, 4000 ft, 6 Aug. 1954, *Stainton, Sykes & Williams* 6720 (BM). BHUTAN. Sarbhang-Chirang road,17 km from Sarbhang, 26°56'N, 90°14'E, mixed forest on steep slope, c. 1000 m, 1 June 1979, *Grierson & Long* 1535 (K). CHINA. Hainan, Ngai distr., Yeung Ling Shan, moist, swampy level land, sandy soil, 19 June 1932, *Lau* 132 (G). TAIWAN. Takow, hanging down from cliff, *Henry* 1144 (K). LAOS. Vientiane (Wiengcham), Nam Yak, in scrub, c. 300 m, 26 April 1932, *Kerr* 21262 (K, L). VIETNAM. Annam, along trail in forest, May – June 1927, *Clemens & Clemens* 3602 (K). THAILAND. Northern, Chiang Mai prov., Doi Angka (Doi Ithanon), c. 1860 m, 6 May 1935, *Garrett* 950 (K, L, TCD). PENINSULAR MALAYSIA. Selangor, Telok Forest Reserve, peat swamp, 6 Nov. 1978, *FRI (Suppiah)* 28198 (K, KEP, L, SAR). SUMATRA. Tapianoeli, Padang Si Dimpoean div., Padang Lawas subdiv., Sopsopan on Aek Si Olip (Topographic Sheet 41, NW quarter), 4 – 20 Sept. 1933, *Rahmat Si Toroes* 5554 (A, K, L, MEL, NY, U, US). BORNEO. Brunei, Temburong distr., Bukit Patoi, primary forest, hillside, yellow clay, 500 ft, 26 May 1958, *BRUN (Ashton)* 3330 (K, KEP, L); Sabah, Mt Kinabalu, Dallas Mt, jungle, 3000 ft, 31 Aug. 1931, *Clemens & Clemens* 26231 (A, BM, K, L, NY); Sarawak, Kuching div., Bau distr., Bau Hills, S of Kuching, on vertically eroded limestone, 10 July 1963, *Fosberg* 43837 (K, L, NY, US); C Kalimantan, Bukit Raya and upper Katingan (Mendawai) R. area, N of Tumbang Samba, on the Samba R., near base camp of Handyani Timber Camp at base of hill, 1°00'S, 113°00'E, primary rainforest, 50 m, 18 Nov. 1982, *Mogea* 3524 (K, KEP, L, US). JAVA. C Java, Gunung Muria, Tjollo, N of Kudus, volcanic, 800 m, 26 Nov. 1951, *Kostermans* 6291 (A, BM, K, L, NY). PHILIPPINES. Balabac Isl., s. loc., Aug.

1913, *BS (Escritor)* 21611 (GH, K, L, NY, US); Luzon, Laguna prov., Los Banos (Mt Maquiling), June – July 1917, *Elmer* 17478 (A, G, GH, K, L, NY, U, US). SULAWESI. Buton Island, SE of Sulawesi, Jismil camp inland from Labuan Tobelo, 4°26'S, 122°59'E, forest in steep valley, upraised coralline limestone, 150 m, 14 Nov. 1989, *Coode* 6242 (AAU, K, L). LESSER SUNDA ISLANDS. Flores, Woro Toro, volcanic breccia, 1000 – 1200 m, 5 Feb. 1910, *Elbert* 4313 (K, L, NY, US). AUSTRALIA. Queensland, Cook district, Prince of Wales Island, *Lindley* 168? (CGE).

33b. var. **microphyllum** *(Hemsl.) Petra Hoffm.*, Kew Bull. 54: 357 (1999). — *A. microphyllum* Hemsl. in H. O. Forbes & Hemsl., J. Linn. Soc., Bot., 26: 432 (1894); Airy Shaw, Kew Bull. 28: 276 (1973). Type: China, Szechuan, Hokiang, *Faber* 97 (K!; NY!).

A. wattii Hook. f., Fl. Brit. India 5: 366 (1887); Mandal & Panigrahi, J. Econ. Taxon. Bot. 4: 256 (1983); Petra Hoffm., Kew Bull. 54: 359 (1999), as synon. nov. Type: Munipur, *Watt* s.n. (K!, holotype; CAL, isotypes).

A. seguinii H. Lév., Repert. Spec. Nov. Regni Veg. 9: 460 (1911), "seguini"; Airy Shaw, Kew Bull. 26: 357 (1972). Syntypes: China, Kouy-Tchéou [Guizhou], Distr. de Tchen-Lin, bords du fleuve à la cascade de Hoang-Ko-Chan, 10 juin 1898, *Séguin* s.n. (E); fleuve Hoa-Kiang, Juin 1905, *Esquirol* 505 (E, K!); Tchai-Choui-Ho, juill. 1909, *Esquirol* 1586 (A!, E, K!).

A. pseudomicrophyllum Croizat, J. Arnold Arbor. 21: 496 (1940); Petra Hoffm., Kew Bull. 54: 359 (1999), as synon. nov. Type: China, Hainan, Po-ting, in forest, Nov. 1936, *Lau* 28228 (A!, holotype). — Note: Holotype and only collection is a poor specimen with one detached fruit. Croizat wrote in the protologue: "...the difference [to *A. microphyllum*] being that in Hemsley's species the primary veins are distinctly ascending, while in the new species they are very broadly spreading, the anastomoses being arranged subparallel with the margins of the leaf". Li Ping Tao (1994: 53) maintained this species and keyed it out against *A. microphyllum* on account of this character as well as the leaf shape and indumentum, which are indistinguishable in *A. microphyllum* and *A. pseudomicrophyllum*. The course of the lateral veins is variable in the examined material and depends much on the width of the leaves.

Spreading shrub, up to 3 m. *Stipules* not foliaceous, linear to deltoid, apically acute, 1.5 – 6 × 0.5 – 1 mm. *Petioles* 2 – 3 × 0.7 – 1 mm. *Leaf blades* lanceolate to linear, basally acute, rarely obtuse to rounded, (1.5 –)3 – 6(– 8) × 0.4 – 1.2(– 1.6) cm, (2.4 –)4 – 6(– 10) times as long as wide, chartaceous to coriaceous, glabrous except along the midvein adaxially, sparsely pilose along the midvein abaxially, sometimes sparsely pilose along the margin. *Staminate inflorescences* 1 – 4 cm long, axillary, sometimes aggregated at the end of the branch, simple or consisting of up to 5 branches. *Stamens* exserted 0.5 – 1 mm from the calyx. *Pistillate inflorescences* c. 2 cm long, axillary, sometimes aggregated at the end of the branch, simple or branched once or twice. *Infructescences* 2.5 – 4 cm long, axes 0.6 – 0.8 mm wide. Fig. 14F.

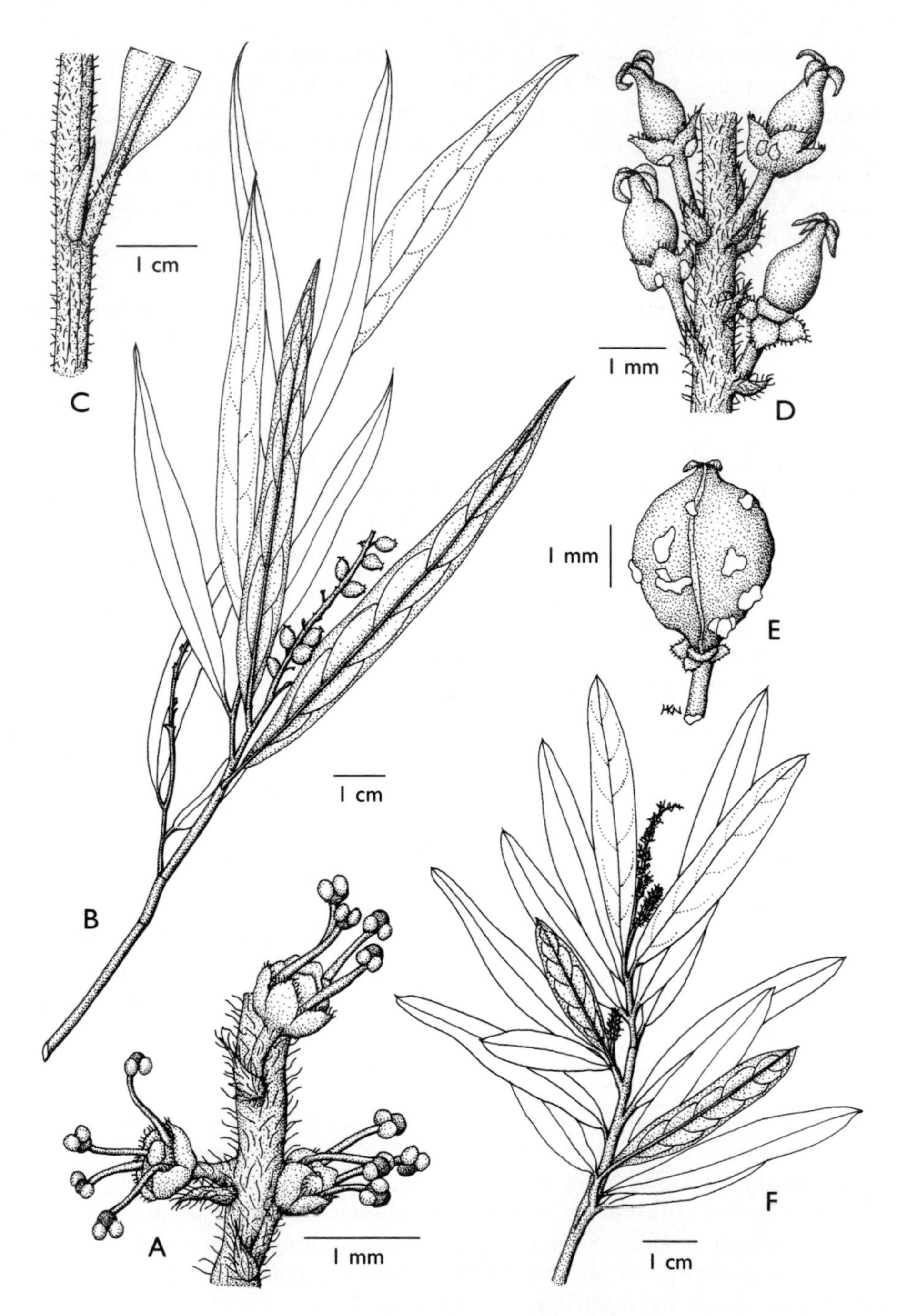

FIG. 14. *Antidesma montanum* var. *microphyllum* and var. *salicinum*. **A – E** *A. montanum* var. *salicinum*. **A** part of staminate inflorescence; **B** habit with infructescence; **C** axis with stipule; **D** part of pistillate inflorescence; **E** fruit in dorsal view. **F** *A. montanum* var. *microphyllum*, habit with staminate inflorescence. **A** from *Stone et al.* 15192 (L); **B – C** from *FRI* 21930 (L); **D** from *FRI* 6424 (L); **E** from *FRI* 15390 (L); **F** from *Kerr* 8636A (L). Drawn by Holly Nixon.

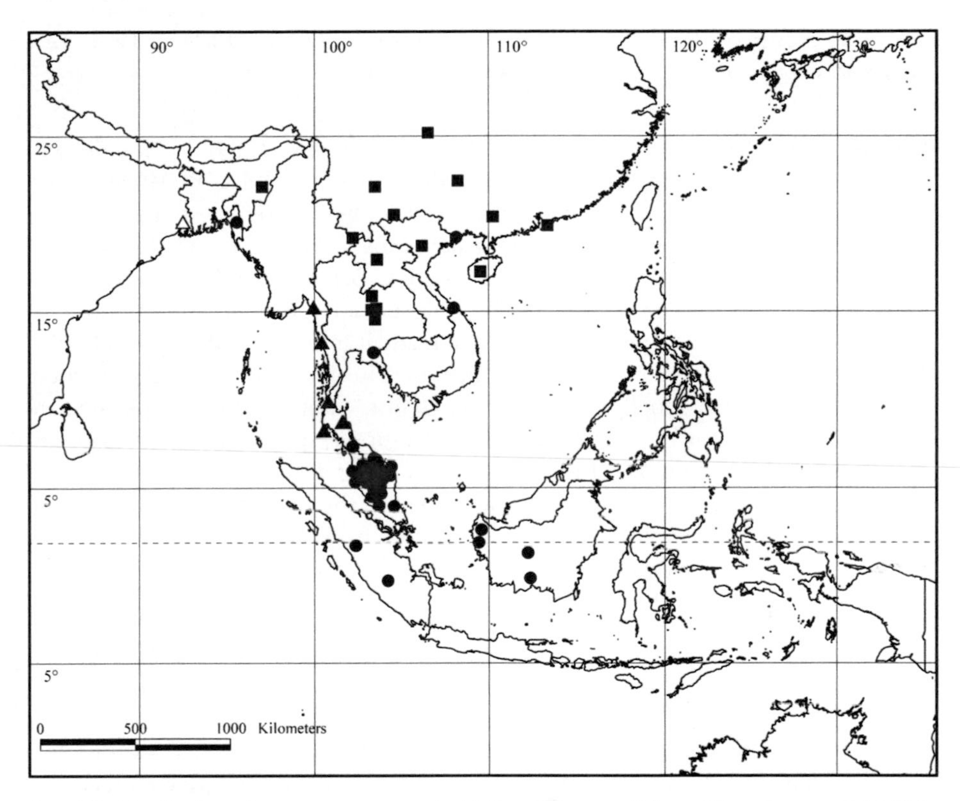

MAP 25. Distribution of *A. montanum* var. *microphyllum* (■), var. *salicinum* (●) and var. *wallichii* (▲), doubtful localities (△).

DISTRIBUTION. India (Assam), southern China (Guangdong, Guangxi, Guizhou, Hainan, Sichuan, Yunnan provinces), Burma (Southern Shan States), Laos, Vietnam, Thailand (Northeast and East). Map 25.

HABITAT. In (evergreen) forest along rivers. On sandstone. 300 – 1000 m altitude.

CONSERVATION STATUS. Least Concern (LC). This variety has a large extent of occurrence. 22 specimens examined.

ETYMOLOGY. The epithet refers to the size of the leaves (Greek, micro- = small; phyllon = leaf).

KEY CHARACTERS. Small-leaved rheophyte.

NOTE. Var. *microphyllum* differs only slightly from var. *salicinum*, mainly in leaf size and shape, and may have to be subsumed under this name when the genus has been revised for India, China, Burma, Laos and Vietnam.

SELECTED SPECIMENS. CHINA. Hong Kong, *Champion* 172 (K). LAOS. Luang Prabang prov., s. loc., 14 Feb. 1932, *Poilane* 20131 (K). VIETNAM. Tonkin, entre Cho-bo et Hoa Binh, 5/6 April 1909, *d'Alleizette* s.n. (L). BURMA. Southern Shan States, Mong Wa, valley of the Nam Live, April 1929, *Kingdon Ward* 8881 (NY). THAILAND. North-eastern, Loei prov., Wang Sapung, stream bed, c. 300 m, 16 March 1924, *Kerr* 8636 (K, L); Eastern, Chaiyaphum prov., Nam Phrom, 16°20'N,

101°45'E, evergreen forest along river, on sandstone, 600 m, 10 Dec. 1971, *Van Beusekom, Geesink, Phengkhlai & Wongwan* 4087 (K, L).

33c. var. **salicinum** *(Ridl.) Petra Hoffm.*, Kew Bull. 54: 359 (1999). — *A. salicinum* Ridl., Fl. Malay Penins. 3: 228 (1924); Airy Shaw, Kew Bull. 28: 276 (1973). Type: Malaysia, Kelantan, Kelantan R., *Ridley* s.n. (K!, lectotype, designated in Hoffmann 1999a: 359). Original Syntypes: Pahang, Kuala Lipis, *Machado* 11556 (K!, SING!); Perak, Plus R., *Wray* 546 (K!, SING!); Dindings, Bruas, *Burn-Murdoch* 261 (K!, SING!). Ridley mixed habitat information with specimen citation and did not cite collection numbers in his protologues which made it somewhat difficult to identify the most suitable specimen for lectotypification. The *Wray* specimen is annotated as "*Antidesma moritzii*, Muell, var? *salicifolia*, Hemsl.", but although duplicated in SING, it does not bear Ridley's handwriting. The *Ridley* specimen bears the genus name and the collection data in Ridley's handwriting, but "*salicifolium* Miq." may have been added in another hand. Ridley stated at the end of the protologue of *A. salicinum* why in his opinion *A. salicifolium* Miq. from Sumatra (which in fact is a synonym of *A. neurocarpum* Miq.) is distinct from his new species. The two remaining syntypes are of poor quality.

Shrub, more rarely tree, usually 1 – 3 m, up to 6 m (*fide Soepadmo* 781: tree 17 m). Bark grey-brown to grey, smooth. *Stipules* not foliaceous, linear to subulate, 2 – 7(– 13) × 0.5 – 1.5(– 3) mm. *Petioles* (0 –)2(– 5) × 1 – 2 mm. *Leaf blades* lanceolate, basally acute, (3 –)5 – 11(– 16) × (0.5 –)1 – 2(– 3.5) cm, (4 –)6(– 10) times as long as wide, membranaceous to chartaceous, more rarely coriaceous, glabrous except along the major veins, on one or both surfaces, rarely slightly pilose abaxially. *Staminate inflorescences* 3 – 13 cm long, slender, axillary. *Stamens* exserted 0.5 – 1 mm from the calyx. *Pistillate inflorescences* c. 2 cm long, axillary, sometimes aggregated at the end of the branch, simple or branched once or twice. *Infructescences* slender. Fig. 14A – E.

DISTRIBUTION. Bangladesh, Vietnam, Peninsular and South-East Thailand, Peninsular Malaysia, Sumatra, Borneo (West and Central Kalimantan). The collection *Alam & Shukla* 7685 (K!) from Rangamati district, south-eastern Bangladesh, so far represents the north-western limit of the distribution area of var. *salicinum.* Map 25.

HABITAT. In riverine rain forest along rivers; often on swampy or seasonally flooded ground; among boulders or in river beds; sometimes partly submerged in river. On sandstone. 0 – 700 m altitude.

CONSERVATION STATUS. Least Concern (LC). This variety has a large extent of occurrence. 62 specimens examined, many recently collected.

VERNACULAR NAMES. Sumatra: krinom masim; Malay Peninsula: penawar (Ba tek), mata pelanduk, mempenai ayer, wamhanu.

ETYMOLOGY. The epithet refers to the similarity with the genus *Salix* L.

KEY CHARACTER. Rheophytic habit.

NOTE. *A. salicinum* was described only because of its lanceolate leaves as opposed to the elliptic leaves of *A. montanum.* In the following paragraph, however,

Ridley described *A. salicinum* var. *latius* with wider leaves. Airy Shaw (1973: 276) called *A. salicinum* "the stenophyllous extreme of the *A. montanum* Bl. complex" and *A. salicinum* var. *latius* "an almost perfect link between the two" (i.e. *A. salicinum* and *A. montanum*).

SELECTED SPECIMENS. BANGLADESH. Rangamati distr., Sitapahar, Kaptai, rain forest, in steep river bank along water edge, 13 May 1996, *Alam & Shukla* 7685 (K). VIETNAM. Tonkin, along Kwangtung-Tonkin border, Ha-coi, Chuk-phai, Taai Wong Mo Shan and vicinity, thicket, 18 Nov. – 2 Dec. 1936, *Tsang* LU 27292 (K). THAILAND. Peninsular, Pattani prov., Ban Kia, Tomo, near river in evergreen forest, 450 ft, 26 April 1931, *Lakshnakara* 794 (K, L). PENINSULAR MALAYSIA. Terengganu, Trengganu Mts, Sungai Kerbat at Jeram Keteh 1.5 miles below Kuala Trengan, boulder stream flood channel, sandstone, 27 June 1971, *FRI (Whitmore)* 20265 (A, K, L); Pahang, Sungai Nipah, 23 June 1932, *SF (Corner)* 25838 (A, K, NY). SUMATRA. Palembang, side of R. Roepit near Soekaradja, 800 ft, 1880, *Forbes* 2930 (BM, GH, K, L). BORNEO. C Kalimantan, near Sampit, swampy forest along river, 5 m, March 1948, *Kostermans* 4713 (A, K, L).

33d. var. **wallichii** *(Tul.) Petra Hoffm.*, Kew Bull. 54: 357 (1999). — *A. oblongifolium* Blume var. *wallichii* Tul., Ann. Sci. Nat. Bot., Sér. 3: 221 (1851); Pax & K. Hoffm. in Engl., Pflanzenr. 81: 107 (1922); Airy Shaw, Kew Bull. 26: 357 (1972). Type: India or., prope Pavoy [probably meaning Tavoy], *Wallich* s.n. (CGE!, holotype).

A. martabanicum C. Presl, Epimel. Bot.: 232 (1849); Airy Shaw, Kew Bull. 26: 357 (1972); Petra Hoffm., Kew Bull. 54: 357 (1999), as synon. nov. Type: Martabania ad Moulmine, *Helfer* s.n. (PRC!, holotype). Possible original material: Tenasserim and Andamans, *Herb. Helfer* Kew Distribution No. 1947 (K!), India or. in Bengalia circa Calcuttam, 1836 – 38, *Helfer* 19 (BO!, G!, ex Prague).

A. oblongum Wall. mss. ex Hook. f., Fl. Brit. India 5: 364 (1887), nom. nud. (non (Hutch.) Keay 1956; basionym: *Maesobotrya oblonga* Hutch. 1912).

Shrub or tree, up to 10 m, diameter up to 6 cm. *Stipules* foliaceous, ovate, apically acuminate, basally nearly symmetrical, parallel- to palmately veined, thin, 7 – 15 × 3 – 7 mm. *Petioles* 3 – 8 × 1 – 2 mm. *Leaf blades* oblong to slightly elliptic, basally rounded, more rarely obtuse or truncate, (8 –)10 – 20(– 25) × (3 –)4 – 7(– 9) cm, 2.5 – 3.4 times as long as wide, membranaceous to chartaceous, more rarely coriaceous, glabrous except along the major veins adaxially, ± spreading-pilose all over abaxially. *Staminate inflorescences* 3 – 13 cm long, axillary but aggregated at the end of the branch, consisting of 10 or more branches. *Stamens* exserted 1 – 1.5 mm from the calyx. *Pistillate inflorescences* 4 – 10 cm long, axillary but aggregated at the end of the branch, simple to consisting of up to 13 branches. *Infructescences* 6 – 20 cm long.

DISTRIBUTION. Peninsular Burma (Tenasserim division) and Peninsular Thailand. The localities "in Bengalia circa Calcuttam" (*Helfer* 19) and "Monte Khasia" (*d'Alleizette* s.n.) are most probably incorrect. Map 25.

HABITAT. In evergreen forest, scrub or along forest edges; usually in damp places, e.g., near waterfalls; in open or shaded, primary or secondary vegetation. 50 – 600 m altitude.

CONSERVATION STATUS. Near Threatened (NT). This variety seems to be more local than the previous three. It is, however, rather vaguely defined taxonomically and very similar to the extremely common var. *montanum*. It may therefore be more common than is indicated by the 14 examined specimens.

ETYMOLOGY. The epithet refers to the collector of the type specimen.

KEY CHARACTERS. Broadly ovate, foliaceous stipules; pistillate inflorescences often aggregated at the end of the branch. Specimens without stipules are very difficult to determine.

SELECTED SPECIMENS. INDIA. Assam, Khasi Hills, 8 – 900 m, July 1909, *d'Alleizette* s.n. (L); West Bengal, in Bengalia circa Calcuttam, 1836 – 38, *Helfer* 19 (BO, G). THAILAND. Peninsular, Nakhon Si Thammarat prov., Khao Luang foothills, evergreen forest along waterfalls, 350 – 400 m, 13 May 1973, *Geesink & Santisuk* 5433 (K, L); Nakhon Si Thammarat prov., Wat Kiriwong, scrub, c. 100 m, 2 May 1928, *Kerr* 15581 (K, L); Nakhon Si Thammarat prov., Khiri Wong trail from village up to Khao Luang area, open, secondary vegetation, forest edge, up to a few 100 m, 15 May 1968, *Van Beusekom & Phengkhlai* 740 (K, L); Ranong prov., Ranong near the hot springs, 9°57'N, 98°40'E, 200 m, 5 July 1992, *Larsen et al.* 43122 (AAU).

34. Antidesma montis-silam *Airy Shaw,* Kew Bull. 28: 269 (1973). Type: Sabah, Lahad Datu district, Mt Silam, Virgin Jungle Reserve, *SAN (Ahmad Datu)* 52772 (K!, holotype; L!, isotype).

Tree, up to 20 m, clear bole up to 10 m, diameter up to 22 cm, straight. Bark whitish, greyish, reddish brown to red, c. 6 mm thick, outer bark 1 – 2 mm thick, fissured or scaly, lenticelled; inner bark greenish, grey, yellow, orange, ochre, brown or red, 2 – 5 mm thick, fibrous; cambium yellow; sapwood white, pink or yellow. *Young twigs* terete to striate, pilose to glabrous, brown. *Stipules* usually caducous, linear to falcate, apically acute, 5 – 10(– 20) × 1 – 2(– 3) mm, pilose to pubescent. *Petioles* flat to channelled adaxially, becoming rugose when old, 3 – 12 × 2 – 3 mm, glabrous or pilose. *Leaf blades* elliptic, oblong or slightly obovate, apically acuminate-mucronate, sometimes caudate, basally acute, 15 – 33 × 6 – 11 cm, 2.6 – 3.5 times as long as wide, eglandular, coriaceous to chartaceous, glabrous, or some hairs only along the major veins abaxially, dull on both surfaces, midvein impressed to flat adaxially, tertiary veins weakly percurrent to reticulate, close together or widely spaced, drying yellowish olive-green. *Staminate plants* unknown. *Pistillate inflorescences* 10 – 15 cm long, axillary, simple, axes 1.5 – 4 mm wide, pilose. *Bracts* deltoid, apically acute, 0.7 – 1.5 × c. 0.5 mm, pilose. *Pistillate flowers* 2.5 – 3 × 1.5 – 2 mm. *Pedicels* 0.2 – 0.5 mm long, pilose to glabrous. *Calyx* 1 – 1.2 × 1.5 – 2 mm, sepals 4 – 5, fused for c. $^{3}/_{4}$ of their length, apically acute, sinuses rounded, shallow, pilose to nearly glabrous outside, glabrous inside. *Disc* slightly shorter than the sepals (in fruit often exserted from the sepals), hirsute. *Ovary* globose, laterally compressed, pilose to pubescent, style lateral to subterminal, stigmas 4 – 5. *Infructescences* 14 – 28 cm long. *Fruiting pedicels* stout, 0.5 – 3 mm long, pilose to glabrous. *Fruits* ellipsoid, laterally compressed, basally symmetrical or asymmetrical, with a lateral to subterminal style, (8 –)10 – 14 × 6 – 10 mm, sparsely pilose, sometimes white-pustulate, reticulate to areolate when dry.

DISTRIBUTION. Borneo: eastern Sabah (Sandakan and Tawau divisions). Map 26.

HABITAT. In primary, sometimes disturbed, dipterocarp forest. On black soil. 70 – 700 m altitude.

CONSERVATION STATUS. Endangered (EN B1ab(i,ii,iii)). Extent of occurrence is under 2500 km^2 and only 15 specimens are known.

ETYMOLOGY. The epithet refers to the type locality.

KEY CHARACTERS. Large, sparsely pilose fruits; long, simple infructescences; wide, relatively short petioles; large leaves; pubescent disc; fused sepals.

SIMILAR SPECIES. *A. brachybotrys* and *A. helferi*, see there. — *A. pendulum* has a less highly fused calyx, longer infructescences, (sub)terminal styles, often apiculate fruits and leaves with adaxially slightly raised to flat midveins. — *A. stipulare* has a less highly fused calyx, usually glabrous disc, ovary and fruits, mango-shaped fruits, usually longer fruiting pedicels, infructescences and leaves with often adaxially raised midveins.

SELECTED SPECIMENS. BORNEO, SABAH. Sandakan, Telupid, mile 87 ½ Hap Sang logging area, flat land, logged over area, 16 July 1976, *SAN (Kodoh & Tuyuk)*

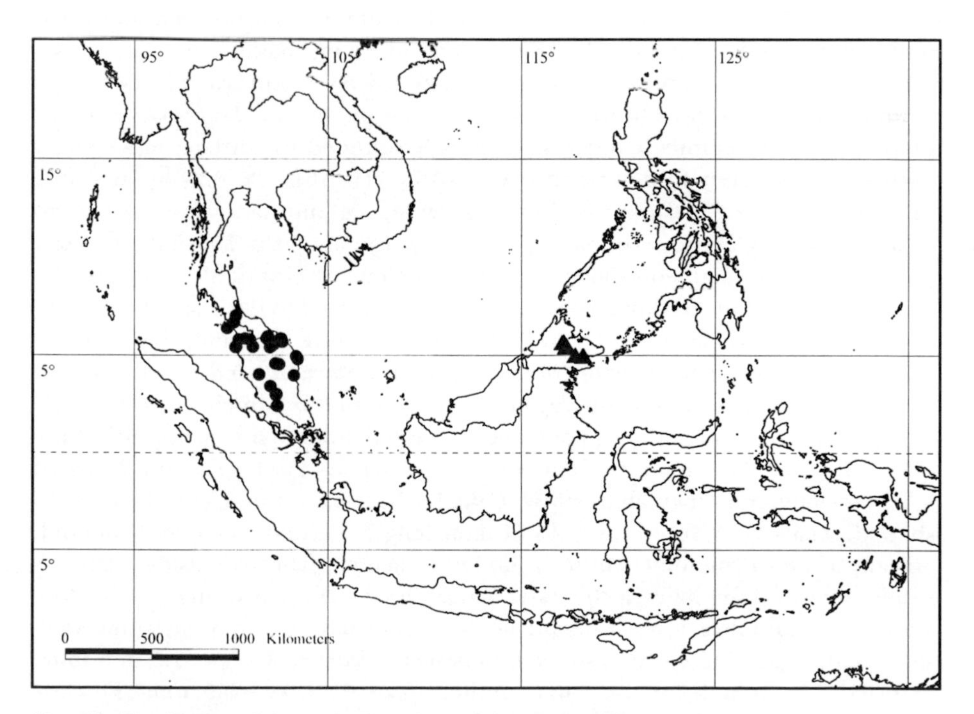

MAP 26. Distribution of *A. montis-silam* (▲) and *A. orthogyne* (●).

83642 (K, KEP, L, SING); Telupid, Kampung Bauto, disturbed dipterocarp forest, flat land, 300 ft, 16 July 1980, *SAN (Dewol Sundaling)* 92191 (K, L, SAR); Sandakan, Bukit Lumisir, primary forest, 1 Feb. 1982, *SAN (Rahim et al.)* 92965 (K, KEP); Telupid, Sungai Ruku-ruku, disturbed forest on hill slope, 6 Aug. 1981, *SAN (Aban Gibot)* 94038 (K, KEP, SAR); Tawau, Lahad Datu distr., K.B.Co. Silam, primary forest, hillside, 900 ft, 19 June 1961, *SAN (Muin Chai)* 25585 (K, L); Lahad Datu distr., K.B.Co. Silam, section 28, primary forest, hilltop, 395 ft, 1 June 1961, *SAN (Muin Chai)* 25061 (K, L); Lahad Datu distr., Mt Silam, foot of hill, 17 June 1984, *SAN (Leopold Madani)* 102183 (K, L, SAR); Lahad Datu distr., Mt Silam, primary forest, hilltop, 2000 ft, 23 April 1962, *SAN (Muin Chai)* 29653 (K, L, SAR); Lahad Datu distr., Mt Silam near Quarry, primary forest, 15 April 1964, *SAN (Meijer & Pereira)* 42240 (K, L, SAR).

35. Antidesma myriocarpum *Airy Shaw,* Kew Bull. 26: 467 (1972). Type: New Guinea, Morobe Distr., Huon Peninsula, near Pindiu, Mongi Valley, in garden regrowth, alt. 900 m, 24 April 1964, *Hoogland* 8829 (K!, lectotype, here designated; A!, CANB!, L!, isolectotypes).

Tree, rarely shrub, up to 23 m, clear bole up to 4 m, diameter up to 34 cm. Bark brown, c. 6 mm thick, flaky, pustular and thinly ridged; inner bark whitish, pink or pinkish brown, fibrous; wood whitish brown, straw, pink or orange brown, hard to moderately hard, rays broad and narrow, pores minute, solitary or in short chains. *Young twigs* terete, shortly pilose to densely ferrugineous-pubescent, brown. *Stipules* early-caducous, linear, apically acute, c. 10 × 1.5 mm, appressed ferrugineous-pubescent. *Petioles* channelled adaxially, basally and, more so, distally often slightly pulvinate and geniculate for 2 – 3 mm, (5 –)7 – 18 × 0.5 – 1.5 mm, almost glabrous to densely ferrugineous-pubescent. *Leaf blades* ovate or oblong, apically acuminate-mucronate, basally rounded to cordate, more rarely obtuse, often slightly folded when dry, (4 –)6 – 11(– 16) × (2 –)3 – 5(– 6.5) cm, (1.5 –)2 – 2.5(– 2.8) times as long as wide, eglandular, chartaceous to subcoriaceous, glabrous, or hairy only along the midvein adaxially, sparsely puberulous to ferrugineous-pubescent along the major veins or all over abaxially, shiny adaxially, shiny or dull abaxially, midvein impressed to flat adaxially, tertiary veins reticulate to percurrent, hardly thicker than fine venation, drying olive-green, lighter abaxially, hairtuft domatia present. *Staminate inflorescences* 4 – 5.5 cm long, axillary, sometimes aggregated at the end of the branch, consisting of 3 – 9 branches, axes ferrugineous-pubescent. *Bracts* deltoid to linear, apically acute, 0.5 – 0.7 × 0.2 – 0.3 mm, pubescent. *Staminate flowers* 1 – 2 × 1 – 1.5 mm. *Pedicels* 0 – 1.2 mm long, not articulated, pilose. *Calyx* 1 – 1.2 × 0.8 – 1 mm, cupular to bowl-shaped, sepals 4 – 5, fused for c. $^2/_3$ of their length, deltoid to narrowly deltoid, apically acute to rounded, sinuses rounded to acute, glabrous outside, glabrous inside but with long hairs at the base, margin erose. *Disc* consisting of 4 – 5 free alternistaminal lobes, lobes ± obconical, 0.3 – 0.5 × 0.3 – 0.5 mm, often not well-separated, appearing cushion-shaped, pubescent. *Stamens* 4 – 5, c. 1.5 mm long, exserted c. 1 mm from the calyx, anthers 0.2 – 0.3 × 0.4 – 0.5 mm. *Pistillode* clavate, 0.4 – 0.7 × 0.2 – 0.3 mm, exserted from or extending to the same length as the sepals, glabrous or very sparsely puberulent. *Pistillate inflorescences* 4 – 7 cm long, axillary or terminal, simple or consisting of up to 10 branches, axes puberulous to pubescent. *Bracts* deltoid, apically acute, 0.4 – 1 × 0.3 – 0.5 mm, pilose. *Pistillate flowers* c. 1.5 × 1 mm. *Pedicels* 0.2 – 1 mm long, glabrous. *Calyx* 0.7 – 1 × 0.7 – 1 mm, cupular to urceolate, sepals 4 – 5(– 6), fused for $^1/_2$ – $^2/_3$ of their length, deltoid, apically acute, sinuses rounded to obtuse, glabrous to pilose outside, glabrous to pilose inside, margin entire. *Disc* shorter than the sepals, shortly pubescent or glabrous outside and at the margin, glabrous inside. *Ovary* ovoid, glabrous, very rarely sparsely pilose, style terminal, stigmas 4 – 6. *Infructescences* 3 – 6 cm long, axes 0.7 – 1.5 mm wide. *Fruiting pedicels* 0 – 1 mm long, sparsely puberulous to glabrous. *Fruits* ellipsoid to globose, laterally compressed, probably terete when fully mature, basally symmetrical, with a terminal style, 3 – 3.5 × 2 – 3 mm, glabrous, very rarely sparsely pilose, sometimes white-pustulate, areolate or fleshy when dry.

35a. var. **myriocarpum**

Tree, rarely shrub, up to 10 m, clear bole up to 4 m, diameter up to 34 cm. *Young twigs* shortly pilose to ferrugineous-pubescent, becoming glabrous when older. *Petioles* shortly puberulous to almost glabrous. *Leaf blades* glabrous, or sparsely puberulous only along the midvein adaxially and along the major veins abaxially, shiny on both surfaces. *Disc* shortly pubescent outside and at the margin, glabrous inside.

DISTRIBUTION. Indonesia: Papua and Papua New Guinea (Central, Eastern Highlands, Madang, Morobe, Western Higlands provinces). Map 19 (p. 139).

HABITAT. In secondary forest; at forest margins leading to grassland; in open scrub; on river banks; in gardens; often in regrowth. On stoney clay or sand. 800 – 2300 m altitude.

CONSERVATION STATUS. Near Threatened (NT). The extent of occurrence (c. 7500 km^2) would justify a higher threat category but might be a collecting artefact. The available habitat information indicates that this species thrives in open habitats such as grassland and other secondary vegetation and is therefore most likely no more threatened than other taxa endemic to New Guinea. 18 specimens examined.

VERNACULAR NAMES. Indonesia: Papua: pupabee or popawee (Dani); PNG: assafa (Dunatina), kinskins (Wahgi: Mini), koril (Chibu: Masul), mupuruma (Asaro: Kefamo).

ETYMOLOGY. The epithet refers to the many small fruits (Greek, myrios = countless; carpos = fruit).

KEY CHARACTERS. Long, thin petioles; shiny leaves with fine, tessellated venation; inflorescences short; small fruits with terminal styles; staminate disc lobes free.

SIMILAR SPECIES. *A. excavatum* and *A. ghaesembilla*, see there. — *A. petiolatum* has wider, ovate stipules, longer, thicker petioles, larger leaves without domatia, almost cauline inflorescences, fewer stamens and smaller fruits. — *A. subcordatum* lacks domatia and has more densely pubescent inflorescences and flowers, larger, always sessile staminate flowers with free sepals, longer infructescences, fruiting pedicels and fruits.

SELECTED SPECIMENS. INDONESIA: PAPUA. Hollandia div., Wiligimaan, Baliem, young secondary growth, on stony sand, 1600 m, 26 June 1961, *BW (Versteegh)* 10494 (L); on stony clay, 1600 m, 28 June 1961, *BW (Versteegh)* 12527 (L); Orion Mts, Tenma R., riverside, 1500 m, 19 May 1959, *Kalkman* 4085 (L); Headwaters of Ok Denim R., 5°02'S, 140°55'E, river bank, 4700 ft, 19 June 1967, *NGF (Henty, Ridsdale & Galore)* 33202 (K, L). PAPUA NEW GUINEA. Madang prov., Ramu subdistr., Dumpu, gully, grassland, 11 Feb. 1989, *Beko* 1 (L); Central div., Mafulu (Mt Mafula), oak forest substage, 1250 m, Sept. – Nov. 1933, *Brass* 5344 (NY, US); Morobe distr., Wantoat (Wantot), 3500 – 6000 ft, 7 Jan. 1940, *Clemens* 10925 (A, US); Eastern Highlands distr., Goroka subdistr., near Dunantina village, low open shrub, 1400 m, 6 June 1956, *Hoogland & Pullen* 5247 (K, L, US); Western Highlands distr., Minj subdistr., Kawimugl, 6°05'S, 145°05'E, regrowth, 7000 ft, 15 Feb. 1965, *NGF (Millar)* 23844 (K, L).

35b. var. **puberulum** *Airy Shaw,* Kew Bull. 33: 17 (1978), "A. myrianthum var. puberulum". Type: North-eastern New Guinea, Morobe Distr., Wau-Edie Creek Road, 1500 m, 4 April 1959, *Brass* 29141 (K!, holotype; A!, CANB!, L!, NY!, US!, isotypes).

Tree, 13 – 23 m, diameter 15 – 30 cm. *Young twigs* densely ferrugineous-pubescent. *Petioles* densely ferrugineous-pubescent. *Leaf blades* glabrous except for the ferrugineous-pubescent major veins adaxially, pilose all over abaxially, densely ferrugineous-pubescent along the veins, shiny adaxially, dull abaxially. *Disc* glabrous.

DISTRIBUTION. Papua New Guinea: Morobe province. All three known specimens come from a small area between 7°10' – 7°20'S and 146°40' – 146°45'E. Map 19 (p. 139).

HABITAT. In rainforest. 1400 – 1600 m altitude.

CONSERVATION STATUS. Critically Endangered (CR B1ab(i,ii)). Extent of occurrence is c. 35 km², and 3 specimens examined. Contrary to var. *myriocarpum*, this local and rarely collected variety seems to prefer primary rain forest and is therefore under much higher threat.

ETYMOLOGY. The epithet refers to the indumentum (Latin, puberulus = minutely hairy).

KEY CHARACTER. Plant with dense ferrugineous indumentum in most parts.

SIMILAR SPECIES. *A. ferrugineum*, see there.

FURTHER SPECIMENS. PAPUA NEW GUINEA. Morobe distr., Edie Creek road, above Wau, Lookout Point, 7°20'S, 146°45'E, 5000 ft, 5 April 1959, *NGF (Womersley)* 11033 (K, L); Morobe distr., Bulolo, upper Nawi Banda logging area, 7°10' S, 146°40'E, in mid-mountain rainforest and sub-canopy tree dominated by *Castanopsis, Canarium*, 4200 ft, 21 Dec. 1962, *NGF (Havel & Kairo)* 17078 (L).

36. Antidesma neurocarpum *Miq.*, Fl. Ned. Ind., Eerste Bijv.: 466 (1861). — *A. microcarpum* Miq., Fl. Ned. Ind., Eerste Bijv.: 184 (1861), nom. nud., printed probably by mistake for *A. neurocarpum.* Type: Sumatra austr., in prov. Lampong, prope Siringkebau, Mangala, *Teysmann* HB 4532 (U!, lectotype, here designated; K!, isolectotype). — Note: Collection number not cited in protologue. The specimen HB 4365 (U!) from Lampong, Mangala, may also be original material.

Shrub, treelet or tree (*fide Wiriadinata* 1324: climber), up to 23 m, clear bole up to 17 m, diameter up to 20 cm (*fide Soepadmo* 626: 50 m, diameter 70 cm). Twigs usually white to grey, also cream, yellowish brown or light brown. Bark brown, grey, white, beige green or reddish brown, 1 – 2 mm thick, smooth, often flaking, non-fissured, soft; inner bark white, green, grey, brown, pink, purple, red or yellow, 1 – 2.5 mm thick, fibrous, soft; cambium yellow, white, green, pink or reddish; wood hard and dense, sapwood white, yellow, brown, pink or red, heartwood pinkish brown, red or red-ochre. *Young twigs* terete, spreading-ferruginous-pubescent but white when becoming glabrous, first brown, whitish grey when older. *Stipules* persistent, foliaceous or not, cordate to lanceolate, apically acuminate to acute, often long mucronate, basally more or less symmetrical, venation more or less reticulate, more rarely linear to subulate, chartaceous, 4 – 15(– 25) × (0.5 –)2 – 10(– 18) mm, pilose, becoming glabrous. *Petioles* slightly channelled to flat adaxially, becoming rugose when old, (0 –)3 – 7(– 10) × c. 1 mm, spreading-ferruginous-pubescent, becoming glabrous when old. *Leaf blades* elliptic, oblong, obovate, lanceolate to linear, sometimes falcate, apically acuminate-mucronate, sometimes caudate, basally acute, (3 –)9 – 14(– 20) × (0.7 –)3 – 4(– 7) cm, (2.2 –)3.3(– 11) times as long as wide, eglandular, chartaceous to coriaceous, glabrous adaxially, ferrugineous-pilose along the veins abaxially, more rarely all over, shiny adaxially, dull to shiny abaxially, midvein shallowly impressed to flat adaxially, tertiary veins weakly percurrent to reticulate, perpendicular tertiary and intersecondary veins conspicuous near the midvein, drying dark reddish brown to grey. *Staminate inflorescences* 2 – 6 cm long, axillary, simple or branched mostly near the base, consisting of up to 7 branches, axes ferrugineous-pubescent. *Bracts* deltoid, apically acute, c. 0.5 × 0.3 mm, ferrugineous-pubescent. *Staminate flowers* c. 1 × 1 – 1.5 mm. *Pedicels* 0 – 4 mm long, not articulated. *Calyx* c. 0.7 × 1 mm, sepals (4 –)5(– 6), nearly free to irregularly fused for up to ½ of their length, c. 0.4 mm wide, more or less reflexed, deltoid, apically acute, pubescent to pilose outside, pilose inside with particularly long hairs at the base, margin entire. *Disc* cushion-shaped, hemispherical, fully or partially enclosing the bases of the filaments and, if present, the pistillode, constricted at the base, glabrous. *Stamens* (3 –)4 – 5, 0.8 – 1 mm long, exserted 0.8 – 1 mm from the calyx, anthers 0.2 – 0.3 × 0.2 – 0.3 mm. *Pistillode* usually absent, rarely hemispherical, up to 0.1 × c. 0.2 mm, glabrous or sparsely pilose. *Pistillate inflorescences* 1.5 – 6 cm long, axillary, simple, rarely once-branched or in fascicles of 2 inflorescences, axes spreading-ferrugineous- to ochraceous-pubescent. *Bracts* deltoid, apically acute, 0.5 – 0.8 × 0.3 – 0.5 mm, pubescent. *Pistillate flowers* 1.5 – 2 × c. 1 mm. *Pedicels* 0.3 – 0.5(– 1.5) mm long, pilose to pubescent. *Calyx* 0.3 – 0.8 × c. 0.8 mm, sepals 3 – 5(– 7), fused only at the base, rarely irregularly fused for up to 1/2 of their length, apically acute, ferrugineous-pubescent to glabrous outside, glabrous or with long hairs inside especially at the base (but not exceeding the calyx). *Disc* exserted from or shorter than the sepals but always conspicuous, glabrous. *Ovary* globose, glabrous, rarely pilose, style terminal to lateral, stigmas 3 – 4. *Infructescences* 3 – 16 cm long. *Fruiting pedicels* 1 – 8(– 20) mm long, ferrugineous-pubescent to almost glabrous. *Fruits* ellipsoid, globose, ovoid or obovoid, dorsiventrally compressed, terete or laterally compressed, basally symmetrical or asymmetrical, with a subterminal to lateral style, 5 – 13 × 3 – 9 mm, glabrous, rarely thinly puberulent, sometimes white-pustulate, reticulate when dry.

NOTE. A very variable species. The fruit shape has caused some taxonomic confusion; this character is constant in almost all other *Antidesma* species. In *A. neurocarpum*, however, there are all intermediates between the dorsiventrally compressed fruits of the typical *A. neurocarpum* on the one hand and the laterally compressed fruits of *A. alatum* on the other (cf. also Airy Shaw 1973: 272). Sometimes both are found on the same specimen (e.g., *Soepadmo* (K) 626 from Pahang, *SAN* 88850 (K, L) from Sabah). No correlation of fruit shape with stipule size, indumentum or other morphological characters could be found.

Throughout the geographic range of the species, some collections have puberulous ovaries and fruits (e.g., *Argent et al.* 93188 (E, K, L) and *Burley et al.* 2983(E, K, L)). This ranges from a very sparse indumentum (e.g., *Church* 571 (L)) to a dense pubescence of the ovary (e.g., *Church et al.* 1825 (L)).

36a. var. **neurocarpum**

A. salicifolium Miq., Fl. Ned. Ind., Eerste Bijv.: 467 (1861), nom. illeg. (non C. Presl 1849). Type: Sumatra occ., in prov. Priaman, *Diepenhorst* HB 2352 (U!, holotype; CAL, K!, MEL?, isotypes). — Note: The collection number is not cited in the protologue. This is a synonym of *A. neurocarpum* and not of *A. forbesii* as erroneously stated by Airy Shaw (1981a: 361). The fruits firmly attached to the holotype in U do not agree with those on the fruiting lectotype of *A. forbesii* but with *A. neurocarpum.* The isotype of *A. salicifolium* Miq. in K has two fruits in a bag which do not agree with either of the two species in question but seem to belong to *A. montanum. Diepenhorst* 2352 is also cited as a paratype of *A. pradoshii* Chakrab. & M. G. Gangop., the type of which is a syntype of *A. forbesii.*

A. alatum Hook. f., Fl. Brit. India 5: 358 (1887); Airy Shaw, Kew Bull. 26: 358 (1972), as synon. nov. Type: Singapore, *Wallich* [no. 8583] (K!, lectotype, here designated; K!, isolectotypes). Original syntypes: Malaysia, Perak, Waterfall Hill, *Wray* [690] (K!); Malacca, *Griffith* Kew Distr. No. 4941 (K!). — Note: Hooker did not cite the numbers in the protologue, only the collectors. However, the localities are correct and all the sheets are annotated in his handwriting.

A. hallieri Merr., Philipp. J. Sci., C, 11: 57 (1916); Airy Shaw, Kew Bull., Addit. Ser. 4: 214 (1975), as synon. nov. e descr. Type: Borneo, s. loc., *Hallier* 1773 (K!, lectotype; here designated; A! [photo of destroyed holotype ex PNH], BO!, L!,

U!, isolectotypes). — Note: The type has long pedicels (up to 9 mm), but is typical *A. neurocarpum.*

A. rubiginosum Merr., Philipp. J. Sci., C, 11: 61 (1916); Airy Shaw, Kew Bull. 26: 358 (1972), as synon. nov. Type: Sarawak, Baram Distr., Baram, 10 Oct. 1894, *Hose* 297 (BM!, lectotype, here designated, A! [photo of destroyed holotype ex PNH], K!, L!, P!, isolectotypes).

A. inflatum Merr., J. Straits Branch Roy. Asiat. Soc. 76: 91 (1917); Airy Shaw, Kew Bull. 26: 358 (1972), as synon. nov. Type: Sabah, Kalabakan watershed, 6 Oct. 1916, along the margins of swamps at sea level, *Villamil* 235 (US!, lectotype, here designated; A! [photo of destroyed holotype ex PNH], K!, isolectotypes).

A. urophyllum Pax & K. Hoffm., Mitt. Inst. Allg. Bot. Hamburg 7: 224 (1931). Type: West-Borneo, auf dem Bukit Raja, um 1250 – 1400 m, im Urwald, 16 Dez. 1924, *Winkler* 883 (HBG, lectotype, here designated). Original syntypes: same locality, 20 Dez. 1924, *Winkler* 967 (BO!, E!, HBG). — Note: Synon. nov. of *A. stipulare* (e descr.) in Airy Shaw (1975: 216). The original syntype in BO is typical *A. neurocarpum*; leaf shape, texture and drying colour as well as flower indumentum and shape differ strongly from *A. stipulare.*

A. cf. *hosei* Pax & K. Hoffm. var. *oxyurum* Airy Shaw, Kew Bull. 36: 362 (1981), **synon. nov.** Type: West Sumatra, Benkoelen, Rimbo-pengaden, 16 March 1931, *De Voogd* 1071 (L!, holotype).

Stipules foliaceous, cordate to lanceolate, apically acuminate to acute, often long mucronate, basally more or less symmetrical, venation more or less reticulate, more rarely linear to subulate, chartaceous, 5 – 15(– 25) × 2 – 10(– 18) mm. *Petioles* 3 – 7 mm long. *Leaf blades* elliptic, oblong or obovate, (6 –)9 – 14(– 20) × (2 –)3 – 4(– 7) cm, (2.2 –)3.3(– 4.6) times as long as wide, without conspicuous stomata abaxially, drying dark reddish brown. *Pedicel* absent. *Infructescences* 4 – 16 cm long. *Fruiting pedicels* 1 – 8(– 20) mm long. *Fruits* dorsiventrally compressed, terete, or laterally compressed, 6 – 10 × 4 – 7 mm, glabrous, rarely thinly puberulent, sometimes white-pustulate.

DISTRIBUTION. Peninsular Thailand, Peninsular Malaysia, Singapore, Sumatra, Borneo. The specimen from Mergui, Peninsular Burma (*Parker* 2741 [CAL]) attributed to this species by Chakrabarty & Gangopadhyay (2000) could not be examined for this study. Map 27.

HABITAT. In lowland to montane forest up to 45 m tall; in dipterocarp forest, associated with *Dryobalanops, Hopea, Lithocarpus, Shorea*; in freshwater swamps; along logging roads and rivers; in open bamboo forest; in submontane mossy forest; in thickets and heath forest; often in dense, humid, shaded habitats; in primary or secondary vegetation. On sand, clay, loam, lateritic or volcanic soil, over sandstone, limestone, shales, basalt or granite. 0 – 1800 m altitude.

USES. The hard wood is used in Java to make walking sticks (Heyne 1917: 77).

CONSERVATION STATUS. Least Concern (LC). This is one of the most common and widespread taxa in western Malesia. 539 specimens examined.

VERNACULAR NAMES. Sumatra: boriengen riembo, ingara lelen, kayu

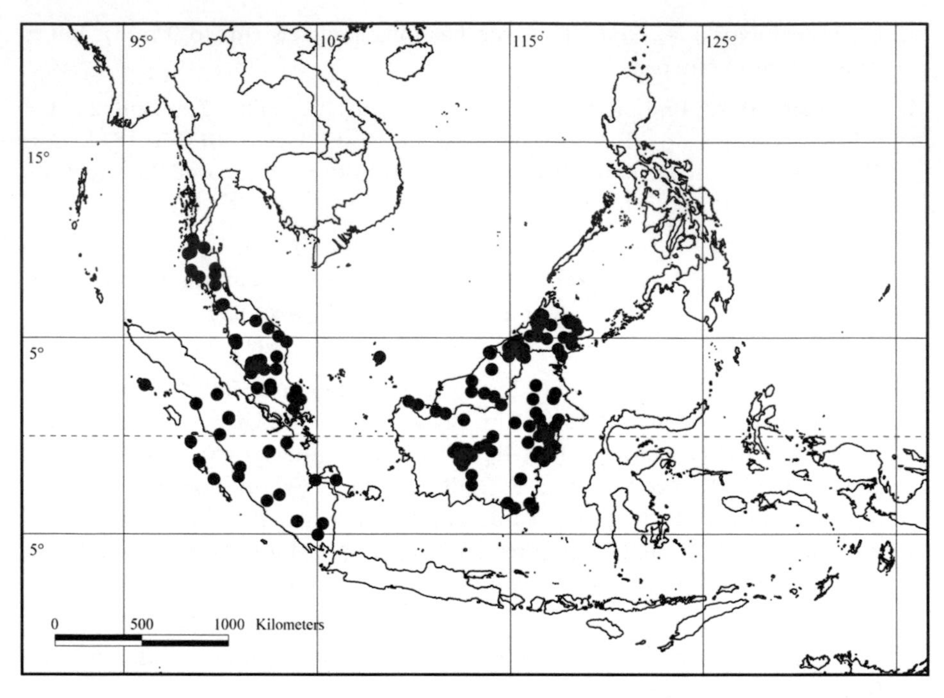

MAP 27. Distribution of *A. neurocarpum* var. *neurocarpum*.

keliengoh, kayu si basa, kayu saber bubu, kayu selipei, useu-useu lutung (Indonesian), sale sale balah; Sabah: kandatan (Dusun), kemuning-kemuning (Kedayan), kilas, lagalagas (Dusun), legas legas, muntinagas (Dusun), prumpung imbaan (Tengara), purak (Dusun), rambai rambai (Dusun), susup (Sungei), tangkukau (Dusun), tutuh (Dusun); Sarawak: bernai betjuping or bertjuping, buah puteh hitam, buti (Iban), mutek (Iban), tulang-patala (Murut), tumas; Kalimantan: beleti limbo, kayu tahum, keteleng (Bassap-Mapulu, teleng = ear, stipule), mayan rimbo, menyalin, patah jarum, pondok, uhai arong.

ETYMOLOGY. The epithet refers to the sculpture of the fruits (Greek, neuron = nerve; carpos = fruit).

KEY CHARACTERS. Ferrugineous indumentum, especially of young parts; stipules foliaceous; leaves drying reddish to grey; sepals free, acute; disc glabrous, as long as sepals.

SIMILAR SPECIES. *A. brachybotrys, A. coriaceum, A. helferi, A. kunsteri*, see there. — *A. riparium* has a hairy disc, highly fused sepals and longer infructescences.

SELECTED SPECIMENS. THAILAND. Peninsular, Ranong prov., Khlong Naka, 9°45'N, 98°40'E, rainforest, on sandstone, 150 – 300 m, 24 Feb. 1974, *Geesink, Hiepko & Charoenphol* 7528 (B, K, L). PENINSULAR MALAYSIA. Johore, Sungai Kayu, 9 March 1937, *SF (Kiah)* 32361 (BM, K, L); Perak, Maxwell's Hill, stream near 3rd mile below Maxwell's Hill, Post Office, 2600 – 2500 ft, 18 Sept. 1945, *SF (Sinclair & Kiah)* 38809 (BM, E, K). SINGAPORE. NE end of MacRitchie Reservoir, 27 Jan. 1949, *SF (Sinclair)* 5535 (E, L). SUMATRA. E coast, Laboehan Batoe subdiv., Bila distr., Hitean Haloban (S of Concession Rantau Parapat B: Topographic Sheet 34, South Center), 17 – 24 May 1933, *Rahmat Si Toroes* 4257 (A, K, L, NY, US). BORNEO. Brunei, Temburong distr., Bukit Patoi, halfway to summit, lowland forest, 26 March 1988, *Wong* 311 (A, AAU, K, L); Sabah, Mt Kinabalu, Penibukan, near camp, jungle hillside, 4000 ft, 9 March 1933, *Clemens & Clemens* 32017 (A, B, K, L, NY); Sabah, Sandakan distr., Botai Forest Reserve, logged area, 21 Nov. 1983, *SAN (Rahim et al.)* 59847 (K, KEP, L, SAR); Sarawak, Miri div., 6 ½ miles Bakam road, river bank, 7 April 1966, *S (Sibat ak Luang)* 24833 (K, KEP, L); Sarawak, Kuching div., Nanga Pelagos, 25 July 1938, *SF (Daud & Tachun)* 35652 (B, K, L, NY, SAR); C Kalimantan, Bukit Raya and upper Katingan (Mendawai) river area, upper Samba R., 60 – 80 km NNW of Tumbang Samba, base camp Tumbang Riang, 0°50'S, 112°50'E, primary rainforest, 150 m, 22 Nov. 1982, *Mogea* 3562 (K, KEP, L, US); E Kalimantan, Samarinda distr., Loa Haur region, sandy loam, 50 m, 30 Aug. 1954, *Kostermans* 10841 (A, K, L).

36b. var. **hosei** *(Pax & K. Hoffm.) Petra Hoffm.*, Kew Bull. 54: 360 (1999). — *A. hosei* Pax & K. Hoffm. in Engl., Pflanzenr. 81: 138 (1922). Type: Sarawak, Baramdistrikt, Mirifluß, *Hose* 549 (K!, lectotype, designated in Hoffmann 1999a: 360; BM!, E!, L!, isolectotypes).

A. plumbeum Pax & K. Hoffm. in Engl., Pflanzenr. 81: 133 (1922). Type: Borneo, Sarawak, Kuching, *Beccari* PB 154 (K!, lectotype, designated in Hoffmann 1999a: 360; P!, isolectotype).

A. hosei Pax & K. Hoffm. var. *microcarpum* Airy Shaw, Kew Bull. 28: 271 (1973). Type: Sarawak, *S (Anderson)* 20204 (holotype K!, isotypes L!, SAR!).

A. hosei Pax & K. Hoffm. var. *angustatum* (Airy Shaw) Airy Shaw, Kew Bull., Addit. Ser. 4 (Euphorb. Borneo): 212 (1975). — *A. neurocarpum* Miq. var. *angustatum* Airy Shaw, Kew Bull. 28: 270 (1973). Type: Sarawak, 3rd Division, Bukit Raya, Kapit, shale ridge, mixed dipterocarp forest, alt. 400 m, 13 Nov. 1964, *S (Suib)* 22269 (K!, holotype, A!, L!, SAR!, isotypes). — Note: Airy Shaw first described this variety for its narrow leaves based on a specimen with dorsiventrally compressed fruits (therefore in *A. neurocarpum*). He later transferred it to *A. hosei* because of the small stipules.

Stipules not foliaceous, lanceolate, linear or subulate, often falcate, 4 – 7 × 0.5 – 1 mm. *Petioles* (0 –)2 – 6(– 10) mm long. *Leaf blades* elliptic, oblong or obovate, (3 –)6 – 10(– 13) × (0.7 –)2 – 3(– 5) cm, (2 –)3.5(– 5.8) times as long as wide, without conspicuous stomata abaxially, drying dark reddish brown to grey. *Pedicels* 0 – 4 mm long. *Infructescences* 6 – 9 cm long. *Fruiting pedicels* 1 – 9 mm long. *Fruits* dorsiventrally compressed, terete, or laterally compressed, 8 – 13 × 6 – 9 mm, glabrous, not white-pustulate.

DISTRIBUTION. Peninsular Malaysia (Kelantan, Terengganu, Pahang), Sumatra (Aceh, one specimen only), Borneo. Map 28.

HABITAT. In lowland to hill (mixed) dipterocarp forest, dipterocarp-nothofagus forest, mixed lowland forest and lower montane forest dominated by

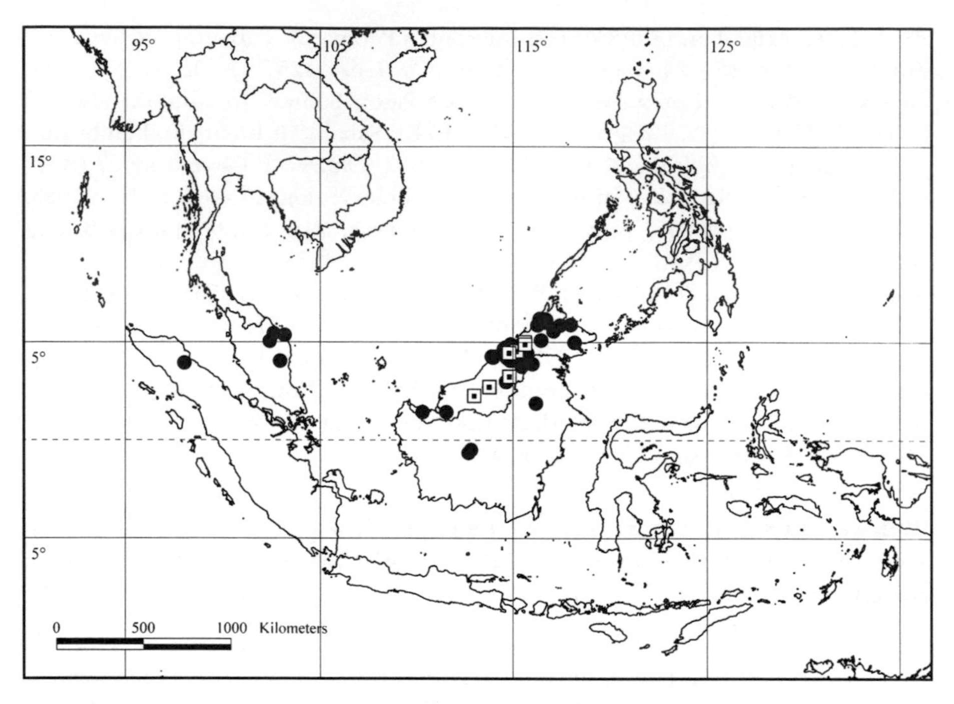

MAP 28. Distribution of *A. neurocarpum* var. *hosei* (●) and var. *linearifolium* (⊡).

Lauraceae, Fagaceae, Myristicaceae; in mossy forest; in heath forest ("kerangas"); in seasonal swamps and peat swamp forest; in riverine forest; in old secondary vegetation around human habitation (Malay: "belukar"); in primary or secondary vegetation; often in shade. On white sand, clay, sandy loam or shallow peat, often alluvial soil, over sandstone, limestone or ultrabasic rock. 0 – 1900 m altitude.

USES. The wood is used as firewood. Reported to be used "as antidote for the species called "Itak". When this is tied onto a string and put across a trail, the enemy named would suffer physical pain leading to paralysis. This plant is carried for protection. Used by Penan in blockade." (*S (Chai et al.)* 68826 from Sarawak).

CONSERVATION STATUS. Least Concern (LC). Like the type variety, var. *hosei* is widespread and frequently collected. 162 specimens examined.

VERNACULAR NAMES. Sabah: kukuleb (Dusun); Sarawak: lapad litak (LB.); ubar-budak (Kelabi); Brunei: tis (Dusun).

ETYMOLOGY. The epithet refers to the collector of the type specimen.

KEY CHARACTER. Stipules not foliaceous; otherwise similar to the type variety.

SIMILAR SPECIES. *A. brevipes*, see there. — *A. stipulare* has more oblong leaves, lacks a rufous indumentum, and the leaves dry olive-green.

NOTE. Differs from the type variety in the narrower, thicker stipules, more coriaceous leaves, weaker rufous indumentum, often pedicellate staminate flowers and less hairy sepals. These differences are not always clear and intergradation occurs (cf. also Airy Shaw 1973: 272).

SELECTED SPECIMENS. PENINSULAR MALAYSIA. Terengganu, Sungai Trengganu at Sungai Panchor, S bank, broad steep rocky hillside, mountains near steep land bondary, granite, 800 ft, 16 Feb. 1972, *FRI* (*Whitmore*) 20571 (K, KEP, L). SUMATRA. Aceh, Gunung Leuser National Park, Sekundur Forest Reserve, upper Besitang R. area, Langkat, base camp at Aras Napal, 3°55'N, 98°00'E, primary forest remnant in logged over area, well drained brown loam soil, 50 – 100 m, 12 Aug. 1991, *De Wilde & De Wilde-Duyfjes* 21367 (L). BORNEO. Brunei, Tutong distr., Kampung Bukit, old secondary forest, 1 May 1989, *Niga Nangkat* 155 (AAU, K, KEP, L, SAR); Sabah, Mt Kinabalu, Penibukan, 4500 ft, 7 March 1933, *Clemens & Clemens* 31940 (A, B, K, NY); Sarawak, Miri div., Bario, Pamerario R., Kalabit Highlands, primary forest, podsol on sandy loam, 1000 m, 1970, *Nooteboom & Chai* 1749 (B, K, KEP, L, SAR, US); Kalimantan, Bukit Raya, 0°39'S, 112°42'E, primary montane forest, c. 1500 m, 24 Jan. 1983, *Nooteboom* 4578 (K, L).

36c. var. **linearifolium** *(Pax & K. Hoffm.) Petra Hoffm.*, Kew Bull. 54: 360 (1999). — *A. linearifolium* Pax & K. Hoffm. in Engl., Pflanzenr. 81: 130 (1922); Airy Shaw, Kew Bull. 28: 275 (1973). Type: Sarawak, *Beccari* PB 3831 (K!, lectotype, designated in Hoffmann 1999a: 360).

Stipules often foliaceous, lanceolate, apically acute-mucronate, basally symmetrical, venation reticulate, chartaceous, 4 – 12 × 1.5 – 4 mm. *Petioles* 2 – 4 mm long. *Leaf blades* lanceolate to linear, sometimes falcate, (5 –)6 – 12(– 13) × (0.7 –)1 – 1.5(– 1.8) cm, (5.5 –)8(– 11) times as long as wide, with conspicuous stomata abaxially (dissecting microscope!), drying dark reddish brown. *Pedicels* 0.5 – 1 mm long. *Infructescences* 3 – 5 cm long. *Fruiting pedicels* 1 – 3 mm long. *Fruits* laterally compressed, 5 – 9 × 3 – 5 mm, glabrous, sometimes white-pustulate.

DISTRIBUTION. Borneo, including Brunei, Sabah and Sarawak. Map 28.

HABITAT. In riverine forest, seasonally flooded areas, usually bordering primary lowland or hill dipterocarp forest. On sandy clay and shales. 20 – 700 m altitude.

CONSERVATION STATUS. Near Threatened (NT). This variety is only known from 13 specimens.

VERNACULAR NAMES. Sarawak: owa.

ETYMOLOGY. The epithet refers to the shape of the leaves (Latin, linearis = linear; folium = leaf).

KEY CHARACTERS. Rheophytic habit.

SELECTED SPECIMENS. BORNEO. BRUNEI. Tutong distr., upstream from Belabau on Tutong R., 4°26'N, 114°48'E, lowland mixed dipterocarp forest, riverside forest, on river bank, probably rheophytic, 20 m, 28 March 1990, *Coode* 6350 (K); Temburong distr., Sungai Temburong at Kuala Belalong, ridge W of river, 4°32'N, 115°09'E, mixed dipterocarp forest, slopes and ridge top, Setap shales, 210 m, 25 June 1989, *Dransfield* 6724 (K, L); Temburong, Amo, Apoi Forest Reserve, Temburong R. catchment, NE spur-ridge of Bukit Belalong opposite Sg. Tulan, 4°32'N, 115°11'E, primary hill dipterocarp forest, steep ridge, grey clay and shale, Setap shale formation, 155 – 180 m, 16 July 1993, *Sands* 5834 (K). SABAH. Sipitang distr., 3 miles N of Mendalong trail to Malaman, relict forest along stream, between boulders, 27 June 1964, *SAN (Meijer)* 43210 (K, L); Sipitang distr., Lumaku Forest Reserve, on foot of Mt Lumaku, primary forest, on river bank, soil formed by rock, sandy, brown-yellow, 14 March 1972, *SAN (Saikeh)* 72101 (K, L); Tongod Kinabatangan, Sungai Enodol, primary forest, 700 m, 26 July 1983, *SAN (Dewol Sundaling)* 99478 (K). SARAWAK. 4th div., Baram, around house compound (Long Selatong Ulu), Kenyah, Cepo Ga', 22 Feb. 1977, *Chin See Chung* 2513 (L); Kapit, Pelagus, near a stream, on sandy clay soil, 8 July 1979, *S (Lee)* 40603 (K, L); Kuching div., Nanga Pelagos, 24 July 1938, *SF (Daud & Tachun)* 35642 (B, K, L, NY).

37. Antidesma orthogyne *(Hook. f.) Airy Shaw,* Kew Bull. 26: 359 (1972); Kew Bull. 26: 459 (1972). — *A. velutinosum* Blume var. *orthogyne* Hook. f., Fl. Brit. India 5: 357 (1887). Type: Malacca, *Griffith* Kew Distr. No. 4928 (K!, lectotype, here designated; GH?, K!, isolectotypes).

Shrub or tree, up to 6 m, diameter up to 5 cm (*fide Alvins* s.n.: large tree 27 m). Bark grey. *Young twigs* terete to obtuse-angled, densely spreading-hirsute, brown. *Stipules* usually persistent, linear to narrowly deltoid, apically acute, (3 –)5 – 7(– 10) × 0.5 – 1.5 mm, hirsute. *Petioles* terete or narrowly channelled adaxially, 3 – 10 × 1 – 2 mm, densely spreading-hirsute. *Leaf blades* elliptic to oblong, apically acuminate-mucronate, basally acute to obtuse, (6 –)15 – 20(– 27) × (3 –)5 – 7(– 10) cm, (2 –)2.8(– 5) times as long as wide, eglandular, chartaceous, glabrous except along the midvein adaxially, spreading-hirsute all over abaxially, especially along the veins, intercostal areas rarely glabrous, dull to moderately shiny on both surfaces, major veins impressed adaxially, tertiary veins weakly percurrent, widely spaced, drying olive-green, lighter abaxially. *Staminate inflorescences* 3 – 6 cm long, axillary, simple or consisting of up to 7 branches, densely set with flowers, axes spreading-hirsute. *Bracts* linear to lanceolate, apically acute to rounded, 0.5 – 0.7 × 0.2 – 0.5 mm, pilose to pubescent, margin fimbriate to glandular-fimbriate. *Staminate flowers* 1.5 – 2 × 1 – 1.5 mm. *Pedicels* 0.2 – 0.5 mm long, not articulated, glabrous to pilose. *Calyx* c. 0.5 mm long, sepals 4, free or nearly so, nearly orbicular, apically rounded to acuminate, pilose outside, glabrous inside, margin erose, sometimes glandular. *Disc* cushion-shaped, enclosing the bases of the filaments and pistillode, hirsute. *Stamens* 4, c. 1.5 mm long, exserted c. 1 mm from the calyx, anthers 0.3 – 0.4 × 0.3 – 0.4 mm. *Pistillode* subulate, c. 0.5 × 0.1 – 0.2 mm, extending to the same length as, to slightly exserted from the sepals, glabrous to hirsute. *Pistillate inflorescences* 4 – 10 cm long, axillary, simple or once-branched at the base, axes densely hirsute. *Bracts* deltoid to linear, 0.5 – 1 × 0.2 – 0.3 mm, hirsute. *Pistillate flowers* 1.5 – 2 × 1 – 1.5 mm. *Pedicels* 0.5 – 0.8 mm long, glabrous. *Calyx* 0.7 – 1 mm long, sepals 4, free or nearly so, c. 0.5 mm wide, narrowly deltoid to orbicular, apically acute to rounded, hirsute outside, glabrous inside but with long hairs at the base, margin sometimes glandular. *Disc* much shorter than the sepals, hirsute. *Ovary* ellipsoid to globose, spreading-hirsute, style terminal to subterminal, stigmas 3 – 6. *Infructescences* 5 – 12 cm long. *Fruiting pedicels* 1 – 2 mm long, spreading-hirsute. *Fruits* ovoid, apiculate, dorsiventrally compressed, basally symmetrical to slightly asymmetrical, with a subterminal to terminal style, 7 – 8 × 4 – 5 mm, pilose, white-pustulate, areolate when dry. Fig. 15A – E.

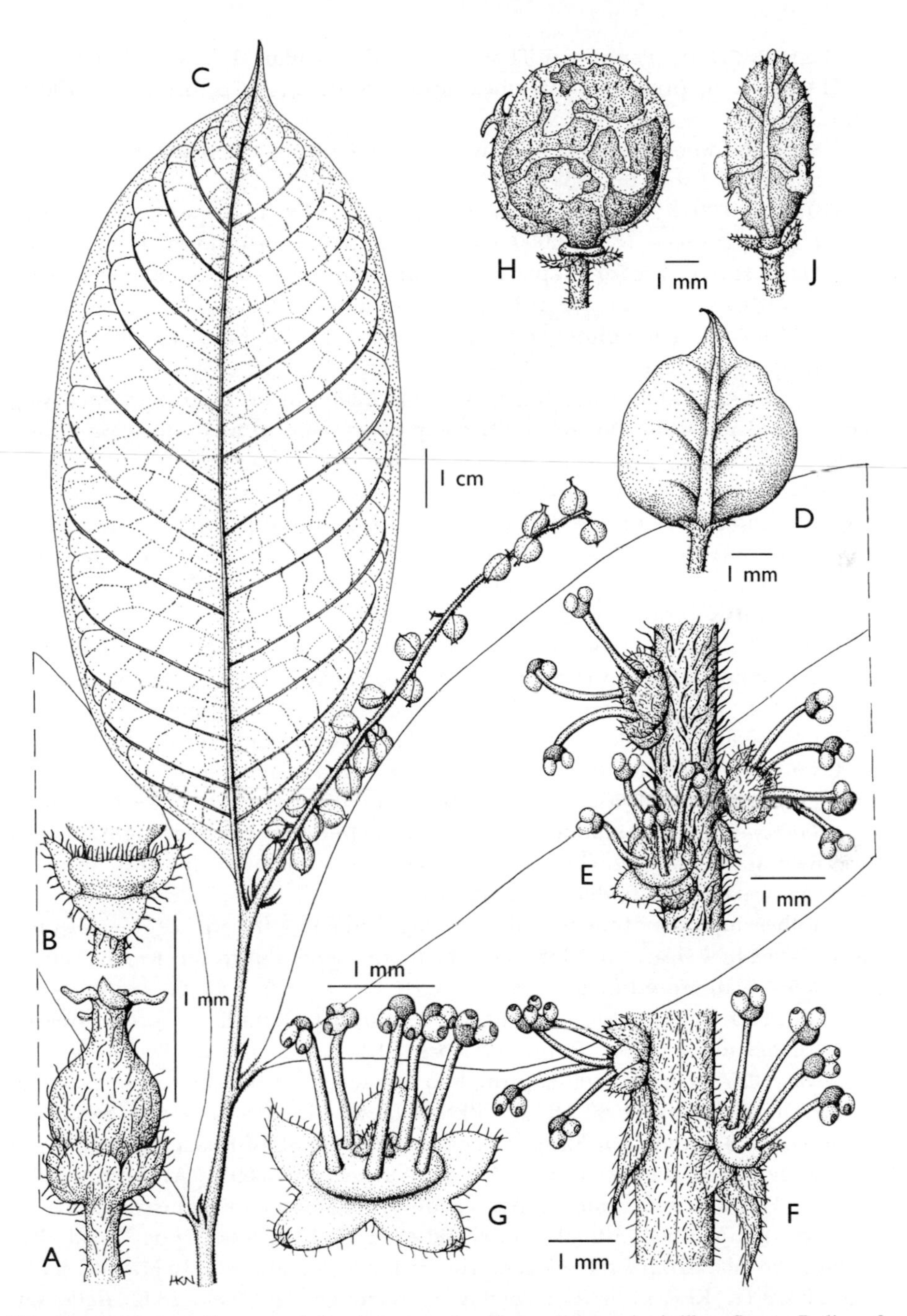

FIG. 15. *Antidesma orthogyne* and *A. velutinosum*. **A – E** *A. orthogyne*. **A** pistillate flower; **B** disc of pistillate flower, calyx partly removed; **C** habit with infructescence; **D** fruit in dorsal view; **E** part of staminate inflorescence. **F – J** *Antidesma velutinosum*. **F** part of staminate inflorescence; **G** staminate flower; **H** fruit in lateral view; **J** fruit in dorsal view. **A – B** from *FRI* 20979 (L); **C – D** from *Sinclair & Kiah* 39949 (L); **E** from *SF (Kiah)* 35033 (L); **F – G** from *Stone & Mahmud Sidek* 12479 (L); **H – J** from *FRI* 4076 (L). Drawn by Holly Nixon.

DISTRIBUTION. Peninsular Thailand and Peninsular Malaysia. Map 26.

HABITAT. In primary, sometimes disturbed, dipterocarp forest. 10 – 700 m altitude.

USES. The wood is used for planks. The juice from the fruits is used as an antiseptic (*FRI (Cockburn)* 7467).

CONSERVATION STATUS. Near Threatened (NT). This species has a large extent of occurrence (c. 100,000 km^2) but intensive agriculture in Peninsular Malaysia causes concern for all species endemic to the area. 45 specimens examined.

VERNACULAR NAMES. beras beras pachat, muranti tukukor.

ETYMOLOGY. The epithet refers to the shape of the ovary (Greek, orthos = straight; gyne = female).

KEY CHARACTERS. Spreading-hirsute indumentum on most parts including the disc and the fruits; dorsiventrally compressed fruits; leaves soft, drying olive-green.

SIMILAR SPECIES. *A. cruciforme* and *A. montanum*, see there. — *A. roxburghii* Tul. from India has larger, reddish drying leaves, more bristly ovaries, numerous stigmas and protruding staminate discs. — *A. velutinosum* has the tertiary veins closer together, longer bracts, glabrous discs, more sepals and stamens, lateral styles and dorsiventrally compressed fruits (Fig. 15).

NOTE. This species has been confused with *A. velutinosum* in the past. Hooker (in the protologue) and Airy Shaw (1972b: 460) recognised it as a separate taxon because of the different fruits but considered staminate specimens indistinguishable from *A. velutinosum.* Airy Shaw even suggested fruit dimorphism in *A. velutinosum.* In fact, the type of *A. orthogyne* is a mixed gathering, consisting of twigs of *A. orthogyne* with very young fruits and one staminate flowering twig of *A. velutinosum.* No staminate specimen had hitherto been recognised and described as *A. orthogyne,* even though there are staminate collections in many herbaria (usually filed under *A. velutinosum*).

Another duplicate of the mixed gathering containing the type of *A. orthogyne* is *Herb. Griffith* Kew Distr. No. 4928 in P! which represents a third species, the similar *A. roxburghii* Tul. from India.

SELECTED SPECIMENS. THAILAND. Peninsular, Narathiwat prov., Waeng, slope of hill in evergreen forest, 17 April 1972, *Sangkhachand, Phusomsaeng & Nimanong* 1057 (K, L); Songkhla prov., Kao Keo Range, evergreen forest, c. 500 m, 12 March 1928, *Kerr* 14503 (K, L). PENINSULAR MALAYSIA. Negeri Sembilan, Pasoh Forest Reserve, near Simpang Pertang, 200 km SE of Kuala Lumpur, 50 ha plot, column 50, 100 m, 30 June 1989, *Gentry & LaFrankie* 66914 (A); Kedah, Bukit Enggang Forest Reserve, catchment area, disturbed hillside forest near stream, low altitude, 13 June 1966, *FRI (Whitmore)* 400 (K, KEP, L); Perak, Upper Perak, EW highway, 26 miles from Grik, hillside primary forest, low altitude, 16 May 1973, *FRI (Ng)* 20979 (K, KEP, L, SING); Penang, Government Hill, Feb. 1892, *Ridley* s.n. (SING); Terengganu, 38th mile state land, Kuala Trengganu-Besut road, W side, 15 July 1953, *SF (Sinclair & Kiah b. Salleh)* 39949 (E, K, L, SING); Kelantan, Kampong Gobek, Kerilla Estates, Tamangan, forest, 100 ft, 1 March 1959, *Shah & Kadim* 497 (K, L, SING); Pahang, Kuala Keniyam, Gua Luas, 100 ft, 4 March 1968, *Shah b. Kadim* 1551 (KEP, L, SING).

38. Antidesma pachystachys *Hook. f.*, Fl. Brit. India 5: 355 (1887). Type: Malaysia, Perak, at Larut, *King's collector* 1934 (K!, lectotype, here designated; P!, isolectotype). Original syntypes: same locality, *King's collector* 2178 (K!); Penang, *Wallich* 8569 (K!). — Note: Hooker gave no collection numbers, but the sheets listed above are annotated by him as *A. pachystachys* in K.

A. pachystachys Hook. f. var. *palustre* Airy Shaw, Kew Bull. 28: 269 (1973), **synon. nov.** Type: Malaya, Johore, mile 13.5, Mawai-Jemulang Road, by swampy stream on hill, 8 Feb. 1935, *SF (Corner)* 29232 (K!, holotype).

Shrub or tree, up to 7 m. *Young twigs* terete to obtuse-angled, ferrugineous-pubescent but soon becoming glabrous, brown. *Stipules* usually persistent, foliaceous, ovate to lanceolate, rarely linear, apically acuminate, basally asymmetrical, chartaceous, reticulately veined, (8 –)15 – 40 × (2 –)5 – 12 mm, pilose to glabrous. *Petioles* slightly channelled adaxially, becoming rugose when old, 3 – 12(– 22) × 2 – 5 mm, ferrugineous-pubescent, glabrescent. *Leaf blades* elliptic-oblong, apically acuminate-mucronate, basally acute to rounded, (15 –)30 – 40(– 45) × (5 –)10 – 12(– 18) cm, (2 –)3.2(– 4.8) times as long as wide, eglandular, coriaceous to chartaceous, glabrous, sometimes pilose only along the midvein adaxially, glabrous to stellate-pilose or with 2-armed hairs all over or only along the major veins abaxially, shiny adaxially, shiny or dull abaxially, major veins sharply raised adaxially, especially in the apical part of the leaf blades, tertiary veins mainly percurrent, close together, drying olive-green to brownish. *Staminate inflorescences* 10 – 27 cm long, axillary, simple, axes glabrous to pubescent. *Bracts* lanceolate, apically acute, 1 – 1.5 × 0.5 – 1 mm, pilose to pubescent. *Staminate flowers* 1.5 – 2 × 1.5 – 2 mm. *Pedicels* absent. *Calyx* 0.8 – 1 × 1 – 1.5 mm, cupular, sepals 4 – 5, fused for $^1/_2$ – $^2/_3$ of their length, nearly orbicular, apically rounded, glabrous to pubescent outside, glabrous inside, margin fimbriate. *Disc* extrastaminal-annular, pilose. *Stamens* 4 – 5, c. 1.5 mm long, exserted 0.5 – 1 mm from the calyx, anthers c. 0.5 × 0.5 mm. *Pistillode* clavate, c. 0.7 × 0.3 – 0.4 mm, extending to the same length as the sepals, pubescent. *Pistillate inflorescences* 15 – 20 cm long, axillary, simple, axes glabrous to ferrugineous-pubescent. *Bracts* linear to lanceolate, 1 – 2 × 0.3 – 0.5 mm, pubescent. *Pistillate flowers* 2 – 2.5 × c. 2 mm. *Calyx* c. 1 × 2 mm, obconical, sepals 5, fused for c. $^1/_2$ of their length, deltoid, apically acute, sinuses acute, glabrous to pilose outside, glabrous inside, margin fimbriate. *Disc* shorter than the sepals but usually visible between them, glabrous or pilose only at the margin. *Ovary* globose, laterally compressed, densely pubescent, style lateral, stigmas 4 – 7, rather long. *Infructescences* 25 – 50 cm long. *Fruiting pedicels* 3 – 6 mm long, glabrous to ferrugineous-pubescent, indumentum sometimes stellate. *Fruits* obliquely ovoid (mango-shaped), laterally compressed, basally symmetrical, with a subterminal style, 12 – 16 × 8 – 9 mm, glabrous to stellate-pilose or with 2-armed hairs, sometimes slightly white-pustulate, reticulate when dry.

DISTRIBUTION. Peninsular Malaysia: Johore, Negeri Sembilan, Pahang, Penang, Perak. Map 29.

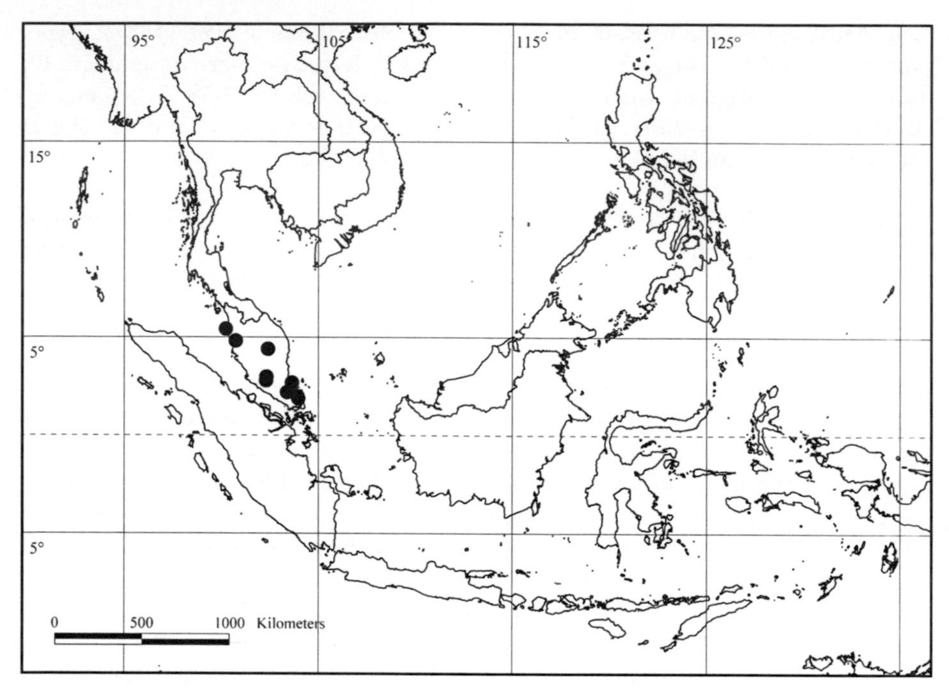

MAP 29. Distribution of *A. pachystachys*.

HABITAT. In dense, shady forest; sometimes in swamp forest. 120 – 500 m altitude.

CONSERVATION STATUS. Near Threatened (NT). Like *A. orthogyne*, the species seems to be widespread in Peninsular Malaysia but a threat category is assigned to it for the same reason. 20 specimens examined.

ETYMOLOGY. The epithet refers to the inflorescence (Greek, pachys = thick, stout; stachys = spike, ear of corn).

KEY CHARACTERS. Veins sharply raised on adaxial leaf surfaces; fruit and leaf indumentum 2-armed to stellate when present; petioles wide; leaves and fruits large; infructescences long.

SIMILAR SPECIES. *A. pendulum* is nearly glabrous and lacks stellate hairs on fruits and abaxial leaf surfaces; it also has adaxially impressed to only slightly raised leaf veins, shorter bracts, fruiting pedicels and fruits, more widely spaced tertiary veins, smaller stipules and glabrous pistillodes. — *A. tomentosum* also lacks stellate hairs and has impressed to flat midveins.

NOTE. The stellate indumentum on the fruits and abaxial leaf surfaces may be 2- to many-armed, sessile or stalked.

SELECTED SPECIMENS. PENINSULAR MALAYSIA. Johore, Labis Forest Reserve, State land 8 mile, S of Labis F. R., secondary forest along extraction road, hillside, 14 April 1966, *FRI (Whitmore)* 151 (K, KEP); Johore, Mersing distr., Labis Forest Reserve, compt. 280, Ulu Endau, lowland forest, 23 March 1968, *FRI (Cockburn)* 7892 (K, KEP, L, SAR, SING); Johore, Sungai Kayu, in swampy forest, 23 Oct. 1936, *SF (Kiah)* 32155 (K, KEP, SING); Johore, Labis, Sg. Juasseh towards Ulu Sg. Kemidak, 1500 ft, 1 Feb. 1971, *Shah & Shukor* 2309 (L, SING); Negeri Sembilan, Pasoh Forest Reserve, 300 m NW of housing, 2°58'N, 102°19'E, on slight slope under dense canopy, 1981, *Rogstad* 631 (A); Pahang, Gemas, in shade, 16 Sept. 1920, *SF (Burkill)* 6391 (SING); Pahang, Pulau Tioman, Bukit Surin, rocky jungle, 1000 – 2000 ft, 19 April 1929, *SF (Henderson)* 21713 (K, L, SING); Perak, s. loc., dense jungle, rich soil, 500 – 800 ft, Nov. 1883, *King's collector* 3778 (G, L, SING).

39. Antidesma pahangense *Airy Shaw,* Kew Bull. 23: 277 (1969). Type: Malaysia, Pahang, Ulu Telom, 17 Aug. 1931, *Dolman* 27623 (K!, holotype; SING! ("Jaamat"), isotype).

Tree, up to 13 m, diameter up to 6 cm. Bark grey, brown or fawn, smooth; inner bark whitish; wood hard, whitish, reddish or brown. *Young twigs* terete to obtuse-angled, ferrugineous-pubescent, brown. *Stipules* often caducous, linear to lanceolate, 5 – 17 × 0.5 – 2(– 4) mm, pubescent. *Petioles* flat to slightly channelled adaxially, 3 – 9(– 13) × 1 – 3(– 5) mm, pubescent, becoming glabrous when old. *Leaf blades* elliptic, more rarely oblong, apically acuminate-mucronate, basally acute, more rarely obtuse or rounded, (10 –)20 – 25(– 33) × (3.5 –)7 – 10(– 17) cm, (2.1 –)2.7(– 3.3) times as long as wide, eglandular, chartaceous to coriaceous, glabrous, or slightly pilose only at the base and along the midvein adaxially, slightly pilose abaxially, more so along the veins, dull on both surfaces, major veins flat or shallowly impressed, tertiary veins reticulate to weakly percurrent, rather widely spaced, drying reddish brown. *Staminate inflorescences* 11 – 25 cm long, axillary, simple, axes ferrugineous-pubescent. *Bracts* linear, 2 – 2.5 × 0.2 – 0.4 mm, pubescent. *Staminate flowers* c. 2 × 2 mm. *Pedicels* 0.5 – 1.5 mm long, not articulated, pilose. *Calyx* c. 1 × 1.5 – 2 mm, bowl-shaped, sepals 5 – 6, fused for $^{1}/_{2}$ of their length, deltoid, apically acute, pilose to pubescent outside, glabrous inside but sometimes with long hairs at the base, margin entire, sometimes fimbriate. *Disc* cushion-shaped, enclosing the bases of the filaments and pistillode, sparsely pilose to densely pubescent. *Stamens* 5 – 6, 1.5 – 2 mm long, exserted 1 – 1.5 mm from the calyx, anthers 0.4 – 0.5 × 0.5 – 0.7 mm. *Pistillode* obconical, sometimes completely sunk into the disc, sometimes crateriform, 0.2 – 0.5 × 0.3 – 0.7 mm, shorter than to exserted from the sepals, pilose to densely pubescent. *Pistillate inflorescences* 10 – 25 cm long, axillary (also cauline?), simple, axes ferrugineous-pubescent. *Bracts* linear, 2 – 2.5 × 0.2 – 0.3 mm, pubescent. *Pistillate flowers* 2 – 3 × 1.5 – 2 mm. *Pedicels* 0.3 – 1 mm long, pubescent. *Calyx* 1 – 1.5 × 1 – 1.5 mm, sepals 5 – 6, fused only at the base, narrowly deltoid, apically acute, pubescent outside, glabrous inside, margin entire. *Disc* shorter than the sepals but usually visible between the sepals, pilose at the margin, otherwise glabrous. *Ovary* ellipsoid, densely ochraceous-pubescent, style subterminal, stigmas 4 – 5. *Infructescences* 20 – 60 cm long. *Fruiting pedicels* 1 – 5 mm long, pubescent. *Fruits* ellipsoid, terete or slightly laterally, or dorsiventrally compressed, basally symmetrical or asymmetrical, with a terminal to distinctly lateral style, 9 – 10 × 7 – 9 mm, pilose, sometimes with stellate hairs, sometimes slightly white-pustulate, reticulate to areolate when dry.

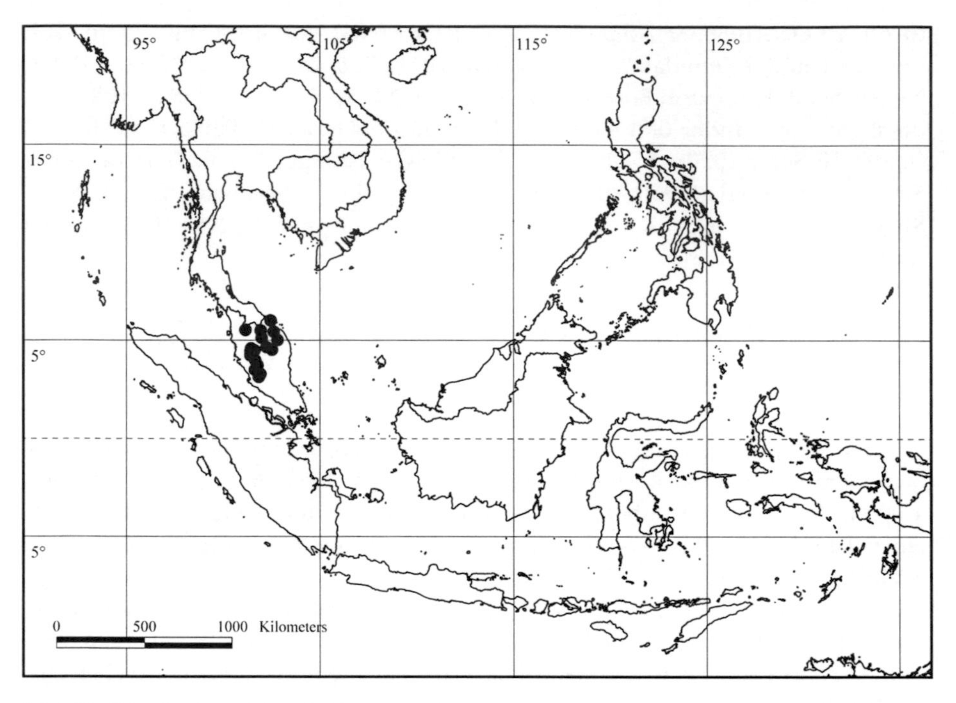

MAP 30. Distribution of *A. pahangense.*

DISTRIBUTION. Peninsular Malaysia: Kelantan, Pahang, Selangor, Terengganu. Map 30.

HABITAT. In mossy forest; in forest with bamboo; often on steep rocky hillsides or streamsides. (70 –)900 – 1400 m altitude.

CONSERVATION STATUS. Near Threatened (NT). This species is also widespread (extent of occurrence c. 35000 km^2) but the general threat of habitat destruction in Peninsular Malaysia causes concern for endemic taxa. 25 specimens examined.

VERNACULAR NAMES. kerdap.

ETYMOLOGY. The epithet refers to the state of Pahang in Malaysia, where the type was collected.

KEY CHARACTERS. Pedicellate flowers; long, narrow bracts; reddish or yellowish, long, shaggy indumentum; sometimes stellate hairs on fruits; leaves drying reddish.

SIMILAR SPECIES. *A. riparium* has shorter bracts, more highly fused, acute sepals in staminate flowers, larger fruits, smaller leaves and a less dense, shorter indumentum. — *A. tomentosum* has sessile staminate flowers, shorter bracts and staminate inflorescences, more percurrent and closer tertiary leaf venation, a shorter and tidier indumentum, and lacks stellate hairs.

SELECTED SPECIMENS. PENINSULAR MALAYSIA. Kelantan, Ulu Kelantan, Sungai Brok, streamside, 10 June 1967, *FRI (Ng)* 5337 (A, K, L, SING); Kelantan, Kuala Limau Nipis, 29 Feb. 1924, *SF (Nur & Foxworthy)* 12161 (BM, K, SING);

Pahang, Fraser's Hill, Big Tree Plot, hill side, 4000 ft, 15 Oct. 1966, *FRI (Kochummen)* 2156 (K, L, SING); Pahang, Fraser's Hill, Big Tree Plot, upper hill forest, 4000 ft, 22 June 1982, *FRI (Kamarudin)* 28750 (K, L); Pahang, Gunong Tahan, 26 July 1936, *SF (Kiah)* 31908 (K, L, SING); Pahang, Jalan Simpon Forest Reserve, Taman Negara, 30 April 1975, *Teo & P* KL 3153 (K, L, SING, U); Selangor, Ulu Batang Kali, very steep hillside, 3000 ft, 26 Nov. 1967, *FRI (Whitmore)* 4547 (A, K, L, SING); Selangor, Ulu Langat, Menuang Gasing, 3°07'N, 101°49'E, Feb. 1912, *Kloss* s.n. (BM, K); Terengganu, Trengganu Mts, Gunong Kerbat, N bank above Kuala Petang, sandstone ridge, narrow rocky ridge crest, 30 June 1971, *FRI (Whitmore)* 20340 (K, L); Terengganu, Ulu Trengganu, summit ridge of Gunong Lawit, 20 ft mossy pole forest, 4300 ft, 28 March 1974, *FRI (Ng)* 22065 (K, L, SING).

40. Antidesma pendulum *Hook. f.*, Fl. Brit. India 5: 356 (1887). Type: Malaysia, Perak, *Scortechini* 818 (K!, holotype; L!, isotype).

A. batuense J. J. Sm., Icon. Bogor 4: 251, t. 380 (1914). Type: Batoe-Inseln (westlich von Sumatra), Feb. 1897, *Raap* 680 (BO!, lectotype, here designated; L!, isolectotype). Original syntype: same locality, *Raap* 659 (BO!).

A. stenophyllum Merr., Philipp. J. Sci., C, 11: 62 (1916) (non Gage 1922); Airy Shaw, Kew Bull. 28: 275 (1973), Kew Bull., Addit. Ser. 4 (Euphorb. Borneo): 216 (1975), **synon. nov.** Type: Sarawak, Mt Sudan, Feb. – June 1914, *BS (Native collector)* 2081 (US!, lectotype, here designated; A!, [photo of destroyed holotype ex PNH], K!, isolectotypes). — Note: Described by Merrill based on a single collection in staminate bud, distinguished from typical *A. pendulum* by the smaller and narrower leaves (15 – 17.5 × 2.2 – 2.7 cm, 6 – 6.6 times as long as wide). This could be a slightly rheophytic ecotype or a poorly developed plant growing under unfavourable conditions.

A. sumatranum Pax & K. Hoffm. in Engl., Pflanzenr. 81: 120 (1922). Type: Sumatra, *Forbes* 1550 (BM!, lectotype, here designated). Original syntype: Sumatra, *Forbes* 1768 (BM!, GH!, "1768 a").

Shrub or tree, up to 6 m, diameter up to 5 cm. Bark very light brown to silvery grey. *Young twigs* terete, glabrous or nearly so, brown. *Stipules* usually persistent, linear to subulate, 6 – 25 × 0.5 – 3 mm, glabrous. *Petioles* channelled to almost flat adaxially, (3 –)5 – 15(– 30) × 1.5 – 4(– 6) mm, glabrous. *Leaf blades* narrowly obovate, more rarely elliptic, apically acuminate-mucronate, sometimes caudate, basally acute, (13 –)25 – 30(– 50) × (4 –)7 – 9(– 14) cm, (2.5 –)3.5(– 5) times as long as wide, eglandular, chartaceous to coriaceous, glabrous on both surfaces, shiny or dull on both surfaces, major veins flat to slightly raised, rarely shallowly impressed adaxially, tertiary veins mainly percurrent, widely spaced, drying yellowish brown, usually lighter abaxially. *Staminate inflorescences* c. 15 cm long, axillary, simple, solitary or 2 per fascicle, axes glabrous. *Bracts* deltoid, apically acute, 0.5 – 0.7 × 0.3 – 0.4 mm, slightly pilose to glabrous. *Staminate flowers* 1 – 1.5 × 1 – 1.5 mm. *Pedicel* absent. *Calyx* c. 0.7 × 1 – 1.5 mm, cupular, sepals 4 – 5, fused

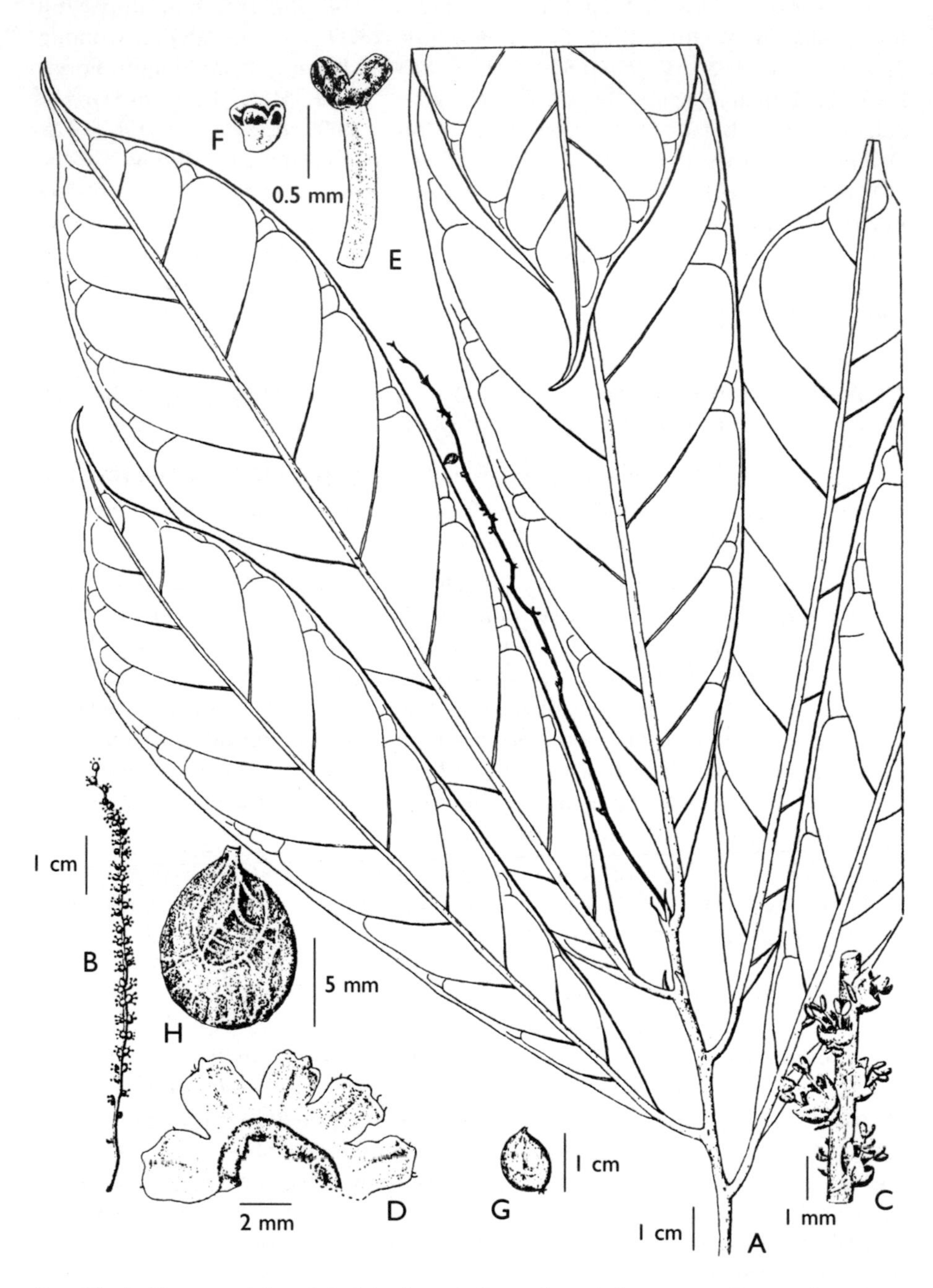

FIG. 16. *Antidesma pendulum.* **A** fruiting branch; **B** staminate inflorescence; **C** part of staminate inflorescence; **D** calyx, removed and spread out; **E** stamen; **F** pistillode; **G** young fruit; **H** fruit. Reproduced from Smith 1914: t. 380 (as *A. batuense* J. J. Sm.). Note that the fruit typically has a more terminal style.

for 1/2 – 3/4 of their length, apically rounded, sinuses rounded, glabrous to pilose outside, glabrous inside, margin erose. *Disc* consisting of 4 – 5 short, free alternistaminal lobes which may appear to be fused extrastaminally, glabrous to pilose. *Stamens* 4 – 5, 1 – 1.5 mm long, exserted c. 0.7 mm from the calyx, anthers c. 0.3 × 0.4 mm. *Pistillode* cylindrical or nearly so, sometimes 2-fid for c. 2/3 of its length, each part 2-fid again apically, 0.5 – 0.7 × 0.2 – 0.5 mm, extending to the same length as or slightly exserted from the sepals, glabrous. *Pistillate inflorescences* 8 – 35 cm long, axillary, simple, axes glabrous to shortly appressed-pilose. *Bracts* deltoid, apically acute, 0.3 – 0.8 × 0.2 – 0.3 mm, glabrous to pilose. *Pistillate flowers* 1 – 1.5 × 1 mm. *Calyx* 0.6 – 0.7 × c. 1 mm, sepals 4 – 5, fused for 1/4 – 1/2 of their length, deltoid, apically obtuse to rounded, sinuses rounded, glabrous to pilose outside, glabrous inside, margin erose, sometimes ciliate. *Disc* shorter than the sepals, glabrous, or pilose only at the margin. *Ovary* ellipsoid, appressed-pilose to appressed-pubescent, style subterminal, stigmas 4 – 5. *Infructescences* 25 – 75 cm long. *Fruiting pedicels* 0 – 1(– 3) mm long, glabrous. *Fruits* ovoid to ellipsoid or lenticular, often apiculate, laterally compressed, basally symmetrical, rarely asymmetrical, with a terminal to subterminal style, 10 – 13 × 7 – 9 mm, glabrous to sparsely pilose (indumentum never stellate), sometimes white-pustulate, reticulate when dry. Fig. 16.

DISTRIBUTION. Peninsular Thailand (Narathiwat), Peninsular Malaysia, Sumatra, Borneo (Kalimantan and Sarawak). Map 31.

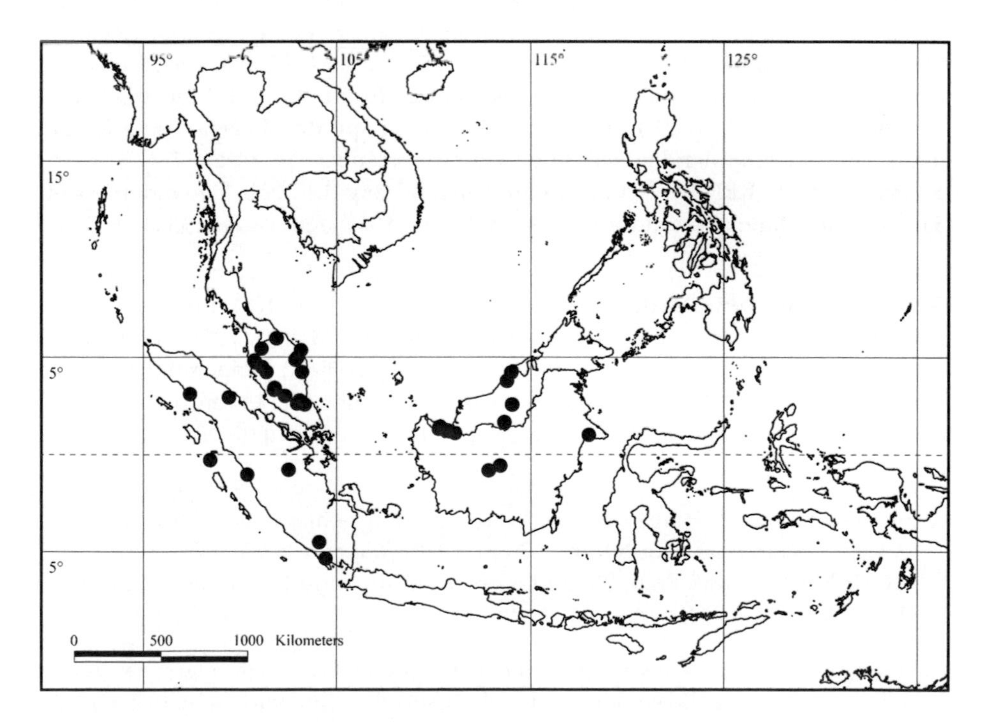

MAP 31. Distribution of *A. pendulum*.

HABITAT. In primary or secondary forest; in mixed dipterocarp forest; in swamp and riverine forest; by waterfalls. On clay-loam soil over limestone, sandstone or granite. 15 – 600(– 1300) m altitude.

CONSERVATION STATUS. Near Threatened (NT). The distribution area of this species may be large enough to classify for Least Concern, but the vast majority of collections of *A. pendulum* were made in primary wet forest, a habitat under ever-increasing threat in Malesia. Nothing is known about the species' adaptability to secondary vegetation. 66 specimens examined.

VERNACULAR NAMES. Sumatra: kayu mangnae.

ETYMOLOGY. The epithet refers to the conspicuously long infructescences (Latin, pendulus = pendulous).

KEY CHARACTERS. Large leaves; thick petioles; plant nearly glabrous; very short fruiting pedicels; long, simple infructescences; fruits hardly longer than wide.

SIMILAR SPECIES. *A. brachybotrys, A. montis-silam, A. pachystachys*, see there. — *A. tomentosum* usually has adaxially impressed to flat major veins, leaves drying reddish to olive-green with the tertiary veins percurrent and close together, cushion-shaped staminate discs, longer fruits and less highly fused sepals.

SELECTED SPECIMENS. THAILAND. Peninsular, Narathiwat prov., Waeng, by stream in tropical evergreen forest, 25 March 1968, *BKF (Phusomsaeng)* 46419 (L). PENINSULAR MALAYSIA. Perak, Gopeng, dense jungle, rocky locality, 300 – 800 ft, June 1883, *King's collector* 4447 (K, L, SING); Perak, Larut, 4 – 600 ft, June 1882, *King's collector* 3865 (3065?) (BM, G, L); Terengganu, Sekayu Forest Reserve, compt. 43, Ulu Trengganu, virgin jungle hillside, 22 Sept. 1969, *FRI (Loh)* 13512 (K, KEP, L). SUMATRA. E coast, Aek Sordang, Loendoet Concession, Koealoe, old jungle and second growth, 1 – 17 April 1927, *Bartlett* 7277 (NY, US); Sumatera Barat, Padang Bungus, Mt Telug Taro, secondary forest, 40 m, 3 June 1953, *Van Borssum Waalkes* 1445 (K, L). BORNEO. Sarawak, Kapit div., Belaga distr., Sungai Iban, Ulu Belaga, lowland mixed dipterocarp forest, near the stream, 5 Nov. 1982, *S (Lee)* 45429 (K, KEP, L); E Kalimantan, Sangkulirang distr., Mt Medandem, N of Sangkulirang, lime and sandstone, 200 m, 7 Aug. 1957, *Kostermans* 13364 (K, L).

41. Antidesma petiolatum *Airy Shaw*, Kew Bull. 33: 16 (1978), nom. nov. — *A. petiolare* Airy Shaw, Kew Bull. 26: 466 (1972), Kew Bull. 28: 276 (1973), nom. illeg. (non Tul. 1851). Type: Territory of New Guinea, Sepik Distr., Aitape subdistr., near Wantipi village (on Bliri R.), steep riverbank in foothills, alt. 195 m, 1 Aug. 1961, *Darbyshire & Hoogland* 8339 (K!, lectotype, here designated; A!, CANB!, L!, isolectotypes).

A. riparium Kaneh. & Hatus., "riparia", nom. nud. in sched.: Dutch New Guinea, Ayerjat, 40 km inward of Nabire, 200 m, 1940, *Kanehira & Hatusima* 12637 (A!, BO!); Nabire, Bivak Prao, 20 km inward from the mouth of Boemi R., 100 m, 1940, *Kanehira & Hatusima* 12836 (BO!).

Tree, up to 10 m, diameter up to 15 cm. Twigs grey. Bark brown, grey or green, smooth or rough, shallowly longitudinally fissured, inner bark salmon, reddish brown, brown cream; wood white, cream or pink. *Young twigs* terete to striate, shortly ferrugineous-pubescent to glabrous, brown, sometimes light brown with a reddish tinge. *Stipules* early-caducous, ovate, apically acuminate, 3.5 – 6 × 2 – 2.5 mm, shortly pubescent. *Petioles* narrowly channelled adaxially to nearly terete, basally and distally sometimes pulvinate up to 10 mm, 15 – 45 × 1.5 – 2 mm, shortly pubescent to glabrous. *Leaf blades* elliptic to ovate, rarely slightly obovate, apically acuminate-mucronate, basally truncate to rounded, more rarely obtuse, sometimes shortly decurrent, (10 –)15 – 20(– 29) × (5 –)7.5 – 10(– 13.5) cm, (1.5 –)1.9 – 2.1(– 2.4) times as long as wide, eglandular, chartaceous, glabrous, or shortly puberulous only along the midvein adaxially and the major veins abaxially, moderately shiny on both surfaces, major veins impressed to flat adaxially, tertiary veins percurrent to weakly percurrent, widely spaced or close together, drying olive-green, lighter abaxially, domatia absent. *Staminate inflorescences* 4 – 7(– 10) cm long, axillary, simple or consisting of up to 10 branches, axes ferrugineous- or ochraceous-pilose to -pubescent. *Bracts* linear, lanceolate or oblanceolate to spathulate, apically acute to rounded, 0.5 – 1.2 × 0.3 – 1 mm, pilose, margin fimbriate. *Staminate flowers* 1 – 2 × 1 – 1.5 mm. *Pedicels* 0.3 – 0.5 mm long, not articulated, glabrous to sparsely pilose. *Calyx* 0.5 – 1 × c. 1 mm, cupular to nearly globose, sepals 3 – 5, fused for $^1/_2$ – $^3/_4$ of their length, deltoid, often unequal, apically acute, sinuses rounded, glabrous outside and inside, margin fimbriate. *Disc* consisting of (2 –)3(– 4) free alternistaminal lobes, lobes ± obconical, c. 0.5 × 0.2 – 0.5 mm, exserted from the sepals, sometimes extrastaminally fused, glabrous at the sides, ferrugineous- to whitish-pubescent apically. *Stamens* 2 – 4, 1 – 1.5 mm long, exserted 0.5 – 1 mm from the calyx, anthers c. 0.4 × 0.5 mm. *Pistillode* subulate, rarely clavate, 0.5(– 1) × 0.1 – 0.3 mm, extending to the same length as or, more rarely, exserted from the disc, ferrugineous- to whitish-pubescent. *Pistillate inflorescences* 2.5 – 6 cm long, axillary to cauline, branched regularly, consisting of 3 – 10 branches, rarely simple, axes pilose to pubescent. *Bracts* linear to lanceolate, apically acute, 0.5 – 0.7 × 0.3 – 0.4 mm, pilose, margin fimbriate. *Pistillate flowers* c. 1 × 0.5 mm. *Pedicels* 0.3 – 0.5 mm long, glabrous to sparsely pilose. *Calyx* 0.4 – 0.5 × 0.4 – 0.8 mm, cupular to nearly globose, sepals 5, fused for $^2/_3$ of their length, deltoid, often unequal, apically acute to obtuse, sinuses narrow, rounded to obtuse, glabrous on both sides, margin fimbriate. *Disc* shorter than the sepals, but indumentum usually exserted from the sepals, ochraceous- to ferrugineous-tomentose at the margin, hairs about as long as the disc. *Ovary* ellipsoid to ovoid, terete, glabrous, style terminal to subterminal, stigmas 3 – 6. *Infructescences* 5 – 7 cm long. *Fruiting pedicels* 0.5 – 1 mm long, glabrous to sparsely pilose. *Fruits* globose to ellipsoid or obovoid, slightly laterally compressed to terete, basally symmetrical, with a subterminal to terminal style, 2 – 3 × 1.5 – 2 mm including the calyx, glabrous, not white-pustulate, wrinkled (neither areolate nor reticulate) when dry. Fig. 17.

DISTRIBUTION. New Guinea. Map 32.

HABITAT. In or at the edges of primary (occasionally disturbed) rainforest; in primary riverine forest; in villages, gardens and regrowth. On old well-drained volcanic soil. 50 – 850 m altitude.

USES. The wood is preferred as taro planting and digging stick; it is also used in bush house construction and as firewood.

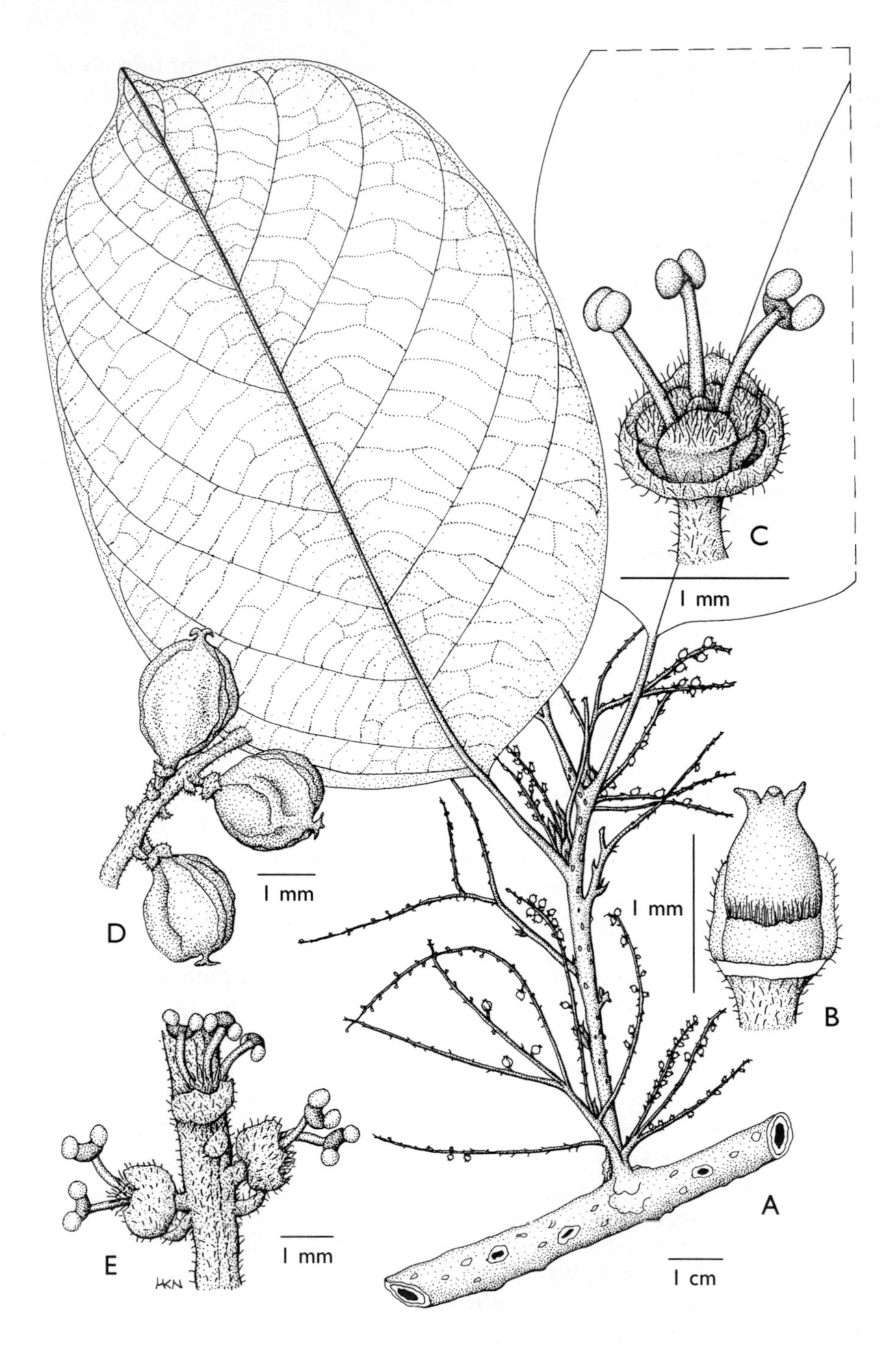

FIG. 17. *Antidesma petiolatum.* **A** habit with hollow branch and infructescences; **B** pistillate flower, part of calyx removed; **C** staminate flower, part of calyx removed; **D** part of infructescence; **E** part of staminate inflorescence. **A – B** from *Jacobs* 9490 (L); **C**, **E** from *Schodde* 3094 (L); **D** from *van Leeuwen* 9870 (L). Drawn by Holly Nixon.

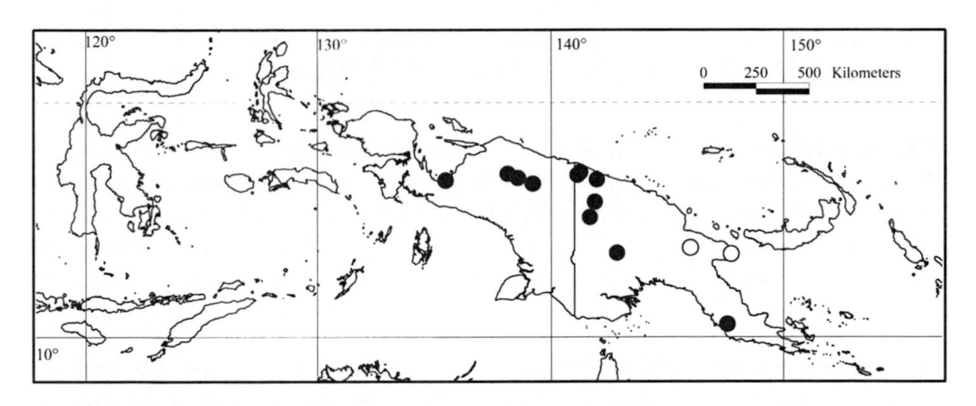

MAP 32. Distribution of *A. petiolatum* (●), doubtful specimens (○).

CONSERVATION STATUS. Near Threatened (NT). This species is widespread in New Guinea (extent of occurrence c. 350,000 km^2) and may be locally common. However, only 18 specimens are known of this rather distinctive species.

VERNACULAR NAMES. as imendal or as imendral (Miyanmin), awilo (Orne), samuli.

ETYMOLOGY. The epithet refers to the long petioles.

KEY CHARACTERS. Ovate, persistent stipules; long petioles; large leaves; fused sepals; hairy disc; staminate flowers on short pedicels; small, glabrous fruits.

SIMILAR SPECIES. *A. excavatum*, *A. ferrugineum* and *A. microcarpum*, see there.

NOTE. The staminate flowering specimen *Brass* 32293 (K, L, NY, US) from Eastern Highlands Province was collected at an unusually high altitude and has consistently extrastaminally fused disc lobes, but shows the characteristic wide stipules, long petioles and large leaves of *A. petiolatum*. In *Clemens & Clemens* 869 (A) from Sattelberg, Morobe province, the staminate inflorescences are like *A. petiolatum* but the petioles are only 6 – 10 mm long and the leaves are only 10 – 15 × 4 – 6 cm and have domatia. This might be a mixed collection as none of the leaves are attached to the flowering twigs; it is provisionally placed here.

SELECTED SPECIMENS. INDONESIA: PAPUA. Bernhard Camp, Idenburg River (Sungai Taritatu), rainforest of lower slopes, 150 m, April 1939, *Brass* 13832 (K, L, US); Ayerjat, 40 km inward of Nabire, in the fringing forest, 200 m, 8 March 1940, *Kanehira & Hatusima* 12637 (A, BO); near Prauwen Bivouac, 10 m, 17 July 1920, *Lam* 660 (K, L); N New Guinea, Rouffaer R. (Sungai Tariku), edge of jungle, hills, 175 m, Aug. 1926, *Van Leeuwen* 9870 (K, L). PAPUA NEW GUINEA. Sandaun prov., Telefomin distr., Miannmin, W of Airstrip on terrace, village areas, gardens and regrowth near airstrip, 760 m, 7 March 1992, *Frodin, Morren & Gabir* 2347 (K); Western section, Musgrave R. c. 4 miles SE of Subitana, 9°27'S, 147°35'E, rainforest, c. 850 ft, 25 Sept. 1962, *Hartley* 10818 (K, L); Mt Bosavi, N Side, N of mission station, 6°26'S, 142°50'E, primary forest, occasionally disturbed or replaced by secondary forest, old well-drained volcanic soil, 600 – 700 m, 29 Oct. 1973, *Jacobs* 9490 (L, US); West Sepik prov., Bewani subprov., Gorge N of Meinat flood plains, 3°07'S, 141°08'E, primary rainforest beside creek, 320 m, 19 Sept. 1982, *LAE (Kerenga)*

56505 (K, L); Sepik distr., Ossima village, 1.5 miles S from village, 3°00'S, 141°15'E, rainforest, 100 ft, 7 March 1964, *NGF (Sayers)* 13237 (K, L, US).

42. Antidesma pleuricum *Tul.*, Ann. Sci. Nat. Bot., Sér. 3: 213 (1851). Type: Phillipines, Calawanio, *Callery* 38 (P! "38bis", lectotype, here designated). Original syntypes: apud Luzonensis, in agro Manillensi, *Cuming* 900 (CGE!, E!, G!, K!, L!, P!, PRC!); Calawanio, *Callery* 14 (P!), 20 (P!, "20bis"). — Note: Staminate collections of *A. pleuricum* cannot be distinguished from those of *A. microcarpum.* The only pistillate element of the protologue is therefore here selected as lectotype.

A. obliquicarpum Elmer, Leafl. Philipp. Bot. 7: 2633 (1915). Type: Philippines, Cabadbaran (Mt Urdaneta), Prov. Agusan, Mindanao, July 1912, humus covered compact red soil on a steep ravine near the forested ridge, 1250 ft, *Elmer* 13277 (NY!, lectotype, here designated; A!, E!, G!, GH!, K!, L!, P!, US!, isolectotypes). — Note: Lectotype accompanied by Elmer's field notes.

?*A. tenuifolium* Pax & K. Hoffm. in Engl., Pflanzenr. 81: 137 (1922). Type: Philippines, Luzon, Camarines, *Curran* 10736 (K!, lectotype, here designated; NY!, US!, isolectotypes). — Note: As staminate collections of *A. pleuricum* and *A. microcarpum* cannot be distinguished, this could also be a synonym of the latter.

Tree, up to 10 m, diameter up to 25 cm. Bark yellowish brown, scaly; inner bark reddish; wood hard, rather brittle and heavy, sapwood whitish, odourless and tasteless. *Young twigs* terete, shortly spreading-pubescent, whitish grey. *Stipules* not seen. *Petioles* channelled adaxially, basally and distally sometimes pulvinate for 2 – 3 mm, 7 – 15(– 22) × 0.8 – 1.2(– 2) mm, pubescent to glabrous. *Leaf blades* ovate to elliptic, more rarely oblong or obovate, apically acuminate, with an obtuse to shortly mucronate apiculum, basally obtuse to truncate, rarely acute, often slightly folded when dry, (5 –)9 – 12(– 22) × (2.5 –)4 – 6(– 9) cm, (1.7 –)2.1(– 2.5) times as long as wide, eglandular, chartaceous to membranaceous, glabrous except for the sparsely pilose major veins, moderately shiny on both surfaces, midvein impressed adaxially, tertiary veins weakly percurrent, drying olive-green. *Staminate inflorescences* 5 – 11 cm long, axillary but aggregated at the end of the branch, consisting of 5 – 50 branches, lax, axes pubescent. *Bracts* lanceolate, apically acute, c. 0.7 × 0.5 mm, pubescent. *Staminate flowers* 1 – 2 × 1 – 2 mm. *Pedicel* absent. *Calyx* 0.5 – 0.7 × c. 1 mm, bowl-shaped, sepals 4 – 5, fused for $^1/_2$ – $^2/_3$ of their length, apically rounded to acute, glabrous to slightly pilose on both sides, margin ciliate. *Disc* extrastaminal-annular (but appearing cushion-shaped), hemispherical, distinctly exserted from the sepals, constricted at the base, 3 – 4(– 7) lobed, the inward pointing lobes filling the space between filaments and pistillode, densely ferrugineous-tomentose. *Stamens* 3 – 4(– 7), c. 1.5 mm long, c. 1 mm long exserted from the calyx, anthers c. 0.4 × 0.3 mm. *Pistillode* globose to subulate, 0.3 – 0.4 × 0.1 – 0.4 mm, slightly exserted from the disc, white-pubescent. *Pistillate inflorescences* 5 – 6(– 12) cm long, axillary but aggregated at the end of the branch, consisting of 3 – 8 branches, axes pubescent. *Bracts* deltoid, apically acute, 0.3 – 0.5 × 0.3 – 0.5 mm, pubescent. *Pistillate flowers* c. 1 × 1 mm. *Pedicels* 1(– 2) mm long, glabrous to pubescent. *Calyx* c.

0.5 × 0.7 – 1 mm, cupular, sepals 4 – 5, fused for $^{2}/_{3}$ – $^{3}/_{4}$ of their length, apically acute, glabrous to sparsely pilose on both sides, margin ciliate. *Disc* exserted from or more rarely shorter than the sepals, densely ferrugineous-tomentose, especially at the margin. *Ovary* lenticular, pubescent to pilose, style lateral, stigmas 3 – 4. *Infructescences* 3 – 12 cm long. *Fruiting pedicels* 1.5 – 3 mm long, pubescent. *Fruits* lenticular to almost bean-shaped, laterally compressed, basally symmetrical, with a lateral style, 3 – 4 × 3 – 4 mm, pilose, white-pustulate, areolate when dry.

DISTRIBUTION. Philippines: Alabat, Catanduanes, Luzon, Mindanao, Mindoro, Samar, Sulu Archipelago, Ticao, Visayas. Map 33.

HABITAT. In forests; often in damp, dense habitats or near streams. On deep, fertile red soil, brown loam, limestone. 10 – 600 m altitude.

USES. The fruits are eaten.

CONSERVATION STATUS. Near Threatened (NT). The species is widespread in the Philippines and seems to be reasonably common, but like all Philippine endemics there is cause for concern over habitat destruction. 25 pistillate specimens examined.

VERNACULAR NAMES. bana (Bilaan); dako-dako; hotot (Manobo).

ETYMOLOGY. The epithet refers to the laterally compressed fruits (Greek, pleuron = side of the body, rib).

KEY CHARACTERS. Long petioles; usually ovate leaves with rounded to truncate base; fused sepals; hairy disc; fruits laterally compressed with lateral styles.

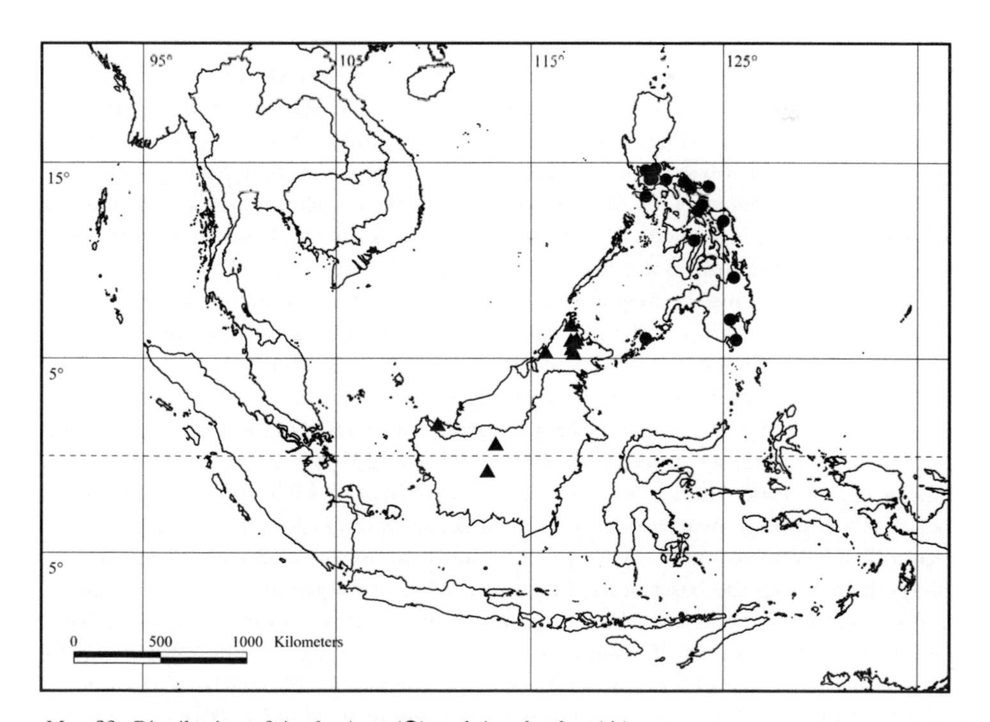

MAP 33. Distribution of *A. pleuricum* (●) and *A. polystylum* (▲).

SIMILAR SPECIES. *A. catanduanense, A. celebicum, A. curranii, A. digitaliforme, A. excavatum, A. microcarpum*, see there. — *A. tetrandrum* has glabrous discs and fruits.

SELECTED SPECIMENS. PHILIPPINES. Alabat Island, s. loc., Sept. – Oct. 1926, *BS (Ramos & Edaño)* 48206 (B, NY); Catanduanes, s. loc., 14 Nov. – 11 Dec. 1917, *BS (Ramos)* 30184 (K, NY, US); Luzon, Sorsogon prov., Irosin (Mt Bulusan), Nov. 1915, *Elmer* 15212 (A, BO, G, K, L, NY, U, US); Luzon, Laguna prov., Los Banos (Mt Maquiling), June – July 1917, *Elmer* 18299 (A, BO, G, GH, K, L, NY, U, US); Mindanao, Davao distr., Todaya (Mt Apo), along the Baracatan creek, in very damp slopes of dense woods, 1750 ft, June 1909, *Elmer* 10936 (BM, E, G, K, L, NY, U, US, WRSL); Mindoro, Mindoro Oriental prov., Lantuyan, N face of Mt Halcon, understorey in submontane forest, canopy 25 m, soil rocky, brown loam, 1 April 1991, *PPI (Stone, Reynoso & Sagcal)* 526 (K); Samar, s. loc., July 1915, *FB (Phasis)* 24232 (US); Sulu Archipelago, s. loc., March 1886, *Vidal* 3694 (K); Ticao Island, s. loc., May – June 1904, *Clark* 1020 (NY, US); Visayas Isl., Brgy. Patag, Paranas, secondary forest, on hill slope, soil brownish, limestone, 22 Oct. 1992, *PPI (Reynoso, Sagcal & Garcia)* 7523 (K).

43. Antidesma polystylum *Airy Shaw*, Kew Bull. 26: 460 (1972). Type: Sabah, Beaufort Distr., Beaufort Hill, sample plot, virgin jungle forest on hill, blackish soil, 17 Oct. 1967, *SAN (Ahmad Talip)* 55652 (K!, holotype; L!, isotype). Paratype: Beaufort Hill, primary forest, brown soil on hillside, alt. 210 m, 22 June 1964, *SAN (Ampuria)* 40238 (K!).

Tree, up to 15 m, clear bole up to 5 m, diameter up to 13 cm. Bark grey, brown or ochreous; inner bark white, pale yellow or brown; sapwood whitish, pink or green. *Young twigs* terete, densely ochraceous-tomentose, brown. *Stipules* caducous, linear to narrowly deltoid, sometimes falcate, 10 – 20 × 1 – 3 mm, densely pubescent. *Petioles* almost terete, narrowly channelled adaxially, 3 – 12 × 2 – 3 mm, densely ochraceous-tomentose. *Leaf blades* oblong to narrowly elliptic, apically acuminate, basally acute to obtuse, 18 – 38 × 5.5 – 13 cm, 2.9 – 3.6 times as long as wide, eglandular, chartaceous, glabrous, or pilose only along the midvein adaxially, pubescent all over abaxially or more rarely only along the major veins, dull on both surfaces, sometimes with minute, regularly spaced white pustules adaxially, midvein impressed adaxially, tertiary veins percurrent, close together (15 – 25 between every two secondary veins), drying olive-green. *Staminate inflorescences* 5 – 9 cm long, cauline, simple or branched once or twice, 5 – 10 per fascicle, axes densely pubescent. *Bracts* lanceolate, apically acute, 0.5 – 1 × c. 0.5 mm, densely pubescent. *Staminate flowers* 2.5 – 3 × 2 – 3 mm. *Pedicels* 0 – 1 mm long, not articulated, glabrous to sparsely pilose. *Calyx* 0.5 – 0.7 × 0.8 – 1.2 mm, sepals 3 – 4, free or nearly so, c. 0.7 mm wide, spreading, deltoid to semiorbicular, apically obtuse to rounded, densely velutinous to pilose on both sides. *Disc* cushion-shaped, enclosing the bases of the filaments, densely ochraceous-tomentose. *Stamens* 3 – 4, 2 – 3 mm long, exserted 1.5 – 2.5 mm from the calyx, anthers 0.3 – 0.4 × 0.4 – 0.6 mm. *Pistillode* absent. *Pistillate inflorescences* 5 – 9 cm long, cauline, simple, 4 – 10 per fascicle, axes densely ochraceous-tomentose. *Bracts* deltoid to narrowly lanceolate, apically acute, 1 – 1.5 × 0.3 – 0.5 mm, densely tomentose. *Pistillate flowers* 2 – 3 × c. 1

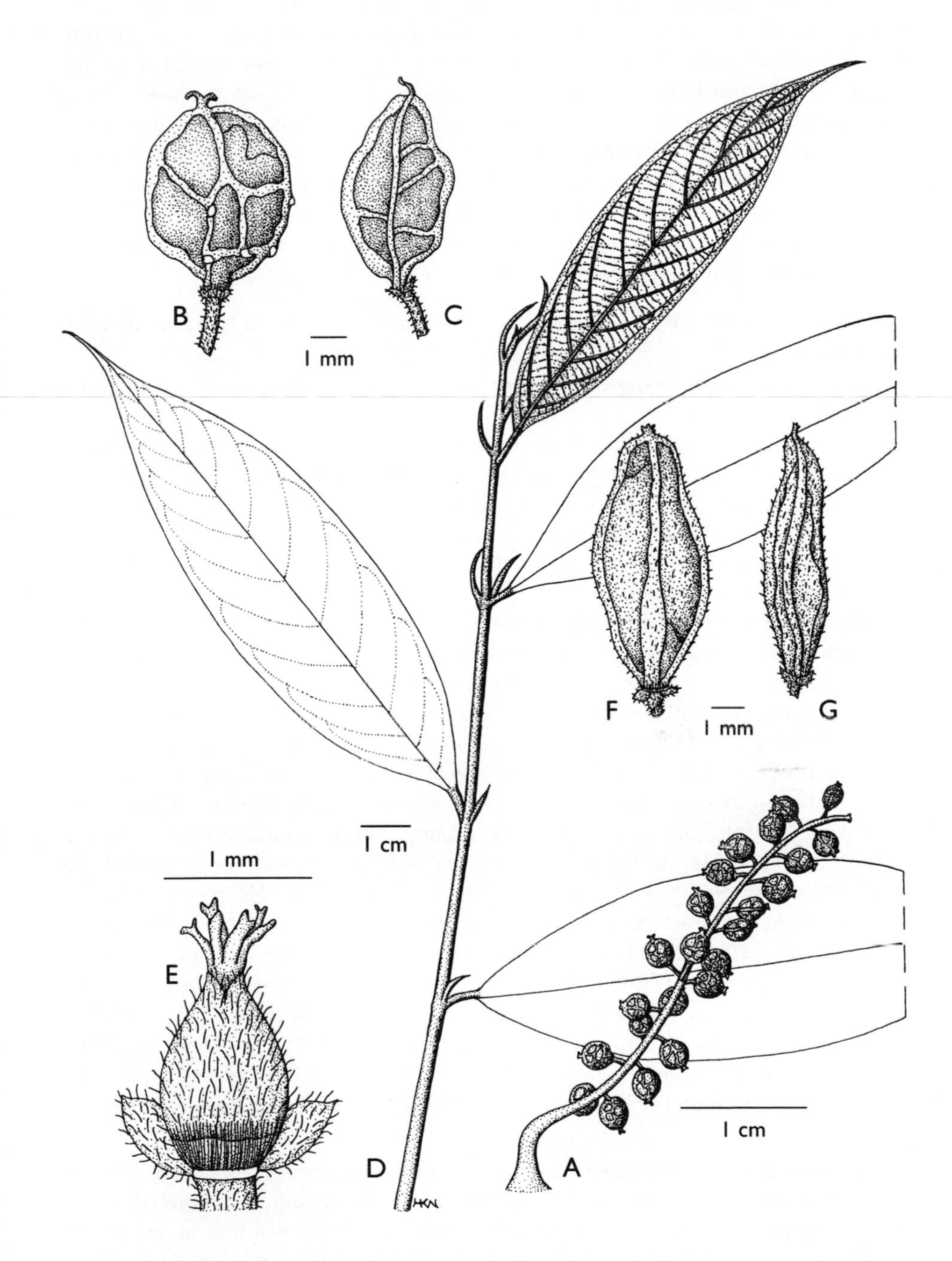

FIG. 18. *Antidesma leucopodum* and *A. polystylum*. **A – C** *A. leucopodum*. **A** infructescence; **B** fruit in lateral view; **C** fruit in dorsal view. **D – G** *A. polystylum*. **D** twig with leaves; **E** pistillate flower; **F** fruit in lateral view; **G** fruit in dorsal view. **A** from *Kostermans* 6952 (L); **B – C** from *S* 19862 (L); **D – E** from *SAN* 107898 (L); **F – G** from *SAN* 60293 (L). Drawn by Holly Nixon.

mm. *Pedicels* 0 – 0.5 mm long, densely tomentose. *Calyx* 0.8 – 1.2 × 1 – 1.5 mm, sepals 4, free or nearly so, c. 0.5 mm wide, deltoid, apically acute, densely velutinous on both sides, individual hairs sometimes longer than the calyx itself. *Disc* shorter than the sepals, densely and long velutinous, hairs sometimes longer than the disc. *Ovary* elongate-ellipsoid, densely ochraceous-velutinous, style terminal, often as thick as the ovary itself, stigmas 6 – 16, sometimes irregularly fused. *Infructescences* 8 – 13 cm long. *Fruiting pedicels* 0.5 – 1 mm long, densely ochraceous-tomentose. *Fruits* elongate-ellipsoid, not or hardly compressed, more or less tetragonal in cross section, basally symmetrical, with a terminal style, 10 – 14 × 3 – 5(– 7) mm, appressed-velutinous, sometimes white-pustulate, reticulate when dry. Fig. 18D – G.

DISTRIBUTION. Borneo, mainly Sabah but also Brunei, Sarawak and Kalimantan. Map 33.

HABITAT. In primary, sometimes disturbed, dipterocarp forest. 130 – 300 m altitude.

CONSERVATION STATUS. Near Threatened (NT). This species is relatively widespread in Borneo, but only 13 specimens are known and there is concern over habitat destruction through logging, burning and intensive agriculture.

VERNACULAR NAMES. Kalimantan: cingkolik.

ETYMOLOGY. The epithet refers to the numerous stylar branches (Greek, poly- = many-; stylos = style).

KEY CHARACTERS. Cauline, densely ochraceous-tomentose inflorescences; elongate-ellipsoid fruits; more or less tetragonal in cross section; 6 – 13 stigmas.

SIMILAR SPECIES. *A. leucopodum*, see there.

SELECTED SPECIMENS. BORNEO. Sabah, Labuk & Sugut, Sungai Soinin, side of Sungai Tongod, *SAN (Aban Gibot & Soinin)* 60293 (L); Labuk & Sugut (Beluran distr.), side of Sg. Sasau, *SAN (Aban & Soinin)* 60393 (K); Labuk & Sugut (Beluran) distr., Hutan Simpan Ulu Tongod, hillside, 19 Jan. 1987, *SAN (Mansus)* 107898 (L); Sandakan, Telupid distr., Karamuak, Kampung Lolou Kapuk, primary forest, hillside, 300 m, 21 March 1985, *SAN (Dewol Sundaling, Longkop & Tuyuk)* 108838 (K, KEP); Pitas distr., Kampung Payas, 100 – 200 m, 13 May 1987, *SAN (George & Amin K.)* 121262 (K); Beluran distr., foot of Bidu-Bidu Forest Reserve, 19 Feb. 1991, *SAN (Julius et al.)*. 131012 (L); Kinabatangan distr., Maliu R. valley ESE of Kampung Enteleben and SW of Bukit Tawai summit, 5°31'N, 117°07'E, in disturbed dipterocarp forest, on a slope in partial shade, 150 m, 7 April 1984, *Sands* 3847 (K); Sarawak, Kuching distr., Gunong Selang, 1000 ft, 9 March 1961, *S (Bujang)* 13432 (K, L); Kalimantan, Liang Gagang, *Hallier* 2906 (BO, K, L, U); Bukit Raya, 0°45'S, 112°45'E, primary dipterocarp forest, c. 130 m, 13 Dec. 1982, *Nooteboom* 4276 (K, L).

44. Antidesma puncticulatum *Miq.*, Fl. Ned. Ind., Eerste Bijv.: 468 (1861). Type: Sumatra austr., in prov. Lampong, ad flumen Tarabangi, *Teysmann* HB 4467 (U!, holotype; BO!, isotype). — Notes: A synonym (with a question mark) of *A. coriaceum* according to Airy Shaw (1975: 210). Examination of the type of *A. puncticulatum*, however, shows that it is clearly conspecific with *A. thwaitesianum* Müll. Arg. Being the older name, it takes priority over *A. thwaitesianum*. Collection number not cited in protologue.

A. thwaitesianum Müll. Arg. in DC., Prodr. 15(2): 263 (1866). — *A. bunius* (L.) Spreng. var. b Thwaites, Enum. Pl. Zeyl.: 289 (1861), nom. inval. according to Art. 24.2. (Greuter *et al.* 2000: 45). — *A. bunius* (L.) Spreng. var. *thwaitesianum* (Müll. Arg.) Trimen, Syst. Cat. Pl. Ceylon: 81 (1885). Type: Sri Lanka, *Thwaites* 2922 (G-DC [microfiche] in staminate flower, lectotype, here designated; BM!, CGE!, FR, G!, G-DC [microfiche], K!, MEL!, TCD!, isolectotypes).

Tree, up to 25 m, clear bole up to 9 m, diameter up to 50 cm. Bark brown, grey, green or red, smooth, sometimes scaly or fissured; inner bark brown, red or yellowish; cambium brown, red or white; sapwood light grey, brown or yellowish. *Young twigs* terete, almost glabrous, more rarely whitish-pubescent, very light grey. *Stipules* early-caducous, hardly ever present on mature leaves, linear to deltoid, brittle, very light grey, c. 2 × 0.7 mm, pubescent. *Petioles* channelled adaxially, striate to rugose when dry, basally and distally usually slightly pulvinate, (3 –)10 – 15(– 23) × 1 – 1.5 mm, glabrous, pilose only when young. *Leaf blades* oblong to elliptic or slightly ovate, apically acuminate, more rarely acute-mucronate, basally acute to obtuse, more rarely rounded (but shortly decurrent at the very base), (7 –)10 – 16(– 21) × (2.5 –)4 – 6(– 21) cm, (2.2 –)2.6(– 3.1) times as long as wide, eglandular, coriaceous, glabrous, shiny on both surfaces, major veins flat to shallowly impressed adaxially, tertiary veins finely reticulate, quarternary veins distinctly prominent, finer venation finely tessellated, drying yellowish brown to yellowish green. *Staminate inflorescences* 3 – 5 cm long, cauline, rarely axillary, simple, up to 10 per fascicle, axes whitish-pubescent. *Bracts* broadly lanceolate, c. 0.7 × 0.5 mm, pubescent. *Staminate flowers* 1 – 2 × 1 – 1.5 mm. *Pedicel* absent. *Calyx* 0.5 – 0.8 × c. 1 mm, globose to cupular, sepals 3 – 5, fused for c. $^1/_2$ of their length, deltoid, apically acute, glabrous to pilose outside, pilose inside, margin entire, fimbriate. *Disc* consisting of 3 – 4 free alternistaminal lobes, lobes ± obconical, c. 0.5 × 0.5 mm, pilose to pubescent. *Stamens* 3 – 4, 1 – 2 mm long, exserted 0.5 – 1.5 mm from the calyx, anthers c. 0.3 × 0.4 mm. *Pistillode* cylindrical to clavate, c. 0.5 × 0.2 mm, extending to the same length or slightly exserted from the disc, ± hidden in the disc indumentum, pubescent. *Pistillate inflorescences* 1 – 4(– 9) cm long, cauline, simple, up to 4 per fascicle, axes whitish-pubescent. *Bracts* deltoid, apically acute, c. 0.7 × 0.3 – 0.5 mm, pubescent. *Pistillate flowers* c. 2 × 1 mm. *Pedicels* 0 – 0.5 mm long, glabrous to slightly pilose. *Calyx* c. 0.5 × 1 mm, shallowly cupular, sepals 4 – 5, fused for c. $^1/_2$ of their length, spreading, deltoid, apically apiculate to acute, glabrous to sparsely pilose outside, glabrous to pilose inside, margin entire, fimbriate. *Disc* shorter than (but visible between) the sepals, shortly pubescent, especially at the margin, in fruit sometimes almost glabrous. *Ovary* ellipsoid, glabrous, style terminal, stigmas 4 – 6. *Infructescences* 4 – 9 cm long. *Fruiting pedicels* 0.5 – 3 mm long, glabrous to sparsely pilose. *Fruits* ellipsoid to slightly oblique or lenticular, distinctly laterally compressed, basally symmetrical to asymmetrical, with a slightly subterminal to lateral style, 6 – 8 × 4 – 6 mm, glabrous, sometimes slightly white-pustulate and often covered with minute white crystals, areolate or fleshy when dry. Fig. 9C – G (p. 93).

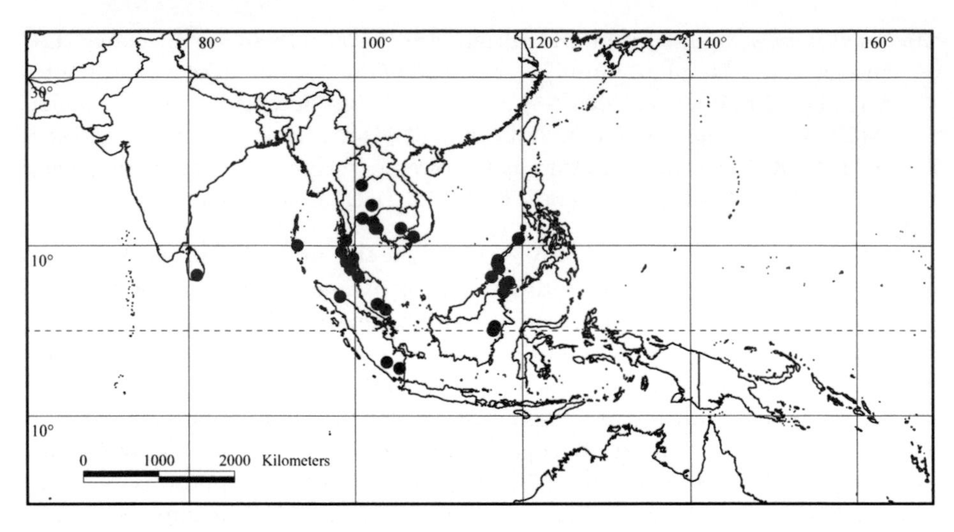

MAP 34. Distribution of *A. puncticulatum*.

DISTRIBUTION. Nicobar and Andaman Islands, Sri Lanka, Vietnam, Cambodia, Laos, Thailand, Peninsular Malaysia, Sumatra, north-eastern Borneo (East Kalimantan, Sabah), south-western Philippines (Balabac, Bancalan, Palawan). Map 34.

HABITAT. In evergreen primary, sometimes disturbed forest; in swamp forest; along river banks; at the inner edge of the mangrove; in secondary vegetation with *Dillenia* and *Melastoma*. On brown to black soil, sandy loam or peat, over sandstone. 0 – 500 m altitude.

CONSERVATION STATUS. Least Concern (LC). This species is classified as Least Concern because of its vast extent of occurrence. It is, however, locally rare (e.g., Mathew & Abraham 1995), and not as frequently collected as other species in this category (68 specimens examined).

VERNACULAR NAMES. Nicobar & Andaman Islands: byee-sane; Thailand: mao luang, mud she, sien; Sumatra: djohan (Malay), kayu uni rawang, lapis putih, niam; Sabah: indorapis; Kalimantan: kayu bei peram puan, maragelang.

ETYMOLOGY. The epithet refers to the finely tessellated leaf venation (Latin, puncticulatus = finely dotted).

KEY CHARACTERS. Higher venation finely tessellated abaxially; leaves coriaceous, shiny, drying yellowish; petioles long; inflorescences short, cauline, whitish-pubescent.

SIMILAR SPECIES. *A. bunius* and *A. coriaceum*, see there.

SELECTED SPECIMENS. ANDAMAN AND NICOBAR ISLANDS. s. loc., 15 May 1915, *Parkinson* 575 (K). VIETNAM. Bao Chang, July 1877, *Pierre* 1971 (K, L, NY). THAILAND. South-eastern, Trat prov., Koh Chang Island, Aw Ong Kang and Salak Koh, 12°00'N, 102°40'E, cleared forest in valley, forest edge, 0 m, 8 May 1974, *Geesink, Hattink & Phengklai* 6635 (K, L); Peninsular, Trang prov., Kantang, 7°25'N, 99°30'E, 30 May 1919, *Haniff & Nur* 4290 (BO, K, SING). PENINSULAR MALAYSIA. Kedah, Bukit Tunjang, 5 April 1924, *Haniff* 13142 (K, SING); Pahang,

Tasek Bera, low altitude, 16 Oct. 1930, *SF (Henderson)* 24060 (K, SING). SUMATRA. Palembang res., Lematang Ilir, c. 75 m, 18 Nov. 1924, *Boschproefstation T (s. coll.)* 883 (BO, L, U). BORNEO. Sabah, Kudat distr., Pulau Banggi (Banguey Island), July – Sept. 1923, *Castro & Melegrito* 1519 (A, BM, BO, US); Sabah, Pantai Barat, Kota Belud distr., Kampung Lubok, Kunonok Ambong, primary forest, swampy flat behind sea shore, 7 June 1961, *SAN (Burgess)* 25181 (K, L, SAR); Sabah, Kinabatangan distr., Batu Puteh camp, primary forest on lowland area, peaty soil, 17 July 1962, *SAN (Singh)* 31066 (K, KEP, L, SAR); E Kalimantan, W Koetai, no. 9, near M. Antjaloeng, forest, rather low country, 15 m, 15 July 1925, *Endert* 2046 (A, BO, K, L). PHILIPPINES. Balabac Island, Cape Melville at S tip of island, just N of the lighthouse, 7°51'N, 117°02'E, logged lowland mixed forest, 30 m, 1 July 1994, *Soejarto, Madulid, Gaerlan, Sagcal & Fernando* 8670 (L); Palawan, Lake Manguao, April 1913, *Merrill* 9450 (BM, K, L).

45. Antidesma rhynchophyllum *K. Schum.* in K. Schum. & Lauterb., Nachtr. Fl. Schutzgeb. Südsee: 294 (1905). Type: New Guinea, Kaiser Wilhelmsland, bei Bivak-Insel, Okt. 1907, *van Leeuwen* 9757 (K!, epitype, here designated; L!, US!, isoepitypes); am Ramuflusse, 22 Juni 1899, *Rodatz & Klink* 12 (WRSL!, lectotype, here designated). — Note: The only extant original material from WRSL consists of two detached fruits without calyx, and two detached leaves, one of which lacks the lower part. As the distinctive characters of *A. rhynchophyllum* are mainly found in the staminate flower, the name is here epitypified with a staminate specimen.

A. obovatum J. J. Sm. in Lorentz, Nova Guinea 8 (1): 230, t. LVII (1910). Type: Niederl. Neu-Guinea, bei Bivak-Insel, Okt. 1907, *Versteeg* 1789 (L!, lectotype, here designated; BO!, K!, P!, U!, isolectotypes). Original syntype: Noord-Fluß am Fuße des Nepenthes-Hügels in Pandanussümpfen, Juni 1907, *Versteeg* 1258 (BO!, K!, L!, P!, U!).

A. cinnamomifolium Pax & K. Hoffm. in Engl., Pflanzenr. 81: 154 (1922), **synon. nov.** Type: Neu Guinea, Kaiser Wilhelmsland, Lordberg, lichter Bergwald, *Ledermann* 10199 (B!, lectotype, here designated). Original syntype: *Ledermann* 10094 (B!).

A. densiflorum Pax & K. Hoffm. in Engl., Pflanzenr. 81: 121 (1922). Type: Neu Guinea, Kaiser Wilhelmsland, Kaiserin Augustafluß-Expedition, Alexishafen, *Schlechter* 19198 (K!, lectotype, here designated; P!, isolectotype). Original syntypes: Lagerplatz 2, *Ledermann* 7456; Hauptlager Malu, *Ledermann* 10792; Pionierlager am Sepik, *Ledermann* 7260 (K!), *Ledermann* 7304 (K!); Maifluß, *Ledermann* 7341.

Tree, up to 12 m, clear bole up to 5 m, diameter up to 25 cm (*fide Aet* 506: liana). Bark brown or grey-brown, finely vertically cracked; inner bark reddish brown, brown or pinkish red, 6 mm thick; wood hard, reddish brown, pink-cream, straw or brownish grey. *Young twigs* terete, pilose to ochraceous-pubescent, brown. *Stipules* usually caducous, narrowly deltoid to subulate, apically acute, 1 – 3 × 0.2 – pustulate, dull to moderately shiny on both surfaces, midvein impressed adaxially, tertiary veins reticulate, widely spaced, drying olive-green, sometimes lighter abaxially, hairtuft domatia always present, usually conspicuous, almost tubular.

Staminate inflorescences 1.5 – 5 cm long, axillary, simple, more rarely consisting of up to 4 branches, weak, axes c. 0.2 mm wide, pilose. *Bracts* ovate to lanceolate, 0.3 – 0.5 × 0.2 – 0.3 mm, pilose, margin fimbriate. *Staminate flowers* 1.5 – 2.5 mm × 1 – 1.5 mm. *Pedicels* 0.3 – 1 mm long, not articulated, glabrous. *Calyx* 0.5 – 1 mm × 1 – 1.2 mm, cupular, sepals 3 – 4, fused for $^1/_4$ – $^1/_3$ of their length, apically rounded to acuminate, sinuses rounded, glabrous outside and inside, margin erose, slightly fimbriate. *Disc* cushion-shaped, enclosing the bases of the filaments, glabrous at the sides, ferrugineous- to ochraceous-pubescent apically. *Stamens* 2 – 3, 1.5 – 2.5 mm long, exserted 1 – 1.5 mm from the calyx, anthers 0.3 – 0.5 × 0.4 – 0.6 mm. *Pistillode* apparently absent, merged with the disc, sometimes visible as swelling between the stamens. *Pistillate inflorescences* 1.5 – 3 cm long, axillary, simple or consisting of up to 4 branches, sometimes in fascicles of 2 inflorescences, very slender, axes pilose to ochraceous-pubescent. *Bracts* ovate to obovate, 0.4 – 0.7 × 0.2 – 0.5 mm, glabrous to pilose, margin fimbriate. *Pistillate flowers* 1 – 2 × 0.5 – 1.2 mm. *Pedicels* 0.3 – 1 mm long, glabrous to sparsely pilose. *Calyx* 0.5 – 1 × 1 – 1.2 mm, cupular to urceolate, sepals 3 – 5, fused for $^1/_2$ – $^2/_3$ of their length, deltoid to semiorbicular, apically acute to rounded, sinuses rounded, glabrous outside and inside, margin fimbriate. *Disc* shorter than the sepals, ferrugineous-tomentose at the margin, hairs about as long as the disc. *Ovary* ovoid, terete, sparsely tomentose to glabrous, style terminal, more rarely subterminal or lateral, stigmas 3 – 6, rather long and thick or irregularly fused. *Infructescences* 2 – 4(– 5) cm long, axes 0.5 – 1 mm wide. *Fruiting pedicels* 0.5 – 1 (– 2.5) mm long, glabrous to sparsely pilose. *Fruits* obovoid, ellipsoid or globose, terete, basally asymmetrical or symmetrical, with a terminal to subterminal style, 2.5 – 4(– 6) × 2.5 – 3 mm, sparsely pilose to glabrous, mostly white-pustulate, areolate or reticulate when dry. Fig. 19.

DISTRIBUTION. New Guinea. Map 35.

HABITAT. In primary and secondary rainforest; in swamp forest and riverine forest; usually in shady understorey; also in scrubby edges and roadside border of forest. On (coral) limestone. 2 – 440 m altitude.

CONSERVATION STATUS. Near Threatened (NT). This species is widespread in New Guinea (extent of occurrence c. 700,000 km^2) but in the absence of

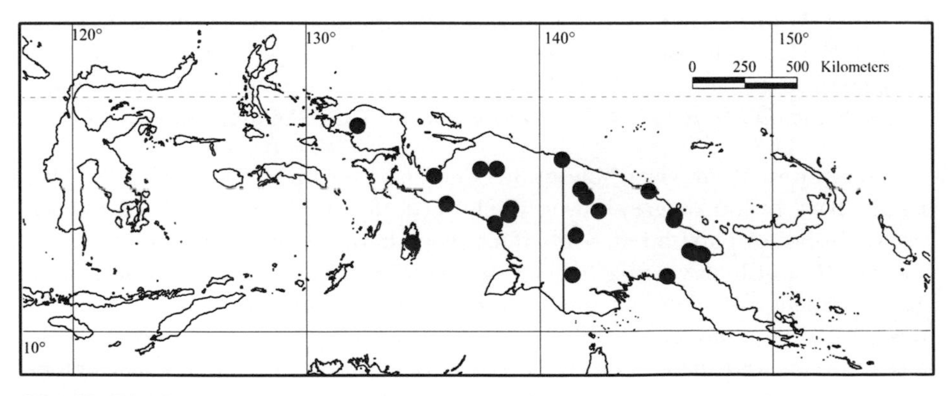

MAP 35. Distribution of *A. rhynchophyllum*.

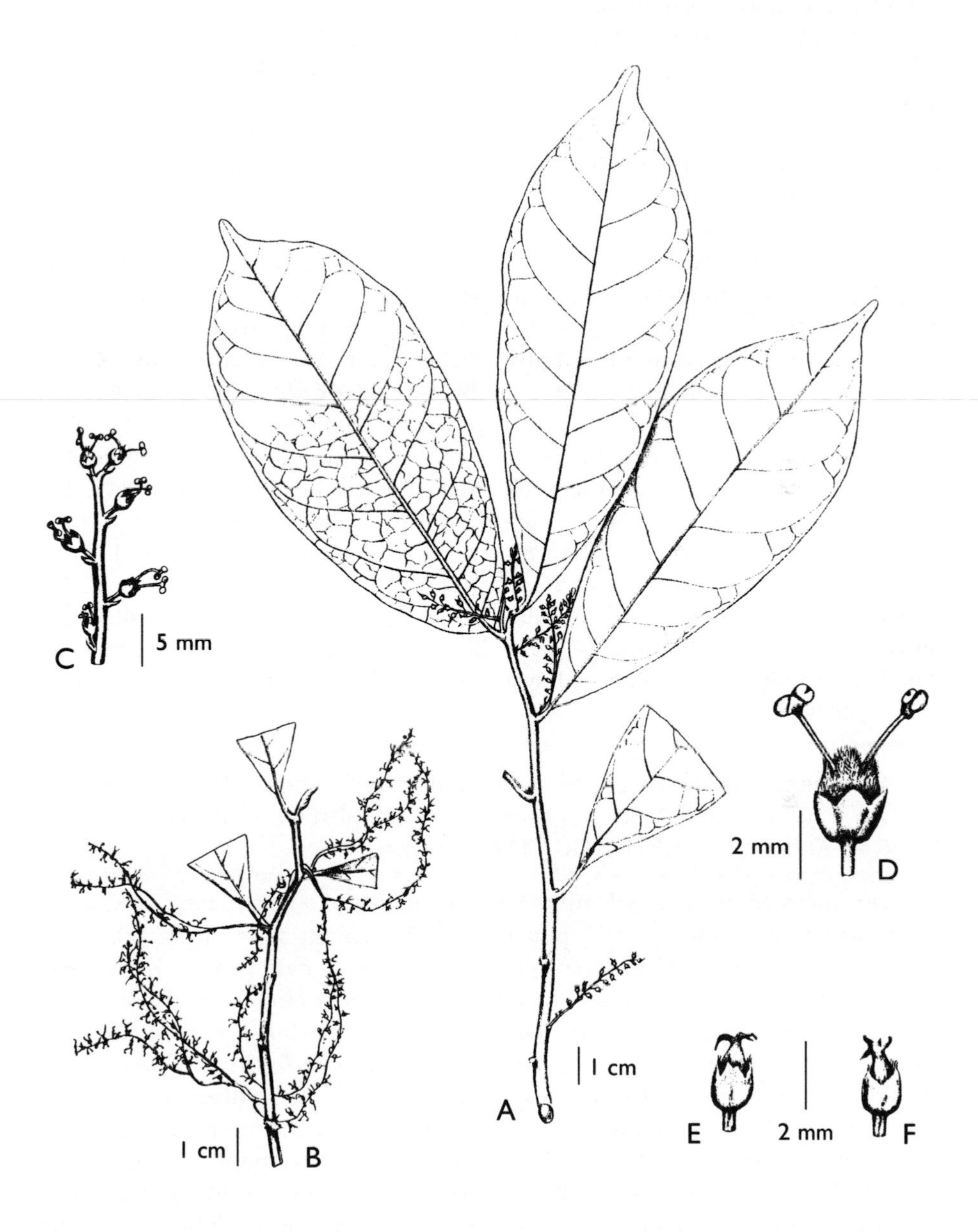

FIG. 19. *Antidesma rhynchophyllum.* **A** pistillate branch; **B** staminate branch; **C** part of staminate inflorescence; **D** staminate flower; **E – F** pistillate flowers. Reproduced from Lorentz 1910: t. LVII (as *A. obovatum* J. J. Sm.).

information about its abundance, it is classified as Near Threatened for the time being. 31 specimens examined.

VERNACULAR NAMES. Moluccas: bima; Indonesia: Papua: tefan kek.

ETYMOLOGY. The epithet refers to the leaf shape (Greek, rhynchos = snout, beak; phyllon = leaf).

KEY CHARACTERS. Thinly pedicellate flowers; fused sepals; hairy disc; 2 – 3 stamens; fruits with (sub)terminal styles; often obovate, thin, abruptly acuminate leaves with domatia.

SIMILAR SPECIES. *A. excavatum*, see there.

SELECTED SPECIMENS. MOLUCCAS. Aru Islands, Kobroor Island, c. 8 km SE of Jirlay, primary rainforest, on coral limestone, low altitude, 29 Oct. 1994, *Nooteboom* 5967 (L). INDONESIA: PAPUA. S New Guinea, Sg. Aendoea near Oeta, 3 m, 9 July 1941, Expedition Lundquist, *Aet* 506 (A, K, L); Hollandia div., Nemo, old secondary forest, level ground, 6 m, 29 March 1956, *BW (Kalkman)* 3461 (K, L); Vogelkop Peninsula, S/SE of Ayawasi, near Tafruwia, 1°14'S, 132°12'E, secondary forest at base of limestone, 440 m, 3 July 1996, *Polak* 1326 (L); Rouffaer R., Motor Bivouac, jungle, 100 m, Nov. 1926, *Van Leeuwen* 11068 (K, L). PAPUA NEW GUINEA. Madang distr. & subdistr., Gogol logging area, 5°15'S, 145°45'E, lowland rainforest, 90 m, 20 June 1974, *LAE (Katik)* 62107 (K, L, US); Sepik distr., Yellow R., near Sepik R., 27 Sept. 1949, *NGF (Womersley)* 3863 (K, L); Morobe distr., Markham road, 20 miles from Lae, 6°35'S, 146°25'E, scrub edge, 50 ft, 2 Feb. 1959, *NGF (Millar)* 10000 (K, L); Gulf distr., W bank, junction of Vailala and Lohiki Rs., primary alluvial forest, c. 70 ft, 25 Jan. 1966, *Schodde & Craven* 4299 (K, L, US).

46. Antidesma riparium *Airy Shaw*, Kew Bull. 23: 282 (1969). Type: Sarawak, 3rd Division, Ulu Bena, Sut, Balleh, on river bank, 22 April 1963, *S (Ashton)* 17802 (K!, holotype; A!, KEP!, L!, MEL!, SAR!, SING!, isotypes).

Tree, up to 15 m, clear bole up to 9 m, diameter up to 20 cm. Bark grey, white or brown, thin, smooth, scaly, papery, not detaching; inner bark yellow or brownish grey, 2 mm thick; sapwood white, yellow or pale brown. *Young twigs* terete, obtuse-angled or flattened, ferrugineous- to ochraceous-pilose, more rarely pubescent or glabrous, first dark brown, light to whitish grey when older, more rarely whitish grey from the beginning. *Stipules* usually persistent, sometimes foliaceous, narrowly deltoid to linear, sometimes falcate or ovate, 6 – 14(– 18) × 1 – 3(– 5) mm, glabrous to sparsely pilose, more rarely ferrugineous-pubescent. *Petioles* channelled or flat adaxially, becoming rugose when old, 2 – 10 × (1 –)1.5 – 3 mm, pilose, more rarely ferrugineous-pubescent, often becoming glabrous when old. *Leaf blades* oblong, more rarely elliptic or slightly obovate, apically acuminate-mucronate, basally acute, more rarely obtuse to rounded, (8 –)14 – 20 (– 26) × (2 –)4 – 6(– 9) cm, (2 –)4(– 6.4) times as long as wide, eglandular, subcoriaceous, more rarely chartaceous, glabrous except for some hairs along the major veins on both surfaces, rarely ferrugineous-pubescent along the major veins adaxially and all over abaxially, shiny or dull on both surfaces, often with conspicuous stomata adaxially (dissecting microscope!), major veins impressed, flat or slightly raised adaxially, tertiary veins percurrent or reticulate, often perpendicular (at right angle to the midvein), but usually parallel to each other, drying reddish brown, slightly lighter to olive-green abaxially, domatia absent. *Staminate inflorescences* 15 – 20 cm long, axillary, simple, more rarely once-branched at the base, rarely consisting of up to 4 branches, lax, axes sparsely pilose. *Bracts* linear, lanceolate or spathulate, apically acute to rounded, 0.7 – 1 × c. 0.5 mm, pilose. *Staminate flowers* 2 – 3 × 2 – 3 mm. *Pedicels* 0 – 1 mm long, not articulated, glabrous. *Calyx* c. 0.5 × 1 – 1.5 mm, bowl-shaped, sepals 5(– 6), fused for ²⁄₃ – ³⁄₄ of their length, apically rounded, glabrous to sparsely pilose outside, glabrous inside, margin erose. *Disc* cushion-shaped, hemispherical, enclosing the bases of the filaments and pistillode, pubescent. *Stamens* 3 – 5, 1.5 – 3 mm long, exserted 1 – 2.5 mm from the calyx, anthers c. 0.4 × 0.4 – 0.6 mm. *Pistillode* clavate, its base buried in the pubescent disc, 0.1 – 0.2 × 0.1 – 0.2 mm, extending to the same length and hardly distinct from the disc, pubescent. *Pistillate inflorescences* 5 – 22 cm long, axillary, simple, solitary or sometimes 2 per fascicle, axes pilose to nearly glabrous, rarely ferrugineous-pubescent. *Bracts* narrowly deltoid to linear, apically acute, 0.5 – 1.2 × 0.2 – 0.7 mm, pilose. *Pistillate flowers* 1.5 – 2.5 × 1 – 1.5 mm. *Pedicels* 0.5 – 2 mm long, glabrous to pilose. *Calyx* 0.7 – 1 × 1.2 – 1.5 mm, shallowly cupular, sepals (4 –)5(– 6), fused for ²⁄₃ – ³⁄₄ of their length, apically acute, sinuses rounded, glabrous or pilose outside, glabrous inside, margin entire. *Disc* shorter than the sepals, but indumentum usually exserted from the sepals, pubescent, especially at the margin. *Ovary* ellipsoid, pilose to almost glabrous, style lateral to subterminal, stigmas 3 – 5. *Infructescences* 18 – 45 cm long. *Fruiting pedicels* 1 – 5 mm long, sparsely puberulous, rarely pubescent. *Fruits* ellipsoid, globose or ovoid, not or hardly compressed (but with ± distinct dorsal and ventral ridges), basally symmetrical, rarely asymmetrical, with a lateral to subterminal, rarely terminal style, 5 – 8 × 4 – 7 mm, pilose to glabrous, sometimes white-pustulate, areolate to reticulate when dry.

46a. subsp. **riparium**

A. globuligerum Airy Shaw, Kew Bull. 36: 635 (1981), **synon. nov.** Type: Sulawesi (Celebes), *Musser* 474 (K!, holotype).

Pistillate inflorescences simple.

DISTRIBUTION. Sumatra (one doubtful specimen from Padang province), Borneo, Philippines (Palawan), Sulawesi. Map 36.

HABITAT. In primary or secondary vegetation; in riverine and swamp forest; in mixed (dipterocarp) forest. On sandy or clay soil, over shales. 60 – 1000 m altitude.

CONSERVATION STATUS. Near Threatened (NT). The species has a fairly wide but scattered distribution. 43 specimens examined.

VERNACULAR NAMES. Brunei: enkunie ai; Sulawesi: lera (Uma).

ETYMOLOGY. The epithet refers to the preferred habitat of the species (Latin, riparius = growing on river banks).

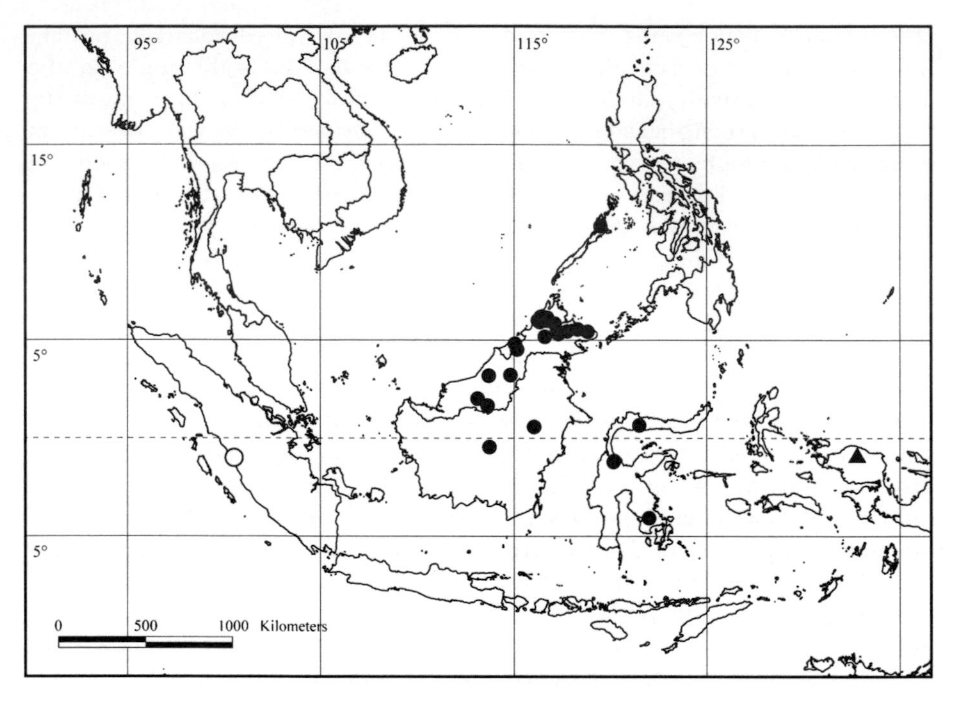

MAP 36. Distribution of *A. riparium* subsp. *riparium* (●), doubtful specimen (○), and subsp. *ramosum* (▲).

KEY CHARACTERS. Ferrugineous indumentum; fused sepals; hairy disc; long inflorescences; leaves drying reddish; fruits ± terete, with dorsal and ventral ridges.

SIMILAR SPECIES. *A. baccatum, A. neurocarpum, A. pahangense*, see there. — *A. tomentosum* has free sepals, shorter inflorescences, smaller staminate flowers and shorter stamens.

NOTE. *Beaman* 9473 from Sabah has atypically large ovate stipules (8 – 20 × 3 – 10 mm), but the young infructescences (46 cm long) exclude *A. neurocarpum*.

SELECTED SPECIMENS. BORNEO. Brunei, Temburong distr., Amo subdistr., Kuala Belalong, Sungai Belalong, 4°32'N, 115°09'E, mixed dipterocarp forest, riverine forest on river bank, Setap shales, 60 m, 15 Feb. 1992, *Dransfield* 7079 (B, K, KEP, L); Sabah, C Sabah, Sandakan, Kinabatangan, Sungai Lokan, side of Lokan R., 6 July 1983, *SAN (Sigin et al.)* 97532 (K, KEP, L, SAR); Sabah, Tawau, Lahad Datu distr., Ulu Segama Forest Reserve, along river, 25 March 1985, *SAN (Leopold Madani et al.)* 108690 (K, L, SAR); Sarawak, Kapit, Sut, Bena, path from Asah's longhouse to Merating, secondary forest, hillside, 215 m, 23 April 1980, *S (Ilias Paie)* 40982 (K, L); Sarawak, Kapit, B7alleh, Mengiong, Sungai Sebatu, on lower valley, rich soil, 500 m, 13 Nov. 1980, *S (Othman et al.)* 41391 (K, L); C Kalimantan, Headwaters of Sungei Kahayan, 5 km NE of Haruwu village, Nyoohoy tributary, 0°28'S, 113°44'E, along river banks, c. 250 m, 24 March 1988, *Burley, Tukirin et al.* 380 (E, K, KEP, L, SAR). PHILIPPINES. Palawan, Mt Kabangaan, on slopes in forest, 2400 ft, 30 April 1929, *BS (Edano)* 77709 (NY). SULAWESI. N Sulawesi, 250 km W of Gorontalo, 75 km inland from Papayuto, on tributary of Sungai

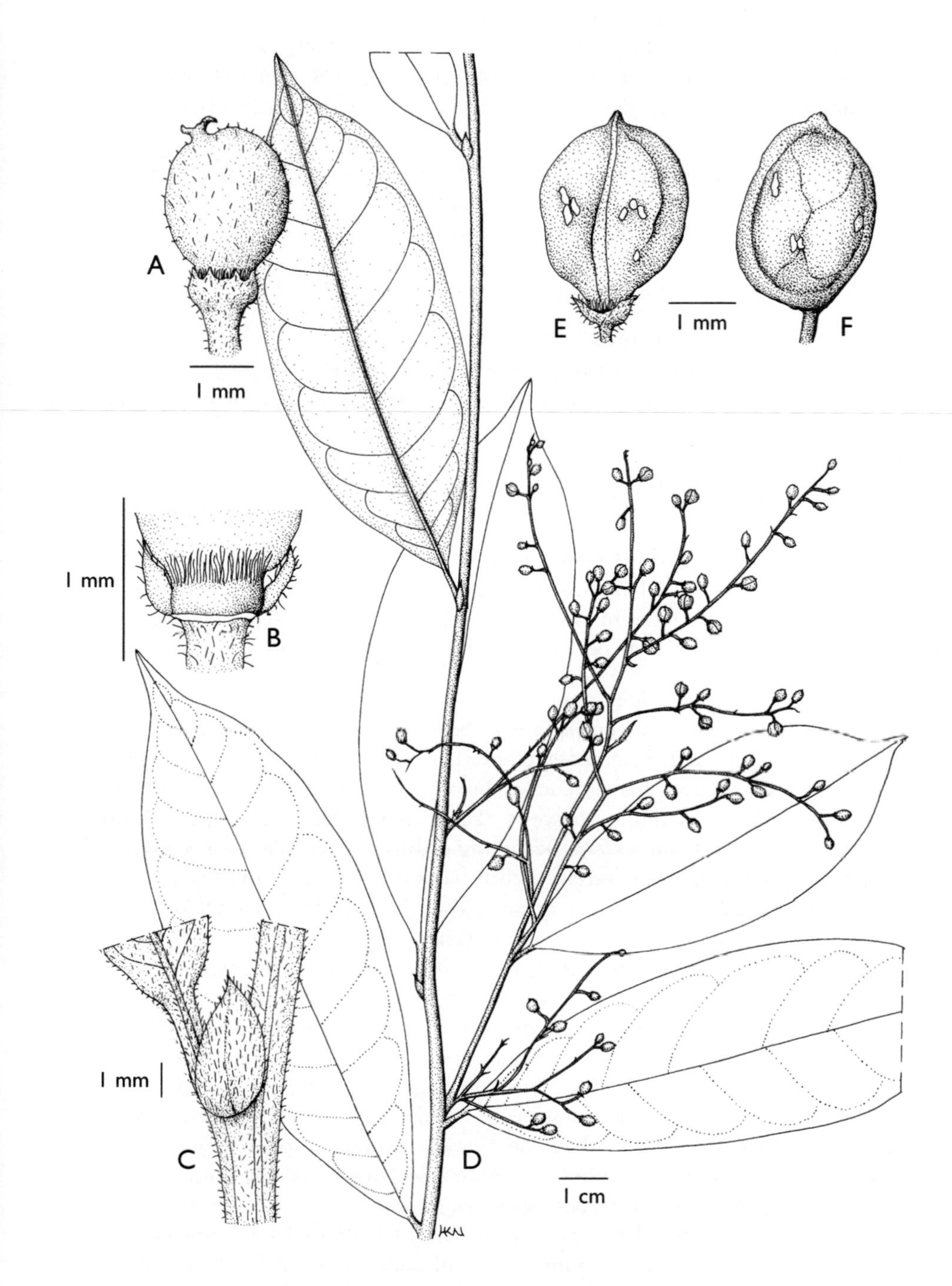

FIG. 20. *Antidesma riparium* subsp. *ramosum.* **A** very young fruit; **B** disc, calyx partly removed; **C** part of branch with stipule; **D** habit with young infructescence; **E** fruit in dorsal view; **F** fruit in lateral view. From *P. van Royen & Sleumer* 7056 (L). Drawn by Holly Nixon.

Papayuto, 0°45'N, 121°30'E, 150 m, 30 March 1990, *Burley et al.* 4209 (A); C Sulawesi, area of Mt Nokilalaki, Paliti, by Lake Lindu, 1°13'S, 120°08'E, swamp forest, 970 m, 1 May 1975, *Meijer* 9933 (L, US); SW of Tongoa, 1°10'S, 120°10'E, partly felled primary forest, 650 m, 15 March 1981, *Johansson, Nybom & Riebe* 385 (K).

46b. subsp. **ramosum** *Petra Hoffm.*, Kew Bull. 54: 355, fig. 5 (1999). Type: Indonesia: Papua, Vogelkop Peninsula, Aifat R. valley, Soererem camp above river, 24 Oct. 1961, *P. van Royen & Sleumer* 7056 (K!, holotype; L!, isotype).

Pistillate inflorescences consisting of up to 4 branches. Fig. 20.

DISTRIBUTION. Indonesia: Papua. Map 36.
HABITAT. In forest, on bank of river. 530 m altitude.
CONSERVATION STATUS. Data Deficient (DD): only known from the type.
ETYMOLOGY. The epithet refers to the inflorescence (Latin, ramosus = branched).
KEY CHARACTERS. Inflorescence branched.

47. Antidesma sootepense *Craib*, Bull. Misc. Inform., Kew 1911(10): 463 (1911), "sootepensis". Type: Thailand, Chiengmai, in deciduous jungle on Doi Sootep, 720 – 750 m, *Kerr* 676 (staminate) (K!, lectotype, here designated; K!, TCD!, isolectotypes). Orignial syntype: same locality, *Kerr* 676a (pistillate) (K!, TCD!).

Shrub or tree (*fide Lakshnakara* 235: woody climber), up to 9 m, diameter up to 20 cm. Twigs brown or greenish brown. Bark grey, brown or tan, thin, smooth, sometimes roughened, sometimes cracked or flaky. *Young twigs* terete, densely, rarely sparsely, ochraceous-tomentose, brown. *Stipules* caducous, linear, apically acute, 1 – 3 × 0.2 – 0.5 mm, pubescent. *Petioles* channelled adaxially, 2 – 11 × 0.7 – 1(– 1.2) mm, densely pubescent, becoming glabrous when old. *Leaf blades* oblong to narrowly elliptic, more rarely slightly obovate or ovate, apically acuminate-mucronate, basally acute, rarely obtuse or rounded, (3 –)5 – 10(– 12.5) × (1 –)1.5 – 3(– 4.2) cm, (1.8 –)2.5 – 3.5(– 4.7) times as long as wide, eglandular, chartaceous, glabrous except along the midvein adaxially, ochraceous-pilose to glabrous all over abaxially, ochraceous-pilose to -pubescent along the veins, dull to moderately shiny on both surfaces, major veins impressed adaxially, tertiary veins reticulate to weakly percurrent, widely spaced (4 – 7 between every two secondary veins), finer venation finely tessellated, drying reddish brown, greyish or olive-green, domatia often present. *Staminate inflorescences* 4 – 11 cm long, axillary, usually aggregated at the end of the branch, simple or consisting of up to 4 branches, axes ochraceous-pubescent. *Bracts* lanceolate, apically acute, 0.3 – 0.6 × 0.2 – 0.3 mm, pilose to pubescent. *Staminate flowers* 1.5 – 2 × 1.5 – 2 mm. *Pedicels* 0 – 0.7 mm long, not articulated, pubescent. *Calyx* 0.3 – 0.5 × 1 – 1.3 mm, cupular to bowl-shaped, sepals 3 – 4, fused for $^1/_2$ – $^3/_4$ of their length, deltoid, apically acute to rounded, sinuses acute, pilose to pubescent outside, glabrous to pubescent inside, with hairs especially at the base, margin erose to entire, fimbriate. *Disc* extrastaminal-annular, sometimes slightly lobed between the stamens, sometimes partly divided into free alternistaminal lobes, glabrous. *Stamens* 3 – 4, 1.5 – 2.5 mm long, exserted 1 – 2 mm from the calyx, anthers 0.3 – 0.5 × 0.4 – 0.5 mm. *Pistillode* clavate, c. 0.5 × 0.2 – 0.5 mm, not exserted or exserted from the sepals, pilose to glabrous. *Pistillate inflorescences* 3 – 8 cm long, terminal, more rarely axillary, simple, rarely branched once or twice, axes ochraceous-pubescent. *Bracts* lanceolate, apically acute, 0.3 – 1 × c. 0.2 mm, pilose. *Pistillate flowers* 1.5 – 2 × 0.8 – 1 mm. *Pedicels* 0.2 – 1 mm long, pilose. *Calyx* c. 0.8 × 0.8 – 1 mm, urceolate, sepals 3 – 5, fused for $^1/_2$ – $^3/_4$ of their length, apically acute to rounded, sinuses acute to rounded, pilose to glabrous outside, glabrous inside but with long hairs at the base, margin erose. *Disc* shorter than the sepals, glabrous. *Ovary* ellipsoid, glabrous, style terminal, thick, stigmas 4 – 8, usually short and thin relative to the style. *Infructescences* 6 – 14 cm long, axes 0.7 – 1.2 mm wide. *Fruiting pedicels* (1 –)2 – 4 mm long, spreading-pilose. *Fruits* ellipsoid, terete (but with ± distinct dorsal and ventral ridges), basally symmetrical, with a terminal to slightly subterminal style, 3 – 5 × 2 – 3.5 mm, glabrous, rarely white-pustulate, areolate when dry.

DISTRIBUTION. Burma (Northern Shan States), Laos, Thailand (excl. Peninsula). Map 37.

HABITAT. In primary or secondary vegetation; in evergreen or deciduous (or mixed evergreen and deciduous) mixed or dipterocarp forest; associated with

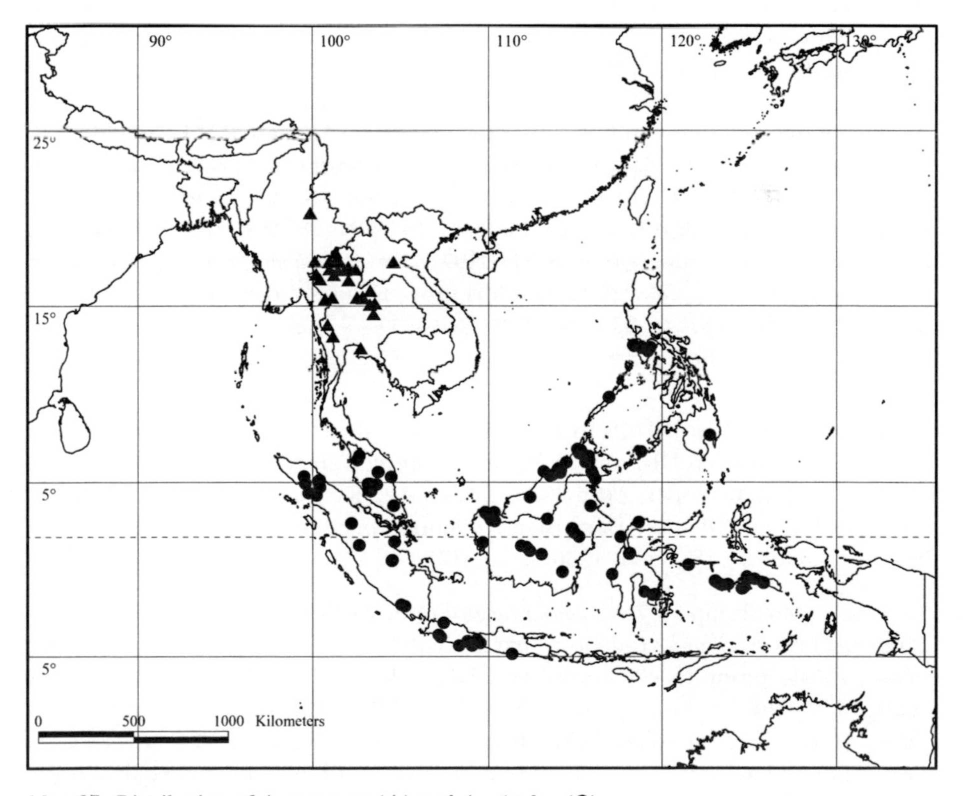

MAP 37. Distribution of *A. sootepense* (▲) and *A. stipulare* (●).

teak, bamboo, pine, oak; in gallery forest; also in gardens; open to shady, sometimes dry, fire-prone habitats. On clay or lateritic soil, over shale, limestone, granite or sandstone bedrock. 120 – 1150 m altitude.

USES. The fruits are eaten.

CONSERVATION STATUS. Near Threatened (NT). The extent of occurrence was calculated as about 500,000 km^2 and 62 specimens have been examined. This species is maybe close to a LC rating as the distribution data suggest that it may be abundant in the undercollected and probably still densely forested mountains of Burma and Laos.

VERNACULAR NAMES. ma-mao duk, mak mao, mak mao sai, sa pho mae.

ETYMOLOGY. The epithet refers to the type locality.

KEY CHARACTERS. Leaves with finely tessellated venation; thin petioles; disc glabrous; staminate disc extrastaminally fused; fruits glabrous with terminal styles.

SIMILAR SPECIES. *A. montanum*, see there. — *A. velutinum* has a ferrugineous indumentum, smaller stipules, thicker petioles, larger leaves, shorter inflorescences, always laterally compressed fruits and usually shorter fruiting pedicels.

SELECTED SPECIMENS. LAOS. Xieng-kouang, recd. 20 May 1921, *Poilane* 2360 (K). BURMA. Northern Shan States distr., Mansam Falls, 2000 ft, 11 Oct. 1911, *Lace* 5480 (K). THAILAND. NORTHERN. Chiang Mai prov., Doi Chiengdao, low slope S of Ban Dam, c. 530 m, 20 Aug. 1935, *Garrett* 987 (K, L, TCD); Mae Hong Son prov., Ban Mae Pang 30 km N of Mae Sariang, 18°28'N, 97°57'E, along stream with evergreen vegetation, 400 m, 11 July 1968, *Larsen, Santisuk & Warncke* 2345 (K, L). NORTH-EASTERN. Phetchabun prov., Lom Kao, along edge of evergreen jungle, c. 1020 m, 8 May 1955, *Smitinand* 11812 (K). EASTERN. Chaiyaphum prov., between Ban Nam Phrom and Tunkamang, 15°40'N, 102°00'E, dry dipterocarp forest, edge with evergreen river board forest, clayey soil, 700 m, 25 May 1974, *Geesink, Hattink & Phengklai* 6984 (K, L). SOUTH-WESTERN. Kanchanaburi prov. & distr., Huay Bankau, 14°55'N, 98°45'E, mixed deciduous forest, on limestone hill, 800 m, 8 Nov. 1971, *Van Beusekom, Phengklai, Geesink & Wongwan* 3529 (K, L). CENTRAL. Saraburi prov., Kang Koi distr., evergreen forest, 6 Oct. 1926, *Lakshnakara* 235 (K). SOUTH-EASTERN. Chon Buri prov., Sriracha distr., Nong Nam Kheo, forest, 15 Nov. 1926, *Collins* 1288 (K).

48. Antidesma spatulifolium *Airy Shaw,* Kew Bull. 23: 283 (1969). Type: Papua New Guinea, Western Division, Mabaduan, common in remnant monsoon-forest patches on granite slopes, April 1936, *Brass* 6512 (K!, holotype; A!, L!, isotypes). Paratype: Western Division, Daru Island, common in edges of rain-forest, scattered in savannah-forest, 18 March 1936, *Brass* 6375 (A!, K!).

Tree or shrub, up to 7 m. *Young twigs* terete, slightly striate, shortly pilose to glabrous, brown. *Stipules* caducous, narrowly deltoid to linear, 1 – 3 × c. 0.5 mm, pilose. *Petioles* terete to channelled adaxially, 1.5 – 4 × 0.5 – 0.7 mm, pilose. *Leaf blades* spathulate, apically retuse, rounded or obtuse, basally acute-cuneate, (1.5 –)2 – 5(– 7) × (0.7 –)1 – 2.5(– 3.5) cm, (1.3 –)1.8 – 2.4(– 2.8) times as long as wide, eglandular, chartaceous, glabrous, or slightly pilose only along the major veins abaxially or on both surfaces, shiny adaxially, moderately shiny abaxially,

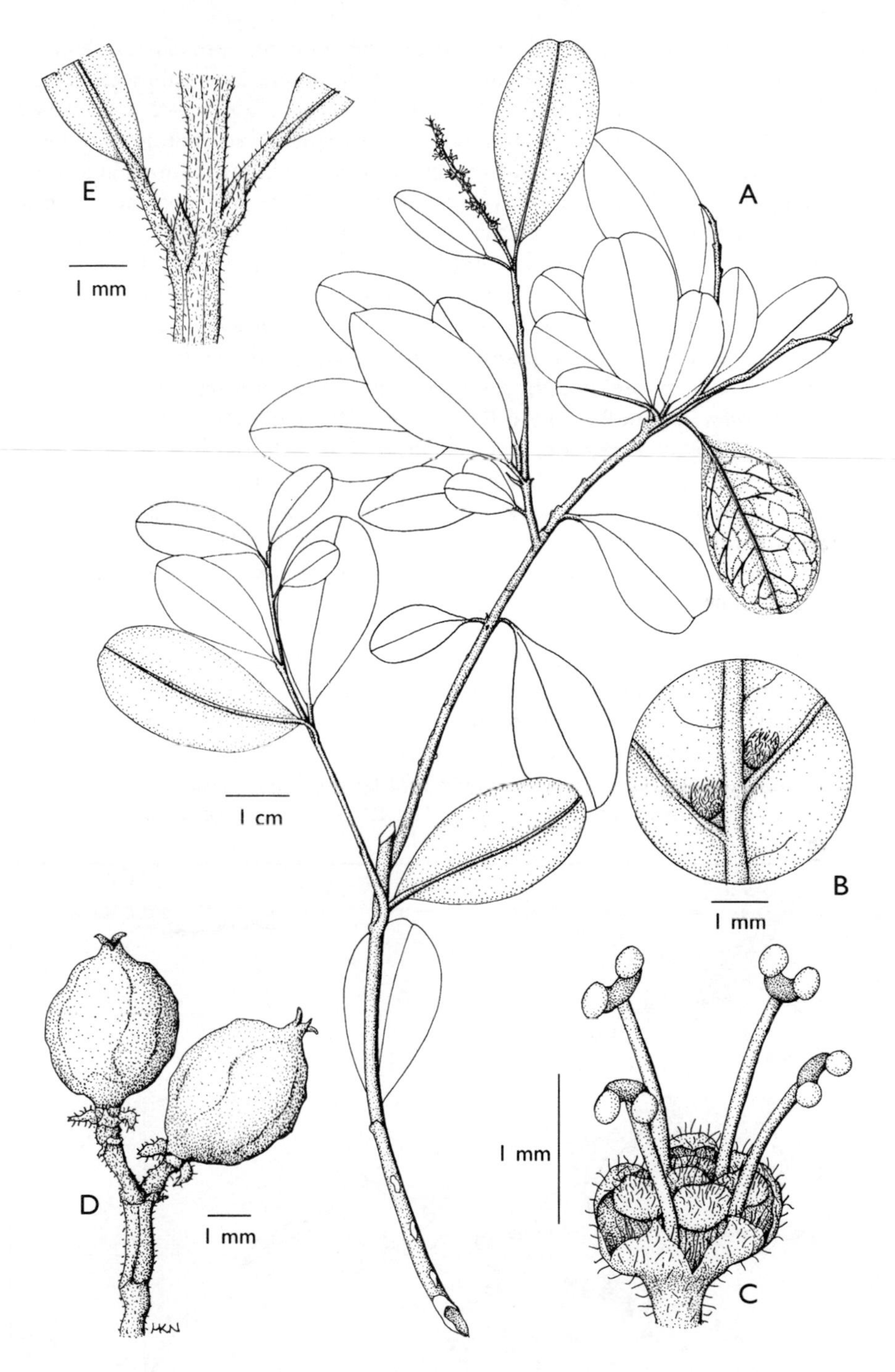

FIG. 21. *Antidesma spatulifolium.* **A** habit with staminate inflorescence; **B** detail from abaxial leaf surface with domatia; **C** staminate flower; **D** part of infructescence; **E** part of branch with stipules, petioles and leaf bases. From *Brass* 6512 (L). Drawn by Holly Nixon.

midvein flat adaxially, tertiary veins reticulate, drying olive-green, hairtuft domatia present. *Staminate inflorescences* c. 3 cm long, axillary, simple, axes pubescent. *Bracts* lanceolate, c. 0.5 × 0.5 mm, pilose. *Staminate flowers* 2 – 2.5 × 1.5 – 2 mm. *Pedicel* absent. *Calyx* 0.6 – 0.8 × 1 – 1.2 mm, sepals 4, free or nearly so, deltoid to oblong, apically acute, glabrous to pilose outside, glabrous inside but with long hairs at the base, margin erose, fimbriate. *Disc* consisting of 2 – 4 free alternistaminal lobes, lobes ± obconical, well-separated from each other, c. 0.5 × 0.2 – 0.5 mm, pilose. *Stamens* 2 – 4, 2 – 2.5 mm long, exserted 1.5 – 2 mm from the calyx, anthers 0.3 – 0.4 × 0.5 – 0.6 mm. *Pistillode* clavate to obconical, 0.3 – 0.5 × 0.2 – 0.3 mm, extending to the same length as the disc, pubescent. *Pistillate inflorescences* axillary, simple, axes sparsely and shortly pilose. *Bracts* deltoid, apically acute, 0.5 – 0.8 × 0.3 – 0.5 mm, sparsely pilose, margin sometimes glandular-fimbriate. *Pistillate flowers* not known; *calyx* in fruit 0.8 – 1 × c. 2 mm, sepals 4 – 6, free or nearly so, deltoid, apically acute, glabrous outside, glabrous inside but with long hairs at the base, margin erose, fimbriate; *disc* shorter than the sepals, pilose, especially at the margin; *stigmas* 4 – 6. *Infructescences* 1.5 – 3 cm long. *Fruiting pedicels* 0.2 – 1 mm long, shortly pilose to glabrous. *Fruits* ellipsoid or lenticular, laterally compressed, basally symmetrical, with a terminal style, 4 – 6 × 3 – 5 mm, glabrous, not white-pustulate, fleshy or areolate when dry. Fig. 21.

DISTRIBUTION. Moluccas (Tanimbar Island), New Guinea (Southern West Papua; Papua New Guinea: Western province). Map 38.

HABITAT. In edges of rain forest; in remnant monsoon forest patches; in *Melaleuca* forest; in savannah. 5 – 100 m altitude.

CONSERVATION STATUS. Vulnerable (VU B1ab(i,ii,iv)). Only 8 specimens of this species are known; it was last collected in 1973. The lack of information may

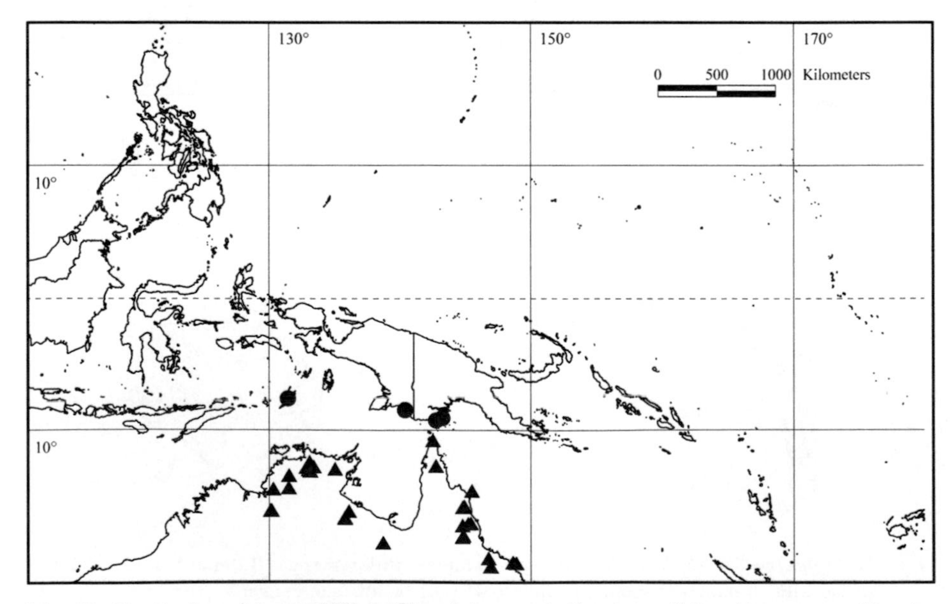

MAP 38. Distribution of *A. spatulifolium* (●) and the very similar *A. parvifolium* (▲, not treated here).

be a collection artefact but in the absence of positive evidence, this distinctive species is classified as Vulnerable.

ETYMOLOGY. The epithet refers to the leaf-shape (Latin, spatha = spathe; folium = leaf).

KEY CHARACTERS. Xerophytic small, spathulate leaves; short, simple, slender inflorescences; hairy disc; free staminate disc lobes; glabrous fruits with terminal styles.

SIMILAR SPECIES. *A. parvifolium* Thwaites & F. Muell. from Australia (see Map 38) has a glabrous disc, terminal to subterminal styles, smaller leaves [(0.5 –)0.8 – 1.8(– 3.5) × (0.3 –)0.5 – 1(– 2) cm] and a more slender habit.

FURTHER SPECIMENS. MOLUCCAS. Tanimbar Isl., Jamdena Isl., Otimmer, c. 100 m, 1 April 1938, *Boschproefstation bb (s. coll.)* 24374 (A, L, SING); Tanimbar Isl., Jamdena Isl., c. 15 km ENE of Otimmer, *Melaleuca* forest, surrounded by primary forest, with wild cattle, 31 March 1938, *Buwalda* 4546 (L). INDONESIA: PAPUA, Merauke area, pr. Kurik, Gali Ephat, forest, 2 June 1961, *Hoogerwerf* 33 (L); Merauke area, pr. Kurik, North Rawah, 3 June 1961, *Hoogerwerf* 73 (L). PAPUA NEW GUINEA, Delta of the Fly R., 1890, *Macgregor in Herb. F. Mueller* s. n. (MEL); Western province, Morehead subdistr., junction of Mai Kussa and Wassi Kussa R., 9°05'S, 142°00'E, 5 m, 11 June 1973, *NGF (Henty)* 49676 (A).

49. Antidesma stipulare *Blume,* Bijdr. Fl. Ned. Ind.: 1125 (1826 – 27). Type: Java, locis umbrosis insulae Nusae Kambangae, *Blume* [1657] (L! herb. no. 903154-342, lectotype, here designated; L!, P!, isolectotypes). — Note: Collection number not cited in protologue.

A. diepenhorstii Miq., Fl. Ned. Ind., Eerste Bijv.: 466 (1861). Type: Sumatra occ., prov. Priaman, *Diepenhorst* [HB 1364] (U!, holotype; BO!, isotype). — Note: Collection number not cited in protologue.

A. amboinense Miq., Ann. Mus. Bot. Lugduno-Batavi 1: 218 (1864). — *A. stipulare* Blume f. *amboinense* (Miq.) J. J. Sm. in Koord. & Valeton, Meded. Dept. Landb. Ned.-Indië 10: 262 (1910). Type: Amboina, *De Fretes* [5566] (U!, holotype; G-DC [microfiche], K!, isotypes s.n.). — Note: Collection number not cited in protologue. Airy Shaw (1975: 217) maintained this taxon on account of the small stipules. The stipules of the type of *A. amboinense* (5 – 10 × 15 – 25 mm) are, however, not unusually small.

A. cordatostipulaceum Merr., Philipp. J. Sci., C, 4: 275 (1909), "cordato-stipulaceum". Type: Philippines, Mindoro, Baco R., *McGregor* 311 (K!, lectotype, here designated). Original syntypes: *Merrill* 1807 (US!), *Merrill* 4048, *McGregor* 179 (K!), *FB (Merritt)* 6794. — Note: The hyphen in the epithet has to be deleted according to Art. 60.9 (Greuter *et al.* 2000: 94).

A. grandistipulum Merr., Philipp. J. Sci., C, 11: 56 (1916). Type: Sarawak, *BS (Native collector)* 1148 (K!, lectotype, here designated; A! and photo ex PNH, isolectotypes). — Note: The duplicate in A was not chosen as lectotype because it has no stipules.

A. sarawakense Merr., Philipp. J. Sci., C, 11: 57 (1916). Type: Sarawak, Rock Road, July 27, *BS (Native collector)* 503 (US!, lectotype, here designated; A! [photo ex PNH only], isolectotype).

A. stenophyllum Gage, Rec. Bot. Surv. India 9: 225 (1922), "tsenophyllum", nom. illeg. (non Merr. 1916). Type: Malaysia, Johore, Gunong Pulai, *Ridley* s.n. (SING!, lectotype, here designated). Original syntypes: Perak, *Scortechini* 638; Larut, *King's collector* 4249, 7880.

A. cordatostipulaceum Merr. var. *lancifolium* Merr., Enum. Philipp. Fl. Pl. 2: 413 (1923), "cordato-stipulaceum". Type: Philippines, Mindoro, *BS (Ramos & Edaño)* 40732 (A!, lectotype, here designated; K!, P!, US!, isolectotypes). Original syntype: Mindoro, *BS (Ramos & Edaño)* 40733 (BO!, L!). — Note: The varietal epithet was changed to "lanceifolium" by Airy Shaw (1983: 5), which is correct according to Art. 60.8 (Greuter *et al.* 2000: 94). However, Stearn (1983: 455 & pers. comm.) considered that this is a word so well established in botanical Latin since Linnean times that it should be applied according to current usage, and this view is followed here.

Tree or shrub, up to 10 m, diameter up to 15 cm. Twigs white to grey. Bark brown, grey, white or green, smooth, soft; inner bark green, grey or yellowish; sapwood white or yellowish. *Young twigs* terete, slightly pilose, becoming glabrous, brown. *Stipules* persistent, usually foliaceous, ovate to cordate, sometimes lanceolate to subulate, apically acuminate-mucronate, rarely obtuse to rounded, basally asymmetrical, parallel-veined, chartaceous, 5 – 60 × 1 – 30 mm, glabrous to very slightly pilose. *Petioles* channelled adaxially, 3 – 12(– 20) × 1 – 3 mm, glabrous to pilose. *Leaf blades* oblong, rarely slightly ovate or obovate, apically long acuminate-mucronate, rarely acute, basally acute, more rarely obtuse to truncate, often shortly decurrent, (9 –)20 – 30(– 40) × (1.5 –)5 – 7(– 10) cm, (3.2 –)4 – 5(– 9.2) times as long as wide, eglandular, coriaceous to chartaceous, glabrous, or slightly pilose along the major veins abaxially, shiny on both surfaces, major veins slightly raised adaxially, secondary veins widely spaced, up to 16 per leaf half, tertiary veins reticulate to percurrent, drying olive-green, lighter abaxially. *Staminate inflorescences* 3 – 13 cm long, axillary, simple, very slender, axes glabrous. *Bracts* ovate, apically acute, c. 0.5 × 0.5 mm, glabrous to pilose, margin sometimes fimbriate. *Staminate flowers* c. 1 × 1.5 mm. *Pedicel* absent. *Calyx* c. 0.7 × 1.5 mm, cupular to bowl-shaped, sepals 5, fused for $^{1}/_{2}$ of their length, apically obtuse to rounded, glabrous on both sides, margin entire to slightly erose. *Disc* cushion-shaped, enclosing the bases of the filaments and pistillode, glabrous, rarely pilose. *Stamens* (4 –) 5, c. 1 mm long, exserted c. 0.5 mm from the calyx, anthers c. 0.3 × 0.5 mm. *Pistillode* flat to cylindrical, 0.1 – 0.2 × 0.3 – 0.4 mm, extending to the same length as the sepals, glabrous, very rarely pilose. *Pistillate inflorescences* 10 – 35 cm long, axillary, simple, slender, axes glabrous. *Bracts* deltoid, c. 0.5 × 0.3 mm, glabrous. *Pistillate flowers* 2 – 2.5 × c. 1.5 mm. *Pedicels* 0.5 – 2 mm long, glabrous. *Calyx* c. 1 × 1.5 mm, shallowly cupular, sepals 5, fused for c. $^{2}/_{3}$ of their length, broadly deltoid, apically obtuse, glabrous on both sides. *Disc* shorter than the sepals, glabrous to pilose. *Ovary* ovoid, glabrous, very rarely pilose, style subterminal, stigmas 3 – 4. *Infructescences* 10 – 50 cm long. *Fruiting pedicels* 1 – 14 mm long, glabrous. *Fruits* obliquely ovoid (mango-shaped), laterally compressed, basally symmetrical, with a lateral to subterminal style, 8 – 15(– 18) × 6 – 11 mm, 1.3 – 1.7 times as long as wide, glabrous, sometimes white-pustulate, reticulate when dry.

DISTRIBUTION. Peninsular Malaysia, Sumatra, Borneo, Java, Philippines (Mindanao, Mindoro, Palawan, Sibuyan?, Sulu Archipelago), Sulawesi, Moluccas (Ambon, Buru, Seram, Sula Islands). The two specimens from Burma (*Russell* 2036 and 2224 [CAL] attributed to this species by Chakrabarty & Gangopadhyay (2000) could not be examined for this study. Map 37.

HABITAT. In primary and secondary vegetation; in lowland to hill dipterocarp forest up to 40 m tall; in mossy upper montane forest; in riverine and swamp forest; in *Agathis* forest; usually in shady understorey. On clay, loam, peat and sandy soil, often acidic, sometimes on ultrabasic soil, over limestone, basalt, sandstone and shale. 0 – 1800 m altitude.

USES. The wood is strong and resistant to termites. It is used for household and construction purposes (Hasskarl 1845: 71).

CONSERVATION STATUS. Least concern (LC). This is one of the most common *Antidesma* species in Malesia. 246 specimens examined.

VERNACULAR NAMES. Malay Peninsula: setundot; Kalimantan: boliti hara, maning; Sulawesi: kau jole (Kulawi); Java (Hasskarl 1845: 80): kisapie, kieueur badak; Moluccas: bua taai - kambing utan (fruits resemble droppings of wild goats), sulamin.

ETYMOLOGY. The epithet refers to the conspicuous stipules (Latin, stipularis = stipulaceous).

KEY CHARACTERS. Plant almost glabrous; leaves shiny, drying olive-green; stipules often foliaceous; inflorescences slender, simple; staminate flowers sessile; stamens very short.

SIMILAR TAXA. *A. kunstleri*, *A. montis-silam*, *A. neurocarpum* var. *hosei*, see there. — *A. tomentosum* has a ferrugineous indumentum at least at the tip of the branches, free, acute sepals, hairy fruits, often reddish drying leaves with more strongly percurrent and closer secondary veins, and usually smaller stipules. Contrary to Airy Shaw (1975: 218), the stipules of *A. tomentosum* can also be almost foliaceous, but they are thicker and usually falcate. The fruits of *A. tomentosum* are usually also falcate and narrower than those of *A. stipulare*, but their length/width ratios overlap.

NOTE. A variable species. The foliaceous stipules are a good diagnostic character if present, but a number of collections throughout the distribution area have small, linear stipules. In *SAN* 25010 (K), for example, the stipules range from 5 × 1 to 17 × 5 mm. There are some collections, mainly from Kalimantan (*Nooteboom* 4485 (K, L), *Veldkamp* 8041 (BO, L)) but also the original syntype of *A. cordatostipulaceum* var. *lancifolium* (*BS* 40733), with pilose flowers, but in the vast majority of collections the flowers are completely glabrous.

SELECTED SPECIMENS. PENINSULAR MALAYSIA. Selangor, Gunong Bunga Bua, high forest on broad ridge, 2800 ft, 28 May 1966, *FRI (Whitmore)* 328 (K, L, SING). SUMATRA. Northern part, Bukit Lawang, Bohorok, Langkat, c. 450 m, 21 Feb. 1973, *Soedarsono* 325 (BO, K, L). BORNEO. Brunei, Belait distr., Labi subdistr.,

Bukit Teraja, valley and ridge to the N of hill rest-shelter, 4°18'N, 114°26'E, hill dipterocarp forest, Lambir formation, sandstone and shale, 250 m, 19 March 1991, *Sands, Awong Kaya & Shanang Fikir* 5456 (K, KEP, L); Sabah, Kinabalu, Penibukan, below Dahobang falls, river bed, 4000 ft, 11 Sept. 1933, *Clemens & Clemens* 40317 (A, K, L, NY); Sarawak, Mt Rayon, old jungle, 400 ft, 19 Jan. 1928, *Native collector* 5002 (A, L, NY, US); E Kalimantan, Mahakam Ulu, Melaham, 26 June 1975, *Wiriadinata* 616 (BO, L, US). JAVA. E Java, Gunung Kumbakarna, Prigi, *Afriastini* 1224 (BO, K, L). PHILIPPINES. Mindoro, Paluan, April 1921, *BS (Ramos)* 39605 (A, BO, K, L, US). SULAWESI. SE Sulawesi, Kolaka, Mt Poli-polia, Tawanggo, primary forest, 200 m, 22 Oct. 1978, *Prawiroatmodjo & Maskuri* 1462 (BO, K, L, SAR). MOLUCCAS. Ambon, s. loc., July – Nov. 1913, *Robinson* Plantae Rumphianae Amboinenses 355 (BM, BO, GH, K, L, NY); Buru, W Buru, Bara, Wae (= river) Duna, Base Camp 9, primary forest on the ridge of the hill, 650 m, 27 Nov. 1984, *Mogea* 5382 (BO, K, L); Seram, Taniwel, Bukit Buria, mixed forest, 300 – 600 m, 1984, *Ramlanto* 317 (BO, K, L).

50. Antidesma subcordatum *Merr.,* Philipp. J. Sci., C, 4: 275 (1909). — *A. subcordatum* Merr. var. *genuinum* Pax & K. Hoffm. in Engl., Pflanzenr. 81: 156 (1922), nom. inval. Type: Philippines, Luzon, Prov. Rizal, Bosoboso, *BS (Ramos)* 1114 (NY!, lectotype, here designated; BO!, GH!, K!, P!, US!, isolectotypes). Original syntypes: *BS (Ramos)* 4564; Prov. Rizal, *Merrill* 2813 (BM!, K!, P!, US!); *FB (Ahern's collector)* 3160 (NY!, P!, US!).

A. fusicarpum Elmer, Leafl. Philipp. Bot. 8: 3081 (1919). Type: Philippines, Luzon, Prov. Laguna, Los Baños (Mt Maquiling), in deeply shaded ravines along the main creek, alt. c. 1000 ft, June – July 1917, *Elmer* 18083 (GH!, lectotype, here designated; BM!, BO!, G!, K!, L!, NY!, US!, isolectotypes).

A. subcordatum Merr. var. *glabrescens* Pax & K. Hoffm. in Engl., Pflanzenr. 81: 157 (1922), **synon. nov.** Type: Philippines, Luzon, Pangasinan, *Merrill* 2860 (NY!, lectotype, here designated; K!, US!, isolectotypes). Original syntypes: Luzon, Bulacan, *Curran* 7213 (US!); Palawan, *Curran* 4492 (NY!, P!, US!).

Tree, up to 8 m, diameter up to 20 cm. *Young twigs* terete, pubescent, brown. *Stipules* early-caducous, subulate, 3 – 6 × 0.5 – 0.8 mm, pubescent. *Petioles* narrowly channelled adaxially, 4 – 10(– 17) × 0.5 – 1(– 1.5) mm, pubescent, glabrescent. *Leaf blades* oblong to ovate, rarely slightly obovate, apically acuminate-mucronate, basally cordate to rounded, (3.5 –)5 – 11(– 18) × (2 –)3 – 5(– 7.5) cm, (1.3 –)1.8 – 2.4(– 4) times as long as wide, eglandular, chartaceous, glabrous with pubescent major veins to pubescent adaxially, pubescent to sparsely pilose all over abaxially, especially along the veins, dull or shiny adaxially, dull abaxially, midvein shallowly impressed adaxially, tertiary veins reticulate to percurrent, drying olive-green, lighter abaxially, domatia absent. *Staminate inflorescences* 2.5 – 9 cm long, axillary, simple or branched at the base, consisting of up to 9 branches, axes densely ferrugineous-pubescent. *Bracts* lanceolate, c. 0.5 × 0.2 – 0.3 mm, densely pubescent. *Staminate flowers* 2 – 3 × 2 – 3 mm. *Pedicel* absent. *Calyx* 0.5 – 1 × c. 1.5 mm, sepals 4

– 5, free, deltoid to oblong, very thin, apically acute to obtuse, pubescent outside, with long hairs inside, especially at the base, margin entire. *Disc* consisting of 4 – 5 free alternistaminal lobes, lobes ± obconical, well-separated from each other, c. 0.5 × up to 0.5 mm, pubescent. *Stamens* 4 – 5, 1.5 – 2.5 mm long, exserted 1 – 1.5 mm from the calyx, anthers c. 0.5 × 0.5 mm. *Pistillode* cylindrical, 0.7 – 1 × 0.2 – 0.4 mm, extending to the same length or exserted from the sepals, pubescent. *Pistillate inflorescences* 4 – 9 cm long, axillary, simple, axes ferrugineous-pubescent. *Bracts* lanceolate, c. 0.7 × 0.3 mm, pubescent. *Pistillate flowers* c. 2 × 0.7 – 1 mm. *Pedicels* 0.5 – 1 mm long, pilose. *Calyx* 0.5 – 1 × c. 1 mm, sepals 5 – 6, fused for $^1/_2$ – $^3/_4$ of their length, deltoid, apically acute, pubescent outside, with long hairs inside, especially at the base, margin entire. *Disc* shorter than the sepals, pubescent, especially at the margin. *Ovary* elongate-ellipsoid, pilose, style terminal, not distinct, pilose, stigmas 4 – 8. *Infructescences* 4 – 9 cm long, axes 0.7 – 1.2 mm wide. *Fruiting pedicels* 0 – 3 mm long, pubescent to glabrous. *Fruits* ellipsoid, laterally compressed, basally symmetrical, with a terminal, more rarely subterminal style, 3 – 6 × 2 – 3.5 mm, pilose, sometimes white-pustulate, areolate when dry. Fig. 13A – D (p. 125).

DISTRIBUTION. Philippines (Coron, Corregidor, Luzon, Masbate, Palawan) and Lesser Sunda Islands (Flores, Timor). The disjunction might indicate an adaptation to drier habitats than its presumed sister taxon *A. ghaesembilla*, or it might simply be a collecting artefact. Map 39.

HABITAT. No data. 20 – 200 m altitude.

CONSERVATION STATUS. Near Threatened (NT). This species has not been frequently collected in recent years, but the distribution area is large even if the disjunction is not a collection artefact. 20 specimens examined.

VERNACULAR NAMES. Philippines: bignay-kalabau (Tagbanua); Lesser Sunda Islands: hau nuif, nuna (Dawan).

ETYMOLOGY. The epithet refers to the shape of the leaf base (Latin, sub- = somewhat; cordatus = heart-shaped).

KEY CHARACTERS. Cordate leaf base; long, slender petioles; densely ferrugineous-pubescent inflorescences; free staminate disc lobes.

SIMILAR SPECIES. *A. edule*, *A. ghaesembilla* and *A. myriocarpum*, see there.

SELECTED SPECIMENS. PHILIPPINES. Coron Island, s. loc., Sept. 1922, *BS (Ramos)* 41160 (A, K, US); Corregidor Isl., s. loc., Sept. 1911, *FB (Curran)* 13212 (A); Luzon, Rizal prov., Bosoboso, Aug. 1907, *BS (Ramos)* 4664 (NY, US); Luzon, Rizal prov., Antipolo, July 1917, *BS (Ramos & Edano)* 29516 (A, US); Luzon, Ilocos Norte prov., Burgos, July 1918, *BS (Ramos)* 32865 (K, US); Luzon, Batangas prov., Mt Lobo, S side, Bo. Ifugom, San Juan, level land near cultivation, low altitude, 24 May 1952, *PNH (Sulit)* 15719 (A, L); Luzon, Rizal prov., Antipolo, June 1910, *Ramos* 379 (FR, G, U, US, WRSL); Masbate Island, s. loc., Aug. 1903, *Merrill* 3051 (BM, K, NY, US). LESSER SUNDA ISLANDS. Flores, Paning-Keka (Cibal), 14 Feb. 1973, *Verheijen* 3317 (L); Flores, W Flores, Wae Longge, 20 m, 20 Dec. 1967, *Schmutz* 2008 (L); Timor, Bocsufa?, c. 200 m, 2 Feb. 1966, *Kooy* 349 (L).

51. Antidesma tetrandrum *Blume*, Bijdr. Fl. Ned. Ind.: 1124 (1826 – 27); Pax & K. Hoffm. in Engl., Pflanzenr. 81: 112, fig. 12 A, D (1922). — *A. blumei* Tul., Ann. Sci. Nat. Bot., Sér. 3: 211 (1851), nom. nov. illeg. — Note: Tulasne referred to

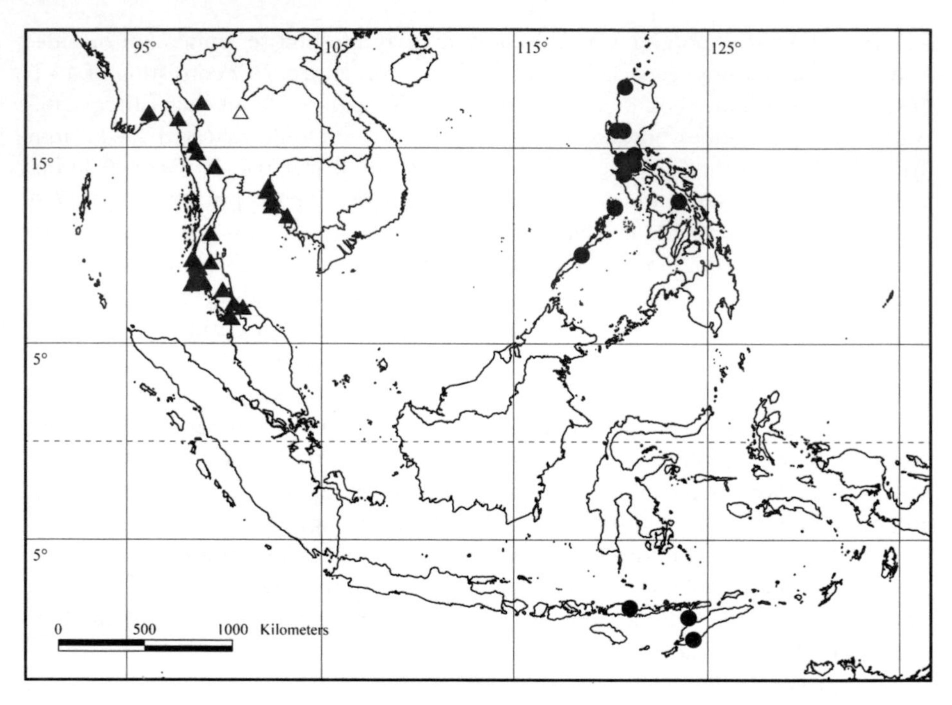

MAP 39. Distribution of *A. subcordatum* (●) and *A. velutinum* (▲), doubtful specimen (△).

Blume's species but disapproved of the epithet "tetrandrum" because he thought that too many *Antidesma* species had four stamens, and replaced it with "blumei". Type: Java, in sylvis montanis, *Blume* [555] (L! herb. no. 910222-1038, lectotype, here designated; BO!, L!, isolectotypes). — Note: Collection number not cited in protologue. Original material also in NY!

A. salaccense Zoll. & Moritzi, Syst. Verz.: 74 (1846). Type: Java, in monte Salak juxta flum. Tjapus alt. 4000 ft, 19 Nov. 1843, *Zollinger* 1784 (P! (fruits & staminate flowers), lectotype, here designated; G!, G-DC [microfiche], MEL!, P!, isolectotypes).

A. auritum Tul., Ann. Sci. Nat. Bot., Sér. 3: 203 (1851). Type: Java, *Zollinger* 2529 (P!, lectotype, here designated; BM!, G!, G-DC [microfiche], MEL!, P!, U!, isolectotypes).

Shrub or tree, up to 15 m, diameter up to 25 cm, often bent, often branching from the base. Bark grey, brown or tan; watery sap reported in *Hassan* 80 and 96. *Young twigs* terete, shortly spreading-pubescent to glabrous, brown. *Stipules* usually persistent (sometimes breaking off early), usually foliaceous, spathulate, cordate or kidney-shaped, sometimes lanceolate, apically rounded to acute, basally symmetrical, reticulately veined, thin, 8 – 15(– 25) × (2 –) 5 – 20 mm, glabrous to sparsely pilose. *Petioles* narrowly channelled adaxially, basally and distally ± pulvinate for 2 – 3 mm, (7 –)10 – 15(– 35) × 0.7 – 1(– 2) mm, spreading-pubescent to glabrous. *Leaf blades* oblong, slightly ovate or obovate, apically long to shortly acuminate-mucronate, basally rounded, more rarely obtuse or truncate, (6 –)10 – 14(– 22) × (2.5 –)4 – 6(– 9) cm, (2 –)2.5(– 3) times as long as wide, eglandular, chartaceous to membranaceous, glabrous, or slightly pilose only along the major veins, dull on both surfaces, major veins impressed adaxially, tertiary veins weakly percurrent, mostly widely spaced, drying olive-green, slightly lighter abaxially. *Staminate inflorescences* 4 – 11 cm long, axillary, branched regularly, consisting of up to 6 branches, rarely simple, axes pilose to pubescent. *Bracts* orbicular to spathulate, 0.3 – 0.5 × c. 0.3 mm, pubescent, margin entire, sometimes fimbriate. *Staminate flowers* c. 1.5 × 1.5 mm. *Pedicel* absent. *Calyx* 0.3 – 0.5 × 0.6 – 0.8 mm, globose to cupular, sepals 4(– 6), fused for c. $^2/_3$ of their length, apically acute to obtuse, pubescent to pilose outside, glabrous inside but with long hairs at the base. *Disc* consisting of 4(– 5) free alternistaminal lobes, lobes ± obconical, with two shallow imprints apically, 0.2 – 0.3 × 0.2 – 0.3 mm, usually exserted from the sepals, glabrous. *Stamens* 4(– 5), 1 – 1.2 mm long, exserted 0.5 – 0.7 mm from the calyx, anthers c. 0.3 × 0.3 mm. *Pistillode* clavate, often crateriform apically, 0.5 – 0.7 × 0.2 – 0.3 mm, extending to the same length or exserted from the sepals, pubescent. *Pistillate inflorescences* 5 – 6 cm long, axillary or terminal, branched regularly, usually consisting of 5 – 8 branches, rarely simple, axes pilose to pubescent. *Bracts* deltoid to lanceolate, apically acute, c. 0.5 × 0.5 mm, pilose, margin fimbriate. *Pistillate flowers* 1 – 1.5 × c. 0.7 mm. *Pedicels* 0.5 – 1 mm long, glabrous to pubescent. *Calyx* c. 0.7 × 0.7 mm, urceolate to cupular, sepals 4 – 5(– 6), fused for c. $^2/_3$ of their length, apically acute, pilose outside, glabrous inside except for some long hairs at the base. *Disc* shorter than the sepals, glabrous. *Ovary* ellipsoid to globose, glabrous, style subterminal, stigmas 4 – 8. *Infructescences* 6 – 13 cm long. *Fruiting pedicels* (1 –)2 – 4 mm long, pilose to pubescent. *Fruits* obliquely ellipsoid, laterally compressed, basally asymmetrical, with a lateral style, 3 – 5 × 2 – 3 mm, glabrous, rarely white-pustulate, areolate when dry.

DISTRIBUTION. Sumatra, Java, Lesser Sunda Islands (Bali). Chakrabarty & Roy (1984: 168) as well as Chakrabarty & Balakrishnan (1992: 19) report *A. tetrandrum* from Great Nicobar Island (*Ahamed Ali* 20 (CAL), *Balakrishnan* 5754 (PBL), 5756 (PBL), *Nair* 7199 (PBL), *Hore* 7774 (PBL)), none of which have been examined for this study. Map 40.

HABITAT. In primary or secondary rainforest up to 40 m tall associated with *Altingia, Castanopsis, Eugenia, Oleandra, Quercus, Schima*; in cloud forest 15 – 25 m tall associated with *Engelhardia, Nauclea, Schima, Wendlandia*; on river banks and marshy ground; at forest edges, in forest remnants, and in secondary vegetation around human habitation (Malay: "belukar"), perhaps planted. On clay soil. 25 – 1800 (– 2300) m altitude.

CONSERVATION STATUS. Least Concern (LC). This species is both common and widespread. 181 specimens examined.

USES. The fruits are sometimes eaten. The hard wood is used to make axe handles in Java (Smith 1910: 282).

VERNACULAR NAMES. Sumatra: awa burunai, burunai alafai, burunai fateoh,

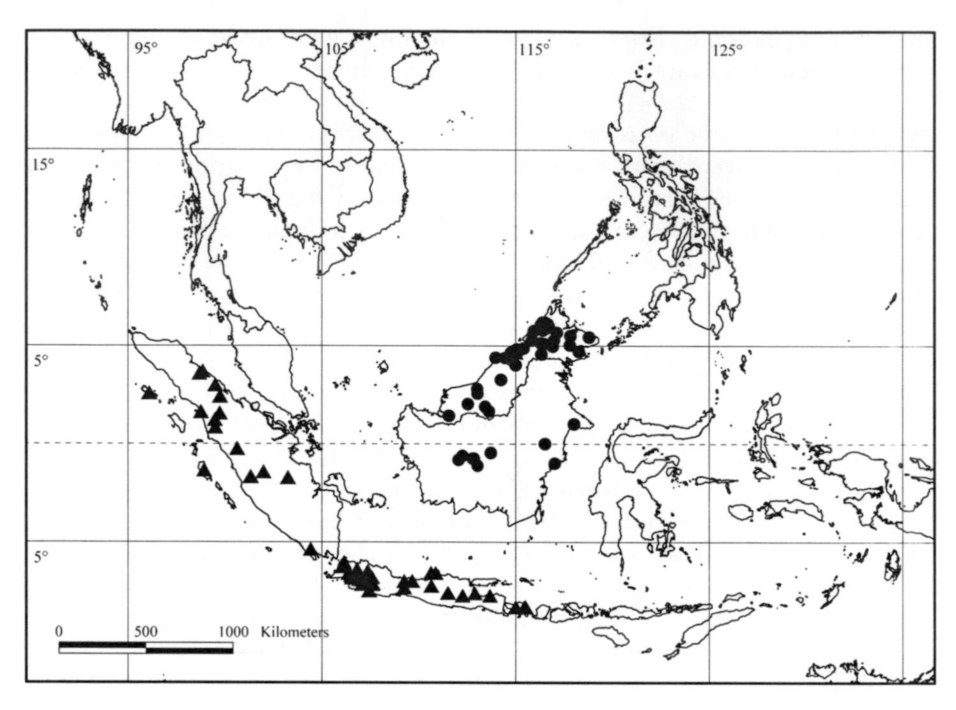

MAP 40. Distribution of *A. tetrandrum* (▲) and *A. venenosum* (●).

bolinai, etna/etwa/awa paudakan/pandakan, kayu si bait ari, kayu simo-simo, selang datan; Java: ande-ande, ande-andejan, andi-andi, anggi-anggi (Javanese), huni potjang or wuni peutjang (Sundanese), ki seueur or ki seueur beureum (Sundanese, seueur = many, refers to the many small edible fruits), kuru mera or huru mera, punai punai, tjungul; Bali: boni-sigium.

ETYMOLOGY. The epithet refers to the number of stamens (Greek, tetra- = four-; -andrus = male).

KEY CHARACTERS. Foliaceous stipules; thin, glabrous leaves; fused sepals; glabrous disc; free staminate disc lobes; very small, glabrous, bean-shaped fruits.

SIMILAR SPECIES. *A. celebicum, A. cuspidatum, A. pleuricum,* see there. — *A. venenosum* has smaller fruits and stipules, however, immature fruits of *A. venenosum* are very similar to those of *A. tetrandrum.* Stipules can also be sometimes reduced or broken off. If in doubt, rely on the locality.

SELECTED SPECIMENS. SUMATRA. Lampung, NW of Kota Agung, 5°23'S, 104°25'E, primary forest recently under trial destruction, 350 – 450 m, 14 May 1968, *Jacobs* 8417 (K, KEP, L); E coast, Laboehan Batoe subdiv., Bila distr., Goenoeng Panjaboengan (in Concession Pangkatan-Zuidwest B: Topographic Sheet 34, S centre), 26 March 1933, *Rahmat Si Toroes* 4361 (A, L, NY, US); Padang Si Dimpoean div., Padang Lawas subdiv., Poelo Liman (Topographic Sheet 41, NW quarter), 29 Aug. – 3 Sept. 1933, *Rahmat Si Toroes* 5328 (A, K, L, NY). JAVA. s. loc., 1846, *Lobb* 245 (BM, CGE, E, K, TCD); C Java, Gunung Muria, Tjollo, N of Kudus, 700 m, 23 Nov. 1951, *Kostermans* 6233 (A, B, K, L, NY); W Java, Tjibodas Forest

Reserve, N side of the Gedeh-Pangrango massif, 30 km SE of Bogor; by the path to Gedeh-Pangrango, Tjibeurum and Kandangbadak, forest dominated by *Altingia excelsa* associated with *Castanopsis* and *Quercus* spp., deep shade, heavy clay soil, 1500 m, 6 Sept. 1969, *Sands* 11 (K, KEP, L, US); Cimonyet, Kiaradua – Sukabumi, Forest Reserve, 700 m, 22 Feb. 1974, *Wiriadinata* 54 (K, L, US); Bantam prov., Gunung Karang, secondary forest, 1600 m, 2 Nov. 1974, *Wiriadinata* 384 (K, L, US). LESSER SUNDA ISLANDS, BALI. Bali Timur, Karangasem, S shoulder of Gunung Agung, 6 km N of Besakih, 8°21'S, 115°26'E, cloud forest in ravines, canopy 15 – 25 m tall, dominants include *Engelhardia, Nauclea, Wendlandia* and *Schima*, 1300 m, 27 June 1994, *McDonald & Ismail* 4713 (K, L); Bali, Gunong Pala, c. 965 m, 19 Sept. 1918, Expedition R. Maier, *Sarip* 318 (L, U); Bali, Bedugul?, 9 Nov. 1966, *Schwabe* s.n. (B).

52. Antidesma tomentosum *Blume*, Catalogus: 109 (1823), "tomentosa"; Bijdr. Fl. Ned. Ind.: 1126 (1826 – 27). Type: Java, *Hb. Blume* s.n. (L! herb. no. 903154-362, lectotype, here designated).

Shrub or tree (*fide Lomudin Tadong* 448: vine), 2 – 6(– 17) m, clear bole up to 6(– 10) m, diameter up to 15 cm, sometimes coppicing at base. Twigs white, brown or greenish brown. Bark brown, grey, white or green, thin, smooth, sometimes flaky, soft; inner bark whitish, yellow, tan, green or red; cambium white; wood hard, heavy, odourless and tasteless, sapwood white, yellow, orange, pink or red-brown. *Young twigs* terete, ferrugineous-pubescent, brown. *Stipules* persistent, sometimes foliaceous, subulate to falcate, rarely lanceolate with an asymmetrical base, strongly parallel-veined, coriaceous, (3 –)6 – 25(– 35) × 1 – 4(– 10) mm, pubescent to pilose, rarely glabrous. *Petioles* terete to 4-angular, channelled adaxially, 1.5 – 15(– 20) × 1 – 5 mm, pubescent to pilose, sometimes becoming glabrous when old. *Leaf blades* elliptic to slightly obovate, ovate, oblong or lanceolate, apically long acuminate-mucronate to caudate, basally acute to cordate, (8 –)15 – 30(– 60) × (1.2 –)5 – 12(– 30) cm, (2 –)2.5 – 3.5(– 13.6) times as long as wide, eglandular, chartaceous, ferrugineous-pubescent along the major veins or all over on both surfaces, adaxially sometimes and abaxially very rarely glabrous, dull to shiny on both surfaces, major veins shallowly impressed to flat adaxially, rarely slightly raised, secondary veins close together, up to 22 per leaf half, tertiary veins mostly percurrent and close together, drying reddish brown to brownish olive-green. *Staminate inflorescences* 4 – 14 cm long, axillary, simple or very rarely once-branched, very rarely 2 – 3 per fascicle, usually densely set with flowers (individual flowers touching each other), axes ferrugineous-pubescent. *Bracts* deltoid to orbicular, apically acuminate to rounded, 0.7 – 1.2(– 1.5) × 0.5 – 1 mm, pubescent. *Staminate flowers* c. 1.5 × 1.5 mm. *Pedicel* absent. *Calyx* 0.5 – 0.7 mm long, sepals 4 – 6, free or nearly so, 0.5 – 1 mm wide, deltoid to oblong, apically acute to obtuse, pubescent to pilose outside, glabrous inside. *Disc* cushion-shaped, enclosing the bases of the filaments and pistillode, rarely only partly (around a few stamens per flower) closed, glabrous to pubescent. *Stamens* (3 –)4 – 6, c. 1 mm long, exserted c. 0.5 mm from the calyx, anthers c. 0.3 × 0.4 mm. *Pistillode* clavate to cylindrical, 0.2 – 0.4 × 0.2 – 0.3 mm, shorter than the sepals, pubescent. *Pistillate inflorescences* (3 –)10 – 30 cm long, axillary, simple, rarely branched once or twice, densely set with flowers (individual flowers touching each other), axes ferrugineous-pubescent. *Bracts* deltoid to linear, 1(– 2) × c. 0.5 mm, pubescent. *Pistillate flowers* 1.5 – 2 × 1 – 1.5 mm. *Pedicels* 0 – 0.5(– 3) mm long, pubescent. *Calyx* 0.5 – 1.3 mm long, sepals (4 –)5(– 6), free or nearly so, c. 0.5 mm wide, narrowly deltoid, apically acute, pubescent outside, glabrous inside. *Disc* shorter than but visible between the sepals, glabrous to pubescent. *Ovary* ovoid, slightly falcate, densely appressed-pubescent, style subterminal, not very distinct, stigmas 3 – 8. *Infructescences* (7 –)10 – 30(– 65) cm long. *Fruiting pedicels* (0.5 –)1 – 7(– 23) mm long, pubescent. *Fruits* obliquely ovoid (mango-shaped) to falcate or elongate-ellipsoid, often slightly beaked, laterally compressed, basally symmetrical to distinctly asymmetrical, with a lateral to subterminal style, (7 –)9 – 15(– 20) × (3 –)5 – 8(– 9) mm, (1.1 –)1.3 – 2(– 3) times as long as wide, thinly puberulous to densely pilose (old fruits sometimes almost glabrous), sometimes white-pustulate, reticulate when dry.

NOTES. *A. tomentosum* is a very variable species. It is here understood in a wide sense, with the single, rheophytic variety, var. *stenocarpum* (Airy Shaw) Petra Hoffm.

52a. var. **tomentosum**

A. cumingii Müll. Arg. in DC., Prodr. 15(2): 249 (1866). Type: Philippines, Luzon, Prov. Albay, *Cuming* 1300 (G-BOISS, holotype; BM!, G!, G-DC [microfiche], CGE!, FHO!, K!, isotypes).

A. persimile Kurz, J. Bot. 13: 330 (1875); Hook. f., Fl. Brit. India 5: 365 (1887), "perserrula"; Mandal & Panigrahi, J. Econ. Taxon. Bot. 4: 256 (1983). Type: Nicobar Islands, Camorta, *Kurz* s.n. (CAL, holotype). — Note: Mandal & Panigrahi (1983: 256) also cited another specimen, Hort. Bot. Bog., *Kurz* 894 (CAL, K!). The Kew duplicate clearly represents *A. tomentosum*, and the description does not provide contradictory evidence. It seems therefore safe to treat this name as a synonym.

A. kingii Hook. f., Fl. Brit. India 5: 356 (1887). Type: Malaysia, Perak, Larut, open jun(gle?), hilly locality, rich soil, 1500 – 2000 ft, Feb. 1883, *King's collector* 3928 (K!, lectotype, here designated; G!, SING!, isolectotypes). — Note: Also original material: Perak, Goping, *King's collector* 844 (K!, P!). Collection numbers not cited in protologue, but sheets are annotated as "A. kingii Hf" by Hooker.

A. longipes Hook. f., Fl. Brit. India 5: 355 (1887), non Pax 1893; Airy Shaw, Kew Bull., Addit. Ser. 4: 218 (1975), as synon. nov. Type: Malaysia, Perak, Larut, Goping, dense jun(gle?), rich soil, rocky bed, 500 – 800 ft, Aug. 1883, *King's collector* 4761 (K!, lectotype, here designated; BM!, G!, L!, SING!, isolectotypes). — Note: Also original material: Goping, *King's collector* 994 (K!). Collection numbers are not cited in the protologue, but sheets are annotated as "A. longipes Hf" and "A. longipes Hookf." by Hooker.

A. membranifolium Elmer, Leafl. Philipp. Bot. 1: 313 (1908), "membranaefolium". Type: Philippines, Luzon, Prov. Tayabas, Lucban, in deeply shaded woody slopes of ravines, 750 m, May 1907, *Elmer* 9088 (L!, lectotype, here designated; A!, E!, G!, K!, isolectotypes). Original syntype: same locality, *Elmer* 7913 (E!, G!, K!, L!, US!). — Note: The original spelling of the epithet is an error to be corrected according to Art. 60.8 (Greuter *et al.* 2000: 94).

A. subolivaceum Elmer, Leafl. Philipp. Bot. 4: 1272 (1911). Type: Philippines, Palawan, Puerto Princesa (Mt Pulgar), in moist fertile soil of humid forests at 750 ft along the trail to Napsan on the west coast of Palawan, March 1911, *Elmer* 12883 (NY!, lectotype, here designated; A!, BM!, BO!, E!, G!, K!, L!, P!, US!, isolectotypes). — Note: The lectotype is accompanied by Elmer's field notes.

A. gibbsiae Hutch. in Gibbs, J. Linn. Soc., Bot., 42: 134 (1914). Type: Sabah, Tenom, secondary jungle, 700 – 1000 ft, Jan., *Gibbs* 2809 (K!, lectotype, here designated; BM!, isolectotype). Original Syntype: same locality, *Gibbs* 2790 (BM!, K!). — Note: These collections may come from an aberrant population with almost glabrous fruits and flowers with 3 stamens and 5 sepals.

A. samarense Merr., Philipp. J. Sci., C, 9: 469 (1914), **synon. nov.** Type: Philippines, Samar, in forests, alt. c. 200 m, April 1914, *Ramos (Phil. Pl.)* 1665 (US!, lectotype, here designated; BM!, BO!, G!, GH!, L!, NY!, P!, SING!, isolectotypes).

A. urdanetense Elmer, Leafl. Philipp. Bot. 7: 2635 (1915). Type: Philippines, Mindanao, Prov. Agusan, Cabadbaran (Mt Urdaneta), Oct. 1912, *Elmer* 13971 (A!, lectotype, here designated; BM!, BO!, E!, G!, K!, L!, NY!, P!, U!, US!, isolectotypes).

?*A. foxworthyi* Merr., Philipp. J. Sci., C, 11: 55 (1916), "foxworthyii". Type: Sarawak, Mt Poe, 26 May 1908, thickets at the edge of clearings, *Foxworthy* 268 (lectotype A! [photo of destroyed holotype ex PNH], here designated). — Note: No specimens of *Foxworthy* 246 could be located. The photograph of the holotype is annotated by Merrill: "*A. foxworthyi* Merr. n. sp.". Both the photograph and Merrill's description correspond with *A. tomentosum*, as was also suspected by Airy Shaw, who listed the name under *A. tomentosum* with a question mark "e descr." (1975: 218). Merrill compared his new species with *A. cumingii*, another synonym of *A. tomentosum*.

A. rivulare Merr., Philipp. J. Sci., C, 11: 60 (1916). — *A. tomentosum* Blume var. *rivulare* (Merr.) Pax & K. Hoffm. in Engl., Pflanzenr. 81: 117 (1922). Type: Sarawak, Sungei Tingei, *Foxworthy* 471 (US!, lectotype, here designated; A! [photo ex PNH], isolectotype). Paratype: Sarawak, Retuh, Sadong, *BS (Native collector)* 2535 (A!, K!).

A. clementis Merr., J. Straits Branch Roy. Asiat. Soc. 76: 90 (1917), nom. illeg. (non Merr. 1914). Type: Sabah, Mt Kinabalu, Lobang, 11 Nov. 1915, *Clemens* 10374 (A!, lectotype, here designated; A! [photo of destroyed holotype ex PNH], isolectotype). — Note: Listed as a synonym (with question mark) of *A. hosei* by Airy Shaw (1975: 211) who had probably not seen the type.

A. ilocanum Merr., Philipp. J. Sci. 16: 549 (1920), **synon. nov.** Type: Phillipines, Luzon, Ilocos Norte Prov., between Bangui and Claveria, 12 Aug. 1918, in thickets and forests at low altitudes, *BS (Ramos)* 32998 (A!, lectotype, here designated; BM!, K!, US!, isolectotypes).

A. impressinerve Merr., Philipp. J. Sci. 16: 548 (1920), **synon. nov.** Type: Philippines, Panay, Capiz Prov., Jamindan, *BS (Ramos & Edaño)* 31409 (A!, lectotype, here designated; K!, US!, isolectotypes). Paratypes: same locality, *BS (Ramos & Edaño)* 31020 (A!, BM!, K!), 31268 (A!, K!, US!). — Note: Although maintained by Airy Shaw (1983: 5), this seems to represent an aberrant population of *A. tomentosum*, distinguished only by the slightly bullate leaves. Only known from the type collection.

?*A. megalophyllum* Merr., Philipp. J. Sci. 16: 551 (1920), **synon. nov.** Type: Philippines, Babuyan Islands, Calayan, 6 June 1917, in forests, alt. c. 40 m, *FB (Velasco)* 26642 (not located). — Note: Airy Shaw (1983: 6) merely listed the name with the remark: "No material at Kew". The name has not otherwise been used in the literature. The original description, however, makes it quite clear that this is a large-leaved form of *A. cumingii*, a synonym of *A. tomentosum*. All measurements in the protologue are well within the limits of the variability of *A. tomentosum*. Merrill even mentioned the typical falcate stipules.

A. bangueyense Merr., Philipp. J. Sci. 24: 114 (1924). — *A. tomentosum* Blume var. *bangueyense* (Merr.) Airy Shaw, Kew Bull., Addit. Ser. 4: 218 (1975), **synon. nov.** Type: Sabah, Banguey Island, Limbuak Valley, at low altitudes, *Castro & Melegrito* 1121 (not located). — Note: A form with smaller, rather wide leaves, particularly frequent on Banguey (Banggi) Island and in the neighbouring district of Kudat. The leaf shape, however, is not discrete from the type variety.

A. tomentosum Blume var. *giganteum* Pax & K. Hoffm., Mitt. Inst. Allg. Bot. Hamburg 7: 224 (1931); Airy Shaw, Kew Bull., Addit. Ser. 4: 218 (1975), as synon. nov. Type: West-Borneo, auf dem Bukit Mehipit, 450 – 500 m, Urwald, 7 Dez. 1924, *Winkler* 632 (HBG, lectotype, here designated; BO!, E!, HBG, isolectotypes). Original syntypes: same locality, 29 Dez. 1924, *Winkler* 1149 (BO!, E!, HBG); am Serawei, oberhalb Lebang Hara, um 180 m, am Ufer, 30 Dez. 1924, *Winkler* 1159 (BO!, HBG).

Petioles 3 – 15(– 20) × 2 – 5 mm. *Leaf blades* elliptic to slightly obovate, ovate or oblong, apically long acuminate-mucronate, (3 –) 5 – 12(– 30) cm wide, (2 –)2.5 – 3.5(– 5.5) times as long as wide. *Staminate inflorescences* 6 – 14 cm long, simple, very rarely 2 – 3 per fascicle. *Pistillate inflorescences* (5 –) 10 – 30 cm long. *Infructescences* 10 – 30 (– 65) cm long. *Fruiting pedicels* 1 – 7(– 23) mm long. *Fruits* obliquely ovoid (mango-shaped) to falcate, often slightly beaked, 5 – 8(– 9) mm wide.

DISTRIBUTION. Peninsular Thailand, Peninsular Malaysia, Sumatra, Borneo, western Java, Philippines, northern and central Sulawesi. Chakrabarty & Balakrishnan (1992: 19 – 20) report *A. tomentosum* from Great Nicobar Island (*Hore* 7594, 8224; *Dwivedi* 7275, 8070, 8072; all CAL & PBL, none of which were examined for this study). Map 41.

HABITAT. In primary and secondary evergreen vegetation; in mixed lowland to hill dipterocarp forest, associated with *Alstonia, Artocarpus, Dipterocarpus, Shorea*; in mossy, montane rainforest; in dry shrubby heath forest on poor sand ("kerangas"); in swamp and riverine forest, sometimes subject to flooding; in plantations; along road sides and forest edges; in thickets; in wet to dry habitats. On clay, sand and ultrabasic soil, over limestone, granite or sandstone. 0 – 1800 m altitude.

USES. The hard wood is used for making ploughs. The ripe fruits are eaten in Java (Smith 1910: 265). The roots are chewed and applied for internal pain (Philippines, *PNH* 38075). The bark is burned and the ash is rubbed on the teeth to colour them (Philippines, *PNH* 13524).

CONSERVATION STATUS. Least Concern (LC). This is a very common and widespread species. 573 specimens examined.

VERNACULAR NAMES. Sumatra: kayu djimang; Malay Peninsula: kayu sireh (Temuan); Sabah: adadsay (Dusun), balinsaay (Dusun), kansali, kobar, kobor or kubor (Duson Banggi), rayan (Kedayan), siop nanah (Dusun); Sarawak: sikandu (Land Dayak); Kalimantan: kosa umpo, kusiro rue, passi haras, sompa; Java: ki seueur, ki seueur lalaki [seueur = many, refers to the many edible fruits], tampar kidang (all Sundanese); Philippines: ata tamsi, balerahay, bongay (Manobo), bungoy, gelebiray (Sub.), malindang (Sub.), padit (Palanan-Agta), tiga (Bukidnon).

ETYMOLOGY. The epithet refers to the indumentum (Latin, tomentosus = covered with short, curled hairs).

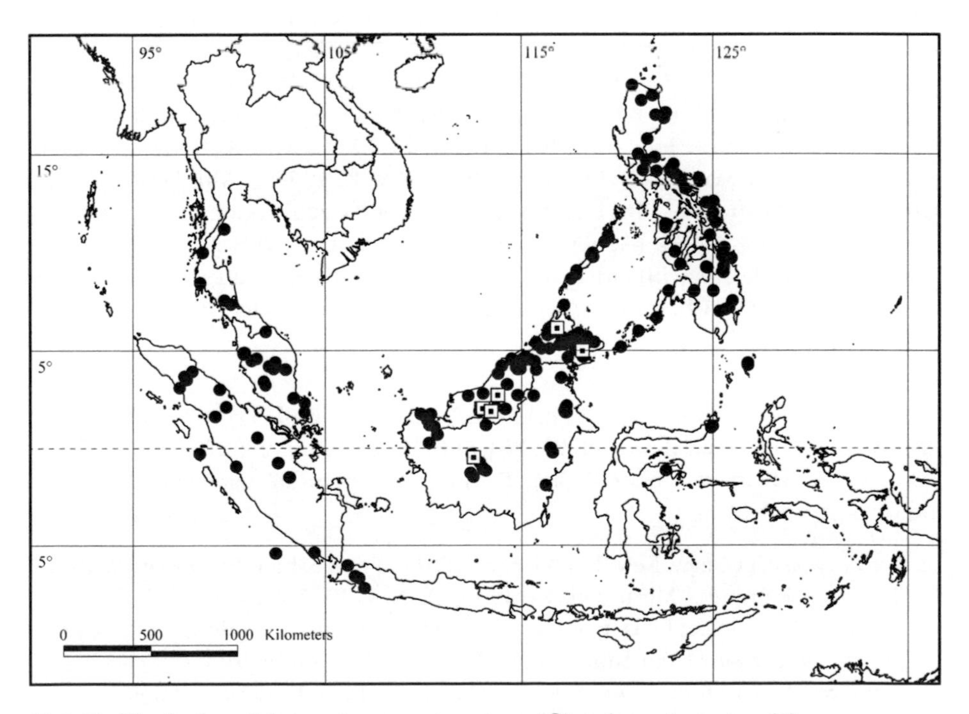

MAP 41. Distribution of *A. tomentosum* var. *tomentosum* (●) and var. *stenocarpum* (▣).

KEY CHARACTERS. Indumentum ferrugineous; petioles wide; stipules persistent; staminate flowers sessile; sepals free, acute; ovaries densely pubescent; fruits mango-shaped.

SIMILAR SPECIES. *A. brachybotrys, A. leucopodum, A. pachystachys, A. pahangense, A. pendulum, A. riparium, A. stipulare,* see there. — *A. velutinosum* has a whitish to ochraceous indumentum, smaller fruits, longer, narrower bracts, pedicellate pistillate flowers and longer sepals. — *A. velutinum* has smaller fruits, shorter, branched inflorescences, less densely pubescent ovaries and an extrastaminal-annular disc.

SELECTED SPECIMENS. THAILAND. Peninsular, Narathiwat prov., Waeng, on slope of hill in evergreen forest, 16 April 1972, *Sangkhachand, Phusomsaeng & Nimanong* 52003 (K). PENINSULAR MALAYSIA. Johore, Gunong Panti, 1°50'N, 103°54'E, on sandstone, 1500 ft, 8 Sept. 1963, *Chew* 721 (A, K, L); Pahang, Taman Negara, Plot 1 along Sungai Tahan trail, slightly disturbed rain forest, c. 90 m, 1975, *Van Balgooy* 2437 (K, L, NY). SUMATRA. Aceh, Sikunder Forest Reserve, WNW of Medan, 3°55'N, 98°05'E, primary rainforest, marshy, 50 – 100 m, 5 Aug. 1979, *De Wilde & De Wilde-Duyfjes* 19422 (K, L, US); Enggano Mt, 1 June 1936, *Luetjeharms* 4037 (K, L). BORNEO, Brunei, Kuala Abang road, mile 6, primary forest, yellow sandy clay on tertiary sandstone hills, 80 ft, 17 June 1957, *BRUN (Ashton)* 86 (K, KEP, L); Sabah, Mt Kinabalu, Dallas, woods, 3000 ft, 18 Aug. 1931, *Clemens & Clemens* 26117 (A, B, BM, K, L, NY); Sabah, Bettokan, near Sandakan, 29 July 1927, *SF (Kloss)* 19012 (BM, BO, K, KEP, NY); Sarawak, Miri div., Baram distr., Entoyut R., Dec. 1894, *Hose* 382 (BM, CGE, K, L); Sarawak, Bintulu div., Tatau distr., Ulu Anap, Bukit Kana, secondary forest, 24 June 1982, *S (Abang Mohtar & Jugah ak Kudi)* 44749 (K, KEP, L, SAR); C Kalimantan, Tumbang Atey, Bukit Raya Expedition, 1°07'S, 113°02'E, much disturbed ex-dipterocarp forest, 75 m, 28 Jan. 1983, *Veldkamp* 8435 (L, SAR, US). JAVA. Preanger res., Tjiastana, Tjisalak, bosch, 600 m, 20 Dec. 1920, *Bakhuizen v. d. Brink Jr.* 651 (U). PHILIPPINES. Leyte, s. loc., June 1913, *Wenzel* 213 (A, E, G, US); Luzon, Apayao subprov., s. loc., May 1917, *BS (Fenix)* 28183 (A, K, P, US); Mindanao, Davao distr., Todaya (Mt Apo), Baruring gorge, on very steep wooded side, 3000 ft, June 1909, *Elmer* 10844 (A, BM, E, G, K, L, NY, US, WRSL). SULAWESI. C Sulawesi, Luwuk area, inland from Batui and Seseba on Batui R., at Sinsing camp, 1°09'S, 122°31'E, forest, steep slopes, 70 – 100 m, 15 Oct. 1989, *Coode* 5948 (AAU, K, L).

52b. var. **stenocarpum** *(Airy Shaw) Petra Hoffm.*, Kew Bull. 54: 361 (1999). — *A. stenocarpum* Airy Shaw, Kew Bull. 23: 281 (1969); Kew Bull. 28: 275 (1973). Type: Sarawak, 3rd Division, Upper Rejang R., Belaga, Nov. 1892, *Haviland* 2186 (K!, holotype).

A. leptodictyum Airy Shaw, Kew Bull 36: 635 (1981). Type: Sabah, Ranau Distr., ab. 13 km from Kampung Merungin, 2 – 30 m, 12 Nov. 1975, *SAN (Leopold & Saikeh)* 82444 (K!, holotype; L!, isotype). — Note: The type and the only other collection of var. *stenocarpum* from Sabah, *SAN (Meijer)* 129649, differ from the specimens from Sarawak in their more reticulate, less percurrent tertiary venation and the basally convex leaf margins. The number of collections of this taxon is, however, too small to ascertain the significance of these characters, and the differences seem insufficient to recognise two distinct rheophytic varieties of *A. tomentosum.*

Petioles 1.5 – 7 × 1 – 3 mm. *Leaf blades* lanceolate to oblong, apically caudate, (1.2 –)2 – 3(– 4) cm wide, (6.5 –)8(– 13.6) times as long as wide. *Staminate inflorescences* 4 – 6 cm long, simple or once-branched. *Pistillate inflorescences* 3 – 5(– 20) cm long. *Ovary* densely (usually appressed-) pubescent. *Infructescences* 7 – 20 cm long. *Fruiting pedicels* 0.5 – 2 mm long. *Fruits* elongate-ellipsoid to mango-shaped, 3 – 6 mm wide.

DISTRIBUTION. Northern Borneo: south-eastern Sarawak (Kapit division), Sabah, East Kalimantan. Map 41.

HABITAT. Rheophyte. In forests along river banks. On sandy alluvial soil. 4 – 120 m altitude.

CONSERVATION STATUS. Near Threatened (NT). Different from the very common type variety, var. *stenocarpum* is only known from 10 specimens.

ETYMOLOGY. The epithet refers to the shape of the fruits (Greek, stenos = narrow; carpos = fruit).

KEY CHARACTERS. Rheophytic habit.

SELECTED SPECIMENS. BORNEO. SABAH. Lahad Datu distr., Mt Silam, 4 km up the telecom road, 3 June 1990, *SAN (Meijer)* 129649 (K, L). SARAWAK. Rapide de Rejang, Sept. 1867, *Beccari* 3829 (L); 3rd div., Nanga Mujong, on a bank sloping to the river, 9 Aug. 1954, *Brooke* 8964 (BM, L, US); Kapit, Nanga Balleh, Upper Rejang R., river bank, low altitude, 4 June 1929, *Clemens & Clemens* 21263 (A, B, BM, K, L, NY); Upper Rejang R., Bet-Gat & Kapit R. margin, 4 m, 26 – 27 June 1929, *Clemens & Clemens* 21667 (K, NY); Belaga, Upper Rejang, 5 Aug. 1927, *Native collector* 5266 (A, NY, US); 7th div., along Batang Rejang, Ulu Kapit, near Rumah Rapek, on river bank, just above water level, on alluvial soil, 19 Sept. 1973, *S (Chai et al.)* 33246 (K, L). KALIMANTAN. W Kalimantan, Serawai, 4 km S of Nanga Jelundung, 0°29'S, 112°32'E, besides Sg. Jelundung, disturbed primary forest, major associates include *Shorea, Lithocarpus, Canarium, Tristania, Calophyllum,* sandy soil, 120 m, 30 Oct. 1995, *Church, Ismail & Ruskandi* 2828 (L).

53. Antidesma vaccinioides *Airy Shaw,* Kew Bull. 28: 280 (1973); Kew Bull., Addit. Ser. 8: 224, plate 2, fig. 4 – 4a (1980). Type: Papua New Guinea, Morobe distr., Angabena ridge, c. 4 miles E of Aseki, 4 April 1966, *Schodde (& Craven)* 4840 (L!, holotype; A!, K!, isotypes).

Tree, c. 5 m. *Young twigs* terete, sparsely and very shortly spreading-pilose, brown. *Stipules* persistent, deltoid, apically acute-mucronate, 0.3 – 0.6 × 0.3 – 0.6 mm, glabrous. *Petioles* channelled adaxially, 3 – 6 × 0.7 – 1 mm, glabrous. *Leaf blades* ovate, apically acuminate or acuminate-mucronate, with a rounded apiculum, basally acute to rounded, decurrent at the very base, 1.5 – 3.5 × 0.7 – 2.2 cm, 1.5 – 2.25 times as long as wide, coriaceous, glabrous, shiny on both surfaces, with (0 –)1 – 2 pairs of elliptic glands (0.6 × 0.3 mm) embedded in the basal third of the revolute leaf margins, midvein flat to slightly raised adaxially, 3 – (5 –)

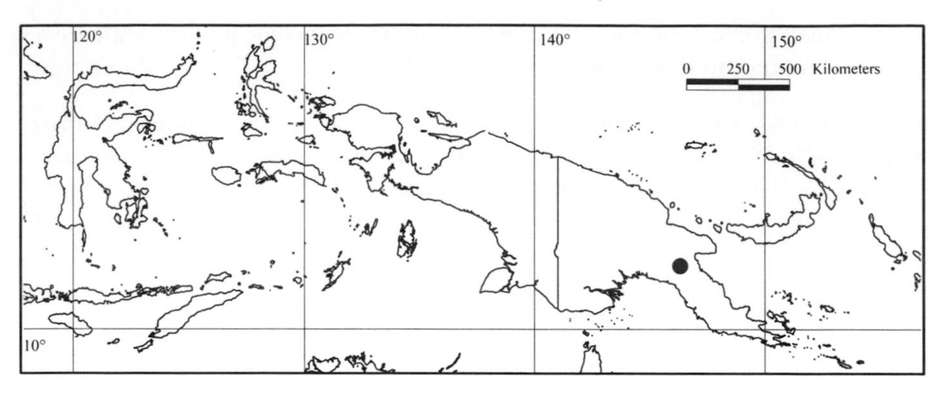

MAP 42. Distribution of *A. vaccinioides*.

veined basally, tertiary veins weakly percurrent, more conspicuous adaxially than abaxially, drying olive-green, lighter abaxially. *Staminate inflorescences* 0.5 – 1 cm long, axillary, simple, sometimes in fascicles of 2 inflorescences, axes c. 0.5 mm wide, minutely and spreading-pilose. *Bracts* deltoid to ovate, apically acute, 0.5 – 1 × 0.5 – 0.8 mm, glabrous. *Staminate flowers* c. 3 × 3 – 4 mm. *Pedicels* 1.5 – 2.5 mm long, articulate at 1 – 1.5 mm from the base, minutely and spreading-pilose. *Calyx* c. 1.5 × 1.5 mm, cupular to nearly globose, sepals 3, fused for $^{2}/_{3}$ – $^{4}/_{5}$ of their length, apically rounded, sparsely and minutely pilose outside, glabrous inside, margin ciliate. *Disc* extrastaminal-annular, strongly erose, thick, glabrous. *Stamens* 7 – 13, 2 – 3 mm long, exserted 1.5 – 2.5 mm from the calyx, anthers 0.3 – 0.4 × 0.4 – 0.5 mm. *Pistillode* absent. *Pistillate plants* unknown. Fig. 22.

DISTRIBUTION. Papua New Guinea. Map 42.

HABITAT. In low ridge mossy forest. 2000 m altitude.

CONSERVATION STATUS. Data Deficient (DD): only known from the type.

ETYMOLOGY. The epithet refers to the similarity with the genus *Vaccinium* L.

KEY CHARACTERS. 1 – 2 pairs of marginal foliar glands; small, coriaceous leaves 3(– 5)-veined basally; very short inflorescences; pedicels articulate; fused sepals; stamens 7 – 13.

SIMILAR SPECIES. This is a highly distinctive plant with no apparent relationship to any other known species of *Antidesma*. The marginal foliar glands, the articulate staminate pedicels and the large number of stamens are unique characters for the entire genus. The typical anther shape and general floral structure, inflorescence and dioecy, on the other hand, place it in *Antidesma*.

54. Antidesma velutinosum *Blume*, Bijdr. Fl. Ned. Ind.: 1125 (1826 – 27); Airy Shaw, Kew Bull. 36: 356, fig. D1 – 7, 365 (1981). Type: Java, July, *Hb. Blume* [1323] (L! herb. no. 903154-370, lectotype, here designated; BO!, L!, isolectotypes). — Note: Lectotype annotated by Blume. Original material also in NY!

A. attenuatum Wall. ex Tul., Ann. Sci. Nat. Bot., Sér. 3: 235 (1851). Type: Malaysia, Penang, 1822, *Wallich* 7286 (K!, lectotype, here designated). Original syntype:

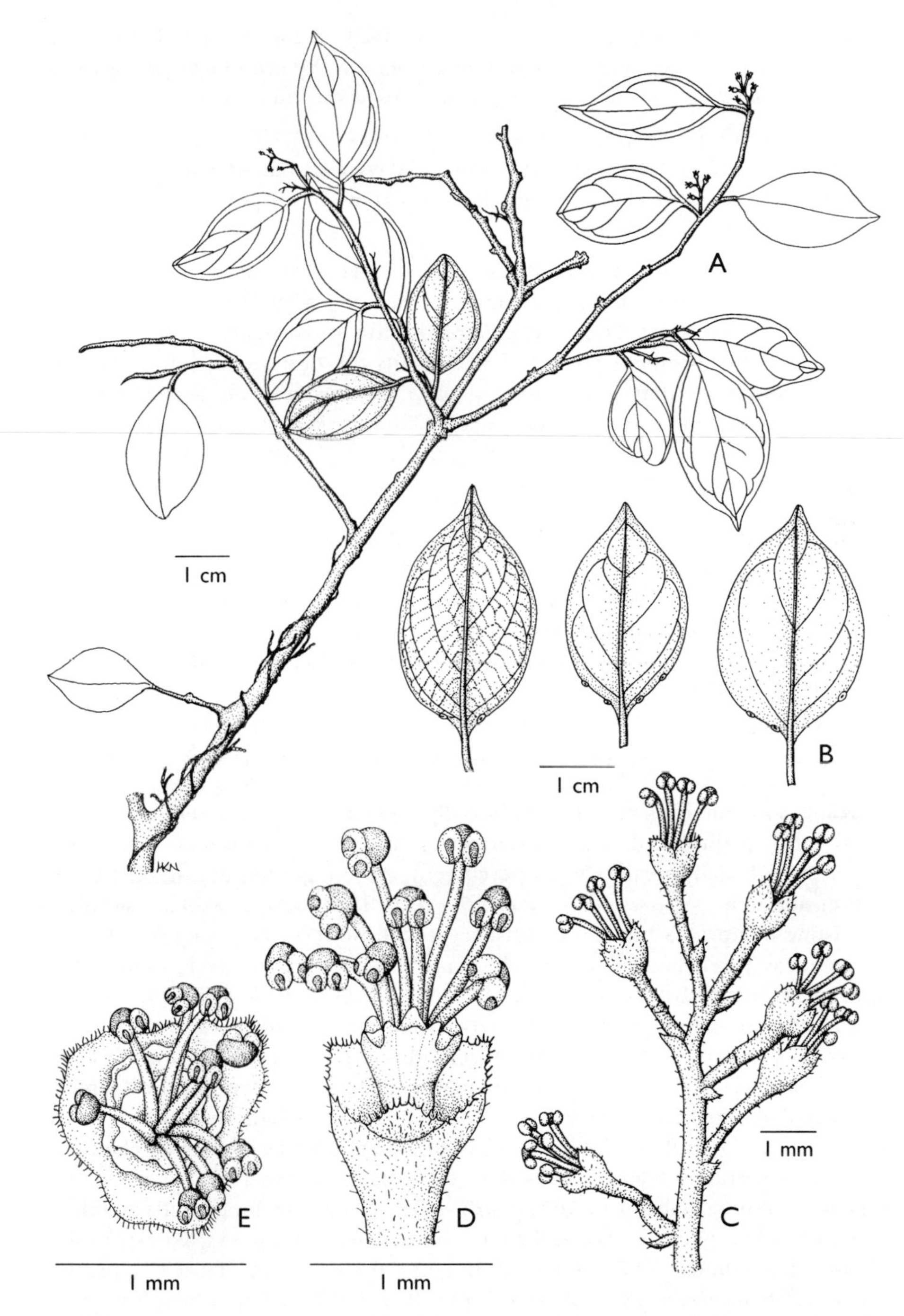

FIG. 22. *Antidesma vaccinioides*. **A** habit with staminate inflorescence; **B** leaves showing the marginal glands; **C** part of staminate inflorescence with articulated pedicels; **D** staminate flower in lateral view, one calyx lobe folded in; **E** staminate flower seen from above. From *Schodde & Craven* 4840 (L). Drawn by Holly Nixon.

Penang, *Wallich* 8582 (K!). — *A. attenuatum* Wall., Num. List: no. 7286 (1832), nom. nud. — Note: Tulasne cited *A. velutinosum* in the protologue as a synonym of *A. attenuatum* without explaining why he used the later name.

A. molle Wall. ex Müll. Arg., Linnaea 34: 67 (1865 – 66). — *A. molle* Wall., Num. List: no. 7287 (1832), nom. nud. Type: Malaysia, in Pulo-Penang, *Hb. Wallich (G Porter coll.)* 7287 (K!, lectotype, here designated). Original syntype: Burma, ad Moulmain, *Helfer* 159.

A. velutinosum Blume var. *lancifolium* Hook. f., Fl. Brit. India 5: 357 (1887), "lancifolia", **synon. nov.** Type: Malaysia, Penang, May 1886, *Curtis* 863 (K!, holotype?). — Note: The type citation in the protologue is only "Penang, Curtis". The specimen *Curtis* 863 at Kew fits the description but there is no other evidence that it is the type of *A. velutinosum* var. *lancifolium.* It is accompanied by a note by Airy Shaw: "Published as *A. velutinosum* Bl. "var. *lancifolia*" Hook. f. in Hook. f., Fl. Brit. India. V. 357 (1887)", which shows that Airy Shaw regarded it as the type.

Shrub or tree, up to 15 m, diameter up to 35 cm. Bark grey, brown or reddish, thin, smooth to slightly cracked; inner bark pink, red or reddish brown; sapwood white or pink. *Young twigs* terete, densely spreading-yellowish to ochraceous-hirsute, brown. *Stipules* usually persistent, lanceolate to linear, (3 –)6 – 10(– 15) × (0.5 –)1 – 3(– 4) mm, densely appressed-hirsute. *Petioles* flat to slightly channelled adaxially, 2 – 6(– 10) × (1 –)2 mm, densely spreading-hirsute. *Leaf blades* oblong to narrowly elliptic, apically acuminate-mucronate, basally acute to rounded, rarely slightly cordate, (8 –)12 – 18(– 27) × (2.5 –)4 – 6(– 10) cm, (2 –)2.6 – 3(– 5) times as long as wide, eglandular, chartaceous, glabrous except along the major veins adaxially, spreading-hirsute all over abaxially, especially along the veins, intercostal areas rarely glabrous, dull to moderately shiny on both surfaces, major veins impressed adaxially, tertiary veins percurrent, close together, drying olive-green, lighter abaxially. *Staminate inflorescences* 5 – 10 (– 15) cm long, axillary, simple or consisting of up to 5 branches, densely set with flowers, axes spreading-hirsute. *Bracts* linear, rarely lanceolate, apically acute, 1.5 – 3 × 0.2 – 0.5(– 1) mm, densely appressed-hirsute. *Staminate flowers* c. 2.5 × 2 – 3 mm. *Pedicels* 0 – 1 mm long, not articulated, spreading-hirsute. *Calyx* 0.5 – 1 mm long, sepals 5 – 7, free or nearly so, sometimes partially fused, narrowly deltoid, sometimes unequal, apically acute, hirsute on both sides. *Disc* cushion-shaped, enclosing the bases of the filaments and pistillode, glabrous. *Stamens* 4 – 8, 1.5 – 2(– 3) mm long, exserted c. 1 mm from the calyx, anthers 0.3 – 0.5 × 0.3 – 0.5 mm. *Pistillode* variable, clavate to globose, sometimes 3-lobed, 0.2 – 0.5 × 0.1 – 0.3 mm, exserted from the disc, hirsute to almost glabrous. *Pistillate inflorescences* 3 – 13 cm long, axillary, simple, rarely branched once or twice at the base, densely set with flowers, axes spreading-hirsute. *Bracts* linear, (1.5 –)2 – 3(– 3.5) × 0.3 – 0.5(– 1) mm, densely appressed-hirsute. *Pistillate flowers* 1.5 – 2.5 × 1 – 2 mm. *Pedicels* 0.2 – 1.5 mm long, spreading-hirsute. *Calyx* (0.8 –)1 – 1.5 mm long, sepals 5 – 8, free, 0.2 – 0.4 mm wide, narrowly deltoid to linear, apically acute, hirsute on both sides. *Disc* much shorter than the sepals, glabrous, very rarely some hairs at the margin. *Ovary* globose, densely to sparsely spreading-hirsute, style subterminal, thick, distinct, stigmas 3 – 5, c. 1 mm long, thin. *Infructescences* 6 – 15 cm long. *Fruiting pedicels* 2 – 5(– 6) mm long, spreading-hirsute. *Fruits* lenticular to obliquely ellipsoid or slightly bean-shaped, laterally compressed, basally distinctly asymmetrical, with a distinctly lateral style, 4 – 7 × 4 – 6 mm, hirsute to glabrous, often white-pustulate, reticulate to areolate when dry. Fig. 15F – J (p. 183).

DISTRIBUTION. Burma, Thailand, Peninsular Malaysia, Singapore, Anambas & Natuna Islands, Sumatra, Java. Map 43.

HABITAT. In primary and secondary evergreen (more rarely semi-evergreen) vegetation; in wet mixed and dipterocarp forest, associated with *Arenga obtusifolia* and other palms, bamboo, dipterocarps; in gallery forest; usually in humid, shady habitats, often close to streams or waterfalls. On volcanic loam, limestone, granitic sand and shale. 20 – 1200 m altitude.

USES. The fruits are eaten locally. The bark is "sold to the Chinese" (*fide Winckel* 311 from Java), probably for medicinal purposes.

CONSERVATION STATUS. Least Concern (LC). This species is widespread and frequently collected. 275 specimens examined.

VERNACULAR NAMES. Burma: kinbalin; Thailand: mount-po-lo; Sumatra: bernai gadja, kayu branei kesuping; Malay Peninsula: ming-baloi = porcupine-cheek (Batek); Kalimantan: kayu sepang; Java: hentjiep or huni serur (Sundanese), ki bulu, ki hilend or ki huni (Sundanese), ki seueur or pohon ki seueur (seueur = many, refers to the many edible fruits, Sundanese), seueur badak, wuni kebo.

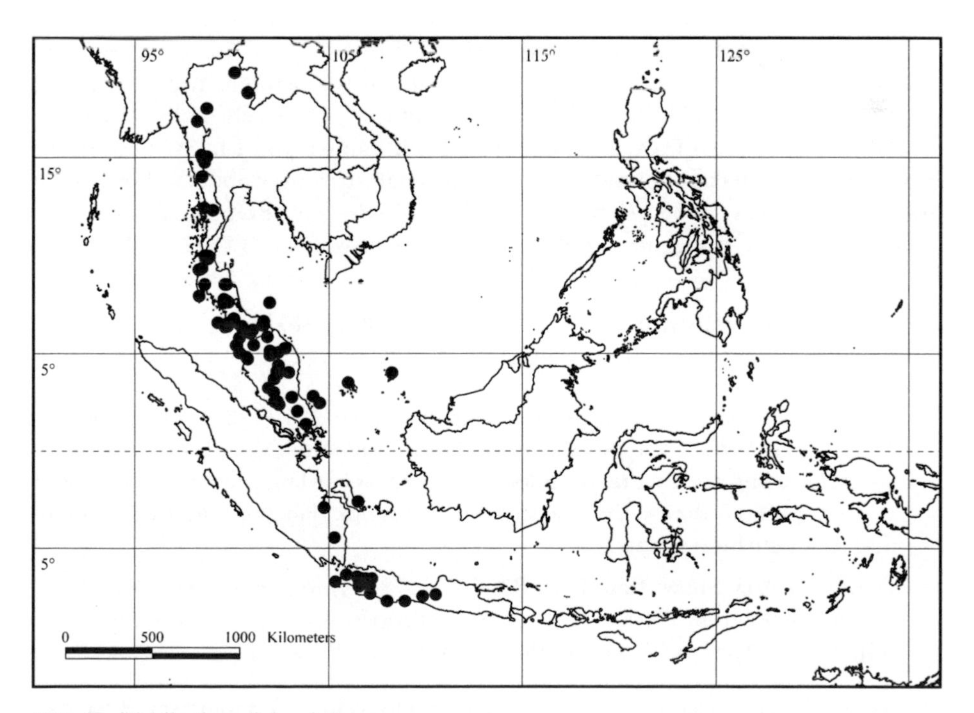

MAP 43. Distribution of *A. velutinosum*.

ETYMOLOGY. The epithet refers to the indumentum (Latin, velutinosus = covered with short, soft, erect hairs).

KEY CHARACTERS. Most parts hirsute; bracts 1.5 mm or longer; fruits lenticular with lateral styles; stigmas long; tertiary veins percurrent, close together; disc glabrous.

SIMILAR SPECIES. *A. elbertii, A. orthogyne, A. tomentosum*, see there. — *A. velutinum* has shorter bracts, shorter, more symmetrical fruits and fewer sepals and stamens.

NOTES. Chakrabarty & Gangopadhyay (1997: 481) reported *A. velutinosum* also from New Guinea, but the cited specimen *Forbes* 617 (CAL) represents *A. excavatum* var. *indutum* in K and L.

Galling of the same kind as in *A. tomentosum* (very dense inflorescences with hundreds of very small, sterile bracts) can be observed in several collections (e.g., *Helfer* 4947 (K), from India).

SELECTED SPECIMENS. BURMA. Tenasserim, Mergui distr., Tharabwin Chaung, 1 Jan. 1932, *Khart* 13271 (K). THAILAND. Northern, Tak prov., Mae Sot, Mussor Village, evergreen forest, 800 – 900 m, 24 July 1959, *Smitinand & Floto* 24343 (K, L); South-western, Kanchanaburi prov., between Kritee and Meung Chah, 15°02'N, 98°45'E, hill evergreen forest, limestone, 1000 m, 9 July 1973, *Geesink & Phengkhlai* 6193 (K, L); Peninsular, Trang prov., Khao Chong, evergreen forest, 16 May 1970, *BKF (Phusomsaeng & Pinnin)* 57534 (K, L). PENINSULAR MALAYSIA. Kelantan, Western part, Sungai Nenggiri near Kuala Jenera, ridge top, dense forest with much bamboo, 17 July 1967, *FRI (Whitmore)* 4076 (K, KEP, L); Penang, 300 ft, April 1881, *King's collector* 1524 (BM, G, K, L). SUMATRA. Bangka, Lobok besar, G. Pading, granitic sand, 20 m, 1 Oct. 1949, *Kostermans & Anta* 1033 (A, BO, K, L). ANAMBAS & NATUNA ISL. Anambas Isl., Siantan, W coast, base of Gunung Manassah, 3°30'N, 106°00'E, low altitude, 30 March 1928, *Henderson* 20130 (BO, CGE, K). JAVA. Banjumas res., Nusa Kambangan Isl., SW part, forest S and W of K. Babakan, humid forest, rich in small streams, dominated by *Arenga obtusifolia*, 21 Nov. 1938, *Kostermans & Van Woerden* 80 (BO, L); Batavia res., Gunung Paniisan, oerwoud, c. 600 m, 28 Oct. 1920, *Van Steenis* 2312 (B, L, NY).

55. Antidesma velutinum *Tul.*, Ann. Sci. Nat. Bot., Sér. 3: 223 (1851). Type: India or., in agro atranico [Lower Burma], *Wallich* s.n. (CGE!, holotype).

A. gymnogyne Pax & K. Hoffm. in Engl., Pflanzenr. 81: 135 (1922). Type: Burma, Tenasserim, *Helfer* 4945 (K!, lectotype, here designated; G!, P!, isolectotypes). — Note: Contrary to the protologue, but in accordance with the accepted concept of *A. velutinum*, the ovary of the type is not glabrous and the staminate disc is not cushion-shaped.

A. spaniothrix Airy Shaw, Kew Bull. 33: 15 (1978), **synon. nov.** Type: Thailand, Peninsular Region, Phuket Circle, Krabi, evergreen forest at foot of limestone hill, 50 m, 4 April 1930, *Kerr* 18852 (K!, holotype; P!, isotype).

Shrub or tree, up to 10(– 20) m, diameter up to 8 cm. Twigs mid-grey. Bark grey or brown, thin, more or less roughened or finely vertically cracked. *Young twigs* terete, densely ferrugineous-tomentose, brown. *Stipules* caducous, linear, apically acute, 3 – 7 × 0.5 – 1 mm, pubescent. *Petioles* channelled adaxially, 2 – 6(– 10) × 1.2 – 1.5 mm, densely pubescent. *Leaf blades* oblong, more rarely slightly obovate, apically acuminate-mucronate, basally acute to obtuse, more rarely rounded, (6 –)8 – 13(– 17.5) × (2.5 –)3 – 5(– 7) cm, (2 –)3(– 4.2) times as long as wide, eglandular, chartaceous, glabrous except along the midvein adaxially, ferrugineous-pubescent all over abaxially, especially along the veins, intercostal areas rarely almost glabrous, rather dull on both surfaces, major veins impressed adaxially, tertiary veins reticulate to weakly percurrent, widely spaced (4 – 7 between every two secondary veins), with strong intersecondary veins, finer venation finely tessellated, drying reddish brown, domatia often present. *Staminate inflorescences* 4 – 7 cm long, axillary, more rarely cauline, consisting of 4 – 11 branches, axes ferrugineous-pubescent. *Bracts* lanceolate, apically acute, 0.3 – 0.8 × 0.3 – 0.5 mm, pubescent. *Staminate flowers* 1 – 2 × 0.7 – 1.5 mm. *Pedicel* absent. *Calyx* c. 0.5 × 0.7 – 1 mm, cupular to bowl-shaped, sepals 3 – 5, almost free to fused for $^1/_2$ of their length, irregularly shaped, pilose to pubescent outside, pilose to pubescent inside, margin erose, fimbriate. *Disc* extrastaminal-annular, lobed, lobes often pointing inwards, filling the space between filaments and pistillode, sometimes appearing cushion-shaped, slightly constricted at the base, glabrous. *Stamens* 3(– 4), 1 – 2 mm long, exserted 0.5 – 1.5 mm from the calyx, anthers 0.2 – 0.3 × 0.3 – 0.4 mm. *Pistillode* clavate, c. 0.5 × 0.2 – 0.3 mm, exserted from the sepals, pubescent. *Pistillate inflorescences* 2 – 4 cm long, axillary, more rarely cauline, consisting of up to 7 branches, more rarely simple, dense (flowers usually touching each other), axes ferrugineous-pubescent to pilose. *Bracts* lanceolate, apically acute, 0.5 – 1 × 0.3 – 0.5 mm, pubescent. *Pistillate flowers* 1.5 – 2 × c. 1 mm. *Pedicels* 0.3 – 0.5 mm long, spreading-pilose to almost glabrous. *Calyx* 0.8 – 1 × 0.8 – 1 mm, urceolate, sepals 3 – 5, fused for $^1/_2$ of their length, apically truncate to acute, sinuses usually rounded, pilose to pubescent, more rarely almost glabrous on both sides, margin fimbriate. *Disc* much shorter than the sepals, glabrous. *Ovary* almost cylindrical, pilose to glabrous, more rarely pubescent, style terminal, usually thick, stigmas 4 – 8, usually short and thin relative to the style. *Infructescences* 3 – 8 cm long, axes 0.7 – 0.8 mm wide. *Fruiting pedicels* c. 1 mm long, spreading-pilose. *Fruits* ellipsoid, laterally compressed, basally symmetrical, with a terminal to slightly subterminal style, 4 – 5 × c. 3 mm, pilose to almost glabrous, often white-pustulate, areolate when dry.

DISTRIBUTION. Burma, Cambodia, Thailand, northern Peninsular Malaysia (Perlis, Kedah). Map 39 (p. 219).

HABITAT. In primary and secondary vegetation; in dry, more rarely wet, evergreen, deciduous or mixed evergreen/deciduous forest; in bamboo forest; often in shady habitats near streams. On limestone, sandstone and granite. 0 – 600 m altitude.

CONSERVATION STATUS. Near Threatened (NT). The distribution area of this species is large, but it occurs in a densely populated and highly cultivated region and is therefore not without concern. 50 specimens examined.

VERNACULAR NAMES. mao kuan or mung mao (Thai), paw mai chaw (Karen).

ETYMOLOGY. The epithet refers to the indumentum (Latin, velutinus = velvety).

KEY CHARACTERS. Dense ferrugineous indumentum; short, strongly branched inflorescences; glabrous disc; leaves drying dark reddish; venation tessellated.

SIMILAR SPECIES. *A. sootepense*, *A. tomentosum*, *A. velutinosum*, see there.

NOTE. Pax & Hoffmann (1922: 136) observed some diandrous flowers among the predominantly triandrous flowers; this could not be confirmed in the present study.

SELECTED SPECIMENS. CAMBODIA. Tatey (Kah Kong), près de village, 18 Feb. 1966, *Martin* 977 (K). BURMA. Tenasserim prov., Moolmyne (Moulmeine), ripa fl. Attran et Choppedong, 1847, *Wallich* 8577 (K, BM). THAILAND. Northern, Tak prov., 20 km E of Mae Sod, 17°20'N, 98°50'E, disturbed mixed deciduous forest, on limestone hills, 500 – 700 m, 30 May 1973, *Geesink, Phanichapol & Santisuk* 5570 (K, L); South-western, Kanchanaburi prov., Kwae Noi R. Basin, near Neeckey, near Wangka, 150 m, 29 April 1946, *Kasim* 172 (K, L); Kanchanaburi prov., Kwae Noi R. Basin, Ka Tha Lai in Pan Paung R. Valley, c. 25 km E of Wangka, 300 m, 2 – 4 June 1946, *Kostermans* 805 (K, L); South-eastern, Chanthaburi prov., Doi Soi Dao, 12°45'N, 102°10'E, evergreen forest, along clearing, on granitic foothill, 300 m, 12 May 1974, *Geesink, Hattink & Phengklai* 6674 (K, L); Peninsular, Chumphon, Sapli, 23 May 1919, *Haniff & Nur* 4249 (K, SING). PENINSULAR MALAYSIA. Kedah, Bukit Tunjang, 950 ft, 5 April 1924, *Haniff* 10406 (SING); Perlis, Mata Ayer Forest Reserve, compt. 14, undulating, bamboo forest, 11 March 1969, *FRI (Kochummen)* 2678 (L).

56. Antidesma venenosum *J. J. Sm.*, Icon. Bogor 4: 41, t. 313 (1914). Type: Borneo, Bukit Mili, Exp. Nieuwenhuis 1898, *Amdjah* 85 (L!, lectotype, here designated). Original syntypes: Rejang Kapit, 1893, *Haviland & Hose* 720 (CGE!, K!, L!, P!); Soengei Saloet Penihin, Exp. Nieuwenhuis 1896 – 97, *Jaheri* 171 (L!), 1548 (A!, BO!, L!); Bloe-oe, Exp. Nieuwenhuis 1896 – 97, 743 (L!, U!); Bukit Liang Karing dibawah, Exp. Nieuwenhuis, 1896 – 97, 1257 (BO!).

Tree, more rarely shrub, up to 18 m, clear bole up to 6 m, diameter up to 25 cm. Twigs green. Bark whitish, grey, greyish yellow, brown, reddish brown or greenish, thin, smooth; inner bark pink, yellow, pale brown, green or grey, 3 mm thick, fibrous; cambium white, yellow or brown; sapwood pink, red, brown, orange, yellow or white, heartwood pinkish brown, *Church* 599: wood with sweet scent. *Young twigs* terete, pubescent, brown. *Stipules* early-caducous, subulate, 5 – 10 × c. 0.5 mm, pubescent. *Petioles* narrowly channelled adaxially, basally and distally usually ± pulvinate for 2 – 3 mm, 5 – 10(– 13) × 0.8 – 1 mm, pilose to pubescent. *Leaf blades* oblong, slightly ovate or obovate, apically (usually long) acuminate-mucronate, basally rounded to truncate, more rarely obtuse, (7 –)10 – 13(– 18) × (2.5 –)3.5 – 5(– 7) cm, (2.3 –)2.8(– 3.8) times as long as wide, eglandular, chartaceous to membranaceous, glabrous, or pilose only along the midvein adaxially, glabrous to pilose (especially along the veins) abaxially, dull to moderately shiny on both surfaces, major veins impressed adaxially, tertiary veins

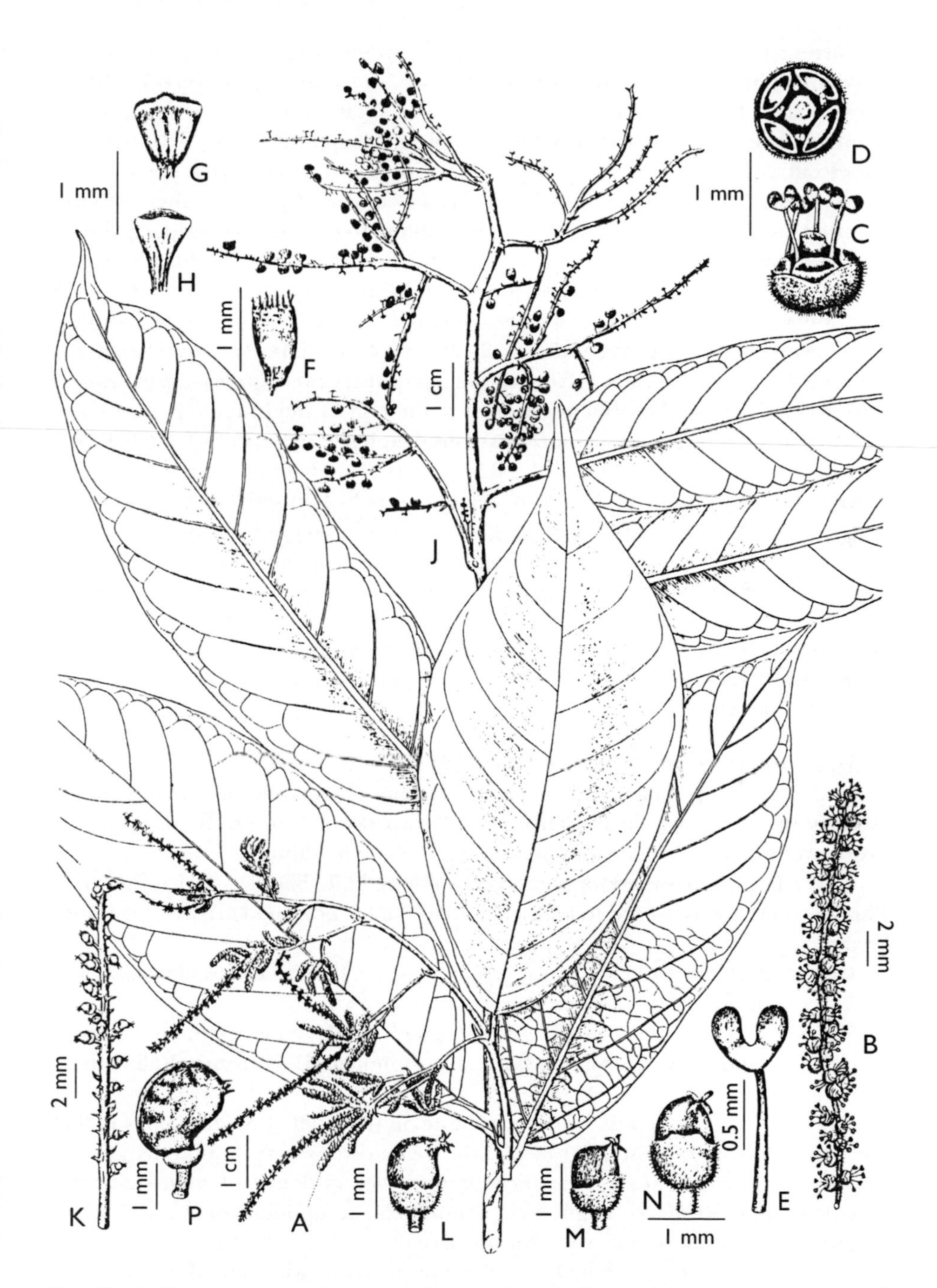

FIG. 23. *Antidesma venenosum.* **A** staminate flowering branch; **B** part of staminate inflorescence; **C** staminate flower; **D** staminate flower seen from above, stamens removed; **E** stamen; **F** pistillode; **G – H** lobes of staminate disc; **J** pistillate flowering branch; **K** part of pistillate inflorescence; **L – M** pistillate flowers in lateral view; **N** pistillate flower in ventral view; **P** fruit. Reproduced from Smith 1914: t. 313.

percurrent to weakly percurrent, medium to widely spaced, drying olive-green adaxially, lighter with dark major veins abaxially. *Staminate inflorescences* 7 – 10 cm long, axillary but aggregated at the end of the branch, consisting of up to 35 branches, axes pubescent. *Bracts* deltoid, apically acute, 0.2 – 0.6 × 0.2 – 0.3 mm, pubescent. *Staminate flowers* 1 – 1.5 × 1 – 1.5 mm. *Pedicel* absent. *Calyx* 0.3 – 0.4 × c. 1 mm, depressed globose, sepals 4(– 5), fused for $^{1}/_{2}$ – $^{2}/_{3}$ of their length, pointing inwards, apically acute to rounded, pilose to pubescent outside, glabrous inside but with long hairs at the base. *Disc* consisting of 4(– 5) free alternistaminal lobes, lobes ± obconical, with two shallow imprints apically, 0.2 – 0.3 × 0.2 – 0.5 mm, longer than the sepals, glabrous. *Stamens* 4(– 5), 0.8 – 1 mm long, exserted 0.6 – 0.8 mm from the calyx, anthers 0.2 – 0.3 × 0.3 – 0.5 mm. *Pistillode* clavate, 0.4 – 0.5 × 0.2 – 0.3 mm, exserted from the sepals, pubescent. *Pistillate inflorescences* 3 – 5(– 8) cm long, axillary but aggregated at the end of the branch, consisting of 4 – 12 branches, axes pubescent. *Bracts* deltoid, apically acute, 0.3 – 0.5 × 0.3 – 0.5 mm, pubescent. *Pistillate flowers* c. 1 × 0.5 – 1 mm. *Pedicels* 0.5 – 1 mm long, pilose. *Calyx* c. 0.5 × 0.8 mm, urceolate to cupular, sepals 4, fused for c. $^{2}/_{3}$ of their length, apically acute, pilose outside, glabrous inside except for long hairs at the base. *Disc* extending to the same length as the sepals, glabrous. *Ovary* ovoid, glabrous, style lateral, stigmas 4 – 6. *Infructescences* 6 – 8 cm long. *Fruiting pedicels* 0.5 – 1(– 1.5) mm long, pubescent to glabrous. *Fruits* bean-shaped to obliquely lenticular, laterally compressed, basally distinctly asymmetrical, with a distinctly lateral style, (1.5 –)2 – 2.5 × 1.5 – 2 mm, glabrous, often white-pustulate, areolate when dry. Fig. 23.

DISTRIBUTION. Borneo. Map 40 (p. 221).

HABITAT. In primary and secondary vegetation; mainly in lowland mixed dipterocarp forest; also in thickets and along roadsides. On sandy soil, clay and loam, over sandstone, shale and limestone. 0 – 1000 m altitude.

USES. The wood is used for handicraft, to make knife handles, and as firewood. The species is also reported to be used as material for poisoned arrows (see under Etymology).

CONSERVATION STATUS. Least Concern (LC). This species is abundant and widespread in Borneo. 217 specimens examined.

VERNACULAR NAMES. Sabah: geruseh puteh (Malay), menggambir (Dusun), tenduripis, tandurupis puru, tendurupis or tondurupis (Dusun); Kalimantan: behuna, uhai.

ETYMOLOGY. The epithet refers to a note on the label of *Nieuwenhuis* s.n. that the plant is used to poison arrows (Latin, venenosus = very poisonous). Smith (1910: 43) stated, however, that there is no other evidence of its toxicity and speculated that *Antidesma* might yield a non-toxic component of the poisonous mixture. It may even be the wood for the arrow.

KEY CHARACTERS. Extremely small, glabrous fruits with lateral styles; many-branched inflorescences; glabrous disc; free staminate disc lobes which are longer than the sepals.

SIMILAR SPECIES. *A. celebicum* and *A. tetrandrum*, see there.

SELECTED SPECIMENS. BORNEO. BRUNEI. Belait distr., Labi subdistr.,

Rampayoh, track through secondary forest along side Sungai Rampayoh c. 2 km from Labi road, 4°22'N, 114°28'E, secondary forest, Lambir formation, sandstone and shale, 20 m, 8 July 1993, *Atkins et al.* 432 (K, L); Tutong distr., Lamunin subdistr., Ladan Hills Forest Reserve, 4°38'N, 114°46'E, lowland dipterocarp forest, Belait Formation, sandstone and clay, 130 m, 30 March 1991, *Sands, Johns, Niga Nangkat, Kalat A., Awong Kaya & Saleh Bat* 5726 (K, L). SABAH. Kinabalu ridges, Dallas, 3000 ft, 20 Oct. 1931, *Clemens & Clemens* 26763 (A, B, AM, K, L, NY); Lamag distr., Sapa Tali, primary forest, flat land, brown sandy soil, c. 130 ft, 30 June 1963, *SAN (Ampuria)* 36412 (K, L, SAR). SARAWAK. Miri div., Baram distr., Long Tarkun, Nov. 1894, *Hose* 317 (BM, CGE, E, K, L); 1st div., Sabal Forest Reserve, 70th mile Serian road, near stream, 300 ft, 15 May 1974, *S (Tong)* 34342 (K, KEP, L, SAR, U). KALIMANTAN. C Kalimantan, Headwaters of Sungei Kahayan, 5 km NE of Haruwu village, 0°28'S, 113°44'E, old secondary forest, open places, 200 m, 4 April 1988, *Burley & Tukirin* 584 (E, K, L, NY, SAR); E Kalimantan, Sangkulirang subdiv. (E Kutei), Sungai Menubar region, ridge, loamy soil containing lime, 15 m, 12 June 1951, *Kostermans* 5164 (A, K, L); C/W Kalimantan, Bukit Raya, 0°45'S, 112°47'E, in primary dipterocarp forest, 130 m, 5 Dec. 1982, *Nooteboom* 4192 (K, L, NY); Tumbang Atey, Bukit Raya Expedition, 1°07'S, 113°02'E, much disturbed ex-dipterocarp forest, 75 m, 31 Jan. 1983, *Veldkamp* 8464 (KEP, L, SAR, US).

UNMATCHED SPECIMENS

The collections listed below were not satisfactorily assigned to any existing taxon but do not show sufficient differential characters to be described as one. They might deserve taxonomic recognition if more material, especially of the opposite sex, is gathered in the future.

Brass 28450. Papua New Guinea, Louisiade Archipelago, Rossel Island, Mt Rossel, Southern slopes, 17 Oct. 1956, *Brass* 28450 (K!, L!, US!).

Undergrowth tree, 3 m high. *Young twigs* terete, densely ochraceous-pubescent, brown. *Stipules* not seen. *Petioles* narrowly channelled adaxially, basally and, more so, distally pulvinate and geniculate for c. 5 mm, 20 – 25 × 1.5 – 2 mm, pubescent. *Leaf blades* elliptic to ovate, apically acuminate-mucronate, basally rounded to obtuse, shortly decurrent, 17 – 18 × 8.5 – 10 cm, 1.7 – 2.1 times as long as wide, eglandular, chartaceous, glabrous except for the puberulous midvein adaxially, pilose all over abaxially, more densely so along the major veins, rather dull on both surfaces, midvein shallowly impressed adaxially, tertiary veins reticulate to weakly percurrent, drying olive-green, domatia present. *Pistillate inflorescences* axillary, simple, axes ochraceous-pubescent. *Bracts* not seen. *Pistillate flowers* not known. *Calyx* in fruit c. 1 × 1.5 – 2 mm, sepals 4 – 5, fused for $^1/_2$ of their length, irregularly shaped, more or less deltoid, apically acute to obtuse, pilose outside, glabrous inside, margin entire to erose. *Disc* shorter than the sepals, but indumentum nearly extending to the same length as the sepals, ochraceous-pubescent at the margin, indumentum longer than the disc, c. 0.5

mm long. *Stigmas* 3 – 6. *Infructescences* c. 5 cm long, axes c. 1.2 mm wide. *Fruiting pedicels* c. 1 mm long, pubescent. *Fruits* ellipsoid, laterally compressed, basally symmetrical, with a terminal style, c. 5 × 3 – 3.5 mm, sparsely tomentose, not white-pustulate, areolate when dry.

HABITAT. In rainforest of a ravine. 750 m altitude.

NOTE. This fruiting collection has large elliptic leaves on long, thick petioles very similar to *A. petiolatum. A. petiolatum,* however, has more or less glabrous leaves and branched, slender infructescences with very small fruits. With regard to leaf indumentum and infructescences, this specimen resembles *A. subcordatum.*

Coode 6138. South-eastern Sulawesi, Kolaka area, Gunung Watiwila foothills, above Sanggona, slopes of Gunung Sopura, 3°49'S, 121°40'E, 3 Nov. 1989, *Coode* 6138 (AAU!, E!, K!, L!).

Straight small tree c. 6 m × 10 cm, with narrow crown. Bark brownish, somewhat flaky, not detaching, inner bark red, thin, hard; sapwood straw, heartwood deep red, hard and dense. *Young twigs* terete, shortly pubescent, brown, with swollen galls but not hollow. *Stipules* early-caducous, linear, c. 2 × 0.3 mm, shortly appressed-pubescent. *Petioles* channelled adaxially, 3 – 5 × 0.8 – 1 mm, shortly appressed-pubescent. *Leaf blades* oblong to elliptic, apically acuminate- to acute-mucronate, basally acute to obtuse, 5 – 8 × 2.5 – 3.5 cm, 2.7 – 3.2 times as long as wide, eglandular, chartaceous, glabrous except for the sparsely pilose midvein abaxially, dull on both surfaces, midvein impressed adaxially, tertiary veins reticulate, widely spaced, hardly prominent, drying greyish green. *Staminate inflorescences* 3 – 4 cm long, axillary and aggregated at the end of the branch, consisting of 2 – 5 branches, axes 0.3 – 0.4 mm wide, shortly pubescent. *Bracts* broadly ovate, 0.7 – 0.8 × 0.7 – 0.8 mm, appressed-pubescent. *Staminate flowers* c. 1.5 × 1.5 mm. *Pedicel* absent. *Calyx* 0.7 – 0.8 × c. 1.5 mm, cupular, sepals 4 – 5, fused for c. $^{3}/_{4}$ of their length, deltoid, thick, apically acute, appressed-pubescent outside, glabrous inside, margin entire. *Disc* consisting of 4 – 5 free alternistaminal lobes, lobes ± obconical, c. 0.5 × 0.4 – 0.5 mm, slightly shorter than the sepals, glabrous at the sides, whitish-pubescent apically. *Stamens* 4 – 5, c. 1.5 mm long, exserted 0.7 – 0.8 mm from the calyx, anthers c. 0.3 × 0.5 – 0.7 mm. *Pistillode* obconical, c. 0.6 × 0.5 – 0.6 mm, hardly exserted from the sepals, densely and shortly pubescent.

HABITAT. In forest with deep leaf litter, many slender smallish trees with scattered emergents, little vegetation on forest floor. 1100 m altitude.

NOTE. This specimen most resembles *A. excavatum* but differs in the short, thin petioles and the disc structure.

Hansen 1391. Central Kalimantan, Kuala Kuayan, camp at logging road c. 9 km W of Pematang logging camp, 2°00'S, 112°28'E, 6 April 1984, *C. Hansen* 1391 (BO!, L!).

Tree, 13 m high, diameter 14 cm. Bark grey. *Young twigs* terete, glabrous, light brown. *Stipules* not seen. *Petioles* widely and shallowly channelled adaxially, c. 2 ×

1.5 – 2 mm, glabrous, rugose. *Leaf blades* elliptic-oblong, apically acuminate-mucronate, basally acute, 7 – 11 × 3 – 4.5 cm, 2.2 – 2.7 times as long as wide, eglandular, chartaceous, glabrous, dull on both surfaces, midvein impressed adaxially, tertiary veins reticulate, widely spaced, hardly visible when dry, drying greyish brown, domatia absent. *Staminate inflorescences* 4 – 6 cm long, cauline, simple, solitary or sometimes 2 per fascicle, axes c. 1 mm wide, sparsely pilose. *Bracts* ovate, apically acute, c. 0.7 × 0.5 – 0.7 mm, pilose. *Staminate flowers* 2.5 – 3 × 2 – 4 mm. *Pedicels* 0 – 0.2 mm long, not articulated, glabrous. *Calyx* c. 0.5 × 1 mm, bowl-shaped, sepals 3, fused for c. $^1/_3$ of their length, deltoid, apically acute, sinuses wide, shallow, pilose outside, glabrous inside, margin fimbriate. *Disc* cushion-shaped, enclosing the bases of the filaments, exserted from the sepals, densely and long tomentose. *Stamens* 3, 2 – 2.5 mm long, exserted 1.5 – 2 mm from the disc, anthers 0.4 – 0.6 × 0.3 – 0.4 mm. *Pistillode* absent.

HABITAT. In primary forest, open area by stream where trees had fallen. 50 m altitude.

NOTE. This staminate flowering collection is most similar to *A. riparium* but differs in persistence of its stipules, shape, shine and venation of its leaves, position and length of its inflorescences, number of sepals and in the absence of a pistillode.

Ridsdale 1377. Philippines, Luzon, Acoje Mine concession area, Santa Cruz, 15°46'N, 120°00'E, 21 May 1988, *Ridsdale* 1377 (K!, L!).

Small tree, 6 m high. *Young twigs* terete, very sparsely pilose, brown. *Stipules* early-caducous, linear, c. 2.5 × 0.5 mm, appressed-pubescent. *Petioles* channelled adaxially, basally and distally slightly pulvinate and geniculate for 2 – 5 mm, 15 – 22 × 0.8 – 1.5 mm, pilose. *Leaf blades* elliptic, apically acuminate-mucronate, basally obtuse to acute, 14 – 18 × 5.5 – 9.5 cm, 1.9 – 2.9 times as long as wide, eglandular, chartaceous, glabrous adaxially, glabrous or sparsely pilose only along the major veins abaxially, shiny on both surfaces, midvein impressed adaxially, tertiary veins weakly percurrent to reticulate, widely spaced, drying olive-green, domatia absent. *Pistillate inflorescences* axillary, simple or consisting of up to 3 branches, sometimes in fascicles of 2 inflorescences, axes pilose. *Bracts* deltoid, c. 0.7 × 0.5 mm, pilose. *Pistillate flowers* c. 2 × 1 – 1.5 mm. *Pedicels* 0.5 – 1.5 mm long, pilose. *Calyx* 1 – 1.2 × 1 – 1.5 mm, cupular to urceolate, sepals 4, fused for $^1/_2$ of their length, deltoid, apically acute, pilose outside, glabrous inside but with hairs at the base extending to the length of the disc indumentum, margin fimbriate. *Disc* shorter than the sepals, ferrugineous-pubescent at the margin, indumentum about as long as the disc. *Ovary* ellipsoid, glabrous, style terminal, stigmas 6 – 8. *Infructescences* 4 – 8 cm long, axes slender, up to 1 mm wide. *Fruiting pedicels* 1 – 1.5 mm long, pilose. *Fruits* ellipsoid, moderately laterally compressed, basally symmetrical, with a terminal style, 5 – 6 × 3 – 4 mm, glabrous, white-pustulate, areolate when dry.

HABITAT. In streamside valley, lightly logged high *Dipterocarpus/ Shorea* forest. On ultrabasic soil.

NOTE. This fruiting specimen from Zambales province, Luzon, with long petioles, large, glabrous leaves, branched, slender inflorescences and a hairy disc, is probably closest to *A. catanduanense*, *A. curranii* and *A. microcarpum*, but differs from these species in its larger fruits and more numerous stigmas. From *A. catanduanense* it also differs in its less highly fused sepals and terminal styles, and from *A. curranii* in its longer petioles and larger leaves. *A. nienkui* Merr. & Chun from Hainan has abaxially pilose leaves, longer, simple, more robust inflorescences, a glabrous disc, a more deeply divided, larger calyx and larger fruits.

Van Beusekom & Phengkhlai 967. Peninsular Thailand, Nakhon Si Thammarat province, Khao Luang, 23 May 1968, *van Beusekom & Phengkhlai* 967 (K!, L!).

Shrub. *Young twigs* terete to striate, ferrugineous-pubescent, glabrescent, light to medium brown. *Stipules* caducous, linear, apically acute, 1 – 1.5 × 0.3 – 0.5 mm, ferrugineous-pilose. *Petioles* channelled adaxially, becoming rugose when old, 2 – 4 × c. 1.5 mm, pilose. *Leaf blades* elliptic to ovate, apically acute- or acuminate-mucronate, basally acute to obtuse, 4.5 – 10.5 × 1.7 – 3.6 cm, 2.7 – 3.4 times as long as wide, eglandular, subcoriaceous, glabrous, shiny on both surfaces, midvein impressed adaxially, tertiary veins reticulate, widely spaced (3 – 5 between every two secondary veins), drying greyish green adaxially, olive-green abaxially, domatia absent. *Staminate inflorescences* 3.5 – 6.5 cm long, axillary, simple or once-branched at the base, axes pilose. *Bracts* deltoid, apically acute, c. 0.7 × 0.5 mm, pilose. *Staminate flowers* c. 2 × 2 mm. *Pedicel* absent. *Calyx* 0.8 – 1 × 1.5 – 2 mm, sepals 4 – 5, free, 0.5 – 1 mm wide, deltoid, apically acute to rounded, pilose outside, glabrous inside, margin entire. *Disc* consisting of 4 – 5 free alternistaminal lobes, lobes ± obconical, well-separated from each other, c. 0.5 × 0.5 mm, glabrous. *Stamens* 4 – 5, 1.5 – 2 mm long, exserted c. 1 mm from the calyx, anthers c. 0.3 × 0.7 mm. *Pistillode* globose, c. 0.5 × 0.5 mm, shorter than the sepals, extending to the same length as the disc, pilose.

HABITAT. In mossy wet evergreen forest. 1500 m altitude.

NOTE. A staminate collection which combines vegetative characters of *A. helferi* with a glabrous disc that consists of free lobes. It much resembles the fruiting specimen *Larsen et al.* 45988 from the same locality (see under *A. helferi*).

DOUBTFUL SPECIES

Antidesma coriifolium *Pax & K. Hoffm.* in Engl., Pflanzenr. 81: 154 (1922). Type: New Guinea, Kaiser Wilhelmsland, Felsspitze, buschwaldähnlicher Gebirgswald, 1400 – 1500 m, *Ledermann* 12950. — Note: Type probably destroyed in Berlin, no isotypes located. The protologue does not contain enough information for neotypification.

Antidesma frutescens *Jack*, Malayan Misc. 2(7): 91 (1822); Merr., J. Arnold Arbor. 33: 216 (1952). Type: Sumatra, Bencoolen, *Jack* s.n., destroyed. — Note: Jack's description points very much to *A. ghaesembilla*, especially the detailed description of the staminate disc. Merrill synonymised the two names on the basis of the vague protologue.

Antidesma pedicellare *Pax & K. Hoffm.* in Engl., Pflanzenr. 81: 162 (1922). Type: New Guinea, Kaiser Wilhelmsland, Sattelberg, s. coll., s.n. — Note: Type probably destroyed in Berlin, no isotypes located. The protologue compares the plant to *A. polyanthum* (syn. *A. excavatum*), differing in its smaller, pedicellate staminate flowers with free disc lobes.

Antidesma perakense *Pax & K. Hoffm.* in Engl., Pflanzenr. 81: 117 (1922). Type: Malaysia, Perak, s. loc., ex Herb. Mus. Perak, s. coll., s.n. Type probably destroyed in Berlin, no isotypes located. Compared to *A. tomentosum* in the protologue.

Antidesma rhamnoides *Tul.*, Ann. Sci. Nat. Bot., Sér. 3: 217 (1851). Type: Stirpis hujus patria incerta est, colitur in caldariis Musei parisiensis, florebatque junio 1842, Brogniart, Msc. in Herb. Mus. par. — Note: No specimen annotated as *A. rhamnoides* could be found in the Paris Herbarium. Pax & Hoffmann (1922: 157) stated that the plant had been cultivated in the Paris Botanical Garden in the past, but could not be found again. The description points to *A. ghaesembilla*, although Tulasne mentioned solitary spikes whereas *A. ghaesembilla* has much-branched inflorescences. There is no description of the pistillate flower or fruit.

Antidesma rumphii *Tul.*, Ann. Sci. Nat. Bot., Sér. 3: 238 (1851), as species non visa; Müll. Arg. in DC., Prodr. 15(2): 267 (1866), as species minus nota; Pax & K. Hoffm. in Engl., Pflanzenr. 81: 166 (1922), as species minus nota; Merr., Interpr. Herb. Amboin.: 316 (1917), pro syn. — *Bunius agrestis* Rumph., Herb. Amboin. 3: 205 (1743), prelinnean. Type: Amboina, Rumph., Herb. Amboin. 3: t. 131 A (1743). — Note: This name is entirely based on *Bunius agrestis* of Rumphius, who described it as the wild species of *Antidesma* in Ambon. The name has very rarely been used after Rumphius. Merrill (1917: 316) reduced it to *A. bunius*. From Rumphius' description, however, this must be *A. ghaesembilla*: growing in more open country, with rounded leaf-tips, smaller fruits and shorter, "many-headed" (many-branched?) infructescences. It is also described as fire-resistant ("arbor salamandra"). *A. ghaesembilla* is frequently found in savannahs, and the only species in the genus for which fire-resistance has been recorded. The only other species recorded on Ambon are *A. heterophyllum* and *A. stipulare*, none of which fits the description given of *Bunius agrestis*. The drawing of an infructescence of *Bunius agrestis* (Rumphius 1743: t. 131 A), however, bears no likeness to that of *A. ghaesembilla* and does hardly differ from that of *Bunius sativa* (t. 131). It is possible that artist or editor made a mistake and included a drawing of the wrong plant.

Antidesma spicatum *Blanco*, Fl. Filip.: 794 (1837), "spicata". Type: Philippines, *Blanco?, Llanos?*, not located. — Note: No original material located (incl. MA, *fide* M. Velayos, curator, in litt. 1995). See under *A. edule*.

Cansjera grossularioides *Blanco*, Fl. Filip.: 73 (1837), "Cansiera"; Fl. Filip., ed. 2: 53 (1845); Merr., Philipp. J. Sci., C, 9: 463 (1914); Sp. Blancoan.: 218 (1918). Type: Philippines, *Blanco? Llanos?*, not located. — Note: No original material located (incl. MA, *fide* M. Velayos, curator, in litt.). Merrill (1918: 218) regarded this species as conspecific to *A. ghaesembilla*, but did not cite any original material. He selected two "illustrative specimens"of *A. ghaesembilla* for his Species Blancoanae (no. 375, Luzon, Laguna Prov., Los Baños, June 1914, *Quisumbing*, K!, and no. 488, Palawan, Taytay, May 1913, K!) to represent *C. grossularioides*. As this term is not equivalent to "type" in the sense of Art. 7.11. (Greuter *et al.* 2000: 9), this did not effect neotypification. Blanco's species maybe identical with *A. ghaesembilla*, but as long as there is no original material, it seems preferable to treat *C. grossularioides* as incompletely known.

Cansjera rheedii *Blanco*, Fl. Filip.: 73 (1837), "Cansiera rheedi"; Fl. Filip., ed. 2: 52 (1845), "Cansiera rhedi"; Merr., Philipp. J. Sci., C, 9: 462 (1914). Type: Philippines, *Blanco? Llanos?*, not located. — Note: No original material could be located. There is one specimen in MA (only photograph seen): "*Llanos* 112, In oppid. Angat, vulgo Bignayrugo(?), A. alexiteria Blanco? Cansiera rheedi Blanco?", designated as "lectotype in sched.", probably by Quisumbing. The specimen on the photograph is most probably *A. ghaesembilla*, and the label itself makes it most unlikely that it is the actual type as Blanco himself would hardly have been so indecisive about his own species. Merrill (1918: 218) regarded this species as conspecific with *Antidesma pentandrum*, the basionym of which he considered to be *Cansiera pentandra* Blanco. *Antidesma* is here treated as a synonym of *A. montanum* (see p. 155).

EXCLUDED NAMES

Antidesma bicolor *Hassk.*, Cat. Pl. Bogor. Alter: 81 (1844), non Pax & K. Hoffm. 1922, "bicoler", nom. nud. is **Excoecaria bicolor** *Hassk.*, Retzia 1: 158 (1855), Euphorbiaceae s.s. Type: *Zollinger* s.n., L (*fide* H.-J. Esser, pers. comm.).

Antidesma crenatum *H. St. John*, Pacific Sci. 26: 279 (1972) is **Xylosma crenatum** (*H. St. John*) *H. St. John*, Phytologia 34: 147 (1976), Salicaceae s.l. Type: Hawaiian Islands, Kauai, 1968, *Hobdy* 8 (BISH, holotype; K!, L!, isotypes).

Antidesma filiforme *Blume*, Bijdr. Fl. Ned. Ind.: 1124 (1826 – 27) is **Galearia filiformis** (*Blume*) *Pax* in Engl. & Prantl, Nat. Pflanzenfam. 3(5): 82 (1890); Forman, Kew Bull. 26: 160 (1971), Pandaceae. Type: Java, in montosis Salak, *Blume* (K!, isotype). — Notes: The sheet in K is labelled as isotype of *A. filiforme* Blume, but bears no original handwriting. There is only one historical sheet of this species from Java in L, labelled: "Blume 43/c, Salak, nov. gen. Rhamnearum" in what could be Blume's handwriting. Although Pax cited the name as "*G. filiformis* (Bl.) Benth.", literature search did not return a reference to that name by Bentham anywhere else. The combination made by Pax is validly published according to Art. 32.4, Ex. 5 (Greuter *et al.* 2000: 54). The author of the new combination is therefore not Boerlage but Pax (see also Forman 1971). Blume's "var. foliis majusculis" in the protologue of *A. filiforme* is not validly published according to Art. 23.6(a) (Greuter *et al.* 2000: 43).

Antidesma litorale *Blume*, Bijdr. Fl. Ned. Ind.: 1123 (1826 – 27) is **Polyosma integrifolia** *Blume*, *fide* Hallier, Meded. Rijks.-Herb.: 7 (1911); Pax & K. Hoffm. in Engl., Pflanzenr. 81: 167 (1922), Polyosmaceae. Type: Java, in maritimis insulae Nusae Kambangae, *Blume* s.n. (L). — Note: The genus is currently under revision for "Flora Malesiana" by Saw Leng Guan (KEP).

Antidesma lunatum *Miq.*, Fl. Ned. Ind., Eerste Bijv.: 467 (1861) is **Aporosa lunata** (*Miq.*) *Kurz*, J. Asiat. Soc. Bengal 42(2): 239 (1873), "lunatum", Phyllanthaceae. Type: Sumatra or., in prov. Palembang, Ogan-ulu, *Teysmann* s.n. (BO, holotype; K, L, isotypes *fide* Schot 2004).

Antidesma megalocarpum *S. Moore* in H. O. Forbes, J. Bot. 61, Suppl.: 46 (1923) is **Rhyticaryum longifolium** *K. Schum. & Lauterb.*, Fl. Schutzgeb. Südsee: 415 (1900), "Rhytidocaryum" *fide* Sleumer in Fl. Males. 1(7): 39 (1971), Icacinaceae. Syntypes: New Guinea, Sogere, *Forbes* 497, 417, 225 (K!).

Antidesma parasitica *Dillwyn*, Rev. Hortus Malab.: 33 (1839) is **Scleropyrum pentandrum** (*Dennst.*) *Mabb.*, Taxon 26: 533 (1977), Santalaceae. Type: Rheede, Hort. Malab. 7: t. 30 (1688).

Antidesma praegrandifolium *S. Moore* in H. O. Forbes, J. Bot. 61, Suppl.: 46 (1923) is **Aporosa praegrandifolia** (S. Moore) Schot, Blumea Suppl. 17: 309 (2004), Phyllanthaceae. Type: New Guinea, Sogere, 1885, *Forbes* 250 (BM!, holotype; K!, L, MEL!, isotypes). — Note: The name was listed as synon. nov. of *Antidesma sphaerocarpum* Müll. Arg. by Airy Shaw (1973: 277).

ACKNOWLEDGEMENTS

This study was financed by a grant from the European Community in the network *Botanical Diversity of the Indo-Pacific Region*, and a grant from the Dutch Science Foundation (NWO). I am thankful to P. Baas and M. Roos at the National Herbarium of the Netherlands, Leiden University Branch (formerly Rijksherbarium Leiden), for obtaining these grants. I would like to thank the management and staff of the Herbarium, Royal Botanic Gardens, Kew, for their help and support, in particular C. Barker, M. Coode, J. Dransfield, K. Ferguson and A. Radcliffe-Smith. P. C. van Welzen (L) and J. M. Lock (K) kindly reviewed the manuscript and made many helpful comments; P. C. van Welzen also tested the key. J. Moat (K) helped to produce the maps, and W. Stuppy (K) assisted with the IUCN ratings. R. Brummitt (K), C. Jarvis (BM), D. Mabberley (L/NSW) and W. Greuter (B) helped to clarify problems regarding typification and etymology. L. G. Saw (KEP) and S. Suddee (BKF) helped with the vernacular names. Thanks go to G. McPherson (MO), S. Dressler (FR), D. Middleton (E) and W. Stuppy (K) for many helpful discussions.

I am indebted to the directors and curators of the following herbaria for making their collections available for study: A, AAU, B, BM, BO, CANB, CGE, E, FHO, G, GH, K, KEP, L, LD, MEL, NY, OXF, P, PRC, SAR, SING, TCD, U, US and WRSL. Fieldwork was conducted with the kind help of J. Beaman (K), A. Ibrahim (SING) and L. G. Saw (KEP). Most of the plates were drawn by H. Nixon (TCD). Thanks go to Herbarium Bogoriense and Nationaal Herbarium Nederland, Leiden for permission to reproduce figures 8, 16, 19 and 23. I would like to thank my two work experience students R. Marchant (University of Reading, U.K.) and M. Schmidt (Göttingen, Germany), for their help with data input and curation.

REFERENCES

Agardh, C. A. (1824). Stilaginae. Aphorismi botanici 14: 199 – 200. Berling, Lund, Sweden.

Airy Shaw, H. K. (1969). New or noteworthy Asiatic species of *Antidesma* L. (Stilaginaceae). Kew Bull. 23: 277 – 290.

—— (1971). CXX. New or noteworthy species of *Aporosa* Bl. Kew Bull. 25: 474 – 481.

—— (1972a). The Euphorbiaceae of Siam. Kew Bull. 26: 191 – 363.

—— (1972b). New or noteworthy species of *Antidesma* L. (Stilaginaceae): II. Kew Bull. 26: 457 – 468.

—— (1973). New or noteworthy species of *Antidesma* (Stilaginaceae): III. Kew Bull. 28: 269 – 281.

—— (1975). The Euphorbiaceae of Borneo. Kew Bull., Addit. Ser. 4: 1 – 224.

—— (1980a). The Euphorbiaceae of New Guinea. Kew Bull., Addit. Ser. 8: 1 – 243.

—— (1980b). A partial synopsis of the Euphorbiaceae-Platylobeae of Australia (excluding *Phyllanthus, Euphorbia* and *Calycopeplus*) [Cover title: The Euphorbiaceae Platylobeae of Australia]. Kew Bull. 35: 577 – 700.

—— (1981a). The Euphorbiaceae of Sumatra. Kew Bull. 36: 239 – 374.

—— (1981b). New species of *Antidesma* (Stilaginaceae) from Malesia and Australia. Kew Bull. 36: 635 – 637.

—— (1982). The Euphorbiaceae of Central Malesia (Celebes, Moluccas, Lesser Sunda Is.). Kew Bull. 37: 1 – 40.

—— (1983). The Euphorbiaceae of the Philippines. Royal Botanic Gardens, Kew, U.K.

Almeida, M. R. & Almeida, S. M. (1987). Correct name for *Antidesma ghaesembilla* Gaertn. J. Bombay Nat. Hist. Soc. 84: 492 – 493.

Angiosperm Phylogeny Group (2003). An update of the Angiosperm Phylogeny Group classification for the orders and families of flowering plants: APG II. Bot. J. Linn. Soc. 141: 399 – 436.

Anonymous (1926). Meeting of the Board of Agriculture. J. Board. Agric. British Guiana 19: 304 – 307.

—— (1929). Notes: The Eradication of *Antidesma* (by E. B. Martyn). Agric. J. British Guiana 2: 52.

Arbain, D. & Taylor, W. C. (1993). Cyclopeptide Alkaloids from *Antidesma montana.* Phytochemistry 33: 1263 – 1266.

Backer, C. A. & Bakhuizen van den Brink, R. C. (1964). Flora of Java 1. P. Noordhoff, Groningen, The Netherlands.

Baillon, H. (1858). Étude générale du groupe des Euphorbiacées. 1 – 684. Victor Masson, Paris, France.

—— (1866). Euphorbiacées australiennnes. Adansonia 6: 337.

Baker, W., Coode, M. J. E., Dransfield, J., Dransfield, S., Harley, M. M., Hoffmann, P. & Johns, R. J. (1998). Patterns of distribution in Malesian vascular plants. In: Hall, R. & Holloway, J. D. (eds.), Biogeography and Geological Evolution of SE Asia: 243 – 258. Backhuys Publishers, Amsterdam, The Netherlands.

Beckett, E. (1927). Noxious Weeds. J. Board Agric. British Guiana 20: 113 – 119.

Bentham, G. (1873). Flora australiensis 6. Lovell Reeve & Co., London, U.K.

Blume, C. L. (1826 – 27). Bijdragen tot de flora van Nederlandsch-Indië. Lands Drukkerij, Batavia (Jakarta), Indonesia.

Bringmann, G., Schlauer, J., Rischer, H., Wohlfarth, M., Mühlbacher, J., Buske, A., Porzel, A., Schmidt, J. & Adam, G. (2000a). Revised Structure of Antidesmone, an Unusual Alkaloid from Tropical *Antidesma* Plants (Euphorbiaceae). Tetrahedron 56: 3691 – 3695.

——, Rischer, H., Wohlfarth, M. & Schlauer, J. (2000b). Biosynthesis of Antidesmone in Cell Cultures of *Antidesma membranaceum* (Euphorbiaceae): An Unprecedented Class of Glycine-Derived Alkaloids. J. Amer. Chem. Soc. 122: 9905 – 9910.

Brouwer, Y. M. (1983). Domatia and their occurrence in the Australian Flora. Austral. Syst. Bot. Soc. Newslett. 34: 6 – 9.

Burkill, I. H. (1935). A dictionary of the economic products of the Malay Peninsula 1: A – H. Crown agents for the colonies, London, U.K.

Burman, J. (1736 or 1737). Thesaurus Zeylanicus. Janssonius-Waesberg & Salomon Schouten, Amsterdam, The Netherlands.

Buske, A., Schmidt, J., Porzel, A. & Adam, G. (1997). Benzopyranones and ferulic acid derivatives from *Antidesma membranaceum*. Phytochemistry 46: 1385 – 1388.

——, Busemann, S., Mühlbacher, J., Schmidt, J., Porzel, A., Bringmann, G. & Adam, G. (1999). Antidesmone, a novel type isoquinoline alkaloid from *Antidesma membranaceum* (Euphorbiaceae). Tetrahedron 55: 1079 – 1086.

——, Schmidt, J. & Hoffmann, P. (2002). Chemotaxonomy of tribe Antidesmeae: Antidesmone and Related Compounds. Phytochemistry 60: 489 – 496.

Chakrabarty, T. & Balakrishnan, N. P. (1992). The Family Euphorbiaceae of Andaman and Nicobar Islands. J. Econ. Taxon. Bot., Addit. Ser. 9: 1 – 122.

—— & Gangopadhyay, M. (1997). Notes on Malesian *Antidesma* L. (Euphorbiaceae). J. Econ. Taxon. Bot. 21: 479 – 481.

—— & —— (2000). The genus *Antidesma* L. (Euphorbiaceae) in the Indian subcontinent. J. Econ. Taxon. Bot. 24: 1 – 55.

—— & Roy, A. K. (1984). Range-Extension of *Antidesma tetrandrum* Bl. (Stilaginaceae). J. Econ. Taxon. Bot. 5(1): 168.

Chase, M. W., Zmarzty, S., Lledó, M. D., Wurdack, K. J., Swenson, S. M. & Fay, M. F. (2002). When in doubt, put it in Flacourtiaceae: a molecular phylogenetic analysis based on plastic *rbcL* DNA sequences. Kew Bull. 57: 141 – 181.

Chun, W. Y. & Chang, C. C. (1965). Flora Hainanica 2. Academia Sinica, China (in Chinese).

Davis, C. C., Webb, C. O., Wurdack, K. J., Jaramillo, C. A. & Donoghue, M. J. (2005). Explosive radiation of *Malpighiales* supports a mid-Cretaceous origin of modern tropical rain forests. Amer. Nat. 165: E36 – E65.

De Jussieu, A. L. (1789). Genera Plantarum. Herissant & Barrois, Paris, France.

De Jussieu, Adr. (1824). De Euphorbiacearum Generibus Tentamen. Didot, Paris, France.

De Padua, L. S., Bunyapraphatsara, N. & Lemmens, R. H. M. J. (eds.) (1999). Plant Resources of South-East Asia (PROSEA) 12(1): Medicinal and poisonous plants I. Backhuys Publishers, Leiden, The Netherlands.

Erdtman, G. (1952). Pollen Morphology and Plant Taxonomy. Angiosperms. Almquist & Wiksell, Stockholm / The Chronica Botanica Co., Waltham, Mass., USA.

Fischer, C. E. C. (1932). VII. - The Koenig collection in the Lund Herbarium. Bull. Misc. Inform., Kew 1932(2): 49 – 76.

Forman, L. L. (1971). A synopsis of *Galearia* Zoll. & Mor. (Pandanaceae). Kew Bull. 26: 153 – 165.

—— (1997). Notes Concerning the Typification of Names of William Roxburgh's Species of Phanerogams. Kew Bull. 52: 513 – 534.

Gagnepain, F. & Beille, L. (1926). Euphorbiacées. In: M. H. Lecomte (ed.), Flore générale de l'Indochine 5(5/6). Masson & Cie., Paris, France.

Gardner, S., Sidisunthorn, P. & Anusarnsunthorn, V. (2000). A Field Guide to Forest Trees of Northern Thailand. Kobfai Publishing Project, Bangkok, Thailand.

Govaerts, R., Frodin, D. G. & Radcliffe-Smith, A. (2000). World Checklist and Bibliography of Euphorbiaceae (and Pandaceae) 1. Royal Botanic Gardens, Kew, U.K.

Greuter, W. R., McNeill, J., Barrie, F. R., Burdet, H. M., Demoulin, V., Filgueiras, T. S., Nicolson, D. H., Silva, P. C., Skog, J. E., Trehane, P., Turland, N. L. & Hawksworth, D. L. (2000). International Code of Botanical Nomenclature (Saint Louis Code). Regnum Veg. 138: 1 – 474.

Gruèzo, W. S. (1991). *Antidesma bunius* (L.) Sprengel. In: E. W. M. Verheij & R. E. Coronel (eds.), Plant Resources of South-East Asia (PROSEA) 2: Edible Fruits and Nuts. Pudoc, Wageningen, The Netherlands.

Hans, A. S. (1970). Polyploidy in *Antidesma*. Caryologia 23: 322 – 327.

—— (1973). Chromosomal Conspectus of the Euphorbiaceae. Taxon 22: 591 – 636.

Hasskarl, J. K. (1845). Aantekeningen over het nut, door de bewoners van Java aan eenige planten van dat eiland toegeschreven. Johannes Müller, Amsterdam, The Netherlands.

Hayden, W. J. & Hayden, S. M. (1996). Two enigmatic biovulate Euphorbiaceae from the Neotropics: relationships of *Chonocentrum* and the identity of *Phyllanoa*. Amer. J. Bot. 83 (Suppl.): 162 (Abstract).

Hegnauer, R. (1966). Chemotaxonomie der Pflanzen, vol. 4: 1 – 551. Birkhäuser, Basel, Switzerland.

—— (1989). Chemotaxonomie der Pflanzen, vol. 8: 1 – 718. Birkhäuser, Basel, Switzerland.

Heyne, K. (1917). De Nuttige Planten van Nederlandsch-Indië, part 3. Ruygrok & Co., Jakarta, Indonesia.

Hickey, L. J. (1979). A revised classification of the architecture of dicotyledonous leaves. In: C. R. Metcalfe & L. Chalk (eds.), Anatomy of the Dicotyledones, 2nd ed., vol. 1: 25 – 39. Clarendon Press, Oxford, U.K.

Hoffmann, P. (1999a). New taxa and new combinations in Asian *Antidesma* (Euphorbiaceae). Kew Bull. 54: 347 – 362.

—— (1999b). The genus *Antidesma* (Euphorbiaceae) in Madagascar and the Comoro Islands. Kew Bull. 54: 877 – 885.

—— (2005). *Antidesma*. In Chayamarit, K. and van Welzen, P.C. (eds), Flora of Thailand 8.1: Euphorbiaceae (Genera A–F): 51 – 81.

——, Kathriarachchi, H. & Wurdack, K. J. (in press). A phylogenetic classification of Phyllanthaceae (*Malpighiales*; Euphorbiaceae *sensu lato*). Kew Bull.

Hooker, J. D. (1887). The Flora of British India 5. Lovell Reeve & Co., London, U.K.

IUCN (2001). IUCN Red List Categories. Prepared by the IUCN Species Survival Commission. IUCN, Gland, Switzerland and Cambridge, U.K.

Jacobs, M. (1966). On domatia — the viewpoints and some facts. Proc. Kon. Ned. Akad. Wetensch. C, 69: 275 – 316.

Kårehed, J. (2001). Multiple origin of the tropical forest tree family Icacinaceae. Amer. J. Bot. 88: 2259 – 2274.

Kathriarachchi, H., Hoffmann, P., Samuel, R., Wurdack, K. J. & Chase, M. W. (2005). Molecular phylogenetics of Phyllanthaceae inferred from 5 genes (plastid *atpB*, *matK*, 3' *ndhF*, *rbcL*) and nuclear *PHYC*. Mol. Phylogen. Evol. 36: 112 – 134.

Köhler, E. (1965). Die Pollenmorphologie der biovulaten Euphorbiaceae und ihre Bedeutung für die Taxonomie. Grana Palynol. 6: 26 – 120.

Lemmens, R. H. M. J. & Wulijarni-Soetjipto, N. (eds.) (1991). Plant Resources of South-East Asia (PROSEA) 3: Dye and tannin-producing plants. Pudoc, Wageningen, The Netherlands.

Levin, G. A. (1986a). Systematic foliar morphology of Phyllanthoideae (Euphorbiaceae). I. Conspectus. Ann. Missouri Bot. Gard. 73: 29 – 85.

—— (1986b). Systematic foliar morphology of Phyllanthoideae (Euphorbiaceae). III. Cladistic analysis. Syst. Bot. 11(4): 515 – 530.

Li, Ping Tao (1994). Flora Reipublicae Popularis Sinicae 44(1): Euphorbiaceae-Phyllanthoideae. Science Press, Beijing, China (in Chinese).

Liddell, H. G. & Scott, K. (1925 – 1940). Greek-English Dictionary, revised and augmented by H. S. Jones, Vol. 1. Clarendon Press, Oxford, U.K.

Linné, C. (1753). Species Plantarum 2. Impensis Laurentii Salvii, Holmiae (Stockholm), Sweden.

—— (1754). Genera Plantarum, ed. 5. Impensis Laurentii Salvii, Holmiae (Stockholm), Sweden.

Lorentz, H. A. (1910). Nova Guinea 8(3): Atlas Botanique. t. 1 – 188. Brill, Leiden, The Netherlands.

Lowry, J. B., Petheram, R. J. & Budi Tangendjaja (1992). Plants fed to Village Ruminants in Indonesia. ACIAR Technical Reports 22. Australian Centre for Agricultural Research, Canberra, Australia.

Mandal, N. R. & Panigrahi, G. (1983). Studies in the systematics and distribution of the genus *Antidesma* L. (Euphorbiaceae) in India. J. Econ. Taxon. Bot. 4: 255 – 261.

Martyn, E. B. (1930). The Eradication of the Weed *Antidesma ghaesembilla*. Agric. J. British Guiana 3: 84 – 87.

Mathew, S. & Abraham, S. (1995). A report on the occurrence of *Antidesma thwaitesianum* Muell.-Arg. (Euphorbiaceae) from South Andaman. J. Bombay Nat. Hist. Soc. 92: 143 – 144.

Meeuse, A. D. J. (1990). The Euphorbiaceae auct. plur. — an unnatural taxon. Eburon, Delft, The Netherlands.

Meeuwen, M. S. van, Nooteboom, H. P. & van Steenis, C. G. G. J. (1960). Preliminary revisions of some genera of Malaysian Papilionaceae I. Reinwardtia 5: 419 – 456.

Mehra, P. N. & Gill, B. S. (1968). In: IOPB chromosome number reports. Taxon 17: 574 – 576.

Mennega, A. M. W. (1987). Wood anatomy of the Euphorbiaceae, in particular of the subfamily Phyllanthoideae. Bot. J. Linn. Soc. 94: 111 – 126.

Merrill, E. D. (1914). Notes on Philippine Euphorbiaceae, II. Philipp. J. Sci., C, 9: 461 – 493.

—— (1917). An interpretation of Rumphius's Herbarium Amboinense. Bureau of Printing, Manila, Philippines.

—— (1918). Species Blancoanae. Bureau of Printing, Manila, Philippines.

—— (1923). An enumeration of Philippine flowering plants 2. Bureau of Printing, Manila, Philippines.

—— (1926). The Flora of Banguey Island. Philipp. J. Sci. 29: 341 – 427.

—— (1952). Partial list of the present locations of authentically named William Roxburgh's Phanerogams. Manuscript at the library of the Royal Botanic Gardens, Kew, U.K.

—— & Chun, W. Y. (1935). Additions to our knowledge of the Hainan flora II. Sunyatsenia 2(3/4): 203 – 332.

Müller Argoviensis, J. (1866). Euphorbiaceae. In: A. de Candolle (ed.), Prodromus systematis naturalis regni vegetabilis 15(2): 189 – 1286. Victor Masson, Paris, France.

Muller, J., Schuller, M., Straka, H. & Friedrich, B. (1989). Palynologia Madagassica et Mascarenica, Fam. 111. Trop. Subtrop. Pflanzenwelt 67: 56 – 98.

Nicolson, D. H., Suresh, C. R. & Manilal, K. S. (1988). An Interpretation of Van Rheede's Hortus Malabaricus. Regnum Veg. 119: 1 – 378. Koeltz Scientific Books, Königstein, Germany.

Obata, J. (1974). Flowering and fruiting observations of *Antidesma pulvinatum* (Hame, Ha'a, Mehame). Newslett. Hawaiian Bot. Soc. 13(5): 20 – 21.

Ochse, J. J. (1931). Vegetables of the Dutch East Indies. Archipel Drukkerij Buitenzorg, Java, Indonesia.

O'Dowd, D. J. & Willson, M. F. (1989). Leaf domatia and mites on Australasian plants: ecological and evolutionary implications. Biol. J. Linn. Soc. 37: 191 – 236.

Pax, F. & Hoffmann, K. (1922). Euphorbiaceae-Phyllanthoideae-Phyllantheae. In: A. Engler (ed.), Das Pflanzenreich 81. 1 – 349. Wilhelm Engelmann, Leipzig, Germany.

Pemberton, R. W. & Turner, C. E. (1989). Occurrence of predatory and fungivorous mites in leaf domatia. Amer. J. Bot. 76: 105 – 112.

Philcox, D. (1997). Euphorbiaceae. In: M. D. Dassanayake & W. D. Clayton (eds.), A revised handbook to the Flora of Ceylon, vol XI: 80 – 283. Oxford & IBH Publishing, New Delhi & Calcutta, India.

Punt, W. (1962). Pollen morphology of the Euphorbiaceae with special reference to taxonomy. Wentia 7: 1 – 116.

Radcliffe-Smith, A. (1996). 153. Euphorbiaceae. In: G. V. Pope (ed.), Flora Zambesiaca 9(4). 1 – 337. Royal Botanic Gardens, Kew, U.K.

—— (2001). Genera Euphorbiacearum. Royal Botanic Gardens, Kew, U.K.

Rheede tot Draakestein, H. A. von (1683). Hortus Indicus Malabaricus 4. Van Someren & van Dyck, Amsterdam, The Netherlands.

Ridley, H. N. (1903). De Maleische Timmerhoutsoorten. Bull. Kolon. Mus. 27: 1 – 108.

Rizvi, S. H., Shoeb, A., Kapil, R. S. & Popli, S. P. (1980). Two diuretic triterpenoids from *Antidesma menasu.* Phytochemistry 19: 2409 – 2410.

Rumphius, G. E. (1743). Herbarium Amboinense III. Amsteldami (Amsterdam), Hagae (Den Haag) & Ultrajecti (Utrecht), The Netherlands.

Savolainen, V., Fay, M. F., Albach, D. C., Backlund, A., van der Bank, M., Cameron, K. M., Johnson, S. A., Lledo, M. D., Pintaud, J.-C., Powell, M., Sheahan, M. C., Soltis, D. E., Soltis, P. S., Weston, O., Whitten, W. M., Wurdack, K. J. & Chase, M. W. (2000). Phylogeny of the eudicots: a nearly complete familial analysis based on *rbcL* gene sequences. Kew Bull. 55: 257 – 309.

Schmid, M. & McPherson, G. (1991). Flore de la Nouvelle-Calédonie et Dépendances 17: Euphorbiacées II, Phyllanthoïdées. Muséum National d'Histoire Naturelle, Paris, France.

Schot, A. M. (2004). Systematics of *Aporosa* (Euphorbiaceae). Blumea Suppl. 17: 1 – 377.

Selling, O. H. (1947). Studies in Hawaiian Pollen Statistics. Part II: The Pollens of the Hawaiian Phanerogams. Bernice P. Bishop Special Publications 38. Bishop Museum, Honululu, Hawaii.

Smith, J. J. (1910). Euphorbiaceae. In: S. H. Koorders & Th. Valeton (eds.), Bijdrage No. 12 tot de kennis der boomsoorten op Java. Meded. Dept. Landb. Ned.-Indië 12: 9 – 637.

—— (1914). Icones Bogorienses IV. Brill, Leiden, The Netherlands.

Sosef, M. S. M., Hong, L. T. & Prawirohatmodjo, S. (eds.) (1998). Plant Resourses of South-East Asia (PROSEA) 5(3). Timber trees: Lesser known timbers. Backhuys Publishers, Leiden.

Stafleu, F. A. & Cowan, R. S. (1976). Taxonomic literature I: A – G. Regnum Veg. 94: 1 – 1136. Bohn, Scheltema & Holkema, Utrecht, The Netherlands.

—— & —— (1983). Taxonomic literature VII: P – Sak. Regnum Veg. 110: 1 – 1214. Bohn, Scheltema & Holkema, Utrecht/Antwerpen; W. Junk, The Hague/Boston.

—— & —— (1988). Taxonomic literature IVV: W – Z. Regnum Veg. 116: 1 – 653. Bohn, Scheltema & Holkema, Utrecht/Antwerpen; W. Junk, The Hague/Boston.

Stearn, W. T. (1983). Botanical Latin, 3rd, rev. ed. David & Charles, Newton Abbot, Devon, U.K.

Steenis, C. G. G. J. Van (1950a). The delimitation of Malaysia and its main plant geographical divisions. Flora Malesiana 1(1): LXX – LXXV. Originally published by Noordhoff-Kolff N.V., Djakarta, Indonesia. Reprint 1985 by Koeltz Scientific Books, Koenigstein, Germany.

—— (1950b). Alphabetical list of collectors. Flora Malesiana 1(1): 5 – 597. Originally published by Noordhoff-Kolff N.V., Djakarta, Indonesia. Reprint 1985 by Koeltz Scientific Books, Koenigstein, Germany.

—— & Schippers-Lammertse, A. F. (1965). Concise plant-geography of Java. In: C. A. Backer & R. C. Bakhuizen van den Brink (eds.), Flora of Java 2: (3) – (72). N. V. P. Noordhoff, Groningen, The Netherlands.

Stuppy, W. (1996). Systematische Morphologie und Anatomie der Samen der biovulaten Euphorbiaceen. Doctoral Dissertation, Univ. Kaiserslautern, Germany.

Tinto, W. F., Blyden, G., Reynolds, W. F. & McLean, S. (1991). Constituents of *Hyeronima alchorneoides*. J. Nat. Prod. 54: 1309 – 1313.

t'Mannetje, L. & Jones, R. M. (eds.) (1992). Plant Resources of South-East Asia (PROSEA) 4: Forages. Pudoc, Wageningen, The Netherlands.

Tulasne, L.-R. (1851). Antidesmata et Stilaginellas, novum plantarum genus. Ann. Sci. Nat. Bot., Sér. 3: 181 – 238.

Vogel, C. (1986). Phytoserologische Untersuchungen zur Systematik der Euphorbiaceae; Beiträge zum infrafamiliären Gliederung und zu Beziehungen im extrafamiliären Bereich. Dissertationes Botanicae 98. Cramer, Berlin & Stuttgart, Germany.

Webster, G. L. (1975). Conspectus of a new classification of the Euphorbiaceae. Taxon 24: 593 – 601.

—— (1989). Revised conspectus of the Euphorbiaceae. Euphorbiaceae Newslett. 2: 4 – 9.

—— (1994a). Classification of the Euphorbiaceae. Ann. Missouri Bot. Gard. 81: 3 – 32.

—— (1994b). Synopsis of the genera and suprageneric taxa of Euphorbiaceae. Ann. Missouri Bot. Gard. 81: 33 – 144.

Whitmore, T. C. (1973). Tree Flora of Malaya 2. Longman, London & Kuala Lumpur.

—— (1975). Tropical rain forests of the Far East. Clarendon Press, Oxford.

Whittaker, R. J. & Jones, S. H. (1994). The role of frugivorous bats and birds in the rebuilding of a tropical forest ecosystem, Krakatau, Indonesia. J. Biogeogr. 21: 245 – 258.

Wight, R. (1844 – 45). Icones plantarum Indiae orientalis 3(2). J. B. Pharoah, Madras, India.

—— (1853). Icones plantarum Indiae orientalis 6. J. B. Pharoah, Madras, India.

Willemstein, S. C. (1987). An evolutionary basis for pollination ecology. Leiden Bot. Ser. 10: 1 – 425.

Willis, J. C. (1966). Willis' dictionary of flowering plants & ferns. 7th ed., revised by H. K. Airy Shaw. The University Press, Cambridge, U.K.

Wurdack, K. J., Hoffmann, P., Samuel, R., de Bruijn, A., van der Bank, M. & Chase, M. W. (2004). Molecular phylogenetic analysis of Phyllanthaceae (Phyllanthoideae *pro parte*, Euphorbiaceae *sensu lato*) using plastid *rbcL* DNA sequences. Amer. J. Bot. 91: 1882 – 1990.

IDENTIFICATION LIST

Only those specimens with clearly identified collector/collection series and collection number are listed. The numbers after the collection numbers correspond to the number of the taxon in the revision.

Aban Gibot SAN 29486: **29**; 29574: **36a**; 29580: **36a**; 31240: **33a**; 31295: **36a**; 32406: **52a**; 32524: **36a**; 32943: **36a**; 34060: **36a**; 56323: **36b**; 57823: **52a**; 65211: **36b**; 66180: **33a**; 66709: **11**; 66715: **11**; 66723: **29**; 66788: **36a**; 70491: **36b**; 74125: **29**; 90036: **36a**; 91251: **52a**; 93029: **33a**; 93995: **36b**; 94010: **52a**; 94038: **34**; 94062: **52a**; 94291: **36a**; 94495: **29**; 94564: **52a**; 97084: **52a**; 97123: **46a**; 99978: **33a** — *Aban Gibot* & 3 UPM SAN 96824: **36b** — *Aban Gibot et al.* SAN 66343: **36b**; 94267: **36b** — *Aban Gibot & Dewol Sundaling* SAN 91188: **29**; 91643: **46a** — *Aban Gibot & Free* SAN 79695: **36a** — *Aban Gibot, Mikil & Free* SAN 79772: **52a** — *Aban Gibot & Petrus* S. SAN 90092: **52a**; 90623: **56** — *Aban Gibot & Saikeh Lantoh* SAN 79399: **52a**; 79411: **49** — *Aban Gibot & Soinin* SAN 60293: **43**; 60393: **43** — *Abang Bujang* 30445: **29**; 36962: **56** — *Abang Mohtar* S 44838: **52a**; 44840: **56**; 48053: **36a** — *Abang Mohtar et al.* S 49499: **36b**; 49523: **52a**; 52718: **52a**; 52776: **36a**; 53915: **3** — *Abang Mohtar & Jugah ak Kudi* S 44749: **52a** — *Abang Mohtar & Othman Ismawi* S 49686: **52a** — *Abang Muas* S 13007: **33a** — *Abbe & Abbe* 10162: **14a** — *Abu* 2270: **28;** 3318: **14a** — *Abubakar* BNB For. Dept. 4100: **29**; KEP 36878: **29**; 36881: **29** — *Achmad* 124: **51**; 284: **51**; 337: **36a**; 409: **51**; 412: **51**; 622: **51**; 1144: **33a**; 1276: **51**; 1339: **51**; 1446: **33a**; 1506: **51**; 1687: **36a** — *Adams* 804: **21** — *Adduru* 23: **33a**; 83: **33a**; 159: **33a** — *Aet Exped. Lundquist* 128: **2**; 449: **10**; 506: **45**; 573: **10** — *Aet & Idjan* 454: **18a**; 850: **18a** — *Afandi Ma'reef* 288: **52a** — *Afriastini* 85: **36a**; 323: **11**; 338: **36a**; 1224: **49**; 2001: **17**; 2002: **49**; 2354: **52a**; 2729: **51** — *Ag. Amin SAN* 105992: **11**; 115571: **33a**; 125960: **11** — *Ag. Amin Sigun* SAN 106133: **56**; 126297: *cf.* **56**; 127344: **56** — *Agam & Aban* SAN 41601: **36a**; 41618: **36a** — *Agama* 418: **36a**; 1067: **33a** — *Aguilar* BS 14336: *cf.* **31** — *Agullana* 3879: **29** — *Ah Wing* SAN 34982: **29** — *Ahern* 5: **21**; 188: **21**; 418: **52a**; 817: **33a** — *Ahern's Collector* 91: **21**; 144: **16a**; 175: **33a**; FB 1111: **21**; 1449: **16a**; FB 2884: **21**; 3160: **50**; FB 3269: **16a**; 3378: **16a** — *Ahmad b. Dewis* S 14327: **11** — *Ahmad Datu* SAN 52772: **34** — *Ahmad Talip* SAN 52783: **36b**; 52825: **52a**; 52952: **33a**; 55002: **33a**; 55652: **43**; 68335: **36a**; 68375: **33a**; 70891: **36a** — *Ahmad Talip & Ally* SAN 32202: **56** — *Ahmad Talip & Ejan* SAN 85702: **52a** — *Ahung* 3265: **24a** — *Aik, Lowa & Kulit* SAN 111927: **36a** — *Akungsai Bakia* 16: **33a**; 557: **52a** — *Alam & Shukla* 7685: **33c** — *Alambra* FB 27824: **42** — *Alcasid et al.* PNH 1588: **33a**; PNH 1590: **33a**; PNH 1603: **33a** — *Alcasid & Edano* PNH 4679: **33a** — *Alejandro* FB 21930: **21** — *Alexius & Dewol Sundaling* SAN 88719: **52a** — *Alston* 12388: **33a**; 12389: **33a**; 12989: **54**; 12996: **33a**; 13015: **54**; 13079: **11**; 13171: **11**; 13377: **36a**; 14694: **21**; 16492: **7** — *Altamirano* FB 31452: **21** — *Altmann* 473: **51**; 600: **1** — *Alvarez* FB 12966: **16a**; FB 18499: **52a**; FB 18585: **5a** — *Alvins* 523: **14a**; 969: **54**; 1048: **54**; 1203: **54**; 1730: **14a**; 1875: **54**; 1877: **54**; 1880: **14a**; 1905: **36a**; 1923: **14a**; 1958: **14a**; 1978: **14a**; 1983: **14a**; 2267: **11**; 2273: **36a**; 3311: **36a** —*Ambri & Arbainsyah* 1730: **36a** — *Ambri & Arifin* W 127: **36a**; AA 133: **11**; W 303: **36a**; AA 438: **21**; W 645: **29**; W 734: **14a**; W 804: **36a**; W 855: **29** — *Ambri, Arifin & Arbainsyah* AA 1265: **36a**; AA 1413: **5a** — *Ambriansyah* AA 666: **29**; AA 765: **36a**; 1540: **52a**; — *Ambriansyah & Arbainsyah* AA1974: **52a** —

Ambriansyah & Arifin AA 64: **29** — *Ambullah* SAN 36006: **33a**; 36096: **34** — *Amdjah* 39: **23**; 74: **49**; 229: **33a**; 249: **49**; 489: **52a**; 643: **56**; 781: **36a**; 801: **36a**; 803: **49**; 1063: **36a** — Amdjah Exped. Nieuwenhuis 85: **56**; 269: **36a** — *Amin* SAN 95491: **21**; 95519: **33a**; 95528: **52a**; 96913: **36a**; 115417: **56**; 115526: **11**; 115596: **11**; 117767: **46a**; 126403: **56**; 126882: **11** — *Amin et al.* SAN 60317: **29**; 68047: **46a**; 93821: **52a**; 94388: **52a**; 117214: **52a**; 118899: **56**; 123230: **36a**; 123428: **56**; 123433: **56** — *Amin & Francis* SAN 114374: **33a**; 115952: **56** — *Amin, Francis et al.* SAN 107803: **36a** — *Amin & Ismail* SAN 107319: **36a** — *Amin, Ismail & Suin* SAN 110433: **36a** — *Amin & Jarius* SAN 109345: **36a**; SAN 115979: **56**; 116063: **33a**; 119815: **56** — *Amin & Lideh* SAN 60472: **56** — *Amin & Soinin* SAN 129204: **33a** — *Amin & Suin* SAN 121635: **56** — *Amin G.* SAN 123479: **21** — *Amin G. et al.* SAN 114337: **21** — *Amin Gambating* SAN 90277: **36a** — *Amin Sigin et al.* SAN 123509: **29** — *Amin Sigon et al.* SAN 59494: **46a**; 60006: **33a**; 67143: **33a**; 68963: **56**; 105613: **21**; 114116: **56**; 114155: **52a**; 116383: **33a**; 116417: **56**; 118126: **33a** — *Amin Sigon & Francis* SAN 116196: **21**; 117915: **21**; 117930: **56** — *Amin Sigon & Heya* SAN 86213: **33a**; 86434: **33a**; 102273: **11** — *Amin Sigon & Jarius* SAN 118119: **21** — *Amin Sigon, Jarius & Francis* SAN 116228: **56** — *Amin Sigon, Mikil et al.* SAN 121349: **52a** — *Amin Sigon & Sigin* SAN 109358: **56** — *Amin Sigon & Suin* SAN 123124: **21** — *Amin Suin* SAN 121531: **56** — *Amir* 16: **21** — *Amiruddin* 14: **14a** — *Ampai Tongseedam* 27: **21** — *Ampuria* SAN 32853: **29**; 34182: **29**; 35317: **29**; 36353: **52a**; 36399: **33a**; 36412: **56**; 36431: **44**; 36439: **44**; 40238: **43**; 40372: **33a**; 40385: **52a**; 40815: **33a**; 40816: **33a**; 41540: **52a**; 41695: **52a**; 42127: **52a** — *Anderson* 163: **51**; 2677: **33a**; 2684: *cf.* **33a**; 3059/1990: **11**; S 3119: **33a**; S 3281: **11**; 3361/5: **11**; 4006: **52a**; 4040: **36b**; 4162/2056: **11**; 4194/2176: **11**; 4584: **36b**; 4631: **36a**; 5112: **11**; 5153: **47**; 8081: **33a**; 8471: **52a**; 8503: **33a**; 9128: **33a**; 9145: **49**; 9762: **11**; S 11773: **11**; S 12416: **11**; S 12423: **33a**; S 14664: **52a**; S 14746: **29**; S 15284: **33a**; S 15477: **52a**; S 16039: **33a**; S 16406: **36a**; S 19147: **33a**; S 20201: **36b**; S 20204: **36b**; S 20946: **33a**; S 25132: **33a**; S 25501: **29**; S 26099: *cf.* **52a**; S 28340: **36a**; S 2873: **11**; S 28376: **36a**; S 31664: **29**; S 31678: **33a** — *Anderson & Ding Hou* 481: **11** — *Anderson & Keng* K 49: **52a** — *Anderson & Sonny Tan* S 26069: **3** — *Andrews* 781: **56** — *Ang* FRI 23313: **52a** — *Ang Khoon Cheng* FRI 23342: **52a** — *Anglade S. J.* 148: **33a**; 149: **33a**; 150: **33a**; 532: **1** — *Anonuevo* PNH 13457: **13**; PNH 13524: **52a**; PNH 13532: **21**; PNH 13568: **31**; PNH 13692: **33a**; PNH 13696: **42** — *Anthony* A 755: **29** — *Anuar & Anthony* 43112: **36a** — *Apostol* 5922: **33a** — *Arbainsyah* 2010: **36a** — *Ardzi b. Arshid* S 16207: **11** — *Argent* 9481: **36a** — *Argent & Amiril Saridan* 9324: **36a**; 9365: **33a** — *Argent, Campbell & Jong* 121987: **36a** — *Argent & Coppins* 1146: **36b**; 1150: **52a**; 1155: **36b**; 1190: **36b** — *Argent, Coppins, Jermy & Chai* 866: **52a** — *Argent, Kamariah, Pendry & Mitchell* 9139: **46a** — *Argent, Pendry & Mitchel* Brunei Rainforest Project 913: **3**; 9128: **49**; 9188: **29** — *Argent & Romero* PPI 9664: **52a** — *Argent, Sidiyasa, Yulita & Wilkie* 93188: **36a**; 93189: **36a** — *Argent & Wilkie* 9436: **33a** — *Ariffin Kalat* 109: **36b**; BRUN 15729: **33a**; BRUN 15758: **11** — *Arifin, Ambri & Arbainsyah* AA 1710: **36a** — *Armstrong* 570: **21** — *Arsat* 644: **36a**; 716: **52a**; 1272: **33a** — *Arsin* 19504: **51** — *Ashton* 7: **49**; 64: **49**; BRUN 86: **52a**; BRUN 110: **36b**; BRUN 622: **36b**; BRUN 1051: **36b**; BRUN 2545: **36a**; BRUN 3330: **33a**; BRUN 5168: **29**; BRUN 5214: **46a**; BRUN 5274: **29**; BRUN 5545: **36b**; S 16482: **36a**; S 16486: **36a**; S 16560: **36a**; S 17779: **33c**; S 17802: **46a**; S 17940: **33a**; S 18060: **29**; S 18111: **29**; S 18166: **29**; S 18226: **36c**; S 18254: **52a**; S 18421: **52a**; S 18622: **49**; S 19605: **36a**; S 21493: **33a** — *Asik J. et al.* SAN 111868: **56**

— *Asik Mantor* SAN 113173: **56**; SAN 116675: **52a**; SAN 121942: **52a**; SAN 123292: **52a**; SAN 127570: **33a**; SAN 128292: **52a**; SAN 135905: **56** — *Aspillero* FB 5225: **52a** — *Atje Exped. van Hustijn* 271: **33a**; 291: **49**; 355: **33a** — *Atkins* 511: **36a** — *Atkins et al.* 432: **56** — *Atkins, Cowley, Sands, Awong, Clayton & Salleh* 599: **33a** — *Atkins & Hyland, Ariffin & Hussain* 506: **49** — *Aumeeruddy* 438: *cf.* **14a** — *Ave* 104: **33a**; 167: **33a**; 4305: **10**; 4355: **10** — *Awang Morshidi* S 24103: **36b** — *Awang Yakup* S 8296: **11** — *Axelius* 164: **14a**; 183: **11**; 281: **36a**; 383: **14a** —

Baba SF 21485: **33a** — *Backer* 300: **23**; 686: **23**; 924: **23**; 2810: **23**; 6083: **54**; 6236: **21**; 7010: **23**; 7224: **33a**; 8749: **32**; 8769: **49**; 8812: **49**; 8822: **54**; 8888: **54**; 8953: **54**; 9020: **54**; 9058: **54**; 10572: **32**; 11074: **32**; 17147: **54**; 18384: **1**; 18871: **23**; 21298: **49**; 21320: **23**; 22558: **32**; 22687: **32**; 23029: **32**; 23241: **32**; 23795: **54**; 25765: **51**; 26132: **51**; 27474: **33a**; 30427: **5a**; 32560: **21**; 32561: **21**; 33913: **5a**; 36061: **33a**; 36732: **33a**; 4516: **23** — *Badal Khan* 90: **1** — *Bahasin* KEP 24926: **14a** — *Bahrgava* 3420: **33a** — *Bakar* 2362: **33a**; 2685: **33a**; 3355: **56**; SAN 25010: **49** — *Bakaroeddin Boschproefstation* bb 68: **11** — *Bakhuizen v. d. Brink* 68: **54**; 626: **51**; 884: **54**; 1390: **51**; 2072: **32**; 2452: **32**; 2473: **32**; 2487: **32**; 3031: **23**; 3036: **23**; 3368: **32**; 4948: **54**; 5054: **52a**; 6084: **54**; 6117: **32**; 6211: **21**; 6792: **21**; 7297: **33a**; 7410: **5a**; 7664: **32** — *Bakhuizen v. d. Brink Jr.* 447: **21**; 651: **52a**; 685: **32**; 873: **5a**; 2406: **33a**; 3134: **33a**; 3423: **54**; 3425: **54** — *Balajadia* KEP 36830: **52a** — *Balakrishnan* 147: **33a** — *Balakrishnan & Chakrabarty* 3091: **33a** — *Balansa* 1494: **5a**; 3222: **21**; 3223: **21**; 3226: **5a**; 3227: **5a**; 3231: *cf.* **33a**; 3239: **21**; 4758: *cf.* **24a**; 4806: **33a**; 4809: **21**; 4818: **5a**; 4834: **5a** — *Baldemor* FB 24562: **16a** — *Bammler* 10: **18a** — *Banang* SAN 51937: **44**; SAN 52062: **44**; SAN 52071: **52a** — *Banlugan et al.* PNH 72730: **31** — *Banyeng et al.* S 64721: **56** — *Banyeng ak Nyudong* S 17222: **52a**; S 22031: **29**; S 24483: **36b**; S 25060: **3**; S 25064: **36a**; — *Banyeng ak Nyudong & Benang ak Bubong* S 26284: *cf.* **29** — *Banyeng ak Nyudong & Dami* S 43694: **29**; S 43916: **29** — *Banyeng ak Nyudong & Ilias Paie* S 45064: **36b** — *Banyeng ak Nyudong & Sibat ak Luang* S 21529: **33a**; S 24475: **40**; S 24915: **14a** — *Barber* 370: **33a**; 391: **56**; 416: **33a**; 1789: **21**; 1807: **21**; 5831: **33a** — *Barbon et al.* PPI 8999: **33a** — *Barbon, Alvarez & Garcia* PPI 1993: **15** — *Barbon, Garcia & Fernando* PPI 12203: **33a**; PPI 12272: **33a**; PPI 12308: **15** — *Barbon, Garcia & Sagcal* PPI 6112: **33a** — *Barbon, Romero & Fuentes* PPI 8139: **33a**; PPI 8411: **33a** — *Barbos* BS 24836: **18a** — *Barker & Verdcourt* LAE 67750: **21** — *Barker & Vinas* LAE 66705: **18b** — *Barnes* FB 167: **16a**; FB 547: **16a** — *Bartlet* 4620: **51**; 4706: **51**; 6830: **33a**; 6924: **28**; 6977: **14a**; 7008: **33a**; 7152: **21**; 7155: **33a**; 7169: **22**; 7237: **22**; 7277: **40**; 7301: **14a**; 7336: **28**; 7390: **33a**; 7640: **40**; 7743: **33a**; 8394: **33a**; 13412: **21**; 13722: **5a**; 14297: **33a**; 14300: **5a**; 14341: **33a**; 14543: **21** — *Bartlet & La Rue* 354: **33a** — *Bateson* 41: **33a** — *Bauerlen in Herb. F. Mueller* 91: **45**; 468: **10**; 475: **10** — *Bawan* FB 24611: **42** — bb series 16634: **21**; 16638: **50**; 20328: **21**; 25456: **21**; 25476: **21**; 28864: **21**; 29851: **21**; 33491: *cf.* **18b** — *Beaman* 136: **36b**; 6958: **36a**; 7269: **36a**; 7662: **36b**; 7784: **36b**; 8312: **36b**; 8456: **29**; 8504: **36a**; 9333: **36a**; 9784: **36a** — *Beaman & Beaman* 7000: **29** — *Beaman et al.* 7275: **52a**; 8978: **49**; 9473: *cf.* **46a**; 9659: **52a**; 9698: **33a**; 9707: **33a**; 9756: **52a**; 10082: **33a** — *Beccari* 55: **20**; 154: **36b**; 197: **49**; 245: **20**; 456: **56**; 635: *cf.* **46a**; 645: **33a**; 664: **51**; 764: *cf.* **33a**; 813: **33a**; 845: **33a**; 1286 or 1256 or 1296: **14a**; 1300: **40**; 1404: **11**; 1464: **36a**; 1717: **36b**; 1852: **49**; 2025: **49**; 2379: **33a**; 2569: **29**; 3142: **11**; 3146: **56**; 3306: **11**; 3371: **33a**; 3590: **33a**; 3829: **52b**; 3831:

36c — *Beddome* 68: **54** — *Beer's* collectors BSIP 6731: **18a**; BSIP 6759: **18a**; BSIP 6827: **18a**; BSIP 7694: **18b**; BSIP 7802: **18a**; BSIP 7823: **18a** — *Beguin* 447: **11**; 1148: **5a** — *Beko* 1: **35a**; 6: **35a** — *Benang ak Bubong* S 24403: **52a**; S 27862: **40** — *Benidick b. Jaibon* A 3229: **33a**; A 3248: **33a** — *Benjamin* LAE 67927: **18a** — *Berhaman et al.* SAN 134455: **36a** — *Bernardo* FB 15481: **5a**; FB 15500: **33a**; FB 24109: **5a** — *Bernstein* 58: **36b** — *Berumai* 14998: **33a** — *Best* SF 13876: **37** — *Beumee* A 567: **23**; 1591: **1**; 2917: **5a**; 5363: **5a** — *Bhargava* 2341: **33a**; 4215: **1** — *Bicknell* 751: **33a**; 865: **33a**; 1270: **33a** — *Binideh* SAN 58440: **11**; 58988: **33a** — *Bish* 25: **21** — BKF series 1068: **33c**; 6206: **5a**; 35156: **21**; 35714: **54**; 40487: **54**; 46419: **40**; 46790: **21**; 48683: **5a**; 51979: **33c**; 52139: **54**; 52178: **33a**; 53513: **36a**; 57534: **54** — *Blake* 16157: *cf.* **21**; 17102: **21** — *Bloembergen* 23: **1**; 3576: **33a**; 4282: **21**; 4461: **21**; 4474: **21** — *Blume* 154: **51**; 209: **32**; 211: **51**; 212: **33a**; 213: **51**; 214: **54**; 555: **51**; 665: **51**; 1323: **54**; 1431: **51**; 1641: **23**; 1657: **49**; 2024: **32**; 2245: **32** — *Boden-Closs* SF 14683: **36a** — *Bodinier* 1128: **24a** — *Boerlage* 41: **49**; 162: **5a**; 289: **49**; 293: **49**; 504: **23**; 574: **33a**; 658: **21** — *Bois* 246: **24a** — *Bon* 2123: **24a** — *Borden* FB 728: **33a**; FB 754: **5a**; FB 756: **5a**; FB 1778: **5a**; FB 1783: **21**; FB 2058: **33a**; FB 2333: **21**; FB 2739: **16a**; FB 3034: **33a** — *Borromeo* 24787: **16a** — Boschproefstation series 257: **29**; 882: **44**; 7530: **7**; 9393: **44**; 17044: **7**; bb 68: **11**; bb 13449: **5a**; bb 14007: **5a**; bb 14329: **7**; bb 14420: **11**; bb 15556: **21**; bb 20264: **5a**; bb 24374: **48**; bb 34074: **11**; bb 34085: **11**; cel/III 32: **17**; cel/V 125: **7**; JA 1846: **5a**; JA 1878: **51**; JA 2088: **33a**; JA 2254: **21**; T 803: **11**; T 883: **44** — *Bourdillon* 165: **33a** — *Bourell* 2242: **33a**; 2375: **33a**; 2379: **33a**; 2434: **33a**; 2458: **21** — *Bourell, Luchavez & Vindum* 2224: **33a** — *Bourne* 211: **33a**; 382: **33a**; 1010: **33a**; 1165: **33a**; 1167: **33a**; 1168: **33a**; 1169: **33a**; 1787: **21**; 2867: **33a**; 2868: **33a** — *Bousi et al.* SAN 123781: **36a**; SAN 123782: **52a** — *Brand* SAN 20066: **36a**; SAN 20839: **29**; SAN 21469: **36a**; SAN 21527: **36b**; SAN 24519: **36a**; SAN 25263: **52a**; SAN 30943: **33a**; SAN 30950: **33a** — *Branderhorst* 214: **21**; 289: **21** — *Brass* 1531: **25**; 2684: **18a**; 2708: **18a**; 3302: **18a**; 3603: **18a**; 3677: **18a**; 3809: **18a**; 3937: **10**; 5344: **35a**; 5633: **18b**; 5746: **21**; 5902: **21**; 6203: **18a**; 6375: **48**; 6512: **48**; 6691: **10**; 7104: **45**; 7294: **18a**; 7333: **45**; 7689: **5a**; 7946: **21**; 8340: **21**; 12866: **18a**; 13832: **41**; 21665: **18a**; 21671: **18a**; 21854: **18a**; 23743: **18a**; 24176: **18a**; 24418: **18a**; 24443: **18a**; 25087: **18a**; 25980: **21**; 27773: **21**; 27776: **18a**; 27790: **18a**; 27947: **18a**; 28118: **18a**; 28414: **18a**; 28430: **18a**; 28516: **18a**; 28661: **18a**; 28905: **18a**; 29141: **35b**; 29501: **18a**; 32293: *cf.* **41**; 32589: **18a** — *Bray* 11625: **14a**; FRI 11717: **54**; FRI 11796: **11** — *Brinkmann* 673: **49**; 828: **33a**; 867: **49** — *Bristol* 1992: **18a**; 2191: **18a**; 2403: **18a** — *Britton* 229: **21** — *Brooke* 8334: **33a**; 8641: **33a**; 8832: **33a**; 8964: **52b**; 9105: **52a**; 9927: **52a**; 10099: **33a**; 10140: **29**; 10160: **52a**; 10274: **56** — *Brown* 33: **18a**; 409: **18b** — *Bruinier/Soeratman* 63: **11** — BRUN series 86: **52a**; 110: **36b**; 622: **36b**; 1051: **36b**; 1379: **36a**; 3330: **33a**; 5168: **29**; 5214: **46a**; 5274: **29**; 5545: **36b**; 15436: **36b**; 15729: **33a**; 15758: **11** — *Brunig* S 8688: **11**; S 8752: **36b**; S 10600: **36b**; S 11998: **11**; S 17524: **11**; S 17535: **11**; S 17544: **33a** — BS series 264: **33a**; 265: **49**; 507: **14a**; 789: **33a**; 808: *cf.* **33a**; 823: **31**; 831: **33a**; 1002: **33a**; 1058: **16a**; 1083: **16a**; 1098: **13**; 1148: **49**; 1231: **21**; 1372: **33a**; 1589: **11**; 1625: **16a**; 2169: **11**; 2183: **21**; 2516: **15**; 3558: **21**; 3656: **33a**; 4079: **52a**; 4492: **50**; 4664: **50**; 5041: **33a**; 5376: **33a**; 5717: **15**; 6158: **33a**; 7213: *cf.* **50**; 7397: **30**; 8079: **33a**; 8371: **50**; 9948: **33a**; 10583: **30**; 10584: **30**; 10956: **52a**; 11114: **33a**; 11445: **5a**; 12168: **5a**; 13339: **52a**; 13868: **52a**; 14085: **15**; 14115: **15**; 14336: *cf.* **31**; 15086: **52a**; 15147: **21**; 15251: **31**; 15335: *cf.* **31**; 15438: **33a**; 15439: **29**; 15455: **33a**; 15644:

21; 15665: *cf.* **33a**; 15731: **31**; 15762: **33a**; 15849: **15**; 16425: *cf.* **42**; 16625: **52a**; 16811: **42**; 16911: **42**; 17528: **52a**; 18337: **33a**; 18731: **16a**; 18874: **21**; 19410: **33a**; 20910: **33a**; 21365: **21**; 21413: **21**; 21476: **33a**; 21611: **33a**; 22506: **5a**; 22556: **33a**; 22729: **21**; 22781: **33a**; 22799: **52a**; 22882: **33a**; 22995: **21**; 23191: **33a**; 23471: **33a**; 23615: *cf.* **42**; 24134: **52a**; 24176: **16a**; 24220: **31**; 24375: **30**; 24415: **52a**; 24436: **15**; 24504: **30**; 24541: **15**; 24836: **18a**; 26069: **21**; 26300: **31**; 26375: **16a**; 26561: **15**; 26597: **16a**; 27302: **21**; 28041: **5a**; 28183: **52a**; 28503: **30**; 28519: **52a**; 28566: **33a**; 28585: **33a**; 29283: **33a**; 29321: **33a**; 29477: **33a**; 29516: **50**; 29838: **30**; 30052: **33a**; 30097: **33a**; 30184: **42**; 30515: **6**; 30533: **52a**; 30542: **30**; 31020: **52a**; 31268: **52a**; 31409: **52a**; 31490: **21**; 31491: **21**; 31949: **33a**; 32262: **21**; 32347: **33a**; 32542: **52a**; 32865: **50**; 32902: **5a**; 32998: **52a**; 33023: **30**; 33057: *cf.* **18a**; 33153: **33a**; 33173: **33a**; 33174: **16a**; 33294: **15**; 33489: **52a**; 33613: *cf.* **30**; 33638: **30**; 33701: **30**; 34371: **33a**; 34694: **33a**; 34803: **52a**; 34927: **33a**; 34928: **33a**; 34970: *cf.* **42**; 35072: **52a**; 35108: **13**; 35715: **31**; 36631: **29**; 36965: **33a**; 37090: **33a**; 37093: **29**; 37337: **29**; 37886: **15**; 38985: **52a**; 39026: **18a**; 39423: **49**; 39605: **49**; 40732: **49**; 40733: **49**; 40835: **33a**; 40899: **49**; 40923: **16a**; 40931: **15**; 41060: **49**; 41084: **49**; 41154: **33a**; 41160: **50**; 41519: **31**; 41841: **30**; 42579: **33a**; 43059: **31**; 43237: **31**; 43582: **33a**; 44006: **33a**; 44013: **52a**; 44014: **52a**; 44532: **16a**; 45189: **24a**; 45205: **52a**; 45303: **52a**; 45360: **52a**; 45926: **33a**; 46159: **52a**; 46234: **52a**; 46539: **21**; 46647: **33a**; 46722: **33a**; 46825: **52a**; 47040: **5a**; 47178: **52a**; 47352: **52a**; 48051: **52a**; 48152: **52a**; 48177: **33a**; 48206: **42**; 48217: **52a**; 48343: **33a**; 48731: **16a**; 49352: **21**; 75109: **33a**; 75150: **52a**; 75342: **42**; 75874: **52a**; 75943: **30**; 75965: **30**; 76137: **52a**; 76335: **33a**; 76429: **52a**; 76906: **21**; 76991: **52a**; 77709: **46a**; 77846: **33a**; 78181: **5a**; 78209: **33a**; 78300: **33a**; 78503: **52a**; 79129: **52a**; 79221: **52a**; 79416: **30**; 79441: **30**; 79466: **52a**; 80399: **33a**; 80687: **33a**; 83664: **52a**; 83820: **13**; 83955: **52a**; 84118: **33a**; 84136: **16a**; 84668: **33a**; 84902: **21** — BSIP series 1228: **18a**; 1351: **18a**; 1969: **18a**; 1976: **18a**; 2233: **18a**; 2247: **18a**; 2480: **18a**; 2672: **18a**; 2726: **18a**; 2856: **18a**; 2860: **18a**; 2916: **18a**; 2927: **18a**; 3351: **18b**; 3497: **18a**; 3502: **18b**; 3607: **18a**; 3893: **18a**; 3941: **18a**; 4970: **18a**; 5002: **18a**; 5420: **18a**; 5463: **18a**; 6731: **18a**; 6759: **18a**; 6827: **18a**; 7694: **18b**; 7802: **18a**; 7823: **18a**; 7869: **18a**; 8397: **18a**; 8700: **18a**; 8807: **18a**; 8934: **18a**; 9649: **18a**; 9793: **18a**; 10115: **18a**; 10186: **18a**; 10483: **18a**; 10496: **18a**; 10769: **18b**; 10830: **18a**; 11002: **18a**; 11240: **18b**; 12030: **18a**; 12415: **18a**; 12691: **18b**; 12860: **18b**; 12982: **18a**; 13434: **18b**; 13508: **18b**; 13573: **18a**; 14307: **18a**; 15714: **18a**; 15727: **18a**; 15787: **18a**; 16444: **18a**; 16995: **18a**; 17260: **18a**; 17497: **18a**; 17536: **18a**; 17849: **18a**; 18007: **18a**; 18034: **18a**; 18453: **18a**; 18506: **18a**; 18537: **18a**; 18843: **18a**; 18966: **18a**; 18979: **18a** — *Buderus* NGF 25804: **21** — *Buennemeijer* 1767: **11**; 2015: **36a**; 4880: **51** — *Bujang* S 12289: **36a**; S 13432: **43**; S 13447: **29** — *Bujang b. Sitam* S 14612: *cf.* **43** — *Bunchuai* 349: **33a**; 11814: **1**; 11825: **21**; 12201: **1**; 15293: **47**; 15294: **33a**; 17127: **21**; 19957: **5a**; 24231: **21**; 25834: **55**; 26410: **44** — *Bunchuai & Nimanong* Fl. Thailand 46297: **5a** — *Bunkird* 32: **33a** — *Bunnab* BKF 35714: **54**; 35740: **28**; 37097: *cf.* **33d** — *Bunpheng* 12442: **21**; 12447: **1**; 15236: **1**; 24176: **47** — *Bunter* SAN 25817: **33a**; SAN 25876: **36a** — *Burgess* SAN 25181: **44** — *Burkill* SF 815: **14a**; 915: **54**; 1358: **37**; SF 2164: **37**; 2639: **21**; SF 2843: **36a**; SF 4489: **36a**; SF 6391: **38**; SF 6541: **21**; SF 6555: **33a**; SF 6611: **33a**; SF 7012: **33a**; SF 8878: **12** — *Burkill et al.* 2344A: **12** — *Burkill & Haniff* SF 13038: **21**; SF 13223: **11**; SF 16803: **54**; SF 16812: **11**; SF16903: **21**; SF 16935: **33c**; SF 17635: **21** — *Burkill & Holttum* SF 8515: **12**; 8655: *cf.* **11** — *Burkill & Shah* 1009:

14a; 1080: **14a** — *Burley* 53: *cf.* **52a**; 68: **5a**; 173: **33a**; 2137: **40** — *Burley & Tukirin* 584: **56** — *Burley, Tukirin et al.* 371: **36a**; 380: **46a**; 408: *cf.* **49**; 477: **29**; 557: **46a**; 620: **36a**; 711: *cf.* **49**; 743: **36a**; 806: **40**; 1096: **14a**; 1134: **52a**; 1261: **14a**; 1344: *cf.* **52a**; 1399: **14a**; 1469: **36a**; 1485: *cf.* **33a**; 1665: **36a**; 1666: **14a**; 1741: *cf.* **36a**; 1761: **51**; 1814: **52a**; 1887: **29**; 1948: **36a**; 2401: **36a**; 2443: **36a**; 2492: **52a**; 2548: **33a**; 2550: **36a**; 2670: **52a**; 2685: **52a**; 2804: **52a**; 2880: **29**; 2886: **40**; 2983: **36a**; 3199: **49**; 3550: **7**; 3558: **7**; 3643: **7**; 4067: **46a**; 4131: *cf.* **4**; 4209: *cf.* **4**; 4219: *cf.* **4**; 4222: **46a** — *Burn-Murdoch* 183: **33c**; 261: **33c** — *Burot Ho* BNB For. Dept. 1767: **33a**; 1884: **33a**; 1899: **33a**; 1902: **33a**; 2685: **33a** — *Burtt* 12654: **29**; 12683: **52a** — *Buwalda* 2981: **5a**; 3004: **5a**; 3022: **33a**; 3479: **51**; 3612: **32**; 3655: *cf.* **5a**; 4546: **48**; 6124: **49**; 6345: **33a**; 6346: **49**; 6433: **11**; 6461: **36a**; 6517: **36a**; 6648: **33a**; 6682: **14a**; 6748: **36a**; 7005: **33a**; 7062: **52a**; 7139: **14a**; 7436: **21**; 7682: **33a**; 7954: **36a**; 8075: **51** — BW series 516: **18b**; 2166: **18b**; 3461: **45**; 3824: **18a**; 3901: **18a**; 4077: **18a**; 4672: **18b**; 4902: **2**; 5518: **18a**; 6740: *cf.* **18b**; 9360: **18b**; 10494: **35a**; 10749: **10**; 12193: **2**; 12527: **35a**; 13308: **10** — *Bygrave* 14: **36b**; 23: **36b** — *Byrnes* 2377: **21**; 2842: **21** —

Callery 14: *cf.* **42**; 20: *cf.* **42**; 38bis: **42**; 42: **15**; 58bis: **33a**; 62bis: *cf.* **42** — *Cammill* FB 116: **5a** — *Campbell* 22/4/7: **36a**; 26/4/7: **36a**; EG 196: **52a**; SAN 112217: **36a**; SAN 112334: **36a** — *Campbell, Aik, Lowa & Kulit* SAN 111905: **36a** — *Canabre BS* 30097: **33a** — *Cantley* 708: **11** — *Carr* 11290: **18a**; 11308: **21**; 11387: **21**; 11968: **18b**; 12265: **21**; 12511: **18a**; 13293: **18a**; 14691: **18a**; 14832: **18a**; 14942: **18a**; 15993: **18b**; SF 26627: **52a**; SF 26812: **36a**; SF 27246: **49** — *Carrick Univ. of Malaya* 3736: **37** — *Carson* SAN 28033: **52a** — *Castillo* 613: **52a**; 615: **36a**; 653: **36a**; BS 22729: **21** — *Castro* 973: **44**; 3791: **36a**; PNH 5695: **13**; PNH 5758: **13** — *Castro & Melegrito* 1494: **33a**; 1519: **44**; 1539: **52a**; 1641: **33a** — *Cavanaugh* NGF 2021: **21** — *Cavanaugh & Fryar* NGF 2086: **18a** — *Celestino & Castro* PNH 1970: **49** — *Celestino & Ramos* PNH 23043: **33a** — *Cenabre* FB 25525: **52a**; FB 31356: **33a** — *Cenabre & Agelar* FB 28868: **30**; FB 28874: **52a** — *Cenabre & Party* FB 28506: **16a**; *FB* 28527: **33a** — *Chai* SAN 18388: **52a**; S 19323: **36a**; S 19339: **36a**; S 19705: **33a**; SAN 21667: **33a**; SAN 25057: **33a**; SAN 29727: **33a**; SAN 29732: **33a**; SAN 29788: **33a**; S 30064: **33a**; S 30072: **52a**; S 33924: **29**; S 34680: **52a**; S 35300: **52a**; S 35364: **36b**; S 35779: **52a**; S 36095: **29**; S 36153: **52a**; S 36745: **36a**; S 37278: **36a**; S 37371: **33a**; S 39509: **33a**; S 39651: **36b**; S 39675: **52a**; S 68826: **36b** — *Chai et al.* S 33246: **52b**; S 37217: **29**; S 37346: **52a** — *Chai & Ilias Paie* S 30679: **56** — *Chakrabarty* 2553: **5a**; 4631: **1** — *Champion* 172: **24a** and **33b** — *Chan* FRI 6706: **11**; FRI 6767: **54**; FRI 6771: **33a**; FRI 11215: **54**; FRI 13054: **14a**; FRI 13184: **36a**; FRI 13212: **54**; FRI 19807: **14a**; FRI 19937: **33a**; FRI 23811: **33c**; FRI 25012: **11**; FRI 25081: **54**; FRI 25116: **11** — *Chantaranothai, Chayamarit, Middleton, Parnell & Simpson* 1320: **54** — *Charington* SAN 24706: **29** — *Charoenmayu* 4379: **55**; 5480: **21** — *Charoenphol, Larsen & Warncke* 3515: **52a**; 3573: *cf.* **54**; 3982: **29**; 4045: **33a**; 4085: **24a**; 4089: **33a**; 4096: **24a**; 4114: **24a**; 5064: **27** — *Chatterjee* 23: **5a**; 181: **5a**; 245: **5a** — *Che Wee Lek* SF 41018: **24a** — *Chelliah* FRI 6903: **54**; FRI 6908: **54**; FRI 6915: **33a**; KEP 104385: **14a**; KEP 104417: **54** — *Chermsirivathana* 1393: **54**; 1394: **54**; 1430: **54** — *Chew* 55: **36b**; 137: **54**; 288: **29**; 363: **36b**; 406: **36b**; 428: **36b**; 456: **33a**; 479: **33a**; 555: **52a**; 713: **33a**; 721: **52a**; 727: **36a**; 763: **14a**; 1031: **52a**; 1344: **11**; SF 41005: **29** — *Chew & Corner* RSNB 4067: **29**; RSNB 4161: **52a**; RSNB 7050: **52a**; SF 31542: **21**; SF 31605: **33a** — *Chew, Corner & Stainton*

RSNB 16: **36b**; RSNB 210: **36a**; RSNB 233: **29**; RSNB 237: **52a**; RSNB 533: **56**; RSNB 1202: **49**; RSNB 1460: **36b** — *Chew & Kiah* SF 40680: **21**; SF 40980: **21** — *Chiao & Fan* 346: **24a** — *Chin* 685: **36a**; 733: **52a**; 1322: **33c**; 1509 A: **24a**; 2513: **36c**; 3392: **33a**; KLU 13382: **14a** — *Chin & Ahmad* 3130: **33a** — *Chin et al.* 4565: **11** — *Ching* 1866: **24a** — *Cho-Ganh* 60: **21** — *Chow* SAN 75336: **52a** — *Chow & Leopold* SAN 76401: **36b** — *Christensen* 564: **52a**; 772: **36a**; 1045: **56**; 1062: **29**; 1180: **56** — *Christophersen* 3240: **18a** — *Chua et al.* FRI 38594: **33a** — *Chung* 2533: *cf.* **29**; 2675: **46a** — *Church* 30: *cf.* **36a**; 108: *cf.* **36a**; 168: **29**; 246: **36a**; 342: **29**; 357: **36a**; 425: **36a**; 496: **33a**; 571: **36a**; 599: **56**; 1187: **33a**; 2293: **49**; 2375: **49**; 2731: **49** — *Church et al.* 709: **36a**; 737: **36a**; 1211: **36a**; 1254: **56**; 1342: **49**; 1556: **52a**; 1620: **33a**; 1652: *cf.* **49**; 1704: **52a**; 1825: *cf.* **36a**; 1864: **29**; 2442: **36a**; 2485: **36a**; 2585: **36a**; 2637: **36a**; 2638: **49**; 2719: **36a**; 2757: **33a**; 2781: **36a**; 2828: **52b**; 2846: **29** — *Clark* FB 987: **33a**; FB 1013: **21**; 1020: **42** — *Clarke* KU 72: **18a**; 14244: **21**; 21645: **21**; 26415: **21**; 26423: **21**; 33506B: **21**; 35143: **33a**; 35355: **1**; 35534: **33a**; 36417C: **5a**; 36633: **1** — *Clarkson* 4036: **21**; 4073: **21**; 4116: **21**; 4117: **21**; 5034: **21**; 5063: **21**; 5637: **21** — *Clarkson & Mc Donald* 6724: *cf.* **21** — *Clarkson & Nelder* 8936: **21** — *Clemens* 339: **29**; 565: **33a**; 623 a: **21**; 649: *cf.* **18a**; 832: **29**; 884: **29**; 1259: **18a**; 1619: **18a**; 2143: **18a**; 4749: **35a**; 9772: **33a**; 9797: **33a**; 9920: **49**; 9944: **29**; 10015: **56**; 10374: **52a**; 10492: **21**; 10690 A: *cf.* **18a**; 10706bis: **18b**; 10790: **29**; 10925: **35a**; 11062bis: **35a**; 17557: **5a**; 17647: **16a**; 18264: **5a**; 18265: **33a**; 50353: **49** — *Clemens & Clemens* 518: **18a**; 869: *cf.* **41**; 1209: **18a**; 3303: **2**; 3402: **21**; 3602: **33a**; 4112: **5a**; 4199: **24a**; 4226: **24a**; 4558: *cf.* **18a**; 4588: **18a**; Field No. 5293: **52a**; Field No. 7203: **49**; Field No. 7824: **49**; 20030: **36b**; 20041: **52a**; 20262: **52a**; 20340: **36a**; 20341: **36a**; 20342: **36b**; 20343: **36b**; 20595: **49**; 21243: **33a**; 21244: **56**; 21245: **56**; 21263: **52b**; 21650: **29**; 21667: **52b**; 21668: **52a**; 21669: **33a**; 21670: **56**; 22116: **33a**; 26117: **52a**; 26120: **33a**; 26231: **33a**; 26257: **49**; 26342: **52a**; 26530: **29**; 26763: **56**; 26789: **52a**; 26863: **49**; 27197: **49**; 27334: **36a**; 29654: **52a**; 29671: **36a**; 30011: **56**; 30274: **36a**; 30275: **49**; 30276: **49**; 30755: **49**; 31940: **36b**; 32017: **36a**; 32198: **36a**; 32854: **49**; 33048: **36b**; 34415: **29**; 40317: **49**; 40394: **36b**; 40572: **36b**; 40686: **36b**; 40878: **29**; 40969: **49**; 50088: **36a**; 50217: **56**; 50255: **36a**; 50267: **36a**; 50326: **36b**; 50399: **29**; 50988: **36a** — *Co* 3139: **16b**; 3142: **16b**; 3420: **52a**; 3452: **52a**; 3455: **52a** — *Cockburn* FRI 7016: **33c**; FRI 7088: **54**; FRI 7101: **36b**; FRI 7109: **14a**; FRI 7124: **54**; FRI 7214: **36b**; FRI 7467: **37**; FRI 7892: **38**; FRI 8180: **36a**; FRI 10622: **33c**; FRI 10930: **36b**; FRI 10959: **36b**; FRI 11069: **14a**; SAN 76810: **36b**; SAN 83092: **36a**; SAN 83302: **36a**; SAN 84868: **36b**; SAN 85057: **52a**; KEP 115971: **54** — *College, Herbar & Shembaganur* 881: **33a** — *Collins* 100: **21**; 198: **21**; 415: **1**; 822: **33a**; 829: **21**; 1016: **1**; 1288: **47**; 1383: **33a** — *Congdon* 220: **54**; 373: **33a**; 427: **54**; 457: **21**; 504: **55**; 610: **33a**; 754: **55**; 1232: **33a** — *Conklin* PNH 17593: **5a**; PNH 17595: **21** — *Conklin & Bawaya* PNH 78678: **5a**; PNH 79569: *cf.* **30**; PNH 80421: **5a**; PNH 80610: **33a**; PNH 80680: **24a** — *Conn & Katik* LAE 66011: **18a** — *Coode* 5838: **17**; 5844: **17**; 5879: *cf.* **33a**; 5948: **52a**; 6060: **52a**; 6068: **17**; 6242: **33a**; 6350: **36c**; 6622: **46a**; 6779: **36b**; 6867: **36b**; 7236: **3**; 7779: **3**; 7945: *cf.* **43**; NGF 32686: **19** — *Coode, Dransfield, Kirkup & Said* 7899: **56** — *Coode, Ferguson et al.* 7185: **29**; 7294: **49** — *Coode & Katik* NGF 32762: **18a** — *Coode & Kirkup* 7010: **29** — *Coode, Kirkup et al.* 6972: **29**; 7040: **29**; 7070: **29** — *Coode & Lelean* NGF 29932: **19** — *Coode & Ridsdale* 5338: **49**; 5408: *cf.* **49**; 5713: **33a** — *Coode & Wong* 6681: **49** — *Coode, Wong et al.* 6356: **29**; 6411: **52a**; 6418: **52a**; 6481: **11**; 6738: **52a**; 6818: **29** — *Copeland*

401: **21**; 6282: **5a** — *Cordero & Espiritu* PNH 91587: **33a** — *Corner* RRS 1526: **18b**; SF 21314: **36a**; 25753: **54**; SF 25838: **33c**; SF 28071: **36a**; SF 28984: **36a**; 29043: **38**; 29232: **38**; SF 29347: **29**; SF 29424: **14a**; SF 29433: **36a**; 29836: **33a**; 30210: **54**; SF 30348: **40**; SF 30423: **49**; SF 30689: **52a**; SF 30872: **52a**; 31573: **33a**; 31607: **21**; SF 31650: **33a**; SF 32534: **14a**; SF 32535: **14a**; 33455: **21**; SF 34532: **14a**; SF 34608: **11**; SF 37080: **52a**; 37692: **33a** — *Corner & Henderson* SF 36602: **33a** — *Cox* 191: **18a** — *Craven & Schodde* 756: **18b**; 901: **18a**; 928: **18a**; 1010: **18a**; 1053: **18a** — *Croft et al.* LAE 68735: **18a**; LAE 68885: **18a** — *Cubitt* 629: **21**; 712: **14a** — *Cuming* 142: **21**; 158: **21**; 474: **5a**; 900: *cf.* **42**; 966: **33a**; 986: **21**; 1203: **5a**; 1246: **33a**; 1300: **52a**; 1303: **5a**; 1316: **33a**; 1348: **16a**; 1354: **52a**; 1426: **33a**; 1446: **5a**; 1511: **33a**; 1513: **33a**; 1650: **21**; 1742: **52a**; 1820: **33a** — *Cunningham* 269: **21**; 290: **21** — *Curran* 3477: *cf.* **30**; BS 4492: **50**; FS 5087: **13**; FB 5150: **21**; FB 5452: **16a**; FB 6365: **5a**; BS 7213: *cf.* **50**; FB 9662: **33a**; FB 10450: **15**; FB 10736: *cf.* **42**; FB 13212: **50**; FB 16617: **33a**; FB 17763: **30**; FB 19142: **16a**; FB 19401: **52a**; FB 19575: **30** — *Curtis* 230: **33a**; 321: **11**; 672: **54**; 702: **11**; 782: **21**; 863: **54**; 984: *cf.* **11**; 1150: **33a**; 1322: **36a**; 1341: **52a**; 1474: **11**; 1487: *cf.* **11**; 1554: **21**; 2278: **11**; 2280: **11**; 2307: **33a**; 2373: **33a**; 2529: **33a**; 2651: **54**; 2842: **54**; 3080: **11**—

Daim Andau 29: **46a**; 163: **52a**; 166: **52a**; 252: **36a**; 265: **36b**; 266: **36a**; 301: **36b**; 320: **46a**; 396: **52a**; 415: **36b**; 469: **49**; 551: **36a**; 664: **52a**; 711: **36a**; 827: **33a**; 905: **36a**; 986: **52a**; 1040: **52a**; 1088: **36a**; 1127: **36b**; 1186: **36a** — *Damanhuri* FRI 36570: **14a** — *Damas* LAE 58888: *cf.* **8** — *Dan b. Bakar* 4352: **36b** — *Danser* 5467: **33a**; 6114: **51**; 6119: **51** — *Darbyshire* 933: **18a** — *Darbyshire & Hoogland* 8339: **41** — *Dardaedi* 703: **33a** — *Darling* FB 14768: **5a**; FB 18675: **42** — *Darnaedi* 703: **33a** — *Daud & Tachun* SF 35642: **36c**; SF 35652: **36a** — *Davidson* 1294: **33a**; 1345: **54**; 1353: **54** — *Davis et al.* 505: **29** — *Dayang Awa & Ilias Paie* S 45670: **49**; S 45682: **52a**; S 46981: **33a** — *Dayang Awa & Lee* S 47673: **36b** — *Dayang Awa & Othman Ismawi* S 47037: **33a**; S 47043: **49** — *Dayang Awa & Yii* S 46636: **52a** — *Dayar Arbain* 240: *cf.* **33a**; 432: *cf.* **33a** — *De Bell* Boschproefstation bb 13449: **5a** — *De Boer* 3513: **1** — *De Gryp* Boschproefstation bb 20264: **5a** — *De Haan* 24: **11**; 102: **11** — *De Jong* 329: **33a**; 463: *cf.* **43**; 644: **52a**; 737: **52a**; 791: *cf.* **52a** — *Demandt & Dillewijn* 4327: **5a** — *De Mesa* FB 27521: **5a**; FB 27664: **52a** — *Dennis* BSIP 7869: **18a** — *Derry* 507: **21**; 521: **14a**; 998: **54** — *De Vogel* 718: **29**; 800: **52a**; 913: **52a**; 1297: **36a**; 2691: **49**; 2961: **49**; 3699: **17**; 3742: *cf.* **7**; 4196: **2**; 4200: **2**; 5612: **18a**; 5625: **17**; 8241: **36a**; 8242: **52a**; 8258: **36a**; 8270: **29**; 8331: **52a** — *De Vogel & Vermeulen* 6508: **17**; 6517: **17**; 6555: **7** — *De Voogd* 212: **36a**; 264: **21**; 1071: **36a**; 2231: **21**; 3118: **51** — *DeVore & Hoover* 103: **21**; 199: **21** — *De Wilde & De Wilde-Duyfjes* 12072: **52a**; 12099: *cf.* **14a**; 12132: **33a**; 12197: **33a**; 12318: *cf.* **14a**; 12372: *cf.* **14a**; 12686: **52a**; 12834: *cf.* **14a**; 13770: **20**; 13878: **33a**; 13961: **33a**; 14341: **20**; 14479: **33a**; 14638: **52a**; 14792: **52a**; 15512: *cf.* **52a**; 15549: **33a**; 15678: **52a**; 16447: **21**; 18024: **33a**; 18150: *cf.* **14a**; 18219: *cf.* **14a**; 18357: **33a**; 18518: *cf.* **14a**; 19211: **52a**; 19265: **33a**; 19309: **52a**; 19317: **49**; 19422: **52a**; 19449: **52a**; 19515: **14a**; 19682: **22**; 19694: **40**; 19771: **52a**; 20032: **52a**; 20051: **52a**; 20111: **33a**; 20209: **49**; 20438: **33a**; 20864: **22**; 20890: **49**; 21089: **14a**; 21158: **52a**; 21193: **52a**; 21331: **52a**; 21367: **36b**; 21375: **33a** — *Dewol & Jumrafiah* SAN 124140: **11** — *Dewol & Karim* SAN 77760: **33a**; SAN 77826: **36b** — *Dewol Sundaling* SAN 55976: **46a**; SAN 78102: **52a**; SAN 80779: **33a**; SAN 80811: **36a**; SAN 80853: **52a**; SAN

80857: **52a**; SAN 83991: **36a**; SAN 84060: **33a**; SAN 90367: **49**; SAN 90514: **36a**; SAN 92191: **34**; SAN 93571: **52a**; SAN 93655: **33a**; SAN 94926: **52a**; SAN 99432: **33a**; SAN 99478: **36c** — *Dewol Sundaling & Harun T.* SAN 89919: **36a** — *Dewol Sundaling, Joseph & Patrick* SAN 69906: **52a** — *Dewol Sundaling & Karim* SAN 78111: **36a** — *Dewol Sundaling & Kodoh T.* SAN 89302: **52a**; SAN 89308: **33a** — *Dewol Sundaling et al.* SAN 74561: **36a**; SAN 74957: **52a**; SAN 108838: **43**; SAN 109231: **52a** — *Dewol Sundaling & Mansus* SAN 107644: **52a** — *Dewol Sundaling & Patrick* SAN 71486: **36a**; SAN 93150: **52a** — *Dewol Sundaling & Petrus S.* SAN 90062: **56** — *Diepenhorst* 1324: **51**; 1364: **49**; 2352: **36a**; 2544: **51** — *Ding Hou* 116: **14a**; 324: **36b**; 437: **52a**; 529: **33a**; 700: **33a**; 824: **33a** — *Djoemadi* 143: **33a** — *Dobremez* NEP 784: **1** — *Doinis Soibeh* 285: **33a**; 301: **33a** — *Dolman* 27623: **39** — *Dolores* PNH 33163: **33a** — *Dransfield* 992: **54**; 1159: **49**; 1177: **33a**; SMHI 1234: **33a**; SMHI 1290: **33a**; 2928: **49**; 6640: **36a**; 6724: **36c**; 7026: **3**; 7032: **52a**; 7079: **46a**; 7102: **36b** — *Dransfield et al.* 7103: **29**; 7458: **49**; 7490: **52a** — *Ducloux* 21?6: **24a** — *Duldulao* FB 25562: **52a** — *Dunlop* 3231: **21** — *Duthie* 2407: **1**; 2408: **1** — *Dyah Noermawati* 9: **5a**; 16: **33a** — *Dyg. Awa & Ilias* S 44331: **36a** — *Dyg. Awa & Lee* S 47545: **36a** — *Dyg. Awa & Paie* S 47401: **33a**; S 48778: **36a** —

Ebalo 590: **21** — *Edano* PNH 357: **52a**; PNH 981: **52a**; PNH 1054: **52a**; PNH 1145: **52a**; PNH 4012: **16a**; PNH 4014: **16a**; PNH 4038: **16a**; PNH 4164: **16a**; PNH 6784: *cf.* **31**; PNH 14081: **33a**; PNH 14117: *cf.* **33a**; PNH 14193: **33a**; PNH 15250: **30**; PNH 15532: **30**; PNH 18052: **5a**; PNH 18053: **5a**; BS 18731: **16a**; 19866: **24a**; PNH 34472: **31**; PNH 34545: **31**; PNH 40164: **30**; BS 41841: **30**; BS 46159: **52a**; BS 46234: **52a**; BS 48731: **16a**; BS 75874: **52a**; BS 75943: **30**; BS 75965: **30**; BS 76137: **52a**; BS 76335: **33a**; BS 76429: **52a**; BS 77709: **46a**; BS 77846: **33a**; BS 78209: **33a**; BS 78300: **33a**; 78305: **52a**; BS 78503: **52a**; BS 79129: **52a**; BS 79221: **52a**; BS 79416: **30**; BS 79441: **30**; BS 79466: **52a** — *Edano & Martelino* BS 35715: **31** — *Eder* FB 28568: **33a** — *Egon* 455: **36a**; A 849: **11** — *Elbert* 372: *cf.* **33a**; 2553: **23**; 3079: **17**; 3139: **17**; 3162: **7**; 3515: **23**; 3608: **21**; 3707: **21**; 3742: **5a**; 3768: **33a**; 3771: **23**; 3775: **21**; 3789: **21**; 3797: **23**; 3798: **21**; 3799: **33a**; 4090: **23**; 4313: **33a**; 4350: **21**; 4564 a: **33a** — *Elgincolin* FB 28574: **5a** — *Elleh* SAN 34377: **29**; SAN 34389: **52a**; 35423: **36a**; 37438: **33a** — *Elmer* 5734: **21**; 6168: **5a**; 6319: **5a**; 6320: **33a**; 6323: **5a**; 6327: **33a**; 6444: **16a**; 7005: *cf.* **15**; 7371: **33a**; 7424: **33a**; 7595: **15**; 7695: **15**; 7913: **52a**; 8136: **21**; 8139: **33a**; 8159: **5a**; 8218: **5a**; 9088: **52a**; 9157: **33a**; 9183: **21**; 9668: **31**; 10312: **5a**; 10374: **33a**; 10844: **52a**; 10936: **42**; 11015: **21**; 11166: *cf.* **31**; 12191: **33a**; 12459: *cf.* **49**; 12519: **15**; 12703: **21**; 12802: **33a**; 12808: **33a**; 12883: **52a**; 13277: **42**; 13549: **33a**; 13971: **52a**; 14508: **31**; 14865: **33a**; 15212: **42**; 15464: **15**; 15601: **15**; 15881: **33a**; 15950: **33a**; 16940: *cf.* **31**; 17264: **31**; 17304: **15**; 17360: **21**; 17478: **33a**; 17516: *cf.* **31**; 17620: **5a**; 17710: **21**; 17783: **33a**; 18083: **50**; 18105: **31**; 18264: **33a**; 18299: **42**; 20036: **36a**; 20189: **36a**; 20856: **36a** — *Endert* 1607: **56**; 1795: **33a**; 1985: **21**; 1986: **21**; 2023: **44**; 2046: **44**; 2119: **36a**; 2301: **56**; 2415: **36a**; 2550: **49**; 2716: **49**; 2776: **36a**; 3052: **33a**; 3064: **56**; 3117: **36a**; 3140: **33a**; 3341: **56**; 3445: **36a**; 3578: **29**; 3666: **36a**; 4025: **33a**; 4187: **36b**; 4349: **36a**; 5249: **52a** — *Enggoh* BNB For. Dept. 10194: **29** — *Enoh* 347: **33a** — *Ernst* 1062: **14a** — *Erwin & Paul* S 27436: **33a** — *Escritor* BS 20910: **33a**; BS 21365: **21**; BS 21413: **21**; BS 21476: **33a**; BS 21611: **33a** — *Esquirol* 505: **33b**; 1586: **33b** — *Eva* AA 40: **29**; AA 42: *cf.* **49** — *Evangelista* 637: **36a**; 698: **36a**; 902: **33a**; 924: **36a**; 980: **33a**;

990: **52a**; 1008: **44**; 1014: **33a** — *Evangelista & Arsat* 965: **36a**; 969: **52a** — *Everett* FB 7264: **52a**; FRI 13557: **52a**; FRI 13592: **14a**; FRI 13614: **49**; FRI 14024: **11**; FRI 14126: **36a**; FRI 14164: **12**; FRI 14184: **54**; FRI 14398: **52a**; KEP 104908 B: **33a**; KEP 104938: **11** — *Evrard* 865: **21** —

Faber 97: **33b** — *Fabia* SAN 25755: **44** — *Fairchild* 315: **23** — *Falconer* 115: **21** — *Fan & Li* 129: **24a** — *Farodo* & collectors BSIP 11240: **18b**; BSIP 12030: **18a** — FB series 30: **21**; 116: **5a**; 167: **16a**; 547: **16a**; 728: **33a**; 754: **5a**; 756: **5a**; 987: **33a**; 1013: **21**; 1111: **21**; 1778: **5a**; 1783: **21**; 2058: **33a**; 2245: **21**; 2333: **21**; 2642: **15**; 2739: **16a**; 2775: **15**; 2884: **21**; 3034: **33a**; 3269: **16a**; 3935: **49**; 4025: **52a**; 5150: **21**; 5225: **52a**; 5452: **16a**; 6115: **33a**; 6365: **5a**; 6560: **21**; 6646: **52a**; 6794: **49**; 7264: **52a**; 8820: **21**; 9604: **33a**; 9662: **33a**; 9848: **5a**; 10450: **15**; 10736: *cf.* **42**; 11291: **33a**; 11882: **33a**; 11949: **5a**; 11976: **21**; 12966: **16a**; 13212: **50**; 13340: **5a**; 14768: **5a**; 15153: **21**; 15389: **21**; 15481: **5a**; 15500: **33a**; 16617: **33a**; 17763: **30**; 18245: **21**; 18258: **5a**; 18499: **52a**; 18585: **5a**; 18675: **42**; 18791: **21**; 19142: **16a**; 19401: **52a**; 19575: **30**; 19802: **42**; 20495: **33a**; 21491: **33a**; 21857: **29**; 21930: **21**; 24109: **5a**; 24232: **42**; 24562: **16a**; 24611: **42**; 24779: **16a**; 24894: **42**; 25525: **52a**; 25562: **52a**; 26663: **52a**; 26692: **52a**; 27253: **33a**; 27293: **16a**; 27521: **5a**; 27664: **52a**; 27824: **42**; 28506: **16a**; 28527: **33a**; 28568: **33a**; 28574: **5a**; 28793: **16a**; 28868: **30**; 28874: **52a**; 29704: **52a**; 29811: **33a**; 29812: **21**; 30312: **42**; 30585: **31**; 30694: **15**; 30706: **30**; 30762: **52a**; 31203: **33a**; 31356: **33a**; 31452: **21**; 31474: **42** — *Fedilis Krispinus* SAN 82159: **36a**; SAN 89889: **52a**; SAN 94726: **36a**; SAN 94757: **52a**; SAN 94769: **36a**; SAN 94785: **36a**; SAN 94900: **52a**; SAN 95416: **49**; SAN 95567: **52a**; SAN 95893: **36a**; SAN 95909: **52a**; SAN 95972: **33a**; SAN 95997: **33a**; SAN 103575: **36a**; SAN 105274: **36a**; SAN 107063: **33a**; SAN 112987: **52a**; SAN 113290: **33a**; SAN 113392: **36a**; SAN 118583: **52a**; SAN 119482: **52a**; SAN 119553: **33a**; SAN 119611: **52a**; SAN 119631: **33a**; SAN 121888: **52a**; SAN 122133: **36a**; SAN 128482: **36a**; SAN 136141: **21**; SAN 136709: **33a** — *Fedilis Krispinus & Asik* SAN 96344: **21**; SAN 110914: **29**; SAN 113014: **52a**; SAN 113255: **46a** — *Fedilis Krispinus & Sumbing* SAN 88220: **36a**; SAN 88336: **36a**; SAN 88489: **33a**; SAN 88498: **33a**; SAN 88963: **36a**; SAN 89748: **36a**; SAN 91341: **33a**; SAN 91460: **36a**; SAN 91757: **29**; SAN 95695: **49** — *Fenix* 111: **33a**; BS 3656: **33a**; BS 4079: **52a**; BS 15147: **21**; BS 15644: **21**; BS 15665: *cf.* **33a**; BS 15731: **31**; BS 15762: **33a**; BS 26069: **21**; BS 28041: **5a**; BS 28183: **52a**; BS 30052: **33a** — *Fernandes* 320: **33a** — *Fernandez* FB 21491: **33a** — *Flemmich & Base* 32603: **11** — *Floto* 7324: **21** — *Floyd* 3471: **18b**; NGF 7260: **18a** — *Floyd & Versteegh* 5775: **45** — F.M.S. Mus. series 10717: **54**; 10757: **36a**; 10931: **39**; 11596: **14a** — *Forbes* 176: **40**; 451: **33a**; 471: **52a**; 482: **52a**; 617: **18b**; 690 or 696: **18b**; 733: **2**; 876: **18a**; 895: **51**; 955: **18b**; 1108: **51**; 1550: **40**; 1610: **40**; 1613: **40**; 1758: **40**; 1768 a: **40**; 1774: **33a**; 1972: **20**; 2451: **20**; 2519: **20**; 2602 a: *cf.* **33a**; 2837: **52a**; 2839: **54**; 2930: **33c**; 3130: **51**; 3180: **36a**; 3300: **49**; 3725: **33a**; 4077: **33a** — For. Dept. F.M.S. series 11550: **14a**; 15605: **44**; 18210: **14a**; 20461: **14a**; 22297: **36a**; 22427: **39**; 22505: **21** and **11**; 22709: **52a**; 23215: **11**; 23318: **33c**; 23719: *cf.* **44**; 23736: **54**; 23812: **14a**; 24142: **52a**; 24227: **36a**; 24417: **54**; 24449: **11**; 29882: **52a**; 33635: **12**; 37631: **36b**; 37641: **36b**; 39496: **14a** — *Foreman* NGF 52051: **18a** — *Foreman & Katik* NGF 48451: **18a** — *Forman* 253: **7**; 316: **4**; 317: **52a**; 335: **7**; 421: **33a**; 496: **36a**; 606: **33a** — *Forman & Blewett* 942: **56** — *Forster* 5944: **21** — *Forsyth* 244: **21** — *Fosberg* 37181: **24a**; 37356: **24a**; 37458: **33a**; 37648: **33a**; 37670: **24a**; 38069: **33a**; 38218: **33a**; 38342: **33a**; 38387: **33a**; 38444: **33a**; 43837:

33a; 43909: *cf.* **36a** — *Foster & Puasa-Angian* 3474: **29** — *Fox* PNH 4573: **33a**; 5022: **33c**; PNH 9828: **33a**; PNH 13337: **33a**; 14156: **33a** — *Foxworthy* 10: **33a**; 32: **11**; 471: **52a**; BS 789: **33a**; BS 808: *cf.* **33a**; BS 1625: **16a**; 1729: **33a** — *Frake* PNH 37987: **33a**; PNH 38024: **52a**; PNH 38059: **33a**; PNH 38075: **52a**; PNH 38372: **33a**; PNH 38400: *cf.* **29** — *Frake & Frake* PNH 36003: **5a** — *Franck* 457: *cf.* **1**; 507: **1**; 530: **33a** — *Franken & Roos* 185: **33a**; 190: **33a**; 308: **33a**; TFB 1683: **28** — *Fraser* 177: **52a**; 704: **52a** — *Free & Sumbing* SAN 79151: **52a** — *Fretes* 5518: **21**; 5566: **49** — FRI series 118: **54**; 151: **38**; 226: **11**; 228: **33a**; 328: **49**; 382: **33a**; 400: **37**; 453: **33a**; 630: **36a**; 697: **40**; 811: **37**; 896: **29**; 916: **11**; 1013: **11**; 1243: **14a**; 1736: **36a**; 1964: **33a**; 1979: **33a**; 2140: **14a**; 2156: **39**; 2503: **52a**; 2678: **55**; 2972: **40**; 3367: **54**; 3368: **36a**; 3450: **36a**; 3456: **14a**; 3459: **33a**; 3466: **54**; 3826: **33a**; 3830: **33a**; 3840: **33c**; 4009: **54**; 4071: **39**; 4076: **54**; 4136: **33c**; 4284: *cf.* **11**; 4308: **28**; 4410: **11**; 4547: **39**; 4760: **49**; 5031: **33a**; 5054: **33a**; 5308: **33c**; 5311: **33c**; 5314: **33c**; 5337: **39**; 5352: **14a**; 5490: **11**; 5707: **11**; 5989 : **12**; 6125: **40**; 6328: **14a**; 6416: **14a**; 6424: **33c**; 6706: **11**; 6767: **54**; 6771: **33a**; 6885: **14a**; 6903: **54**; 6908: **54**; 6915: **33a**; 6978: **54**; 7016: **33c**; 7088: **54**; 7101: **36b**; 7109: **14a**; 7124: **54**; 7214: **36b**; 7467: **37**; 7892: **38**; 8180: **36a**; 8589: **33c**; 8860: **33a**; 8907: **33c**; 10622: **33c**; 10930: **36b**; 10959: **36b**; 11069: **14a**; 11215: **54**; 11281: **54**; 11296: **11**; 11326: **14a**; 11350: **14a**; 11377: **33c**; 11396: **54**; 11455: **49**; 11583: **40**; 11717: **54**; 11739: **12**; 11796: **11**; 11837: **40**; 12184: **39**; 12317: **36a**; 12705: **36a**; 12818: *cf.* **51**; 13054: **14a**; 13184: **36a**; 13212: **54**; 13398: **49**; 13405: **14a**; 13426: **28**; 13453: **54**; 13512: **40**; 13557: **52a**; 13592: **14a**; 13614: **49**; 14024: **11**; 14126: **36a**; 14164: **11**; 14184: **54**; 14398: **52a**; 14629: **11**; 14830: **22**; 14893: **54**; 15205: **36a**; 15390: **33c**; 15876: **54**; 15979: **54**; 15992: **28**; 16143: **36a**; 16237: **14a**; 16288: **54**; 16290: **54**; 16460: **36a**; 17102: **14a**; 17275: **36a**; 17286: **14a**; 17295: **11**; 17397: **33a**; 17872: **33a**; 17974: **29**; 18376: **36a**; 18402: **49**; 19807: **14a**; 19937: **33a**; 20054: **33c**; 20069: **40**; 20073: **36a**; 20074: **54**; 20075: **52a**; 20265: **33c**; 20340: **39**; 20497: **33c**; 20571: **36b**; 20664: *cf.* **11**; 20757: **54**; 20979: **37**; 21642: **36a**; 21930: **33c**; 22041: *cf.* **37**; 22044: **36b**; 22046: **36b**; 22065: **39**; 23313: **52a**; 23342: **52a**; 23631: **29**; 23811: **33c**; 25012: **11**; 25081: **54**; 25116: **11**; 25233: **14a**; 25566: **36a**; 25644: **14a**; 26207: **14a**; 26851: **29**; 28169: **11**; 28198: **33a**; 28750: **39**; 29088: **14a**; 29343: **54**; 31289: **33a**; 31472: **39**; 31956: **33c**; 33668: **54**; 33987: **21**; 34230: **33c**; 34392: **54**; 34585: **21**; 34603: **33a**; 35002: **21**; 36280: **54**; 36290: **33c**; 36570: **14a**; 38311: **54**; 38594: **33a**; 39431: **54**; 39909: **36b**; 44632: **33a**; 44650: **54**; 98990: **36a**; 99149: **36a** — *Frodin* UPNG 501: **21**; UPNG 2010: **18a** — *Frodin & Hill* NGF 26330: *cf.* **18a**; NGF 26360: **18a** — *Frodin & Morren* 3112: **18a**; 3362 A: **41** — *Frodin, Morren & Gabir* 2347: **41**; 2377: **18a**; 2396: **41**; 2483: **41** — *Fukuoka* 11425: **24a**; 62027: **1**; 62065: **1**; 62238: **33a**; 62354: **33a**; 62525: **1** — *Fukoka & Na Nakhon* T 36098: **21**; T 36099: **33a** — *Fung* LU (Lingnan Univ.) 20102: **33a**; LU 20103: **33a**; LU 20157: **21** — *Furtado* SF 37905: **5a**; SF 37906: **5a**; SF 37907: **33a** —

Gachalian PNH 15526: **16a** — *Gaerlan et al.* PPI 2766: **52a**; PPI 2957: **52a**; PPI 3053: **30**; PPI 4509: **33a**; PPI 4571: **13**; PPI 4587: **33a**; PPI 4689: **13**; PPI 10351: **33a**; PPI 10382: **33a**; PPI 10386: **5a**; PPI 10465: **33a** — *Gafui* & collectors BSIP 8397: **18a**; BSIP 8934: **18a**; BSIP 10115: **18a**; BSIP 10186: **18a**; BSIP 10483: **18a**; BSIP 10496: **18a**; BSIP 10769: **18b**; BSIP 10830: **18a**; BSIP 11002: **18a**; BSIP 12691: **18b**; BSIP 12860: **18b**; BSIP 12982: **18a**; BSIP 16444: **18a**; BSIP 16995: **18a**; BSIP 17260: **18a**; BSIP 17497: **18a**; BSIP 17536: **18a**; BSIP 18453: **18a**; BSIP 18506: **18a**; BSIP 18537: **18a**; BSIP

18843: **18a**; BSIP 18966: **18a**; BSIP 18979: **18a** — *Galau* S 15647: **29** — *Gallatly* 176: **1** — *Galoengi* 210: **33a**; 257: **33a** — *Galore* NGF 17569: **18a** — *Gamble* 19: **33a**; 874: **21**; 9059: **21**; 11695: **33a**; 12816: **33a**; 14272: **33a**; 20557: **33a**; 20968: **21** — *Garcia et al.* PPI 15141: **16a** — *Garrett* 143: **54**; 232: **21**; 712: **5a**; 765: **33a**; 950: **33a**; 987: **47**; 1134: **5a**; 1330: **33a**; 1394: **1**; 1492: **47** — *Gaudichaud* 65 or 69: **54** — *Gaudichaud* in Herb. Wallich 327: **21** — *Gauth* 10698: **21** — *Geesink* 9182: **29**; 9269: **36b**; 9288: **33a**; 9322: **52a** — *Geesink et al.* 3348: **33a**; 4872: **55**; 4944: **36a**; 4985: **1**; 5019: **44**; 5029: **21**; 5166: **22**; 5289: *cf.* **20**; 5345: **36a**; 5358: **52a**; 5427: **54**; 5433: **33d**; 5472: **36a**; 5520: **1**; 5570: **55**; 5572: **33a** (L) and **55** (K); 5576: **55**; 5596: **21**; 5597: **21**; 5614: **5a**; 5672: **47**; 6075: **1**; 6193: **54**; 6195: **5a**; 6251: **33a**; 6555: **33a**; 6635: **44**; 6674: **55**; 6749: **44**; 6772: **1**; 6811: **33a**; 6894: **33b**; 6984: **47**; 7204: **54**; 7213: **36a**; 7214: **33a**; 7291: **33a**; 7361: **54**; 7399: **36a**; 7456: **33a**; 7528: **36a**; 7602: **1**; 7635: **22**; 7636: **22**; 7647B: **22**; 7927: **47** — *Gentry et al.* 66666: **21**; 66914: **37**; 67004: **14a**; 67121: *cf.* **39**; 67223: **39** — *George* 13629: **21** — *George et al.* S 42888: **36a**; S 42889: **36a**; SAN 117638: **44**; SAN 120951: **36a**; SAN 120970: **33a**; SAN 120988: **52a**; SAN 121784: **34**; SAN 123721: **36a** — *George & Amin K.* SAN 121262: **43** — *Gerus* KEP 99477: **49** — *Gibbs* 2790: **52a**; 2809: **52a** — *Gideon* LAE 57480: **10**; LAE 57542: **41**; LAE 76946: **25**; LAE 76970: **18a**; LAE 77029: **21**; LAE 78653: **18a** — *Gideon & Croft* LAE 76118: **18a** — *Gideon & Kainwaka* LAE 76936: **18a** — *Giesen* 34: **33a**; 145: **11**; 146: **11** — *Gillison* NGF 22024: **21**; NGF 22310: **18a**; NGF 22452: **18b**; NGF 22492: **18a**; NGF 25323: **18a** — *Glassman* 2686: **18a**; 2697: **18a** — *Godefroy-Lebeuf (comm.)* 332: **44** — *Gojar* FB 31474: **42** — *Goklin T.* 1306: **33a**; 2057: **33a**; 2090: **33a**; BNB For. Dept. 2157: **21**; BNB For. Dept. 2248: **21**; 2285: **29** — *Gomez* in Herb. Wallich 8578: *cf.* **33d** — *Goodenough* 410: **36a**; 1483: **44**; 1716: **44**; 1855: **54**; 1906: **36a** — *Goverse & Adriansyah Berau* 478: **52a** — *Govindarajalu* 9339: **33a** — *Graeffe* 1399: **18a**; 1592: **18a** — *Grasshoff* 833: **36a** — *Grierson & Long* 1535: **33a** — *Griffith* 1845: **14a**; 4922: **21**; Kew distr. no. 4923: **14a**; Kew distr. no. 4924: **1**; Kew distr. no. 4927: *cf.* **33d**; Kew distr. no. 4928: **37**; Kew distr. no. 4932: **21**; Kew distr. no. 4935: **33a**; Kew distr. no. 4941: **36a**; Kew distr. no. 4955: **11** — *Griffith* in Herb. Mergin 1109: **55** — *Griswold* 84: **29** — *Groenhart* 211: **33a** — *Groff* 6083: **47**; 6138: **21**; 6157: **1** — *Guard & Shaw* For. Dept. F.M.S. 23736: **54**; For. Dept. F.M.S. 23812: **14a** — *Guard & Syed Ali* For. Dept. F.M.S. 23719: *cf.* **44** — *Gusdorf* 68: **44**; 116: **54** — *Gutierrez* PNH 78036: **5a**; PNH 78089: **52a**; PNH 78159: **24a** — *Gutierrez et al.* 117037: **52a**; PNH 117087: **52a**; PNH 117297: **30**; PNH 117611: **52a**; PNH 117617: **30** — *Gwynne-Vaughan* 304: **33a**; 455: **21** —

Haegens 290: **33a** — *Haenke* 448: **21**; 449: **21**; 540: **30** — *Haines* 150: **21**; 3526: **21**; 3527: **21**; 5823: **21** — *Haji Suib* S 23461: **29** — *Halle* TFB 50: **49**; TFB 344: **33a** — *Hallier* 200a: **16a**; 620: *cf.* **36a**; 701: **33a**; 738: **33c**; 925: **33c**; 1126: **49**; 1237: **33a**; 1773: **36a**; 1830: *cf.* **36a**; 1901: **36a**; 2104: **33a**; 2574: **36a**; 2724: **56**; 2906: **43**; 4043: **16a**; 4200: **16a**; 4701: **33a** — *Hallier f.* 345: **5a** — *Hamid* 10574: **33a**; 10588: **36a**; 11578: **40** — *Hamid* (forester) C.P. 3860: **55** — *Hamzah* Boschproefstation 9393: **44** — *Hance* 1807: **33b** — *Haniff* SF 6926: **24a**; 10406: **55**; 13142: **44**; SF 14348: **21**; 14990: **52a**; SF 15525: **54**; SF 15922: **21**; 21011: **49** — *Haniff & Nur* SF 2094: **36a**; SF 2725: **36a**; 3904: **33a**; 4249: **55**; SF 4267: **55**; 4290: **44**; 4709: **21**; 8078: **33c**; 10166: **33c**; SF 10259: **37** — *Hansen* 257: **36b**; 898: **33a**; 1112: **29**; 1255: **11** — *Haniff & Lelai* 10456: **33a** — *Hansen & Smitinand* 10852: **54**; 11952: **54**; 37260: **22** — *Hara*

94/38*: **24a** — *Hardial* 351: **14a**; 650: **33a** — *Hardial Singh & Noor* 65: **54**; 73: **52a** — *Harmand* in Herb. Pierre 2909: **5a** — *Harsukh* 22485: **21** — *Hartley* 9853: **18a**; 9877: **21**; 10508: **18a**; 10604: **21**; 10761: **18a**; 10794: **18a**; 10810: **41**; 10818: **41**; 10874: *cf.* **45**; 10951: **45**; 10953: **18a**; 10968: **18a**; 10991: **18a**; 11012: **18a**; 11336: **18a**; 11366: **18a**; 11476: **18a**; 11525: **18a**; 11530: **18a**; 11589: **18a**; 11908: **18a**; 11963: **18a**; 12261: **18a**; 12451: **18a**; 12582: *cf.* **19**; 13082: **21**; 13305: **21** — *Harvey* BNB For. Dept. 10127: **36a** — *Hasan & Nurta* 80: **51**; 96: **51** — *Haslani Abdullah* 50: **11**; 66: **49**; 73: **49** — *Hassan* 780: **33a** — *Hassan & Kadim* 83: **11** — *Hasskarl* 13: **51**; 564: *cf.* **33a** — *Hasskarl* or *Bickbier* 395: **52a** — *Hatusima* 17388: **33a**; 17687: **33a**; 19118: **24a** — *Havel* NGF 9170: **18a** — *Havel & Kairo* NGF 11141: **18a**; NGF 17078: **35b** — *Haviland* 303: **33a**; 719: **33a**; 720: **56**; 725: **33a**; 726: **29**; 727: **33a**; 793: **33a**; 794: **33a**; 1337: **49**; 1374: **56**; 1731: **40**; 1737: **40**; 1762: **33a** and **14a**; 1844: **29**; 1889/1327: **49**; 1900: **49**; 2033/1537: **33a**; 2186: **52b**; 2221/1728: **52a**; 3102: **33a**; 3104: **33a**; 3105: **33a**; 3263: **29**; 5002: **52a** — *Haviland & Hose* 720: **56**; 730: **29**; 3104: **33a**; 3247: **52a**; 3251: **11**; 3262: **11**; 3264: **29**; 3265: **33a**; 3664: **49**; 3668: **49**; 3671: **11**; 3674: *cf.* **33a**; 3675: **33a** — *Hayata* 127: **33a** — *Heaslett* 8: **52a** — *Helfer* 18: **21**; 19: **33d**; Kew distr. no. 4922: **21**; Kew distr. no. 4942: **22**; Kew distr. no. 4944: **54**; Kew distr. no. 4945: **55**; Kew distr. no. 4946: **54**; Kew distr. no. 4947: **33d** and **54** — *Henderson* 327: **18a**; 10423: **14a**; 10477: **14a**; 10556: **36a**; F.M.S. Mus. 10717: **54**; F.M.S. Mus.10757: **36a**; 10758: **52a**; F.M.S. Mus. 10931: **39**; 11035: **14a**; 11036: **14a**; 11183: **33a**; F.M.S. Mus. 11596: **14a**; 18214: **33a**; SF 18230: **33a**; 18352: **54**; 18379: **54**; SF 19571: **33c**; 20130: **54**; SF 20373: **29**; SF 21713: **38**; SF 21722: **33a**; SF 21767: **54**; SF 21786: **54**; SF 21787: **52a**; SF 21875: **52a**; SF 21986: **29**; SF 21991: **29**; SF 22150: **33c**; SF 22212: **36a**; SF 22348: **52a**; SF 22353: **52a**; SF 22390: **33a**; SF 22591: **33c**; SF 23011: **24a**; SF 23660: **52a**; SF 23816: **21**; SF 24060: **44**; SF 25019: **33a**; 25092: **24a**; 29515: **54**; SF 35755: **11**; SF 38956: **21** — *Henry* 780: **33a**; 1144: **33a**; 1885: **33a**; 8099: **21**; 8100: **21**; 8371: **21**; 8488: **21**; 9530: **33b**; 9530 A: **33b**; 9530 B: **33b**; 13032: **1**; 13667: **33a**; 13667 A: **33a**; 13688: **33a** — *Henty* NGF 9809: **21**; 11543: **18b**; NGF 11594: **21**; NGF 11601: **18a**; NGF 12418: **18a**; NGF 14355: **18a**; NGF 14853: **18a**; NGF 16726: **18a**; NGF 16729: **18a**; NGF 28061: **18a**; NGF 29154: **5a**; NGF 29385: **18a**; NGF 38521: **25**; NGF 38595: **18a**; NGF 49676: **48** — *Henty & Coode* NGF 29180: **19** — *Henty & Katik* NGF 38729: **21**; NGF 42936: **9** — *Henty & Lelean* NGF 41887: **25**; NGF 41900: **18a** — *Henty, Ridsdale & Galore* NGF 31867: **18a**; NGF 33202: **35a** — *Herbst* 680: *cf.* **5a** — *Hernaez* 1406: **15** — *Herre* 176: **18a**; 1062: **21**; 1167: **33a** — *Hewitt* 8: **36b**; 17: **56**; 19: **49**; 314: **29**; 498: **29** — *Heyligers* 1608: **21** — *Heyne* 1374: **1**; 7282 F: **1** — *Hillebrand* 1889: **5a** and **21** — *Hochreutiner* 772: **51**; 1114: **5a**; 2530: **33a**; 2589: **5a** — *Hoed & Kostermans* 885: **47** — *Hoeft* 2140: **18b** — *Hoffman?* 4347: **18a** — *Hoffmann* 3: **5a**; 4: **5a**; 5: **37**; 6: **37**; 8: *cf.* **38**; 9: **37**; 13: **49**; 14: **33a** — *Hohenacker* Pl. Indiae or. (Terr. Canara.) 167: **1**; 459 a: **33a** — *Hole* 660: **21** — *Hollrung* 734: **18a**; 757: **18b**; 766: **18a**; 882: **18a** — *Holmberg* 852: **54** — *Holttum* 9396: **36a**; SF 9455: **33a**; SF 9607: **14a**; 9620: **54**; SF 9839: **54**; SF10640: **52a**; SF 10978: **52a**; SF 15202: **33a**; SF 15328: **33c**; SF 18094: **36a**; 19832: **21**; SF 20821: **33c** — *Hommel* 109 d: **54** — *Hoogerwerf* 19: **21**; 33: **48**; 57: **21**; 73: **48**; 112: **21** — *Hoogland* 3889: **21**; 4528: **18a**; 4621: *cf.* **25**; 4712: **18a**; 4714: **21**; 4783: **18b**; 8829: **35a**; 8977: **18b** — *Hoogland & Craven* 10179: **18b**; 10381: **18a** — *Hoogland & Macdonald* 3457: **21** — *Hoogland & Pullen* 5247: **35a** — *Hoogland & Taylor* 3734: **18a** — *Hooker* 36: **21**; 53: **33a**; 57: **33a**; 2555: **21** — *Hooker*

& Thompson 229: **21** — *Horgen et al.* 389: *cf.* **52a** — *Horsfield* 165: **21**; *Antid* 3: **33a**; *Antid* 4: **33a** — *Hose* 69: **11**; 139: **11**; 297: **36a**; 317: **56**; 382: **52a**; 391: **33a**; 469: **33a**; 549: **36b**; 758: **56**; C F Field No 4901: **14a** — *Hosokawa* 9470: **18a** — *Hotta* 12637: **36a**; 13539: **56**; 13838: **3**; 14295: **36a** — *Hou* 343: **11**; 645: **36a** — *How* 70549: **21**; 70973: **5a**; 71576: **33a**; 72748: **33a**; 73029: **33a**; 73679: *cf.* **24a** — *Howroyd* SAN 29353: **36a**; SAN 29362: **36a** — *Hu* 5976: **21**; 5977: **21**; 8366: **21**; 12207: **21**; 12917: **21** — *Huc* 124: **49**; 189: **33a**; 569: **36a**; 584: **14a**; TFB 1352: **11** — *Hullett* 359: **33a**; 629: **54**; 861: **54** — *Hume* 7251b: **33a**; 7433: **21**; 7523: **14a**; 7635: **33a**; 8192: **52a**; 8431: **33a**; 8476: **36a**; 8604: **11**; 9172: **14a**; 9200: **49**; 9355: **54** — *Huq* 10481: **33c** — *Huq & Mia* 10280: *cf.* **33a** — *Hutchinson* FB 3935: **49**; FB 4025: **52a**; FB 6115: **33a**; FB 6560: **21** — *Hyland* 7819: **21**; 8563: **21**; 9143: **21**; 10260: **18a**; 10265: **18a**; 11538: **18a** —

Iboet 14: **51**; 62: **40**; 119: **21**; 135: **51**; 346: **36a**; 367: **33a**; 450: **33a** — *Idjan & Mochtar* 382: **21** — *Ijiri & Niimura* 624: **10** — *Ilias & Azahari* S 35706: **36b** — *Ilias & Yeo* S 38337: **36b** — *Ilias Paie* S 13602: **29**; S 16373: **36b**; S 16605: **36a**; S 16628: **36b**; S 19350: **56**; S 19862: **29**; S 22942: **29**; S 24261: **52a**; S 24949: **14a**; S 25297: **52a**; S 25850: **52a**; S 26027: **36a**; S 26340: **36b**; S 26949: **52a**; S 27933: **36a**; S 28004: **36a**; S 28020: **29**; S 36278: **29**; S 36975: **40**; S 39006: **33a**; S 39025: **36a**; S 40982: **46a**; S 42496: **40**; S 42707: **40**; S 42728: **33a**; S 45137: **36c** — *Ilias Paie & Jugah ak Kudi* S 35980: **52a**; S 38690: **11** — *Inagat* 22484: **21**; 25988: **21** — *Indir Alam* Boschproefstation bb 14007: **5a** — *Inpirat* 9738: **21** — *Ismael* 70: *cf.* **33a** — *Ismail* 28129: **39** — *Iwatsuki & Fukuoka* T 10300: **1** — *Iwatsuki et al.* S 59: **33a**; S 60: **33a**; A 83: **49**; A 89: **49**; A 197: **49**; 252: **33a**; 301: **24a**; B 7158: **33a**; 8447: **54**; T 10956: **1**; T 14543: **54** —

Ja series 2515: **33a**; 2807: **5a**; 6534: **51** — *Ja'amat* KEP 35928: **33c**; KEP 36597: **14a**; 39260: 52a — *Jacob* 17625: **33a** — *Jacobs* 4531: **51**; 4616: **33c**; 5685: **11**; 7641: **30**; 7916: **15**; 7989: **15**; 7991: **15**; 8417: **51**; 8445: **40**; 8456: **52a**; 8505: **33a**; 8714 -/A: **10**; 9005: **10**; 9490: **41**; 9587: **18a**; 9647: **18a**; 9674: **18a** — *Jacobson* 2425: **11**; 2428: **11** — *Jagor* 303: **14a** — Jaheri Exped. Nieuwenhuis 16: **33a**; 72: *cf.* **36a**; 133: **33a**; 149: **52a**; 171: **56**; 248: *cf.* **40**; 371: **49**; 674: **29**; 743: **56**; 812: **36a**; 911: **29**; 1257: **56**; 1456: *cf.* **29**; 1544: **49**; 1548: **56**; 1569: **49**; 1609: **36a**; 1716: **33a**; 1893: **33a** — *Jahluhur* 47: **33a** — *Jamatdom Awang* 18231: **14a** — *James Ahwing* SAN 33988: **36a** — *James D.* S 29840: **52a** — *James D. Mamit* S 34390: **33a**; S 35155: **49**; S 37662: **33a** — *James D. Mamit et al.* S 34438: **49**; S 34638: **52a**; S 35080: **52a** — *Jaray* 18: **21**; 51: **33a** — *Jarvie* 5400: **49** — *Jarvie & Ruskandi* 5455: **52a**; 5738: **29**; 5862: **52a**; 6009: **36a**; 6187: **29** — *Jaswir Singh* SAN 24177: **56**; SAN 30704: **52a** — *Jibrin Sibil* 133: **33a** — *Jimpin S.* SAN 118576: **56** — *Johansson, Nybom & Riebe* 62: **49**; 123: **49**; 385: **46a**; 386: **7** — *Johns* 6605: **36a**; 6730: **36a**; 7478: **52a**; 7638: **36b** — *Johns & Sands* 6804: **49** — *Johns et al.* 7052: **49** — *Joseph, Madilil & Ahad* SAN 116947: **36a** — *Joseph B. et al.* SAN 122469: **33a**; SAN 124080: **21**; SAN 124090: **21** — *Joseph Radin* SAN 102213: **33a** — *Jugah ak Kudi* S 23770: **52a** — *Jugah ak Kudi, Jegong & Johnny* S 57625: **29** — *Jugah ak Kudi & Sangat* S 32717: **11** — *Julius et al.* SAN 131012: **43** — *Jumatin & Tuyuk* SAN 92458: **33a** — *Junghuhn* 87: **51**; 89: **51**; 94 or 911: **51**; 106: **51**; 109: **51**; 220: **51**; 875: **51** — *Jusimin Duaneh* 293: **29**; 394: **36a** —

Kadir & Enggoh BNB For. Dept. 10329: **36a** — *Kadir b. Abdul* SAN 16888: **36a** — *Kairo & Streimann* NGF 35615: **18a**; NGF 35720: **35a** — *Kajang* For. Dept. F.M.S. 18201: **14a** — *Kajewski* 2208: **18a**; 2308: **18a**; 2348: **18a**; 2398: **18a**; 2450: **18a**; 2468: **18b**; 2569: **18a**; 2593: **18a** — *Kalkman* BW 3461: **45**; 4085: **35a** — *Kalong* For. Dept. F.M.S. 20270: **54**; For. Dept. F.M.S. 20461: **14a**; For. Dept. F.M.S. 22427: **39** — *Kalshoven* 25: **5a** — *Kamarudin S.* FRI 28750: **39**; FRI 31289: **33a**; FRI 31472: **39**; FRI 34585: **21**; FRI 34603: **33a** — *Kambira* SAN 117538: **56**; SAN 117539: **33a** — *Kamis* 4283: **44**; 4758: **36a** — *Kanehira* 1325: **18a**; 1433: **18a**; 2481: **21**; 2528: **52a**; 2629: **52a** — *Kanehira & Hatusima* 11526: **45**; 11699: **10**; 12057: **18a**; 12637: **41**; 12836: *cf.* **41** — *Kanis* 1030: **9** — *Kanis & Ding Hou* SAN 49330: **36a** — *Karim Momin* SAN 78115: **36b** — *Karim & Shah* 43: **54** — *Karta* 48: **33a**; 211: **33a**; 237: **33a**; 271: **33a** — *Kartawinata* 251: **5a**; 681: **36a**; 758: **52a**; 817: **52a**; 1176: **11**; 1180: **36a**; 1193: **36a**; 1261: **29**; 1411: **33a** — *Kasim* 172: **55**; 743: **33a** — *Kasim*, AR & ZAI (281) 914: **33a**; 1990: **14a** — *Katik* NGF 46964: **18a**; LAE 62107: **45**; LAE 62148: **25**; LAE 70798: **18a** — *Katik et al.* LAE 70851: **18a**; LAE 70937: **18a**; LAE 74727: **18a** — *Katik & Henty* LAE 74792: **18a** — *Kato, Okamoto & Walujo* B 9042: **52a**; B 9125: **36a**; B 10191: **36b**; B 10493: **36b** — *Kato, Okamoto & Ueda* B 11736: **18a** — *Kato, Sunarno & Akiyama* C 3360: **2** — *Kato et al.* C 2033: **49**; C 2050: **49**; C 6438: **49** — *Kato & Wiriadinata* B 5310: **29**; B 5316: **36a**; B 6136: **29** — *Keenan, Tun Aung & Rule* 937: **54**; 960: **54**; 1173: **54**; 1417: *cf.* **33d**; 1420: **54** — *Keith* 4522: **36a**; 6229: **36a** — *Keng* 1349: **24a** — *Keng et al.* 11: **33c**; 48: **52a**; 99: **12**; 6117: **33a** — *Kenneally* 8158: **21**; 8598: **21**; 8633: **21**; 8635: **21**; 8780: **21** — KEP series 24926: **14a**; 29792: **14a**; 34633: **28**; 35928: **33c**; 36597: **14a**; 36830: **52a**; 36878: **29**; 36881: **29**; 36930: **44**; 38600: **33a**; 41459: **33a**; 48859: **33a**; 55036: **21**; 71037: **33a**; 71948: **33c**; 77826: **54**; 78634: **12**; 79142: **14a**; 80181: **36a**; 80307: **52a**; 80933: **28**; 93138: **11**; 94034: **11**; 94399: **52a**; 98022: **11**; 98036: **11**; 98378: **14a**; 99477: **49**; 99581: **11**; 104385: **14a**; 104417: **54**; 104564: **21**; 104806: **54**; 104908B: **33a**; 104938: **11**; 108851: **54**; 108855: **54**; 115971: **54** — *Kere* BSIP 4970: **18a**; BSIP 5002: **18a** — *Kerenga* LAE 56505: **41**; LAE 73837: **18a** — *Kerenga & Lelean* LAE 73950: **2** — *Kerenga & Symon* LAE 56821: **8** — *Kern* 8475: **33a** — *Kerr* 618: **33a**; 618 A: **33a**; 676: **47**; 676 A: **47**; 1115: **5a**; 1117: **5a**; 1253: **5a**; 2060: **33a**; 2429: **54**; 3750: **21**; 4134: **33a**; 4135: **33a**; 4140: **21**; 4143: **33a**; 4144: **1**; 4146: **1**; 4148: **33a**; 4149: **44**; 4210: **21**; 4264: **21**; 5039: **54**; 5279: **5a**; 5460: **33a**; 5631: **33a**; 5649: **47**; 5742: **47**; 5845: **44**; 5853: **33a**; 5865: **44**; 6049: **33a**; 6097: **33a**; 6283: **5a**; 6466: **33a**; 6825: **55**; 6875: **44**; 6969: **27**; 7217: **54**; 7313: **33a**; 8636: **33b**; 8636 A: **33b**; 8783: **33a**; 9673: **27**; 10352: **33a**; 10791: **24a**; 10854: **24a**; 10912: **21**; 11995: **54**; 12165: *cf.* **52a**; 12190: **28**; 12211: **54**; 12221: **36a**; 12588: **33a**; 12757: **33a**; 13019: **33a**; 13634: **21**; 13985: **54**; 14486: **36a**; 14503: **37**; 14592: **33a**; 14820: **55**; 14870: **24a**; 15341: **24a**; 15375: **33a**; 15378: **33a**; 15467: **22**; 15471: **54**; 15581: **33d**; 15626: **33a**; 16003: **37**; 16061: **33a**; 16516: **28**; 16882: **54**; 16884: **36a**; 16923: **22**; 16928: **22**; 16955: **22**; 17467: **54**; 17468: **33d**; 17884: **33a**; 17891: **28**; 18006: **24a**; 18012: **27**; 18076: **27**; 18334: **21**; 18523: **54**; 18621: **36a**; 18703: *cf.* **22**; 18778: *cf.* **52a**; 18796: **44**; 18821: **44**; 18852: **55**; 18927: **33a**; 19055: **44**; 19147: **22**; 19548: **1**; 19821: **1**; 21262: **33a**; 21392: **5a**; 21470: **1**; 21485: **33a**; 21595: **24a**; 21677: **54**; 21684: **33a** — *Keßler* 1564: **33a** — *Keßler et al.* Berau 12: **52a**; Berau 16: **52a**; Berau 92: **36a**; Berau 109: **52a**; Berau 146: **52a**; Berau 268: **29**; 1466: **36a**; 1705: **33a**; 1731: **21**; 1787: **21**; 1790: **21**; 1886: **33a**; 1906: **21**; 2125: **36a**; 2207: **33a**; 2282: **29** — Kew distr. no. 1341: **14a**; 1342: **11**; 1346: **11**; 2655: **1**; 2655/1: **21**; 2655/2: **47**; 4922: **21**; 4923: **14a**; 4924:

1; 4927: **55**; and *cf.* **33d**; 4928: **37**; 4932: **21**; 4935: **33a**; 4941: **36a**; 4942: **22**; 4944: **54**; 4945: **55**; 4946: **54**; 4947: **33d** and **54**; 4955: **11** — *Khairuddin* FRI 38311: **54** — *Khairuddin & Damanhuri* FRI 31956: **33c** — *Khan, Huq & Rahman* K 5467: **1** — *Khart* 13271: **54** — *Khoon Winit* 632: **44** — *Kiah* S 240: **33a**; S 294: **21**; 517: **33a**; SF 23939: **33c**; 24276: **54**; SF 24284: **24a**; 24302: **54**; SF 24304: **24a**; 24392: **54**; SF 31908: **39**; SF 32155: **38**; SF 32361: **36a**; SF 35019: **54**; SF 35033: **37**; SF 35136: **33a**; SF 35150: **54**; SF 35191: **11**; SF 35233: **24a**; SF 37228: **54**; SF 38964: **33a** — *Kiah & Strugnell* SF 23998: **54** — *Kiladja*: 104: **52a** — *King & Darrow* 5605: **1** — King's Collector 143: **21**; 350: **14a**; 395: **21**; 402: **21**; 574: **21**; 759: **28**; 844: **52a**; 994: **52a**; 1429: **33a**; 1439: **21**; 1440: **21**; 1503: **28**; 1524: **54**; 1565: **33a**; 1666: **54**; 1685: **33a**; 1934: **38**; 2031: **33a**; 2178: **38**; 2211: **36a**; 2300: **52a**; 2314: **14a**; 2378: **40**; 2804: **52a**; 2831: **54**; 2896: **33a**; 3029: **33a**; 3073: **36a**; 3460: **40**; 3464: **14a**; 3778: **38**; 3845: **28**; 3865 *or* 3065: **40**; 3928: **52a**; 3936: **11**; 4046: **33a**; 4056: **22**; 4212: **11**; 4300: **33a**; 4447: **40**; 4466: **40**; 4535: **52a**; 4761: **52a**; 4808: **52a**; 5010: **36a**; 5372: **11**; 5422: **11**; 5509: **11**; 5553: **38**; 5598: **11**; 5637: **22**; 5969: **21**; 7601: **54**; 8329: **52a**; 8336: **36a**; 8394: **11**; 8430: **36a**; 8470: **11**; 8556: **36a**; 8626: **28**; 10840: **33a**; 10961: **36a** — *Kingdon Ward* 8881: **33b**; 22165: **5a** — *Kinsun Bakia* 25: **33a**; 285: **33a**; 471: **21**; 510: **33a**; 539: **33a** — *Kinted* SAN 15870: **11** — *Kirkup* 268: *cf.* **36a**; 723: **52a**; 759: **52a** — *Kjellberg* 3947: **7**; 4132: **7** — KL series 1015: **54**; 1304: **52a**; 1307: **11**; 1461: **54**; 1462: **14a**; 2621: **14a**; 2686: **21**; 2839: **37**; 2943: **21**; 3009: **14a**; 3142: **33c**; 3146: **54**; 3153: **39**; 3527: **14a**; 3609: **33a**; 3961: **14a**; 4008: **21** — *Klemme FB* 6646: **52a**; FB 11291: **33a** — *Kloss* 6687: **55**; 6941: **55**; 14462: **51**; SF 18686: **36a**; 18974: **52a**; SF 19012: **52a** — *Kloss & Robinson* 6831: **54**; 6897: **54** — *Koch* 78: **10** — *Kochummen* FRI 2140: **14a**; FRI 2156: **39**; 2271: **28**; FRI 2503: **52a**; FRI 2678: **55**; FRI 2972: **40**; FRI 11455: **49**; FRI 16143: **36a**; FRI 16237: **14a**; FRI 16288: **54**; FRI 16290: **54**; FRI 16460: **36a**; FRI 18376: **36a**; FRI 18402: **49**; FRI 26207: **14a**; FRI 29088: **14a**; FRI 29343: **54**; KEP 77826: **54**; KEP 78634: **12**; KEP 79142: **14a**; KEP 80933: **28**; KEP 93138: **11**; KEP 94034: **11**; KEP 94399: **52a** — *Kodoh & Tuyuk* SAN 83642: **34** — *Koelz* 18941: **1**; 19174: **21**; 25131: **1**; 26156: **33a**; 28180: **1** — *Kohyama, Tukirin & Yamada* 10596: **33a** — *Koie & Olsen* 1497: **18a** — *Kokawa* 6245: **36a**; 6360: **11** — *Kokawa & Hotta* 72: **52a**; 394: **49**; 1126: **33a**; 5563: **36b**; 5973: **52a** — *Kollmann* 707: **33a** — *Kondo & Edano* PNH 36550: **33a** — *Konta et al.* T 29544: **33a**; T 29619: **33a** — *Kooper* 730: **21**; 1406: **33a**; 1817: **51**; 1844: **51** — *Koorders* 1814 b: **5a**; 1815 b: **5a**; 1818 b: **5a**; 1820 b: **5a**; 1821 b: **5a**; 1823 b: **5a**; 1824 b: **5a**; 1825 b: **5a**; 1826 b: **51**; 1828 b: **51**; 1829 b: **51**; 1830 b: **51**; 1833 b: **51**; 1835 b: **51**; 1836 b: **51**; 1837 b: **54**; 1840 b: **33a**; 1841 b: **5a**; 1842 b: **32**; 1843 b: **51**; 1844 b: **54**; 1845 b: **52a**; 1846 b: **5a**; 1847 b: **21**; 1848 b: **33a**; 1849 b: **33a**; 1851 b: **33a**; 1852 b: **51**; 1855 b: **51**; 1856 b: **21**; 1858 b: **51**; 1859 b: **33a**; 1860 b: **5a**; 1861 b: **51**; 1862 b: **54**; 1863 b: **5a**; 1864 b: *cf.* **5a**; 1865 b: **21**; 1866 b: **5a**; 1867 b: **5a**; 1868 b: **33a**; 1869 b: **33a**; 1871 b: **33a**; 1872 b: **5a**; 2253 b: **21**; 2307 b: **21**; 2308 b: **21**; 2310 b: **21**; 2311 b: **21**; 5390 b: **54**; 9920 b: **51**; 9949 b: **33a**; 10011 b: **5a**; 10092 b: **51**; 10302 b: **33a**; 10303 b: **49**; 10953 b: **5a**; 11170 b: *cf.* **54**; 11171 b: **51**; 11438 b: **5a**; 11440 b: **21**; 11445 b: **5a**; 11446 b: **21**; 11447 b: **21**; 11449 b: **33a**; 11450 b: **33a**; 11458 b: **21**; 12411 b: **5a**; 12739 b: **5a**; 12908 b: **33a**; 13094 b: **5a**; 13581 b: **51**; 14182 b: **5a**; 14413 b: **5a**; 14415 b: **51**; 14815 b: **21**; 14824 b: **33a**; 15584 b: **51**; 16790 b: **7**; 16791 b: **52a**; 16792 b: **7**; 16793 b: **7**; 16795 b: **7**; 16797 b: **17**; 16799 b: **7**; 16800 b: **7**; 16854 b: **52a**; 19851/13851 b: **51**; 20496 b: **51**; 20585 b: **5a**; 20591 b: **51**; 21032 b: *cf.* **5a**; 21753 b: **33a**; 21806 b: **33a**; 22671 b: **5a**; 22721 b: **33a**; 22927 b: **33a**; 22946 b: **33a**; 23065 b:

33a; 24026 b: **33a**; 24195 b: **54**; 24197 b: **32**; 24205 b: **54**; 24215 b: **51**; 24220 b: **32**; 24541 b: **5a**; 24665 b: **5a**; 24681: **5a**; 25041 b: **21**; 25222 b: **21** and **33a**; 25235 b: **5a**; 25543 b: **51**; 25662 b: **32**; 25766 b: **54**; 25915 b: **51**; 26113 b: **33a**; 26885: **49**; 26887 b: **5a**; 26892 b: **54**; 27075 b: **51**; 27123 b: **5a**; 27370 b: **33a**; 27374 b: **5a**; 27731 b: **51**; 27741 b: **5a**; 28098 b: **5a**; 28100 b: **33a**; 28101 b: **5a**; 28262 b: **5a**; 28268 b: **5a**; 28914 b: **33a**; 28933 b: **33a**; 28977 b: **1**; 29225 b: **5a**; 29226 b: **51**; 30013 b: **33a**; 30030 b: **5a**; 30168 b: **33a**; 30293 b: **23**; 30571 b: **5a**; 30572 b: **51**; 30574 b: **33a**; 30576 b: **52a**; 30905 b: **49**; 30983 b: **5a**; 31003 b: **52a**; 31100 b: **5a**; 32146 b: **51**; 32896 b: **32**; 32910 b: **54**; 33431 b: **52a**; 33435 b: **52a**; 33503 b: **5a**; 33529 b: **33a**; 33537 b: **1**; 33542 b: **1**; 34518 b: **23**; 36773 b: **21**; 37073 b: **33a**; 38103 b: **51**; 38110 b: **51**; 38580 b: **5a**; 38690 b: **51**; 38808 b: **5a**; 39179 b: **33a**; 39270 b: **21**; 39276: **49**; 41566 b: **32**; 41711 b: **32**; 44009 b: **51** — *Kooy* 349: **50**; 356: **21** — *Kornassi* Exped. Rutten 1326: **21**; 1333: **5a**; 1412: **49** — *Korthals* 3: **52a**; 7: **52a**; 158?: **33a**; 507: **21**; 610: **1** — *Koster* BW 6740: *cf.* **18b** — *Kostermans* 5: **51**; 22: **32**; 26: *cf.* **29**; UNESCO 47: **5a**; UNESCO 123: **33a**; S 142: **14a**; UNESCO 151: **54**; 283: **54**; 624: **18a**; 683a: **18a**; 802: **33a**; 805: **55**; 973: **18a**; 1523: **18a**; 1658: **18a**; 4118: **33a**; 4322: **36a**; 4464 a: **36a**; 4503: **36a**; 4511: **36a**; 4713: **33c**; 4754: **18a**; 5164: **56**; 5506: **33a**; 6204: **51**; 6233: **51**; 6291: **33a**; 6296: **5a**; 6797: **29**; 6836: **29**; 6863: **36a**; 6952: **29**; 7008: **33a**; 7632: **36a**; 7830: **18a**; 9612: **11**; 9623: **36a**; 9710: **36a**; 9783: **29**; 9987: **21**; 10408: **49**; 10553: **52a**; 10565: **52a**; 10728: **36a**; 10841: **36a**; 12545: **36a**; 12744: **49**; 12761: **52a**; 12996: **49**; 13010: **49**; 13150: **46a**; 13364: **40**; 13740: **36a**; 13788: **33a**; 14009: **52a**; 18191: **33a**; 18877: **21**; 19305: **33a**; 19344: **33a**; 19360: **54**; 21025: **33a**; 21074: **36a**; 21154: **33a**; 21164: **49**; 21393: **52a**; 21603: **52a**; 21659: **33a**; 21868: **54**; 22003: **52a**; 22018: **52a**; 22217: **21**; 23288: **5a**; 23867: **49**; 24853: **21**; 26037: **33a**; 26052: **33a**; 27406: **5a**; 27749: **5a**; 28216: **5a**; 28233: **5a**; bb 33491: *cf.* **18b**; Boschproefstation bb 34074: **11**; Boschproefstation bb 34085: **11** — *Kostermans & Anta* 567: **36a**; 689: **11**; 815: **11**; 863: **11**; 1033: **54** — *Kostermans & Den Hoed* 229: **1** — *Kostermans et al.* 409: **33a** — *Kostermans & Soegang* 77: **18a** — *Kostermans & Wirawan* 83: **33a**; 113: **5a**; 246: **5a**; 834: **33a** — *Kostermans & Van Woerden* 32: **23**; 66: **49**; 80: **54** — *Koyama* T 61132: **5a**; T 61197: **1** — *Koyama et al.* T 33738: **54**; T 33882: **54**; T 33923: **54** — *Koyama & Nantasan* 49876: **1** — *Kramadibrata* 204: **49** — *Kramer* 1: **32** — *Kuhl & Van Hasselt* 39: **54** — *Kulip* SAN 109547: **36a** — *Kulip et al.* SAN 133306: **34** — *Kumin Muroh* SAN 69377: *cf.* **52a** — *Kumul* NGF 36283: **21** — *Kunstler* 6: **54**; 9: **14a**; 123: **14a**; 3460: **40** — *Kuntze* 5160: **51**; 5812: **33a**; 6176: **33a** — *Kurata & Nakaike* 855: **24a**; 870: **24a**; 1914: **24a** — *Kurz* 1598: **1** — *Kuswata* 251: **5a** —

Lace 4231: **21**; 4617: **24a**; 4825: **21**; 5480: **47** — LAE series 51290: **18a**; 51739: **10**; 51901: **10**; 51912: **18a**; 52532: **18a**; 52609: **18a**; 53583: **10**; 53861: **18a**; 53993: **18a**; 55363: **21**; 56505: **41**; 56821: **8**; 57480: **10**; 57542: **41**; 58888: *cf.* **8**; 62107: **45**; 62148: **25**; 66011: **18a**; 66705: **18b**; 67750: **21**; 67927: **18a**; 68735: **18a**; 68885: **18a**; 70517: **18a**; 70798: **18a**; 70851: **18a**; 70937: **18a**; 73395: **35a**; 73837: **18a**; 73950: **2**; 74727: **18a**; 74792: **18a**; 75163: **18a**; 76118: **18a**; 76936: **18a**; 76946: **25**; 76970: **18a**; 77029: **21**; 78653: **18a** — *LaFrankie* 2106: **40**; 2163: **40**; 2434: **37**; 2533: **54**; 3032: **37**; 3229: **37**; 3368: **37** — *Lagrimas* PNH 34953: **52a**; PNH 41775: **52a** — *Lagrimas et al.* PNH 39434: **52a** — *Lajangah* SAN 32173: **36b**; SAN 32174: **33a**; SAN 32273: **56**; SAN 44561: **21** — *Lake & Kelsall* in Herb. Ridley 4019: **44**; 4020: **33c** — *Lake & Viebak* 4021: **33a** — *Lakshnakara* 235: **47**; 365: **21**; 397: **21**; 472: **5a**; 684: **29**; 794: **33c**; 899:

21 — *Lam* 660: **41**; 707: **10**; 2693: **52a**; 2785: **7**; 3298: **52a**; 3379: **7**; 3424: **21** — *Laman et al.* 70: **36a**; 95: **29**; 272: **29**; 605: **36b**; 607: **36a**; 623: **49**; 684: **29**; 697: **36b**; 947: **33a**; 954: **33a** — *Lambert & Brunson* 2: **21** — *Lamsudin* For. Dept. F.M.S. 23215: **11** — *Langlasse* 26: **16a**; 38: **33a**; 302: **14a** — *Lanjouw* 152: **33a** — *Larivita & Katik* LAE 70517: **18a** — *Larsen* 9464: **54**; 10599: **47** — *Larsen et al.* 40946: **33a**; 41216: **33a**; 41416: **33a**; 41659: **52a**; 41776: **33a**; 42132: **33a**; 42154: **33a**; 42703: **33a**; 42714: **54**; 42715: **33a**; 42767: **55**; 42797: **21**; 42800: **21**; 42830: **33a**; 42882: **33a**; 42941: **29**; 43086: **21**; 43122: **33d**; 43259: **55**; 43297: **33a**; 43476: **33a**; 43594: **54**; 44075: **24a**; 44153: **24a**; 44256: **33c**; 44257: **33a**; 44396: **47**; 44409: **47**; 44736: **1**; 45470: **24a**; 45988: *cf.* **22**; 46114: **1**; 46268: **33a**; 46384: **54** — *Larsen & Larsen* 31058: **33a**; 32629: **37**; 32685: **54**; 32842: **29**; 32909: **36a**; 33084: **21**; 33163: **28**; 33398: **36a**; 33497: **22**; 33908: **1**; 33929: **5a**; 34146: **21**; 34267: **1**; 70318: **33a** — *Larsen, Larsen, Nielsen & Santisuk* 30637: **55**; 30646: **33a**; 31342: **33b**; 31427: **47** — *Larsen, Santisuk & Warncke* 1951: **1**; 1984: **1**; 2080: **47**; 2127: **21**; 2345: **47**; 2637: **33a**; 2652: **1**; 2672: **1**; 2856: **5a**; 2989: **33a**; 3049: **47**; 3088: **33a**; 3118: **24b**; 3137: **24b**; 3293: **24a** — *Larsen, Smitinand & Warncke* 157: **33a**; 485: **24a**; 497: **47**; 843: **1** — *Lasquety* FB 31203: **33a** — *Lassan* SAN 70669: **44** — *Lassan & Ampon* SAN 71521: **52a** — *Lassan & Fox* SAN 72825: **52a** — *Latiff* 3835: **36a** — *Latiff et al.* 2863: **33a**; 933: **14a**; 4000: **54** — *Lau* 90: **21**; 132: **33a**; 393: **5a**; 397: **5a**; 498: **5a**; 3934: **24a**; 4482: **24a**; 20232: **5a**; 28228: **33b** — *Laumonier* 208: **33a**; TFB 178: *cf.* **20**; TFB 199: *cf.* **20**; 892: **49**; 1381: **51**; 1429: *cf.* **54**; TFB 1530: **49**; TFB 3498: **36a**; 6403: *cf.* **20**; 6589: *cf.* **20** — *Laumonier, Franken & Roos* TFB 1383: **52a** — *Lauterbach* 102: **18a**; 215: **5a**; 1434: **18a**; 3132: **18a**; 6390: **5a** — *Lawson* 107: **33a** — *Lazarides* 7853: **21**; 7914: **21** — *Leach* 3856: **21** — *Leavasa* 12: **35a** — *Ledermann* 6599: **18b**; 6982: **18b**; 7260: **45**; 7304: **45**; 7769: **18b**; 7980: **18a**; 9143: **18a**; 9280: **18a**; 9375: *cf.* **18b**; 9415: **18a**; 9547: **18a**; 9551: **18b**; 9800: **18b**; 9891: **18a**; 10022: **18a**; 10094: **45**; 10199: **45**; 10410: **18a** — *Lee* S 38119: **36b**; S 38226: **36b**; S 38617: **52a**; S 38854: **36b**; S 40603: **36c**; S 40658: **33a**; S 40676: **36a**; S 41986: **56**; S 41993: **36b**; S 43243: **36a**; S 43249: **33a**; S 44320: *cf.* **29**; S 44505: **36a**; S 45429: **40**; S 45503: **29**; S 52312: **52a**; S 52325: **29**; S 53823: **29**; S 54053: **29**; S 54200: **36a**; S 54590: **56**; S 54600: **40** — *Lee & Pius* SAN 119789: **36a** — *Leeuwenberg* 13238: **23**; 13244: **5a** — *Leeuwenberg & Rudjiman* 13096: **36a**; 13430: **36a** — *Lehmann* S 30140: **52a** — *Lei* 549: **21**; 788: **21** — *Leiberg* 6085: **16a** — *Leighton* 94: **29**; 298: **36a**; 759: **36a** — *Lelean & Stevens* LAE 51290: **18a** — *Lelean & Streimann* LAE 52532: **18a** — *Leopold Madani* SAN 33199: **44**; SAN 33231: **52a**; SAN 35045: **29**; SAN 35076: **11**; SAN 40458: **56**; SAN 63514: **52a**; SAN 81188: **52a**; SAN 81596: **49**; SAN 88719: **52a**; SAN 88850: **36a**; SAN 88880: **33a**; SAN 88890: **56**; SAN 89367: **36b**; SAN 90110: **56**; SAN 90756: **33a**; SAN 90788: **33a**; SAN 91120: **33a**; SAN 91680: **29**; SAN 91700: **29**; SAN 102183: **34**; SAN 124437: **56**; SAN 124456: **36a** — *Leopold Madani et al.* SAN 74312: **52a**; SAN 74383: **56**; SAN 108690: **46a**; SAN 114410: **36a**; SAN 114434: **36a**; SAN 132688: **36a**; SAN 133486: **36a** — *Leopold Madani & Amin* SAN 75364: **36a** — *Leopold Madani & Dewol Sundaling* SAN 60176: **33a**; SAN 60437: **36a**; SAN 74334: **33a** — *Leopold Madani & Ismail* SAN 108585: **29**; SAN 108587: **36a**; SAN 108907: **52a**; SAN 111279: **29**; SAN 111303: **36a** — *Leopold Madani & Kodoh* SAN 81369: **36a**; SAN 81453: **29** — *Leopold Madani & Petrus S.* SAN 92524: **33a**; SAN 92526: **56** — *Leopold Madani & Saikeh* SAN 82444: **52b**; SAN 82681: **33a** — *Leopold Madani & Taha* SAN 83516: **36b** — *Leschenault* 58: **1** — *Lesmy Tipot* FRI 33668: **54**; FRI 33987:

21 — *Lestari & Arifin* 14: **36a** — *Liang* 61735: **33a**; 61914: **21**; 66550: **21** — *Liborio Ela Ebalo* 871: **29**; 1040: **29** — *Lindley* 100: **21**; 168?: **33a** — Lingnan University 12059: **24a** — *Linsley Gressitt* 212: **24a**; 538: **24a**; 770: **21**; 798: **33a**; 923: **21**; 1136: **21**; 1262: **24a**; 1420: **24a**; 1542: **24a** — *Lipaqeto* BSIP 3351: **18b**; BSIP 3497: **18a**; BSIP 3502: **18b** — *Lobb* 225: **51**; 245: **51**; 248: **51**; 460: **33a**; 1125: **54** — *Loerzing* 5312: **52a**; 6383: **5a**; 6684: **52a**; 6856: **52a**; 11424: **51**; 13129: **21**; 14009: **5a**; 14438: **33a**; 14564: **33a**; 14738: **52a**; 15242: **52a**; 16122: **51**; 16884: **51**; 16885: **51** — *Loh* FRI 6885: **14a**; FRI 6978: **54**; FRI 13398: **49**; FRI 13405: **14a**; FRI 13426: **28**; FRI 13453: **54**; FRI 13512: **40**; FRI 17102: **14a**; FRI 17275: **36a**; FRI 17286: **14a**; FRI 17295: **11**; FRI 17397: **33a** — *Loher* 4640: **5a**; 4646: **16a**; 4647: **16a**; 4648: **16a**; 4650: **15**; 4651: **15**; 4652: **5a**; 4653: **33a**; 4654: **33a**; 4655: **33a**; 4656: **33a**; 4657: **21**; 4658: **21**; 4659: **21**; 4660: **21**; 4661: **21**; 4662: **21**; 4829: **16a**; 6817: **21**; 6853: **16a**; 6863: **16a**; 14915: **15**; 14945: **16a** — *Lomudin Tadong* 159: **46a**; 191: **33a**; 205: **36b**; 334: **33a**; 384: **36a**; 385: **36a**; 429: **36b**; 442: **33a**; 448: **52a**; 454: **36a**; 522: **52a**; 529: **49** — *Lorence Lugas* 231: **56**; 437: **56**; 469: **46a**; 1390: **56**; 1834: *cf.* **46a**; 1871: **36b**; 1974: **49**; 2063: **46a**; 2103: **46a**; 2161: **49**; 2337: **33a**; 2357: **56** — *Lowry et al.* 5245: 2 — *Luetjeharms* 3755: **52a**; 4037: **52a** — *Lumb* for *W Meijer* ALFB 106/87: **33a** —

Mabesa FB 24894: **42** — *MacGregor* 659: **21** — *Machado* 11556: **33c** — *Madulid et al.* PPI 6647: **21**; PNH 117785: **15**; PNH 117959: **30**; PNH 118196: **21** — *Madulid & Majaducon* 8359: **33a** — Maengkom Boschproefstation 7530: **7** — *Mahyar* 908: **56**; 972: **46a** — *Maidin* 1505: **52a**; 1561: **33a**; BNB For. Dept. 7327: **33a**; KEP 41459: **33a** — *Maikin et al.* SAN 129772: **36a**; SAN 133172: **36a** — *Main* (Exped. Polak) 93: **51**; 518: **10**; 2141: **33a**; 2190: **36a**; 2195: **3** — *Maingay* Kew distr. no. 1341: **14a**; Kew distr. no. 1342: **11**; 1343: **54**; 1344: **21**; Kew distr. no. 1346: **11**; 3057: **54** — *Mair* NGF 1862: **18b** — *Majaducon* 8480: **33a** — *Majawat & Lassan* SAN 88024: **52a** — *Maji* 1403: **21** — *Majumber & Islam* 123: **21** — *Mangold* BW 2166: **18b** — *Mansus* SAN 107898: **43**; SAN 117155: **56** — *Mansus & Aban Gibot* SAN 69245: **33a** — *Mansus et al.* SAN 122202: **36b** — *Mansus, Amin & Jarius* SAN 118902: **52a**; SAN 118905: **33a** — *Mansus, Lideh & Donggop* SAN 111841: **44** — *Mansus & Tuyuk* SAN 113500: **33a** — *Mansus, Tuyuk & Good* SAN 109221: **52a** — *Maradjo* (Ass. *W Meijer*) 92: *cf.* **33a** — *Marcal & Heya* SAN 86204: **11** — *Marcan* 142: **33a**; 181: **44**; 1196: **33a**; 1226: **33a**; 1291: **44**; 1300: **55**; 1387: **33a**; 2211: **21** — *Martin* 19: **21**; 977: **55**; S 36974: **40**; S 38007: **36a**; S 38158: **36b**; S 38173: **36b** — *Marvis* or *Maroris* 1172: **27** — *Mashiah & Meijer* SAN 141747: **52a** — *Maskuri* 184: **33a**; 212: **33a**; 280: **33a**; 785: **36a**; 790: **36a**; 791: **36a**; 795: **36a** — *Mat Asri* FRI 21642: **36a**; FRI 25566: **36a**; FRI 25644: **14a**; FRI 26851: **29** — *Matamin Rumutom* 60: **52a**; 118: **33a**; 121: **52a**; 143: **36a**; 145: **52a**; 159: **36a**; 265: **33a**; 371: **33a**; 426: **52a**; 493: **33a** — *Matin* SAN 106677: **36b** — *Matthew* 23359: **33a**; 41456: **33a** — *Maung* 5442: **21**; 13002: **54** — *Mauriasi & collectors* BSIP 8700: **18a**; BSIP 8807: **18a**; BSIP 9793: **18a**; BSIP 12415: **18a**; BSIP 13434: **18b**; BSIP 13508: **18b**; BSIP 13573: **18a**; BSIP 14307: **18a**; BSIP 15714: **18a**; BSIP 15727: **18a**; BSIP 15787: **18a**; BSIP 17849: **18a**; BSIP 18007: **18a**; BSIP 18034: **18a** — *Maxwell* 71-248: **28**; 71-255: **55**; 71-315: **21**; 74-309: **1**; 75-238: **1**; 75-470: **33a**; 75-473: **33a**; 75-549: **33a**; 75-557: **21**; 75-598: **47**; 75-614: **33a**; 75-788: **33a**; 75-892: **54**; 75-921: **24b**; 75-940: **24a**; 76-444: **5a**; 76-776: **14a**; 77-60: **11**; 77-368: **54**; 77-391: **14a**; 78-77: **14a**; 78-232: **14a**; 81-42: **11**; 81-60: **14a**; 81-110: **14a**; 81-248: **14a**; 82-223:

33a; 82-278: **11**; 82-284: **11**; 82-289: **14a**; 83-28: **5a**; 84-88: **54**; 84-192: **33a**; 84-223: **33a**; 84-315: **33a**; 84-386: **33a**; 84-489: **33a**; 85-103: **33a**; 85-276: **54**; 85-313: **28**; 85-373: **33a**; 85-423: **33a**; 85-443: **21**; 85-481: **33d**; 85-490: **33d**; 85-508: **36a**; 85-808: **47**; 85-852: **24a**; 85-927: **24a**; 85-1011: **33a**; 85-1047: **33d**; 85-1064: **36a**; 86-21: **28**; 86-209: **54**; 86-244: **55**; 86-285: **54**; 86-289: **54**; 86-338: **21**; 86-363: **24a**; 86-513: **54**; 86-883: **21**; 86-1103: **24a**; 87-454: **36a**; 87-535: **33a**; 87-560: **55**; 87-572: **33a**; 87-693: **1**; 87-758: **47**; 87-984: **5a**; 87-1081: **47**; 88-584: **47**; 88-642: **5a**; 88-696: **1**; 88-699: **1**; 88-801: **5a**; 89-435: **33a**; 89-711: **33a**; 89-728: **47**; 89-740: **33a**; 89-752: **33a**; 89-772: **21**; 89-808: **1**; 89-1055: **33a**; 89-1201: **1**; 89-1210: **47**; 90-522: **47**; 90-701: **47**; 90-726: **47**; 90-727: **47**; 90-807: **1**; 91-450: **33a**; 91-458: **33a**; 91-486: **33a**; 91-735: **33a**; 91-831: **1**; 91-933: **47**; 92-167: **5a**; 92-256: **1**; 92-337: **33a**; 92-444: **5a**; 92-675: **47**; 93-526: **5a**; 93-560 or 568: **1**; 93-595: **1**; 93-694: **47**; 93-767: **5a**; 93-918: **33a**; 93-925: **55**; 93-954: **1**;93-1218: **1**; 94-210: **54**; 94-224: **54**; 94-437: **5a**; 94-491: **33a**; 94-505: **33a**; 94-509: **33a**; 94-527: **55**; 94-600: **5a**; 94-610: **33a**; 94-695: **21**; 94-709: **21**; 94-735: **47**; 94-742: **33a**; 95-465: **33a**; 95-470: **1**; 95-514: **33a**; 95-555: **47**; 95-820: **33a**; 95-847: **47**; 95-999: **1**; 96-593: **5a**; 96-608: **33a**; 96-679: **33a**; 96-724: **47**; 96-848: **5a**; 96-1042: **1** — *McClure* CCC 8152: **21**; CCC 8980: **33a** — *McDonald* NGF 8155: **18b**; NGF 8206: **18b**; NGF 8221: **18a** — *McDonald & Afriastini* 3305: **33a** — *McDonald & Ismail* 3607: **52a**; 4713: **51** — *McDonald & Sunaryo* 4304: **33a**; 4507: **23** — *McGregor* 179: **49**; 311: **49**; BS 1231: **21**; 10279: **16a**; 10280: **30**; BS 10583: **30**; BS 10584: **30**; BS 11445: **5a**; BS 18874: **21**; BS 22781: **33a**; BS 22799: **52a**; BS 22882: **33a**; BS 22995: **21**; BS 23191: **33a**; BS 32262: **21**; BS 32347: **33a**; BS 32542: **52a**; BS 43582: **33a**; BS 45926: **33a**; BS 47352: **52a** — *McKee* 8510: **21**; 8512: **21** — *Mead* S 17: **11** — *Mearns* BS 2516: **15** — *Meijer* 1936: **36a**; 1959: **49**; 3787: **32**; 5256: **33c**; 7657: **14b**; 9307: *cf.* **46a**; 9719: **7**; 9933: **46a**; 10129: **49**; 10414: **51**; 10651: *cf.* **5a**; 10686: **7**; 10811: **17**; 11289: **18a**; SAN 19164: **36a**; SAN 19279: **33a**; SAN 19852: **36a**; SAN 19865: **21**; SAN 20124: **36a**; SAN 20134: **36a**; SAN 20243: **33a**; SAN 20994: **36b**; SAN 21079: **36b**; SAN 21270: **52a**; SAN 21325: **36b**; SAN 21335: **52a**; SAN 22097: **52a**; SAN 23410: **33a**; SAN 24081: **36b**; SAN 25401: **33a**; SAN 27648: **33a**; SAN 28715: **36a**; SAN 29521: **36a**; SAN 36688: **52a**; SAN 37948: **33a**; SAN 41014: **36a**; SAN 42242: **33a**; SAN 43210: **36c**; SAN 43865: **44**; SAN 121368: **33a**; SAN 122624: **36a**; SAN 129649: **52b**; SAN 136596: **21** — *Meijer & Aban* SAN 128794: **36a** — *Meijer et al.* 12008: **14a** — *Meijer & Leopold Madani* SAN 131953: **36a** — *Meijer & Pereira* SAN 42240: **34** — *Melegrito* 2488: **11**; 3362: **21** — *Mendoza* PNH 12217: **33a**; PNH 18546: *cf.* **31**; 20478: **33a**; PNH 37430: **21**; PNH 41920: **13**; PNH 41962: **52a**; PNH 41972: *cf.* **31**; PNH 42266: **33a**; PNH 42275: **52a**; PNH 42467: **52a**; PNH 42513: **33a**; 97496: **15**; PNH 97740: **21**; PNH 98591: **33a** — *Mendoza & Convocar* PNH 10273: **13**; PNH 10463: **52a**; PNH 10622 a: **33a**; PNH 10759: **52a** — *Mergin* in Herb. Griffith Kew distr. no. 4927: **55** — *Merrill* Species Blancoanae 8: **5a**; Sp. Blanco. 31: **33a**; Sp. Blanco. 272: **33a**; Sp. Blanco. 375: **21**; Sp. Blanco. 718: **16a**; Species Blancoanae 915: **16a**; 661: **21**; 861: **52a**; 893: **15**; 939: **21**; 1492: **21**; 1691: **21**; 1807: **49**; 2108: **21**; 2193: **21**; 2194: **5a**; 2498: **33a**; 2559: **5a**; 2576: **21**; 2636: **21**; 2678: **33a**; 2813: **50**; 2860: **50**; 2908: **21**; 2920: **5a**; 2943: **33a**; 2975: **33a**; 3051: **50**; 3148: **16a**; 3477: **33a**; 3640: **21**; 3784: **16a**; 4048: **49**; BS 5376: **33a**; BS 5717: **15**; 5734: **21**; 8274: **33a**; 9294: **33a**; 9295: *cf.* **33a**; 9336: *cf.* **33a**; 9450: **44**; 9546: **52a**; 11605: **33a** — *Merrill King et al.* 5433: **21**; 5448: **21**; 5491: **21** — *Merritt* 4051: **16a**; FB 6794: **49**; FB 8820: **21**; FB 9848: **5a**; BS 15849:

15 — *Merritt & Curran* BS 8371: **50** — *Metzner* 148 a: **33a**; 182: **33a** — *Meyer* FB 2245: **21**; FB 2642: **15**; FB 2775: **15** — *Mikil* SAN 30217: **49**; SAN 30313: **21**; SAN 34538: **56**; SAN 38542: **49**; 41776: **36a**; SAN 41924: **52a**; SAN 41938: **36b**; 42093: **29** — *Mikil et al.* SAN 117685: **52a**; SAN 120954: **44** — *Milford* 409: **24a** — *Millar* NGF 9941: **45**; NGF 10000: **45**; NGF 14507: **18a**; NGF 14549: **18a**; NGF 14588: **18a**; NGF 23003: **18a**; NGF 23458: **19**; NGF 23844: **35a**; NGF 35125: **18b**; NGF 35126: **18b**; NGF 35400: **18a** — *Millar & Vandenberg* NGF 35225: **35a** — *Miller* 9: **14a** — *Milliken* 917: **7** — *Minjulu* SAN 77025: **33a** — *Miranda* FB 11882: **33a**; FB 18245: **21**; FB 18258: **5a** — *Mitsuru Hotta* 13629: **11** — *Miyoshi Furuse* 812: **33a**; 813: **33a**; 865: **24a**; 905: **24a**; 1019: **24a**; 1173: **24a**; 1957: **33a**; 2352: **33a**; 2519: **24a**; 2535: **24a**; 2536: **24a**; 2660: **33a**; 2671: **24a**; 2943: **24a**; 2944: **24a**; 3340: **33a**; 3385: **24a**; 3518: **24a**; 3548: **24a**; 3549: **24a**; 3714: **33a**; 3767: **33a**; 3802: **24a**; 3850: **24a**; 4029: **24a**; 4219: **33a**; 4220: **33a**; 4601: **33a**; 7987: **24a**; 10216: **24a** — *Mochtar* 81 A: **36a**; 108 A: **36a** — *Mogea* 3473: **36a**; 3524: **33a**; 3542: **40**; 3562: **36a**; 3631: **36a**; 3663: **49**; 3785: **29**; 3796: **36b**; 3915: **36b**; 4005: **29**; 4039: **36a**; 4124: **36a**; 4165: *cf.* **52a**; 4278: *cf.* **49**; 4435: **36a**; 5288: **49**; 5382: **49** — *Mohidin* S 21618: **52a**; S 21646: **29**; S 21647: **52a**; S 21689: **3** — *Mohtar & Othman Ismawi* S 49707: **36a** — *Mohzan* KEP 99581: **11** — *Moi & Inu* NGF 25955: **45**; NGF 25971: **21**; NGF 25977: **21** — *Molesworth Allen* 4819: **24a** — *Mondi* 16: **33a**; 62: **11** — *Mooney* 67: **21**; 68: **21**; 217: **21**; 358: **5a**; 1110: **33a**; 3714: **33a** — *Morley & Kardin* 589: **33a** — *Morren* 277: **41** — *Morse* 621: **33a**; 676: **21** — *Motley* 72: **36a**; 403: **21**; 694: **33a** — *Moulton* 66: **56** — *Moysey & Kiah* SF 31870: **11**; SF 33604: **29**; SF 33633: **52a**; SF 33762: **29**; SF 33862: **33c**; SF 33915: **11**; SF 33943: **40** — *Mueller-Dombois & Balakrishnan* 68091212: **21** — *Muin Chai* SAN 18388: **52a**; SAN 21639: **36a**; SAN 21685: **33a**; SAN 25061: **34**; SAN 25585: **34**; SAN 25974: **52a**; SAN 25983: **36b**; SAN 26067: **36a**; SAN 26108: **36a**; SAN 29653: **34**; SAN 29729: **33a**; SAN 29801: **52a**; SAN 31617: **52a**; SAN 31680: **36a**; SAN 31701: **33a**; SAN 33403: **36a** — *Mujia & Chong* SAN 33689: **21** — *Mujin* SAN 26757: **56** — *Mulyati Rahayu* 603: **29**; 694: **36a** — *Mulyati Rahayu & Maskuri* 550: **36a**; 551: *cf.* **33a** — *Murata* 17050: **33a** — *Murata et al.* B 1205: **49**; B 1216: **33a**; B 1268: **52a**; B 4458: **33a**; B 4606: **33a**; B 4614: **33a**; T 14934: **1**; T 16318: **33a**; T 16708: **21**; T 16709: **21**; T 16764: **5a**; T 17048: *cf.* **55**; T 17480: **33a**; T 17777: **55**; T 37332: **21**; T 37333: **21**; T 37361: **1**; T 37394: **1**; 37674: **1**; T 37821: **1**; T 37833: **21**; T 43130: **1**; T 50009: **1** — *Murdoch* 98: **33c** — *Muroh K.* SAN 69377: **52a** — *Murton* 34: **33a**; (com.) 106: **14a** — *Musser* KAN-1: *cf.* **46a**; S-3a: *cf.* **46a**; S-4: **7**; S-13: **7**; 19: **5a**; 119: **5a**; 474: **46a**; 502: *cf.* **46a**; 519: *cf.* **46a**; 572: **49**; 668: **49**; 834: **7**; 901: *cf.* **46a**; 1013: *cf.* **46a**; 1105: **7** —

Nagata 234: **5a** — *Nahar* 12667: **36a** — *Nair* 2626: **33a**; 2663: **5a**; 3560: **21**; 3661: **33a**; 3705: **33a** — *Nakhan* 3482: **21**; 14521: **1** — Native collector 32: **33a**; 92: **40**; C 133: **29**; 147: **33a**; BS 264: **33a**; BS 265: **49**; 503: **49**; 504: **14a**; 506: **33a**; BS 507: **14a**; 508: **14a**; 509: **33a**; BS 1148: **49**; 1151: **33a**; 1384: **33a**; 1506: **33a**; BS 1589: **11**; 1805: **49**; 1915: **11**; 2081: **40**; BS 2169: **11**; 2418: **29**; 2433: **33a**; 2535: **52a**; 2841: **5a**; 2859: **1**; NGF 4645: **18a**; 5000: **52a**; 5001: **49**; 5002: **49**; 5189: **49**; 5192: **52a**; 5265: **56**; 5266: **52b**; 5941: **21** — *Nedi* 727: **11** — Neth. Ind. For. Service Ja 2515: **33a**; Ja 2807: **5a**; bb 16634: **21**; bb 16638: **50**; bb 20328: **21**; bb 25456: **21**; bb 25476: **21**; bb 28864: **21**; bb 29851: **21** — *Newman* 50: **33a** — *Ng* FRI 1013: **11**; FRI 1243: **14a**; FRI 1736: **36a**; FRI 1964: **33a**; FRI 1979: **33a**; FRI 5031: **33a**; FRI 5054: **33a**; FRI 5308: **33c**; FRI

5311: **33c**; FRI 5314: **33c**; FRI 5337: **39**; FRI 5352: **14a**; FRI 5490: **11**; FRI 5707: **11**; FRI 5989: **12**; FRI 6125: **40**; FRI 6328: **14a**; FRI 20757: **54**; FRI 20979: **37**; FRI 22041: *cf.* **37**; FRI 22044: **36b**; FRI 22046: **36b**; FRI 22065: **39**; KEP 98022: **11** — *Ng & Beltran* FRI 6416: **14a**; FRI 6424: **33c** — *Ngadiman* SF 34540: **11**; SF 35922: **14a**; SF 36630: **33a**; SF 36849: **14a** — NGF series 1862: **18b**; 2021: **21**; 2086: **18a**; 2111: **18a**; 3125: **18a**; 3746: **18b**; 3863: **45**; 4442: **18a**; 4645: **18a**; 7260: **18a**; 8155: **18b**; 8206: **18b**; 8221: **18a**; 8257: **18a**; 8314: **18a**; 9022: **18a**; 9170: **18a**; 9809: **21**; 9941: **45**; 10000: **45**; 10103: **18a**; 10344: **18a**; 10455: **18a**; 10848: **18a**; 11033: **35b**; 11141: **18a**; 11543: **18b**; 11594: **21**; 11601: **18a**; 12418: **18a**; 12769: **18a**; 13237: **41**; 14355: **18a**; 14507: **18a**; 14549: **18a**; 14588: **18a**; 14853: **18a**; 15309: **45**; 16328: **18a**; 16337: **18a**; 16726: **18a**; 16729: **18a**; 17078: **35b**; 17569: **18a**; 19672: **18b**; 19675: **25**; 22024: **21**; 22310: **18a**; 22452: **18b**; 22492: **18a**; 23003: **18a**; 23458: **19**; 23844: **35a**; 25323: **18a**; 25804: **21**; 25912: **18a**; 25955: **45**; 25971: **21**; 25977: **21**; 26330: *cf.* **18a**; 26360: **18a**; 27636: **2**; 27964: **21**; 28061: **18a**; 29154: **5a**; 29180: **19**; 29385: **18a**; 29932: **19**; 31019: **18a**; 31867: **18a**; 31944: **18a**; 32686: **19**; 32762: **18a**; 33202: **35a**; 33970: **18a**; 35125: **18b**; 35126: **18b**; 35225: **35a**; 35400: **18a**; 35615: **18a**; 35720: **35a**; 36283: **21**; 36744: **18a**; 37095: **18a**; 38068: **18a**; 38521: **25**; 38595: **18a**; 38729: **21**; 39231: **18a**; 41887: **25**; 41900: **18a**; 42490: **18a**; 42936: **9**; 43886: **18a**; 44196: **18a**; 45167: **18a**; 45413: **18a**; 46807: **18a**; 46964: **18a**; 47800: **18a**; 48451: **18a**; 49676: **48**; 52051: **18a** — *Nielsen & Balslev* 992: **29** — *Niga Nangkat* 155: **36b**; 226: **36b**; 353: **36b**; BRUN 15436: **36b** — *Nilphanit* 10517: **1**; 10524: **5b** — *Nima for Womersley* NGF 37095: **18a** — *Nimanong* 41850: **1** — *Nimanong et al.* 1609: **33a**; 1838: **6a** — *Nimanong & Phusomsaeng* 41862: **54** — *Niniek & Wardi* 531: **33a** — *Niniek Mulyati Rahayu* 124: **29** — *Nitta* 15079: **51** — *Nitta & Yoshida* 57: **33a** — *Niyomdham et al.* 212: **33d**; 219: **55**; 357: **44**; 951: **33a**; 986: **33a**; 1112: **33a**; 1214: **33a** — *Niyomdham & Ueachirakan* 1750: **33a** — *Noe* 20: **55**; 38: **33a**; 64: **33a** — *Noerkas* 217: **21**; 498: **7** — *Noltee* 4619: **5a** — *Noorsiha et al.* FRI 39431: **54** — *Nooteboom* 1432: **36a**; 4020: **33c**; 4182: **36a**; 4183: **49**; 4192: **56**; 4205: **33c**; 4276: **43**; 4394: **36a**; 4404: **49**; 4406: **14a**; 4482: **14a**; 4485: *cf.* **49**; 4514: **14a**; 4578: **36b**; 4629: **29**; 4709: **36a**; 4851: **36a**; 5152: **49**; 5176: **18a**; 5918: *cf.* **18b**; 5967: **45** — *Nooteboom & Chai* 1749: **36b** — *Nordin* 46132: **36a**; SAN 54598: **29** — *Nordin Abas* SAN 46268: **52a**; SAN 84207: **11**; SAN 84309: **11**; SAN 85631: **21**; SAN 85658: **52a** — *Nordin Abas & Ali* SAN 54438: **36a** — *Nuhamara* 13: **11** — *Nur* 158: **51**; 744: **5a**; 1415: **33a**; 1605: **33a**; 1608: **33a**; SF 6100: **33a**; 6584: **54**; SF 6890: **54**; 7414: **5a**; 7415: **33a**; SF 11109: **39**; SF 11276: **12**; SF 11774: **36b**; 11874: **14a**; SF 18561: **33a**; SF 18849: **54**; 21744: **33a**; 25166: **52a**; SF 32682: **14a**; SF 32732: **39** — *Nur & Foxworthy* SF 12161: **39** — *Nurta, Hasan & Sukandi* 143: **51** —

Oates 39: **33a** — *Ogata* 10333: **54**; 10672: **52a**; 10679: **36a**; 10708: **36a**; 10797: **36a**; 10918: **33a**; 10962: **33a**; 11452: **36a**; 11492: **52a**; 11498: **36a**; 11627: **52a**; 11633: **33a**; 11635: **33a** — *Ogata et al.* B 40: **11**; B 65: **11**; B 88: **29**; B 92: **36b**; B 107: **29**; B 248: **56**; B 277: **36a**; 297: **33a**; B 307: **11** — *Okada* 3432: **51**; 3492: **54** — *Okada et al.* 29: **33c**; 23511: **36a**; 23903: **29**; 25412: **52a** — *Okada & Komara* 31765: **33c**; 32262: **29** — *Okunara & Sunagawa* 116: **33a** — *Oldham* 364: **24a**; 744: **24a**; 939: **24a** — *Omar* 8540: **14a** — *Omar Musi* SAN 106912: **36a** — *Ong & Soepadmo* 172: **33c** — *Oro* FB 30694: **15**; FB 30706: **30**; FB 30762: **52a** — *Orolfo* 1815: **52a**; 3076: **36a**; 3077: **52a**; KEP 36930: **44** — *Orstroom* 13130: **51** — *Osman* 7: **36b**; 39: **56**; For. Dept. F.M.S.

11550: **14a** — *Otanes* BS 18337: **33a** — *Othman et al.* S 41391: **46a**; S 43536: **36c**; S 48846: **36a**; S 62130: **29**; S 62150: **56** — *Othman & Rantai* S 57482: **36a** — *Othman Haron* S 19939: **29**; S 19976: **36a**; S 19980: **52a**; S 21331: **36b**; S 21397: **56**; S 29976: **36a** — *Othman Ismawi* S 37568: **33a**; S 56618: **11** — *Othman Ismawi et al.* S 37501: **49**; S 37509: **52a**; S 37527: **52a**; S 43526: **33a**; S 43733: **29**; S 43746: **33a**; S 43861: **29**; S 43878: **52a**; S 56498: *cf.* **46a** — *Othman Ismawi & Munting* S 54347: **52a** —

Paduada FB 29811: **33a**; FB 29812: **21** — *Palee* 340: **1** — *Pancho* 62: **33a** — *Panigrahi* 8411: **1**; 11188: **21**; 15188: **1**; 20842: **1** — *Panoff & Panoff* 451: **18a** — *Parish* 280: **1** — *Parker* 2632: **54**; 2734: **54**; 3131: **22** — *Parkinson* 435: **33a**; 575: **44**; 5241: **54**; 14433: **55** — *Parnell et al.* 95-066: **5a**; 95-118: **33a**; 95-232: **1**; 95-318: **1**; 95-632: **1** — *Pascual* 2362: **56**; *FB* 28793: **16a** — *Patrick Lassan* SAN 91740: **36a**; SAN 111575: **46a** — *Paymans* 43: **29** — *Penas* FB 26692: **52a** — *Pereira* SAN 29768: **36a**; SAN 41596: **33a** — *Perumal & LaFrankie* 734: **36b** — *Petelot* 758: **33a**; 1428: **24a** — *Pham Hoang Ho* 5005: **1** — *Phasis* FB 24232: **42** — *Phengklai* 3215: **21**; 3973: **33a**; 4270: **21**; 6545: **21**; 6586: **21**; BKF 6206: **5a** — *Phengklai et al.* 3746: **1**; 4182: **1**; 6206: **5a**; 6963: **5a** — *Phengnaren* 36332: **24b**; 37453: **33a**; 40401: **44**; 41133: **47** — *Phromdej* 32: **33a** — *Phuakam* 14: **1** — *Phusomsaeng* 82: **33a**; 134: **36a**; BKF 46419: **40**; BKF 52139: **54** — *Phusomsaeng et al.* 1587: **33a** — *Phusomsaeng & Pinnin* BKF 53513: **36a**; BKF 57534: **54** — *Pickles* SAR 3652: **36b**; 3742: **36a** — *Pierre* 190: **21**; 1971: **44** — *Pikkoh* SAN 68833: **36a** — *Pingkun* 10250: **11** — *Pinuim et al.* 449: **33a** — *Piper* 354: **21**; 445: **21** — *Pitty & Ogata* SAN 63256: **52a** — *Playfair* 163: **21** — *Pleyte* 19: **21**; 184: **18a**; 288: **18a**; 619: **10** — PNH series 357: **52a**; 981: **52a**; 1054: **52a**; 1145: **52a**; 1588: **33a**; 1590: **33a**; 1603: **33a**; 1970: **49**; 3618: **31**; 4012: **16a**; 4014: **16a**; 4038: **16a**; 4164: **16a**; 4573: **33a**; 4679: **33a**; 5695: **13**; 5758: **13**; 6087: **13**; 6162: **30**; 6203: **30**; 6366: **13**; 6641: **33a**; 6784: *cf.* **31**; 6911: **31**; 8034: **24a**; 8052: **52a**; 8369: **33a**; 9828: **33a**; 9933: **5a**; 10032: **52a**; 10273: **13**; 10463: **52a**; 10622a: **33a**; 10759: **52a**; 12217: **33a**; 12297: **33a**; 13337: **33a**; 13457: **13**; 13524: **52a**; 13532: **21**; 13568: **31**; 13692: **33a**; 13696: **42**; 14081: **33a**; 14117: *cf.* **33a**; 14193: **33a**; 14387: **52a**; 14420: **13**; 15250: **30**; 15526: **16a**; 15532: **30**; 15719: **50**; 16784: **5a**; 17028: **21**; 17593: **5a**; 17595: **21**; 18052: **5a**; 18053: **5a**; 18546: *cf.* **31**; 21591: **31**; 22852: **5a**; 23043: **33a**; 33163: **33a**; 34472: **31**; 34545: **31**; 34953: **52a**; 36003: **5a**; 36466: **5a**; 36550: **33a**; 37430: **21**; 37987: **33a**; 38024: **52a**; 38059: **33a**; 38075: **52a**; 38372: **33a**; 38400: *cf.* **29**; 39434: **52a**; 40164: **30**; 41775: **52a**; 41920: **13**; 41962: **52a**; 41972: *cf.* **31**; 42266: **33a**; 42275: **52a**; 42467: **52a**; 42513: **33a**; 72730: **31**; 78036: **5a**; 78089: **52a**; 78159: **24a**; 78678: **5a**; 79569: *cf.* **30**; 80421: **5a**; 80610: **33a**; 80680: **24a**; 87809: **33a**; 91587: **33a**; 97740: **21**; 98591: **33a**; 117087: **52a**; 117297: **30**; 117611: **52a**; 117617: **30**; 117785: **15**; 117959: **30**; 118196: **21**; 118666: **33a**; 118752: **30** — *Podzorski* 508: **52a**; SMHI 741: **44**; 753: **52a**; SMHI 892: **52a** — *Poilane* 1234: **5a**; 2360: **47**; 6206: **24a**; 7698: **5a**; 7925: **5a**; 8294: **24a**; 8795: **1**; 8843: **1**; 10126: **5a**; 10502: **5a**; 10636: **5a**; 10901: **5a**; 12322: **5a**; 12485: **1**; 13407: **21**; 15135: **1**; 15329: **21**; 16413: **21**; 16439: **44**; 20131: **33b**; 21287: **33a** — *Polak* 493: **33a**; 539: **33a**; 642: **33a**; 709: **33a**; 917: **2**; 1196: **18a**; 1326: **45** — *Poore* 195: **14a**; 364: **14a**; 1024: **21**; 4640: **33a**; 6043: **37**; 6188: **14a** — *Popta* 622/101: **5a**; 861: **5a** — *Porter* in Herb. Wallich 8582: **54**; 8584: **11**; 9101: **11** — *Posthumus* 1024: **33c**; 1037: **21** — *Powell* 341: **18a**; 344: **18a**; 1176: **18a** — *Powell & H'ng Kim Chey* 708: **5a**; 842: **5a** — PPI series 526: **42**; 835: **5a**; 847: **5a**; 863: **21**; 866: **21**; 1091: **15**; 1092: **15**; 1095: **52a**; 1262: **52a**;

1344: **30**; 1993: **15**; 2766: **52a**; 2957: **52a**; 3053: **30**; 4119: **16a**; 4131: **21**; 4189: **16a**; 4214: **21**; 4455: **21**; 4509: **33a**; 4571: **13**; 4587: **33a**; 4689: **13**; 6112: **33a**; 6597: **21**; 6647: **21**; 6750: **33a**; 6833: **33a**; 6926: **15**; 7100: **31**; 7167: **5a**; 7179: **31**; 7243: **33a**; 7357: **33a**; 7523: **42**; 8139: **33a**; 8411: **33a**; 8999: **33a**; 9664: **52a**; 10351: **33a**; 10382: **33a**; 10386: **5a**; 10465: **33a**; 12203: **33a**; 12272: *cf.* **33a**; 12308: **15**; 15141: **16a** — *Prance* 30105: **33a**; 30569: **36b**; 30691: **36b**; 30693: **36b** — *Prapat* 57: **55** — *Prawiroatmodjo & Maskuri* 1201: **21**; 1220: **49**; 1233: **17**; 1417: **17**; 1424: **33a**; 1462: **49** — *Prawiroatmodjo & Soewoko* 1887: **17**; 1891: **49**; 1979: **46a** — *Prazer* 193: **21** — *Pringgo Admodjo* 1: **54**; 2: **54**; Exped. *van Daalen* 184: **21** — *Puasa-Angian* 1547: **29**; 3776 or 36353: **56**; BNB For. Dept. 3929: **21**; 3986: **36a**; 7725: **33a**; 7737: **33a**; 7768: **21**; 10471: **33a**; 10497: **33a**; 10084: **44**; 10087: **33a**; BNB For. Dept. 36706/3929: **21**; BNB For. Dept. 48856: **44**; KEP 48859: **33a**; KEP 55036: **21** — *Pullen* 456: **8**; 978: **2**; 1790: **18a**; 5581: *cf.* **18a**; 6033: **2**; 7277: **18a**; 7278: **18a**; 7344: **10**; 7541: **9**; 8163: **25**; 8272: **18a**; 8280: **18a** — *Pulsford* UPNG 167: **21** — *Purseglove* 4390: **33a**; 4476: **33a**; 4932: **33a**; 4980: **49**; 4981: **33a**; 5109: **56**; 5140: **56**; 5173: **52a**; 5319: **52a**; 5372: **36a** — *Purseglove & Shah* 4391: **33a**; 4496: **33a** — *Put* 555: **27**; 695: **33a**; 1038: **33a**; 1512: **33a**; 1691: **21**; 1761: **33a**; 2765: **44**; 2949: **27**; 3820: **5a**; 3886: **33a**; 3893: **33a**; 3897: **5a** — *Putz* FRI 21930: **33c**; FRI 23631: **29** —

Quisumbing Species Blancoanae 488: **21**; Sp. Blanco. 488: **21**; PNH 8034: **24a**; PNH 8052: **52a**; BS 78181: **5a**; BS 84668: **33a** —

Raap 171: **36a**; 213: **36a**; 509: **5a**; 659: **40**; 680: **40**; 722: **52a**; 729: **52a**; 830: **5a**; 860: **54** — *Rabil* 58: **33a**; 266: **33a**; 351: **54** — *Rachmat* 575: *cf.* **51**; 617: **5a**; 641: **18a** — *Raghavan* 97258: **33a**; 97360: **33a** — *Rahim et al.* SAN 59847: **36a**; SAN 92965: **34**; SAN 92967: **49**; SAN 101804: **52a** — *Rahman & Paul* SAN 90059: **29** — *Rahmat Si Boeea* 6078: **33a**; 6480: **33a**; 6665: **33a**; 6820: **33a**; 7536: **33a**; 7726 or 7736: **33a**; 7802: **33a**; 8127: **33a**; 9136: **52a**; 9612: **33a** — *Rahmat Si Toroes* 44: **33a**; 248: **36a**; 505: **33a**; 846: **33a**; 1054: **21**; 2238: **33a**; 2259: **14a**; 2471: **33a**; 2510: **33a**; 3038: **21**; 3219: **33a**; 3321: **21**; 3531: **21**; 3550: **33a**; 3708: **33a**; 3842: **21**; 3907: **22**; 3928: **14a**; 4105: **21**; 4155: **21**; 4185: **21**; 4251: **52a**; 4255: **14a**; 4257: **36a**; 4361: **51**; 4424: **33a**; 4461: **14a**; 4489: **36a**; 4493: **14a**; 4534: **52a**; 4554: **52a**; 4620: **51**; 4623: **33a**; 4639: **33a**; 4692: **33a**; 4706: **51**; 4766: **33a**; 4828: **21**; 4888: **33a**; 4933: **51**; 4969: **33a**; 4987: **51**; 4994: **33a**; 5051: **33a**; 5147: **33a**; 5201: **33a**; 5235: **33a**; 5296: **33a**; 5328: **51**; 5349: **33a**; 5445: **51**; 5554: **33a**; 5561: **33a**; 6276: **14a** — *Ramamoorthy & Ganghi* 2699: **33a** — *Ramaswami* 1471: **21** — *Ramlanto* 21: **36a**; 279: **54**; 317: **49**; 454: **49** — *Ramlanto & Zainal Fanani* 675: **17**; 732: **49** — *Ramli* KEP 98378: **14a** — *Ramos* 37: **33a**; 372: **21**; 379: **50**; BS 1002: **33a**; BS 1058: **16a**; BS 1083: **16a**; 1114: **50**; 1360: **36a**; BS 1372: **33a**; 1381: **36a**; 1510: **33a**; 1555: **15**; 1657: **52a**; 1665: **52a**; 1707: **36a**; BS 2183: **21**; BS 3558: **21**; BS 4664: **50**; BS 5041: **33a**; BS 7397: **30**; BS 8079: **33a**; BS 10956: **52a**; BS 11114: **33a**; BS 12168: **5a**; BS 13339: **52a**; BS 13868: **52a**; BS 15086: **52a**; BS 15251: **31**; BS 15335: *cf.* **31**; BS 16625: **52a**; BS 17528: **52a**; BS 19410: **33a**; BS 23471: **33a**; 23524: **31**; BS 23615: *cf.* **42**; BS 24134: **52a**; BS 24176: **16a**; BS 24220: **31**; BS 24375: **30**; BS 24415: **52a**; BS 24436: **15**; BS 24504: **30**; BS 24541: **15**; BS 27302: **21**; BS 30184: **42**; BS 30515: **6**; BS 30533: **52a**; BS 30542: **30**; BS 32865: **50**; BS 32902: **5a**; BS 32998: **52a**; BS 33023: **30**; BS 33057: *cf.* **18a**; BS 33153: **33a**; BS 33173: **33a**;

BS 33174: **16a**; BS 33294: **15**; BS 39423: **49**; BS 39605: **49**; BS 40835: **33a**; BS 40899: **49**; BS 40923: **16a**; BS 40931: **15**; BS 41060: **49**; BS 41084: **49**; BS 41154: **33a**; BS 41160: **50**; 41183: **33a**; BS 41519: **31**; BS 42579: **33a**; S 43059: **31**; BS 43237: **31**; BS 76906: **21**; BS 76991: **52a**; BS 80399: **33a**; BS 80687: **33a** — *Ramos & Convocar* BS 83664: **52a**; BS 83820: **13**; BS 83955: **52a**; BS 84118: **33a**; BS 84136: **16a** — *Ramos & Deroy* BS 22506: **5a**; BS 22556: **33a** — *Ramos & Edano* BS 26375: **16a**; BS 26561: **15**; BS 26597: **16a**; BS 28503: **30**; BS 28519: **52a**; BS 28566: **33a**; BS 29283: **33a**; BS 29321: **33a**; BS 29477: **33a**; BS 29516: **50**; BS 31020: **52a**; BS 31268: **52a**; BS 31268: **52a**; BS 31409: **52a**; BS 31490: **21**; BS 31491: **21**; BS 33489: **52a**; BS 33613: *cf.* **30**; BS 33638: **30**; BS 33701: **30**; BS 36631: **29**; BS 36965: **33a**; BS 37090: **33a**; BS 37093: **29**; BS 37337: **29**; BS 37886: **15**; BS 38985: **52a**; BS 39026: **18a**; BS 40732: **49**; BS 40733: **49**; BS 44006: **33a**; BS 44013: **52a**; BS 44014: **52a**; BS 44532: **16a**; BS 45189: **24a**; BS 45205: **52a**; BS 45303: **52a**; BS 45360: **52a**; BS 46539: **21**; BS 46647: **33a**; BS 46722: **33a**; BS 46825: **52a**; BS 47040: **5a**; BS 47178: **52a**; BS 48051: **52a**; BS 48152: **52a**; BS 48177: **33a**; BS 48206: **42**; BS 48217: **52a**; BS 48343: **33a**; BS 49352: **21**; BS 75109: **33a**; BS 75150: **52a**; BS 75342: **42**; BS 84902: **21** — *Ramos & Pascasio* BS 34371: **33a**; BS 34694: **33a**; BS 34803: **52a**; BS 34927: **33a**; BS 34928: **33a**; BS 34970: *cf.* **42**; BS 35072: **52a**; BS 35108: **13** — *Ramsri* 28: **36a** — *Rananand* 11901: **1** — *Rankin* 1436: **21** — *Rant* 544: **21** — *Rao* 9063: **33a**; 38898: **1**; 39008: **1** — *Rastini* 123: **33a**; 146: **33a**; 173: **7**; 244: **21** — *Rau* 29: **18a**; 220: **21**; 3809: **1** — *Raub* For. Dept. F.M.S. 22505: **21** and **11**; — *Razali Jaman* 166: **52a** — *Reillo* BS 15438: **33a**; BS 15439: **29**; BS 15455: **33a**; BS 16425: *cf.* **42** — *Reinecke* 139: **18a**; 343: **18a**; 408: **18a**; 512: **18a**; 513: **18a** — *Reinwardt* 17: **51**; 45: **51**; 52: **33a**; 63: **52a**; 1816: **51** — *Rena George* S 43408: **52a** — *Reynoso* PNH 87809: **33a** — *Reynoso et al.* PNH Field No. 512: **33a**; PPI 835: **5a**; PPI 847: **5a**; PPI 863: **21**; PPI 866: **21**; PPI 1091: **15**; PPI 1092: **15**; PPI 1095: **52a**; PPI 1262: **52a**; PPI 1344: **30**; PPI 4119: **16a**; PPI 4131: **21**; PPI 4189: **16a**; PPI 4214: **21**; PPI 4455: **21**; PPI 6597: **21**; PPI 7100: **31**; PPI 7167: **5a**; PPI 7179: **31**; PPI 7243: **33a**; PPI 7357: **33a**; PPI 7523: **42** — *Reynoso & Espiritu* PNH 118666: **33a**; PNH 118752: **30** — *Richards* 1173: **56** — *Ridley* 2: **21**; 521: **14a**; 1352: **21**; 1657: **14a**; 1716?: **11**; 2286: **52a**; 2287: **54**; 2337: **36a**; 2338: **33a**; 2339: **33a**; 2340: **33c**; 2345: **33c**; 2677: **14a**; 2975: **26**; 2978: **36a**; 3167: **36a**; 3430: **11**; 3440: **54**; 3449: **54**; 4154: **36a**; 4159: **29**; 5046: **14a**; 5500: **21**; 5501: **33a**; 5502: **37**; 5503: **36a**; 6943: **33a**; 7894: **52a**; 7960: **11**; 8014: **33a**; 8016: **36a**; 8178: **33a**; 8344: **12**; 9312: **33a**; 9451: **33a**; 9581: **29**; 11036: **33a**; 11045: **11**; 11059: **54**; 11504: **38**; 11670: **33a**; 13061: **33c**; 13067: **14a**; 13075: **14a**; 13076: **14a**; 13253: **33a**; 13714: **39**; 14045: **36a**; 14540: **33c**; 14906: **33a**; 14911: **33a**; 14912: **33a**; 14913: **21**; 14914: **21**; 15194: **33a**; 15458: *cf.* **1**; 15525: **54**; 15734: **33a** — *Ridsdale* Project Barito Ulu 8: **36a**; 59: **33a**; Project Barito Ulu 358: *cf.* **36c**; Project Barito Ulu 464: **36a**; 689: **33a**; 747: **33a**; 999: **33a**; 1006: **33a**; 1021: **33a**; SMHI 1544: **21**; SMHI 1577: **33a**; SMHI 1774: **33a**; SMHI 1861: **49**; SMHI 1873: **33a**; 1943: **36a**; 2209: **10**; NGF 33970: **18a**; NGF 36744: **18a** — *Ridsdale & Baquiran* ISU 386: **30** — *Ridsdale et al.* ISU 20: **52a**; ISU 270: **30**; ISU 450: **33a**; ISU 480: **52a**; 1242: **49**; 1300: *cf.* **49**; 1670: **15**; NGF 31944: **18a** — *Ridsdale & Katik* NGF 38068: **18a** — *Ridsdale & Lavarack* NGF 31019: **18a** — *Ritchie* FB 30: **21** — *Robinson* Plantae Rumphianae 334: **5a**; Pl. Rumph. 355: **49**; Pl. Rumph. 356: **49**; 1709: **49**; 1710: **21**; 1799: *cf.* **18a**; 5740: **33a**; 5759: **21**; BS 6158: **33a**; BS 9948: **33a**; BS 14085: **15**; BS 14115: **15** — *Rock* 1625: 1 — *Rodatz & Klink* 12: **45** — *Roemer* 262: **45** — *Rogstad* 588:

37; 631: **38** — *Rola* FB 26663: **52a** — RSNB series 16: **36b**; 210: **36a**; 233: **29**; 237: **52a**; 533: **56**; 1202: **49**; 1460: **36b**; 4067: **29**; 4161: **52a**; 7050: **52a** — *Rutten* 98: **33a**; 188: **56**; 245: **33a**; 251: **33a**; 1637: **18a**; 1691: **21** —

S series 17: **11**; 59: **33a**; 60: **33a**; 142: **14a**; 240: **33a**; 294: **21**; 329: **52a**; 449: **11**; 2873: **11**; 3119: **33a**; 3281: **11**; 5371: **40**; 8296: **11**; 8688: **11**; 8752: **36b**; 10600: **36b**; 11773: **11**; 11998: **11**; 12289: **36a**; 12416: **11**; 12423: **33a**; 13007: **33a**; 13162: **36b**; 13432: **43**; 13447: **29**; 13602: **29**; 14327: **11**; 14612: *cf.* **43**; 14664: **52a**; 14746: **29**; 15284: **33a**; 15477: **52a**; 15647: **29**; 16039: **33a**; 16207: **11**; 16373: **36b**; 16406: **36a**; 16482: **36a**; 16486: **36a**; 16560: **36a**; 16605: **36a**; 16628: **36b**; 17222: **52a**; 17524: **11**; 17535: **11**; 17544: **33a**; 17779: **33c**; 17802: **46a**; 17940: **33a**; 18060: **29**; 18111: **29**; 18166: **29**; 18226: **36c**; 18254: **52a**; 18421: **52a**; 18622: **49**; 19147: **33a**; 19323: **36a**; 19339: **36a**; 19350: **56**; 19605: **36a**; 19705: **33a**; 19862: **29**; 19939: **29**; 19976: **36a**; 19980: **52a**; 20201: **36b**; 20204: **36b**; 20946: **33a**; 21331: **36b**; 21397: **56**; 21493: **33a**; 21529: **33a**; 21618: **52a**; 21646: **29**; 21647: **52a**; 21689: **3**; 21795: **36a**; 22031: **29**; 22119: **49**; 22122: **36a**; 22269: **36b**; 22942: **29**; 23204: **52a**; 23461: **29**; 23770: **52a**; 23900: **33a**; 24103: **36b**; 24140: **36b**; 24261: **52a**; 24372: **3**; 24403: **52a**; 24475: **40**; 24483: **36b**; 24619: **36b**; 24819: **36b**; 24833: **36a**; 24915: **14a**; 24949: **14a**; 25052: **52a**; 25060: **3**; 25064: **36a**; 25071: **36b**; 25132: **33a**; 25297: **52a**; 25501: **29**; 25850: **52a**; 26027: **36a**; 26069: **3**; 26099: *cf.* **52a**; 26284: *cf.* **29**; 26340: **36b**; 26949: **52a**; 27101: **33a**; 27436: **33a**; 27862: **40**; 27933: **36a**; 27972: **36a**; 28004: **36a**; 28020: **29**; 28340: **36a**; 28376: **36a**; 29840: **52a**; 29976: **36a**; 30064: **33a**; 30072: **52a**; 30140: **52a**; 30679: **56**; 31664: **29**; 31678: **33a**; 32717: **11**; 33246: **52b**; 33588: **29**; 33924: **29**; 34165: **33a**; 34256: **36a**; 34313: **29**; 34318: **36a**; 34333: **40**; 34342: **56**; 34390: **33a**; 34438: **49**; 34638: **52a**; 34680: **52a**; 34750: **52a**; 34923: **36b**; 34943: **52a**; 35080: **52a**; 35155: **49**; 35300: **52a**; 35364: **36b**; 35706: **36b**; 35779: **52a**; 35980: **52a**; 36095: **29**; 36153: **52a**; 36278: **29**; 36745: **36a**; 36974: **40**; 36975: **40**; 37217: **29**; 37278: **36a**; 37346: **52a**; 37371: **33a**; 37501: **49**; 37509: **52a**; 37527: **52a**; 37568: **33a**; 37662: **33a**; 38007: **36a**; 38119: **36b**; 38158: **36b**; 38173: **36b**; 38226: **36b**; 38337: **36b**; 38401: **56**; 38617: **52a**; 38690: **11**; 38854: **36b**; 39006: **33a**; 39025: **36a**; 39509: **33a**; 39651: **36b**; 39675: **52a**; 40603: **36c**; 40658: **33a**; 40676: **36a**; 40982: **46a**; 41391: **46a**; 41986: **56**; 41993: **36b**; 42154: **33a**; 42272: **33a**; 42496: **40**; 42707: **40**; 42728: **33a**; 42888: **36a**; 42889: **36a**; 43243: **36a**; 43249: **33a**; 43408: **52a**; 43526: **33a**; 43536: **36c**; 43694: **29**; 43733: **29**; 43746: **33a**; 43861: **29**; 43878: **52a**; 43916: **29**; 44320: *cf.* **29**; 44331: **36a**; 44505: **36a**; 44749: **52a**; 44838: **52a**; 44840: **56**; 45064: **36b**; 45137: **36c**; 45429: **40**; 45503: **29**; 45670: **49**; 45682: **52a**; 45941: **29**; 46176: **36a**; 46636: **52a**; 46981: **33a**; 47037: **33a**; 47043: **49**; 47401: **33a**; 47545: **36a**; 47673: **36b**; 48053: **36a**; 48778: **36a**; 48846: **36a**; 49499: **36b**; 49523: **52a**; 49686: **52a**; 49707: **36a**; 50664: **11**; 52312: **52a**; 52325: **29**; 52718: **52a**; 52776: **36a**; 53524: **52a**; 53535: *cf.* **29**; 53823: **29**; 53915: **3**; 54053: **29**; 54200: **36a**; 54347: **52a**; 54590: **56**; 54600: **40**; 55652: **43**; 56498: *cf.* **46a**; 56537: **52a**; 56618: **11**; 56731: **36a**; 56742: **36a**; 57482: **36a**; 57625: **29**; 57631: **36a**; 58247: **36b**; 61435: **49**; 62130: **29**; 62150: **56**; 64721: **56**; 68826: **36b** — *Saigol* SAN 93062: **36a** — *Saikeh Lantoh* SAN 67227: **29**; SAN 68068: **33a**; SAN 72101: **36c**; SAN 72164: **36b**; SAN 72404: **33a**; SAN 73311: **36a**; SAN 82391: **52a**; SAN 83178: **52a** — *Sain* 16572: **1** — *Sajor* FB 27253: **33a** — *Sako* 95: **24a** — *Salvoza* BS 28585: **33a**; FB 29704: **52a** — *Samsuri Ahmad* S 329: **52a**; 745: **33a**; 1026: **14a**; 1120: **33a**; 1128: **11**; 1222: **11**; 1889:

36a — *Samsuri Ahmad & Ahmad Shukor* 480: **36a**; 515: **40**; 686: **40** — SAN series 4192: **34**; 4793: **36a**; 15870: **11**; 16259: **11**; 16282: **36a**; 16308: **36b**; 16888: **36a**; 17089: **29**; 17119: **11**; 17368: **46a**; 17469: **36b**; 18388: **52a**; 19164: **36a**; 19279: **33a**; 19852: **36a**; 19865: **21**; 20066: **36a**; 20124: **36a**; 20134: **36a**; 20243: **33a**; 20839: **29**; 20994: **36b**; 21079: **36b**; 21270: **52a**; 21325: **36b**; 21335: **52a**; 21469: **36a**; 21527: **36b**; 21639: **36a**; 21667: **33a**; 21685: **33a**; 22097: **52a**; 22561: **52a**; 23410: **33a**; 24011: **46a**; 24050: **52a**; 24081: **36b**; 24177: **56**; 24240: **29**; 24519: **36a**; 24706: **29**; 25010: **49**; 25057: **33a**; 25061: **34**; 25181: **44**; 25263: **52a**; 25401: **33a**; 25585: **34**; 25755: **44**; 25817: **33a**; 25876: **36a**; 25974: **52a**; 25983: **36b**; 26067: **36a**; 26108: **36a**; 26252: **36a**; 26538: **52a**; 26757: **56**; 27421: **52a**; 27648: **33a**; 27958: **11**; 28033: **52a**; 28715: **36a**; 29353: **36a**; 29362: **36a**; 29486: **29**; 29521: **36a**; 29574: **36a**; 29580: **36a**; 29653: **34**; 29727: **33a**; 29729: **33a**; 29732: **33a**; 29768: **36a**; 29788: **33a**; 29801: **52a**; 30217: **49**; 30313: **21**; 30704: **52a**; 30943: **33a**; 30950: **33a**; 31066: **44**; 31240: **33a**; 31295: **36a**; 31617: **52a**; 31680: **36a**; 31701: **33a**; 32173: **36b**; 32174: **33a**; 32202: **56**; 32273: **56**; 32406: **52a**; 32524: **36a**; 32943: **36a**; 33199: **44**; 33231: **52a**; 33403: **36a**; 33689: **21**; 33988: **36a**; 34060: **36a**; 34182: **29**; 34377: **29**; 34389: **52a**; 34538: **56**; 34982: **29**; 35045: **29**; 35076: **11**; 35317: **29**; 36006: **33a**; 36096: **34**; 36353: **52a**; 36399: **33a**; 36412: **56**; 36431: **44**; 36439: **44**; 36688: **52a**; 37948: **33a**; 38542: **49**; 40238: **43**; 40372: **33a**; 40385: **52a**; 40458: **56**; 40815: **33a**; 40816: **33a**; 41014: **36a**; 41540: **52a**; 41596: **33a**; 41618: **36a**; 41695: **52a**; 41924: **52a**; 41938: **36b**; 42127: **52a**; 42240: **34**; 42242: **33a**; 43210: **36c**; 43865: **44**; 44561: **21**; 46268: **52a**; 49112: **36a**; 49330: **36**; 50644: **11**; 51855: **33a**; 51937: **44**; 52062: **44**; 52071: **52a**; 52772: **34**; 52825: **52a**; 52952: **33a**; 54438: **36a**; 54598: **29**; 54608: **36a**; 55976: **46a**; 56323: **36b**; 56808: **36a**; 56829: **46a**; 56843: **49**; 57154: **36a**; 57202: **36a**; 57290: *cf.* **33a**; 57299: **34**; 57823: **52a**; 58440: **11**; 59494: **46a**; 59847: **36a**; 60006: **33a**; 60176: **33a**; 60293: **43**; 60317: **29**; 60393: **43**; 60437: **36a**; 60472: **56**; 63514: **52a**; 65211: **36b**; 66180: **33a**; 66312: **33a**; 66343: **36b**; 66692: **52a**; 66709: **11**; 66715: **11**; 66723: **29**; 66788: **36a**; 67143: **33a**; 67227: **29**; 68047: **46a**; 68068: **33a**; 68335: **36a**; 68375: **33a**; 68833: **36a**; 68963: **56**; 69245: **33a**; 69377: **52a**; 69906: **52a**; 70491: **36b**; 70669: **44**; 70891: **36a**; 71486: **36a**; 71521: **52a**; 72101: **36c**; 72164: **36b**; 72404: **33a**; 72825: **52a**; 73311: **36a**; 74125: **29**; 74312: **52a**; 74334: **33a**; 74383: **56**; 74561: **36a**; 74957: **52a**; 75336: **52a**; 75364: **36a**; 75732: **52a**; 76063: **52a**; 76169: **33a**; 76277: **33a**; 76297: **21**; 76401: **36b**; 76810: **36b**; 77025: **33a**; 77155: **33a**; 77760: **33a**; 77826: **36b**; 78102: **52a**; 78111: **36a**; 78115: **36b**; 78556: **36a**; 79151: **52a**; 79399: **52a**; 79411: **49**; 79695: **36a**; 79772: **52a**; 80779: **33a**; 80811: **36a**; 80853: **52a**; 80857: **52a**; 81188: **52a**; 81369: **36a**; 81453: **29**; 81596: **49**; 82159: **36a**; 82391: **52a**; 82444: **52b**; 82681: **33a**; 83092: **36a**; 83178: **52a**; 83302: **36a**; 83516: **36b**; 83642: **34**; 83991: **36a**; 84060: **33a**; 84189: **52a**; 84207: **11**; 84309: **11**; 84868: **36b**; 85057: **52a**; 85631: **21**; 85658: **52a**; 85702: **52a**; 86146: **33a**; 86204: **11**; 86213: **33a**; 86434: **33a**; 87559: **52a**; 88024: **52a**; 88220: **36a**; 88336: **36a**; 88489: **33a**; 88498: **33a**; 88719: **52a**; 88850: **36a**; 88880: **33a**; 88890: **56**; 88963: **36a**; 89302: **52a**; 89308: **33a**; 89367: **36b**; 89748: **36a**; 89889: **52a**; 89919: **36a**; 90036: **36a**; 90059: **29**; 90062: **56**; 90092: **52a**; 90110: **56**; 90277: **36a**; 90367: **49**; 90514: **36a**; 90623: **56**; 90756: **33a**; 90788: **33a**; 91120: **33a**; 91188: **29**; 91251: **52a**; 91341: **33a**; 91460: **36a**; 91643: **46a**; 91680: **29**; 91700: **29**; 91740: **36a**; 91757: **29**; 92191: **34**; 92458: **33a**; 92524: **33a**; 92526: **56**; 92965: **34**; 92967: **49**; 93029: **33a**; 93062: **36a**; 93150: **52a**; 93571: **52a**; 93655: **33a**; 93821: **52a**; 93995: **36b**; 94010: **52a**; 94038: **34**; 94062: **52a**;

94267: **36b**; 94291: **36a**; 94388: **52a**; 94495: **29**; 94564: **52a**; 94726: **36a**; 94757: **52a**; 94769: **36a**; 94785: **36a**; 94900: **52a**; 94926: **52a**; 95416: **49**; 95491: **21**; 95519: **33a**; 95528: **52a**; 95567: **52a**; 95695: **49**; 95893: **36a**; 95909: **52a**; 95972: **33a**; 95997: **33a**; 96344: **21**; 96824: **36b**; 96913: **36a**; 97084: **52a**; 97123: **46a**; 97171: **36a**; 97390: **33a**; 97532: **46a**; 99432: **33a**; 99478: **36c**; 99886: **49**; 99978: **33a**; 101500: **36a**; 101804: **52a**; 101922: **33a**; 102183: **34**; 102213: **33a**; 102273: **11**; 103575: **36a**; 103610: **36a**; 105274: **36a**; 105287: **11**; 105613: **21**; 105992: **11**; 106133: **56**; 106677: **36b**; 106912: **36a**; 107063: **33a**; 107200: **52a**; 107319: **36a**; 107644: **52a**; 107803: **36a**; 107898: **43**; 108585: **29**; 108587: **36**; 108690: **46a**; 108838: **43**; 108907: **52a**; 109221: **52a**; 109231: **52a**; 109345: **36**; 109358: **56**; 109547: **36a**; 109722: **49**; 110433: **36a**; 110610: **36a**; 110640: **52a**; 110827: **36a**; 110914: **29**; 111279: **29**; 111303: **36a**; 111575: **46a**; 111841: **44**; 111868: **56**; 111905: **36a**; 111927: **36a**; 112217: **36a**; 112334: **36a**; 112987: **52a**; 113014: **52a**; 113173: **56**; 113255: **46a**; 113290: **33a**; 113392: **36a**; 113500: **33a**; 114116: **56**; 114155: **52a**; 114337: **21**; 114374: **33a**; 114410: **36a**; 114434: **36a**; 115417: **56**; 115526: **11**; 115571: **33a**; 115596: **11**; 115952: **56**; 115979: **56**; 116063: **33a**; 116196: **21**; 116228: **56**; 116383: **33a**; 116417: **56**; 116675: **52a**; 116947: **36a**; 117155: **56**; 117214: **52a**; 117538: **56**; 117539: **33a**; 117638: **44**; 117685: **52a**; 117767: **46a**; 117915: **21**; 117930: **56**; 118119: **21**; 118126: **33a**; 118576: **56**; 118583: **52a**; 118899: **56**; 118902: **52a**; 118905: **33a**; 119480: **36a**; 119482: **52a**; 119553: **33a**; 119611: **52a**; 119631: **33a**; 119789: **36a**; 119815: **56**; 120951: **36a**; 120954: **44**; 120970: **33a**; 120988: **52a**; 121262: **43**; 121349: **52a**; 121368: **33a**; 121531: **56**; 121635: **56**; 121784: **34**; 121888: **52a**; 121942: **52a**; 122114: **52a**; 122133: **36a**; 122202: **36b**; 122469: **33a**; 122624: **36a**; 123124: **21**; 123230: **36a**; 123292: **52a**; 123428: **56**; 123433: **56**; 123479: **21**; 123509: **29**; 123721: **36a**; 123781: **36a**; 123782: **52a**; 124080: **21**; 124090: **21**; 124140: **11**; 124437: **56**; 124456: **36a**; 125304: **36a**; 125312: **52a**; 125960: **11**; 126297: *cf.* **56**; 126403: **56**; 126882: **11**; 127344: **56**; 127570: **33a**; 128292: **52a**; 128482: **36a**; 128794: **36a**; 129106: **36b**; 129204: **33a**; 129649: **52b**; 129772: **36a**; 131012: **43**; 131953: **36a**; 132688: **36a**; 133172: **36a**; 133306: **34**; 133486: **36a**; 134455: **36a**; 135905: **56**; 136141: **21**; 136596: **21**; 136709: **33a**; 141747: **52a** — *Sands* 11: **51**; 1348: **10**; 2974: **18a**; 3847: **43**; 5834: **36c**; 5854: **52a**; 5893: **36b**; 5970: **36b** — *Sands et al.* 2733: **18a**; 5456: **49**; 5458: **29**; 5510: **49**; 5670: **29**; 5726: **56**; 5942: **29** — *Sands & Johns* 5447: **29** — *Sangkhachand* 825: **33a**; 958: **47**; 1566: **24a**; 1640: **33a**; 1787: **54**; 1939: **36a**; 9767: **47**; 9781: **47**; 12411: **33a**; 17139: **36a**; 23859: **1**; 23895: **1**; 23922: **47**; BKF 40487: **54**; 45576: **33a**; 46202: **1** — *Sangkhachand, Phusomsaeng & Nimanong* 1057: **37**; BKF 1068: **33c**; BKF 51979: **33c**; BKF 52003: **52a**; BKF 52178: **33a** — *Sani Sambuling* 383: **33a**; 695: **52a**; 719: **56** — *Santisuk* 673: **1**; 957: **1**; 1103: **5a**; 1117: **5a**; 1119: **1**; 1269: **33a** — *Santos* 4166: **33a**; 4353: **29**; 4665: **5a**; 4977: **21**; 5264: **21**; 5724: **5a**; 5729: **21**; BS 26300: **31**; BS 31949: **33a** — *Sanusi b. Tahir* 5207: **33a**; 9176: **33a**; 9762: **11** — *Sapiin* 144: **51**; 199: **51** — Sarawak Museum Series 461: **49**; 462: **49** — Sarip Exped. *R. Maier* 136: *cf.* **33a**; 318: **51** — *Sarkat Danimihardja* 2119: **17** — *Sathaphen et al.* 201: **33a** — *Sauliere* 27: **33a**; 56: **33a**; 122: **33a**; 272: **33a** — *Saunders* 72: **21**; 171: **18a** — *Sauveur* K 25: **36a** — *Saveur* 147: **50** — *Saw* FRI 34392: **54**; FRI 35002: **21**; FRI 36280: **54**; FRI 36290: **33c**; FRI 39909: **36b**; FRI 44632: **33a**; FRI 44650: **54**; KEP 80181: **36a** — *Sawan Tangki*

SAN 125304: **36a**; SAN 125312: **52a** — *Sayers* 165: **18a**; NGF 13237: **41**; NGF 19672: **18b**; NGF 19675: **25** — *Schiffner* 2164: **33a**; 2165: **5a**; 2167: *cf.* **52a** — *Schlechter* 13513: **33a**; 14311: **18a**; 16826: **18b**; 18398: **18a**; 18465: **18a**; 19198: **45** — *Schmutz* 115: **33a**; 562: **21**; 618: **21**; 680: **5a**; 683: **21**; 762: **21**; 809: **21**; 1144: **21**; 2008: **50**; 2751: **5a**; 2888: **33a**; 3096 A: **33a**; 3761: **21**; 4464: **33a**; 4513: **33a** — *Schodde* 2619: **21**; 3094: **41**; 3105: **41** — *Schodde & Craven* 4072: **18a**; 4074: **18a**; 4213: **18b**; 4226: **2**; 4286: **18b**; 4299: **45**; 4565: **21**; 4615: **18b**; 4796: **18a**; 4840: **53**; 5025: **2**; 5034: **2** — *Scholler & Teo* KL 3961: **14a** — *Schram* BW 516: **18b**; BW 9360: **18b**; BW 10749: **10**; BW 13308: **10** — *Schultz* 694: **21**; 748: **21** — *Schwabe* 184: **18a**; 185: **45** — *Scortechini* 29: **33a**; 699: **36a**; 818: **40**; 955: **33a** — *Seidenfaden* 2679: **33a** — *Seimund* 54: **33a**; 120: **33a**; 124: **33a**; 838: **33c**; 840: **33c** — *Senada* S 810108: **46a**; 810108 (or 310108): *cf.* **46a** — *Servinas* BS 16811: **42**; BS 16911: **42** — *Setiabudi* 29: **11**; 146: **36a**; 156: **33a**; 157: **11** — SF series 915: **54**; 1358: **37**; 2094: **36a**; 2164: **37**; 2725: **36a**; 2843: **36a**; 4267: **55**; 4489: **36a**; 5371: **14a**; 5506: **14a**; 5535: **36a**; 6100: **33a**; 6391: **38**; 6541: **21**; 6890: **54**; 6926: **24a**; 8515: **12**; 8878: **12**; 9455: **33a**; 9607: **14a**; 9839: **54**; 10259: **37**; 10640: **52a**; 10978: **52a**; 11109: **39**; 11276: **12**; 11774: **36b**; 12161: **39**; 13038: **21**; 13223: **11**; 13876: **37**; 14348: **21**; 14683: **36a**; 15202: **33a**; 15328: **33c**; 15525: **54**; 15922: **21**; 16803: **54**; 16903: **21**; 16935: **33c**; 17635: **21**; 18094: **36a**; 18230: **33a**; 18686: **36a**; 18849: **54**; 19012: **52a**; 19571: **33c**; 20373: **29**; 20821: **33c**; 21314: **36a**; 21485: **33a**; 21713: **38**; 21722: **33a**; 21767: **54**; 21786: **54**; 21787: **52a**; 21875: **52a**; 21986: **29**; 21991: **29**; 22150: **33c**; 22212: **36a**; 22348: **52a**; 22353: **52a**; 22390: **33a**; 22591: **33c**; 23011: **24a**; 23660: **52a**; 23816: **21**; 23939: **33c**; 23998: **54**; 24060: **44**; 24284: **24a**; 24304: **24a**; 25019: **33a**; 25838: **33c**; 26627: **52a**; 26812: **36**; 28071: **36a**; 28793: **54**; 28819: **36a**; 28829: **52a**; 28984: **36a**; 29347: **29**; 29424: **14a**; 29433: **36a**; 30348: **40**; 30423: **49**; 30689: **52a**; 30872: **52a**; 31542: **21**; 31605: **33a**; 31650: **33a**; 31870: **11**; 31908: **39**; 32155: **38**; 32361: **36a**; 32534: **14a**; 32535: **14a**; 32682: **14a**; 32732: **39**; 33297: **21**; 33604: **29**; 33633: **52a**; 33762: **29**; 33862: **33c**; 33915: **11**; 33943: **40**; 34532: **14a**; 34540: **11**; 34608: **11**; 35019: **54**; 35033: **37**; 35136: **33a**; 35150: **54**; 35191: **11**; 35233: **24a**; 35642: **36c**; 35652: **36a**; 35755: **11**; 35922: **14a**; 36602: **33a**; 36630: **33a**; 36849: **14a**; 37080: **52a**; 37228: **54**; 37905: **5a**; 37906: **5a**; 37907: **33a**; 38809: **36a**; 38956: **21**; 38964: **33a**; 39832: **21**; 39949: **37**; 40033: *cf.* **11**; 40113: **11**; 40680: **21**; 40803: **33a**; 40806: **33a**; 40862: **36b**; 40947: **54**; 40980: **21**; 41005: **29**; 41018: **24a** — *Shaari A.* 94292: **33a** — *Shah* 124: **14a**; 284: **22**; 301: **22**; 494: **11**; 497: **37**; 1329: **33c**; 1377: **37**; 1551: **37**; 1791: **37**; 1974: **36b**; 2581: **39**; 2590: **54**; 2712: **33c**; 2732: **39** — *Shah et al.* 2064: **36a**; 2068: **14a**; 3325: **52a**; 3714: **52a**; 3720: **38**; 3779: **54**; 4896: **29** — *Shah & Ali* 2850: **33c**; 2928: **52a** — *Shah & Mahmud* 40: **52a**; 4887: **52a** — *Shah & Lee* 2695: **54** — *Shah & Nur* 569: **33a**; 741: **49**; 908: **39**; 1773: **37**; 1798: **33c**; 1841: **52a**; 1848: **54**; 1871: **52a**; 1897: **54**; 1975: **49** — *Shah & Samsuri* 2426: **54**; 3813: **36a** — *Shah & Sanusi* 2125: **49** — *Shah & Shukor* 2251: **40**; 2309: **38**; 2337: **52a**; 3154: **33a**; 3169: **36b** — *Shah & Sidek* 1138: **52a** — *Shah & Tan* 5038: **55** — *Shaik Mokim* 798: **1**; 1134: **1** — *Shea & Aban* SAN 77155: **33a** — *Shea & Chow* SAN 75732: **52a** — *Shea & Minjulu* SAN 76063: **52a**; SAN 76169: **33a**; SAN 76277: **33a**; SAN 76297: **21** — *Shepherd & Rieley* 945: **33a**; 9476: **33a** — *Shimizu et al.* 7992: **33a**; T 8796: **1**; T 8914: **5a**; T 10453: **1**; T 10822: **1**; T 14698: **55**; T 17944: **1**; T 19025: **1**; T 21015: **1**; T 21836: **1**; T 22699: **1** — *Shimizu & Stone* T 14438: **54**; — *Shiu Ying Hu* 7220: **24a**; 10080: **24a**; 10254: **21**; 10265: **21**; 10286: **21**; 13722: **24a** —

Sibat ak Luang S 5371: **40**; S 21795: **36a**; S 22119: **49**; S 22122: **36a**; S 23204: **52a**; S 24140: **36b**; S 24372: **3**; S 24619: **36b**; S 24819: **36b**; S 24833: **36a**; S 25052: **52a**; S 25071: **36b** — *Sidiyasa* Project Barito Ulu 533: **29**; 607: **29**; 1004: **36a**; 1020: **49**; 1065: **21**; 1420: **33a** — *Sigin et al.* SAN 56808: **36a**; SAN 56829: **46a**; SAN 56843: **49**; SAN 66692: **52a**; SAN 97171: **36a**; SAN 97532: **46a**; SAN 99886: **49**; SAN 107200: **52a**; SAN 109722: **49**; SAN 110610: **36a**; SAN 110640: **52a** — *Sigin Gambukas* SAN 97390: **33a** — *Simbut* SAN 78556: **36a** — *Simpson* 2091: **3**; 2456: **3** — *Simpson & Mardh* 2071: **29** — *Simpson et al.* 1511: **55** — *Sinanggul* SAN 54608: **36a**; SAN 57154: **36a**; SAN 57202: **36a**; SAN 57290: *cf.* **33a**; SAN 57299: **34** — *Sinclair* SF 5371: **14a**; SF 5506: **14a**; SF 5535: **36a**; 6558: **54**; 7672: **37**; 7678: **37**; 9865: **24a**; 10678: **29**; SF 40033: *cf.* **11**; SF 40113: **11**; SF/S 40113/2776: **11**; SF 40947: **54** — *Sinclair & Edano* 9587: **33a** — *Sinclair & Kadim b. Tassim* SF? 10247: *cf.* **29** — *Sinclair & Kiah* SF 38809: **36a**; SF 39832: **21**; SF 39949: **37**; SF 40803: **33a**; SF 40806: **33a**; SF 40862: **36b** — *Singghatsathit* 4405: **5a** — *Singh* SAN 22561: **52a**; SAN 24011: **46a**; SAN 24050: **52a**; SAN 24240: **29**; SAN 27421: **52a**; SAN 31066: **44** — *Singh & Aban* SAN 26252: **36a** — *Singh et al.* SAN 49112: **36a** —*Singh & Eging* SAN 51855: **33a** — *Singh & Nordin* 48466: **36a** — *Singh & Talip* SAN 50644: **11**; SAN 50664: **11** — *Sirirugsa* 655: **21**; 829: **21**; 926: **54**; 975: **36a** — *Sirute'e & collectors* BSIP 9649: **18a** — *Smith* 250: **33a**; 687: **54**; 926: **32**; 19670: **51**; 71948: **33c** — *Smitinand* 6228: **33a**; 7305: **5a**; 7918: **33a**; 9743: **47**; 11761: **5b**; 11812: **47**; 11833: **47**; 12840: **33a**; 15622: **21**; 21479: **54**; 22276: **54**; BKF 48683: **5a** — *Smitinand & Floto* 24343: **54** — *Smitinand & Nalamphum BKF* 46790: **21** — *Smitinand & St. John* 25082: **33a** — *Smythies* S 13162: **36b** — *Smythies et al.* SAN 17089: **29**; SAN 17119: **11**; SAN 17368: **46a**; SAN 17469: **36b** — *Soedarsono* 263: **49**; 325: **49** — *Soegeng Reksodihardjo* 230: **21**; 333: **18a**; 645: **29** — *Soejarto & Fernando* 7375: **52a** — *Soejarto & Madulid* 6120: **33a**; 6123: **49**; 9012: **21** — *Soejarto & Reynoso* 6263: **33a** —*Soejarto et al.* 5878: **55**; 5890: **55**; 6454: *cf.* **33a**; 6504: **21**; 6510: **21**; 6512: **21**; 7520: **21**; 7790: *cf.* **15**; 7937: **15**; 8105: *cf.* **52a**; 8260: **33a**; 8285: **33a**; 8317: **33a**; 8328: **33a**; 8406: **33a**; 8455: **33a**; 8574: *cf.* **33a**; 8670: **44**; 8744: **33a**; 8752: **33a** — *Soenarko* 318: **5a** — *Soepadmo* 120: **52a**; 325: **5a**; 626: **36a**; 781: **24a**; 863: **33c**; HUM 9036: **14a** — *Soepadmo & Mahmud* 1223: **33a**; 9191: **36a** — *Soepadmo & Suhaimi* S 9: *cf.* **20**; S 46: **20**; S 261: **33c**; 285: **33a**; 318: **11** — *Sohmer* 12419: **33a** — *Sohmer & Katik* LAE 75163: **18a** — *Soinin et al.* SAN 101922: **33a** — *Soinin & Suin* SAN 129106: **36b** — *Soinin Satman* SAN 66312: **33a** — *Song Xianghou* 47: **24a**; 690: **33b**; 955: **24a**; 965: **24a** — *Sore et al.* BSIP 2672: **18a** — *Sorensen* 479: **27**; 2663: **1**; 2983: **33a**; 3110: **1**; 3290: **1**; 3356: **1**; 3423: **1**; 3833: **21**; 3895: **1**; 4345: **47**; 5154: **47**; 5386: **5a** — *Sow* For. Dept. F.M.S. 39496: **14a**; 80036: **33a** — *Spare* SF 33297: **21**; 36702: **21** — *Squires* 114: **33c**; 201: **33c**; 882: **1** — *St John* 19035: **18a**; 19036: **18a** — *Stainton et al.* 6720: **33a** — *Stauffer & Sayers* 5602: **18a** — *Steiner* PNH 22852: **5a**; PNH 36466: **5a** — *Steup* Boschproefstation 17044: **7** — *Stevens et al.* 75: **52a**; 210: **36b**; 283: **29**; 334: **36a**; 430: **36a** — *Steward & Cheo* 538: **33b**; 854: **24a** — *Stone* 6286: **36a**; 7489: **54**; 8772: **33a**; 8776: **33a**; 10883: **33c**; 11799: **37**; 13836: **52a**; 14392: **26** — *Stone et al.* PPI 526: **42**; PPI 6750: **33a**; PPI 6833: **33a**; PPI 6926: **15**; 13783: **49**; 15192: **33c** — *Stone & Sidek* 12452: **33c**; 12479: **54** — *Stone & Streimann* LAE 53583: **10** — *Streimann* 8715: *cf.* **2**; NGF 25912: **18a**; NGF 44196: **18a**; NGF 45167: **18a**; NGF 45413: **18a**; NGF 47800: **18a**; LAE 51739: **10** — *Streimann et al.* LAE 51901: **10**; LAE 51912: **18a** — *Streimann & Kairo* NGF 27636: **2**; NGF 27964: **21**; NGF 39231: **18a**; NGF 42490: **18a** — *Streimann & Katik* NGF 46807: **18a** — *Streimann &*

Lelean LAE 52609: **18a** — *Streimann & Stevens* LAE 53861: **18a**; LAE 53993: **18a** — *Strugnell* 12146: **49**; 12898: **11**; 20299: **14a** — *Subramanian* 1513: **1** — *Sugau* 47: **33a** — *Suib* S 22269: **36b** — *Sulit* PNH 3618: **31**; PNH 6087: **13**; PNH 6162: **30**; PNH 6203: **30**; 6257: **13**; PNH 6366: **13**; PNH 6911: **31**; PNH 8369: **33a**; PNH 9933: **5a**; PNH 10032: **52a**; PNH 12297: **33a**; PNH 14387: **52a**; PNH 14420: **13**; PNH 15719: **50**; PNH 17028: **21**; PNH 21591: **31**; FB 24779: **16a**; FB 30312: **42** — *Sulit & Conklin* PNH 16784: **5a** — *Sulit & Party* PNH 6641: **33a** — *Sumbing & Asik* SAN 110827: **36a**; SAN 122114: **52a** — *Sumbing & Martin* SAN 103610: **36a** — *Sumbing Jimpin* SAN 101500: **36a**; SAN 105287: **11**; SAN 119480: **36a** — *Suppiah* FRI 11281: **54**; FRI 11296: **11**; FRI 11326: **14a**; FRI 11350: **14a**; FRI 11377: **33c**; FRI 11396: **54**; FRI 11583: **40**; FRI 11739: **11**; FRI 11837: **40**; FRI 14830: **22**; FRI 14893: **54**; FRI 28169: **11**; FRI 28198: **33a**; 98969: **36a**; FRI 98990: **36a**; KEP 104564: **21**; KEP 104806: **54**; KEP 108851: **54**; KEP 108855: **54** — *Suvarnakoses* 699: **33a**; 2833: **21**; 9286: **54**; 11577: **22**; 17581: **33a**; 25265: **21**; 29547: **21**; 38821: **22** — *Sylvester Tong* S 34923: **36b** — *Symington* 21074: **14a**; For. Dept. F.M.S. 22297: **36a**; 22619: **49**; For. Dept. F.M.S. 22709: **52a**; 23227: **36a**; For. Dept. F.M.S. 24142: **52a**; For. Dept. F.M.S. 24227: **36a**; 24237: **37**; For. Dept. F.M.S. 24417: **54**; For. Dept. F.M.S. 24449: **11**; 246191: **49**; KEP 29792: **14a**; For. Dept. F.M.S. 29882: **52a**; For. Dept. F.M.S. 37631: **36b**; For. Dept. F.M.S. 37641: **36b** — *Symington & Kiah* SF 28793: **54**; SF 28819: **36a**; SF 28829: **52a** —

Taam 706: **24a**; 948: **24a**; 1340: **5a**; 1459: **5a**; 1757: **5a**; 28821: **24a** — *Tafuku* 11: **24a** — *Tagawa et al.* T 2332: **1**; FRI-T 8620: **54**; T 9170: **47**; T 9797: **1**; T 9824: **47** — *Tahir* 805: **33a**; 1196: **21** — *Takahashi* 62750: **1** — *Takeuchi* 4388: **18a**; 4444: **35a**; 4605: **45**; 6196: **18a**; 7048: **18a**; 9146: **10**; 9175: **18a** — *Talib B. & Marsal* SAN 86146: **33a** — *Tamesis* FB 11949: **5a**; FB 11976: **21**; FB 13340: **5a** — *Tanaka & Shimada* 13570: **24a**; 13571: **24a** — *Tandom* BNB For. Dept. 2828: **21**; 4223: **33a**; KEP 38600: **33a** — *Tanosa* FB 15153: **21** — *Tantra* 1537: *cf.* **46a** — *Tappenbeck in Herb. Lauterbach* 140: **18a** — *Tarmiji & Dewol Sundaling* SAN 84189: **52a** — *Tarmiji & Paul* SAN 87559: **52a** — *Tarrosa Miranda & Rafael* FB 18791: **21** — *Tay & Tan* 94-0043: **18a** — *Taylor* 301 or 105: **21**; 395: **21**; 2681: *cf.* **7** — *Teo* 343: **21** — *Teo & P* KL 2621: **14a**; KL 2686: **21**; KL 2839: **37**; KL 2943: **21**; KL 3009: **14a**; KL 3142: **33c**; KL 3146: **54**; KL 3153: **39**; KL 3527: **14a**; KL 3609: **33a** — *Teo & Remy* KL 4008: **21** — *Teruya* 2656: **14a** — *Tessier-Yandell* 180: **21** — *Teysmann* 257: **33a**; 266: **33a**; 567?: **54**; 751: **21**; 3661: **29**; 3733: **33a**; 3738: **29**; 3816: **29**; 4431: **33a**; 4467: **44**; 4532: **36a**; 5283: **7**; 5589: **18a**; 8276: **49**; 8278: **52a**; 8416: *cf.* **36a**; 8417: **33a**; 8421: **11**; 8424: **33a**; 10727: **5a**; 10927: **33c**; 11327: **33a**; 11330: **33c**; 11623: **21**; 12362: **21**; 12388: **5a**; 12582: **5a**; 12661: **21**; 12693: **7**; 12798: **7**; 17528: **49**; 17529: **21** — *Thakur Rup Chand* 1702: **33a**; 3079: **21**; 4686: **1**; 4781: **1**; 5089: **1**; 6292: **33a**; 6293 A: **33a**; 8287: **33a** — *Thaufeck (Taufik?)* SAN 27958: **11** — *Thaworn* 12227: **44**; 12527: *cf.* **33d**; 12882: **36a**; 12895: **21**; 14850: **21**; 14862: **52a**; 15309: **21**; 15325: **24a**; 15447: **33a**; 17165: **36a**; 17363: **36a**; 17555: **55**; 18162: **33a** — *Thorel* 684: **5a**; 1139?: **1**; 1201: **21** — *Thorenaar* 257: **29**; 260: **21**; 273: **21** — *Thorne & Henty* 27478: **18a** — *Thwaites* C.P. 660: **5a**; C.P. 773: **21**; C.P. 2922: **44** — *Ting & Shih* 588: **24a** — *Tingguan* 37376: **36a** — *Tirvengadum & Nanakorn* 1987: **24a** — *Tong* S 33588: **29**; S 34165: **33a**; S 34333: **40**; S 34342: **56**; S 34750: **52a**; S 34943: **52a** — *Tong et al.* S 34256: **36a**; S 34313: **29**; S 34318: **36a** — *Torquebiau et al.* 578: **11** — *Treub* 1893: *cf.* **49** — *Troth* 874: **1**; 875: **1** — *Tsang* LU

15564: **21**; LU 15601 20: **33a**; LU 16085: **21**; LU 16163: *cf.* **33a**; 16796: **21**; 20571: **24a**; 21017: **24a**; 22894: **24a**; LU 27292: **33c**; 29038: **24a**; 30196: **5a** — *Tsiang Ying* 3171: **5a** — *Tsugaru* T 61688: **1**; T 61697: **1**; T 61772: **24a**; T 61847: **1** — *Tuke I. Djuda* 94221: *cf.* **36a**; 94230: **52a** — *Tukirin Partomihardjo* 568: **11** —

Udasco FB 27293: **16a** — *Umbai* for *A H Millard* KL 1015: **54**; KL 1304: **52a**; KL 1307: **11**; KL 1461: **54**; KL 1462: **14a** — Univ. of San Carlos 313: **33a** — *Upik Rosalina* TFB 2675: *cf.* **11**; TFB 2762: **14a**; TFB 2862: *cf.* **20** — *Utja & Wasijat* Ja 6534: **51** —

Valera 4734: **52a**; 4835: **52a** — *Valeton* 6: **49**; 7: **54**; 51: **51**; 192: **49**; 254: **23** — *Van Balgooy* 1940: **5a**; 2099: **33a**; 2105: **33a**; 2132: **49**; 2334: **33a**; 2414: **33a**; 2437: **52a**; 2454: **54**; 2511: **33c**; 2514: **33c**; 2630: **28**; 3130: *cf.* **46a**; 3936: **21**; 4715: **49**; 4933: **49**; 4972: **49**; 5068: **33a**; 5235: **32**; 6531: **33a**; 6787: **18b**; 7281: **51** — *Van Balgooy & Van Setten* 5449: **29**; 5680: **32**; 5688: **33a** — *Van Balgooy & Wiriadinata* 2864: **51**; 2867: **51**; 2870: **32**; 2894: **51**; 2897: **51** — *Van Beusekom & Charoenphol* 1670: **33a**; 1877: **24b**; 1947: **33a** — *Van Beusekom & Geesink* 3322: **24b** — *Van Beusekom et al.* 3529: **47**; 3646: **1**; 4087: **33b**; 4137: **47** — *Van Beusekom & Phengkhlai* 64: **24a**; 177: **33a**; 209: *cf.* **54**; 245: *cf.* **33a**; 283: **33a**; 304: **33a**; 305: **54**; 503: **5a**; 538: **54**; 547: *cf.* **52a**; 574: **52a**; 604: **55**; 704: *cf.* **52a**; 740: **33d**; 792: **54**; 1081: **1**; 1107: **1**; 2318: **1**; 2658: **47** — *Van Beusekom & Santisuk* 2802: **52a**; 3163: **27**; 3264: **5a** — *Van Beusekom & Smitinand* 2107: **1**; 2123: **21** — *Van Borssum Waalkes* 793: **23**; 1445: **40**; 2298: *cf.* **33a**; 2429: **21** — *Van der Meer & Den Hoed* 1536: **21** — *Van der Meer Mohr* 15: **33a**; 16: **33a**; 17: **33a**; 71: **33a** — *Van der Sijde* BW 4077: **18a**; BW 5518: **18a** — *Van Heel* 46: **33a** — *Van Leeuwen* 7540: **32**; 9164: **10**; 9757: **45**; 9870: **41**; 11068: **45**; 12679: **1** — *Van Leeuwen-Reijnvaan* 11749: *cf.* **5a**; 14191: **23** — *Van Niel* 3873: **11**; 4246: **11**; 4254: **11**; 4303: **11**; 4598: **36b** — *Vanoverbergh* 1154: **5a**; 1194: **15** — *Vanpruk* 732: **33a**; 800: **54** — *Van Royen* 4635: **21**; 4840: **21**; 5258: *cf.* **33a**; NGF 16328: **18a**; 20074: **45** — *Van Royen & Sleumer* 6913: **10**; 7056: **46b**; 7263: **10** — *Van Slooten* 2124: **36a**; 2187: **36a**; 2204: **36a** — *Van Steenis* 1115: **54**; 1164: **36a**; 1211: **36a**; 1220: **54**; 2312: **54**; 2330: **51**; 2345: **33a**; 2414: **32**; 3233: **5a**; 3289: **5a**; 3421: **33a**; 3427: **29**; 3593: **49**; 3605: **49**; 3798: **49**; 6163: **33a**; 7498: **1**; 8922: **49**; 9210: **49**; 9402: **33a**; 10096: **49**; 10582: **5a**; 12674: **21**; 17353: **51**; 17380: **32**; 17403: **54**; 17407: **52a**; 17508: **33a**; 17546: **51**; 18078: **23**; 18545: **11** — *Van Valkenburg* 1014: **29**; 1049: **56**; 1108: **36a** — *Van Valkenburg & Stockdale* 1075: **36a** — *Van Welzen* 810: **36a** — *Varadarajan et al.* 1501: **5a**; 1528: **21**; 1539: **21**; 1573: **33a** — *Vaupel* 216: **18a** — *Veldkamp* 8041: *cf.* **49**; 8054: **29**; 8063: **52a**; 8094: **52a**; 8184: **56**; 8218: **36a**; 8329: **49**; 8344: **36a**; 8346: **52a**; 8435: **52a**; 8445: **33a**; 8464: **56**; 8496: **49**; 8548: **33a** — *Venugopal* 14565: **21** — *Verheijen* 1075: **33a**; 1076: **33a**; 1077: **33a**; 1510: **33a**; 1511: **33a**; 1512: **33a**; 2263: *cf.* **5a**; 2937: **23**; 3063: *cf.* **5a**; 3100: **5a**; 3106: **33a**; 3202: **33a**; 3231: *cf.* **5a**; 3285: **21**; 3305: **33a**; 3317: **50**; 4528: **5a** — *Verhoef* 87: **36a** — *Vermeulen* 784: **36b**; 867: **52a**; 897: **36b**; 1229: **33a** — *Vermeulen & Duistermaat* 938: **36a**; 1143: **36b** — *Versteegh* 1258: **45**; 1730: **10**; 1734: **10**; 1770: **10**; 1789: **45**; BW 3824: **18a**; BW 3901: **18a**; BW 4672: **18b**; BW 4902: **2**; BW 10494: **35a**; BW 12527: **35a** — *Vethevelu* FRI 25233: **14a** — *Vians & Nagari* UPNG 4853: **18a** — *Vidal* 335/1664: **21**; 564: **5a**; 564 c, d: **52a**; 565: **5a**; 565 bis: **5a**; 565 c: **5a**; 566: **33a**; 567: **21**; 568: **33a**; 569: **33a**; 570: **21**; 570 bis, c, d, e, f: **21**; 587: **16a**; 590: **42**; 591: *cf.* **42**; 875: **52a**; 876: **16a**; 903: **33a**; 904: **33a**; 1293: **1**; 1740: **52a**; 1747: **33a**; 1751: **33a**; 1762: **5a**; 1762 bis: **15**; 1763: *cf.* **33a**; 1790: **1**; 1796: *cf.* **42**; 1802:

33a; 3692: **33a**; 3694: **42**; 3695: **49**; 3696: **52a**; 3697: **33a**; 3698: **33a**; 3751: **52a**; 3891: *cf.* **42**; 5299: **5a** — *Vidal et al.* 6122: **47**; 6209: **1** — *Villamil* 235: **36a**; 349: **21**; FB 20495: **33a**; FB 21857: **29** — *Villan* BS 29838: **30** — Villar FB 30585: **31** — *Vinas* 134: **18a** — *Vinas & Nagari* UPNG 4866: **18b**; UPNG 7611: **18b** — *Vink* BW 12193: **2** — Vogt BU-37: **1** — *Vreeken-Buijs* 16: *cf.* **52a** —

Waas 625: **21** — *Walker* 8445: **24a**; For. Dept. F.M.S. 23318: **33c** — *Walker et al.* 6199: **24a** — *Walker & Tawada* 6643: **24a**; 7164: **24a**; 7192: **33a**; 7275: **24a** — *Wallace* 118: **21** — *Wallich* 7280 c, d, f: **21**; 7281: **1**; 7282: **5a**; 7282 A: **28**; 7282 B: **33a**; 7282 F: **33a**; 7284: **1**; 7285: **1**; 7285 B: **1**; 7288: **33a**; 8569: **38**; 8577: **55**; 8583: **36a** — *Wallich* in Herb. Heyne 7282: **33a** — *Wang* 33766: **21** — *Warburg* 1325: **33a**; 11869: **52a**; 11874: **33a**; 11875: **5a**; 11999: **33a**; 12315: **33a**; 13137: **52a**; 13604: **21**; 13607: **33a**; 13975: **52a**; 13978: **33a**; 13979: **5a**; 13980: **33a**; 13981: **33a**; 13982: **33a**; 13983: **33a**; 13984: **33a**; 13986: **33a**; 17297: **49**; 20520: *cf.* **18a**; 20640: **33a**; 20650: **33a** — *Waterhouse* Y 118: **18a**; 139 B: **18a**; Y 147: **18a**; 365 B: **18a**; 771 B: **18a**; 791 B: **18a** — *Watt* 11850: **21** — *Waturandang* Boschproefstation cel/V 125: **7** — *Weber* 11: **33a**; 33: **44**; 1492: **24a**; 1577: **30** — *Wen Ho Qun* W 135: **33a** — *Wenzel* 34: **33a**; 46: **33a**; 106: **33a**; 107: **33a**; 115: **13**; 116: **15**; 213: **52a**; 319: **13**; 555: **33a**; 705: **52a**; BS 823: **31**; BS 831: **33a**; 995: *cf.* **42**; BS 1098: **13**; 1204: **52a**; 1242: **33a**; 1266: **33a**; 1327: **15**; 1335: *cf.* **30**; 1347: **15**; 1458: **15**; 1707: **52a**; 2792: **52a**; 3114: **52a**; 3128: **33a**; 3147: **33a**; 3194: **16a**; 3353: **33a** — *Werrachai Nanakorn* 518: **21** — *Whistler* 150: **18a**; 151: **18a**; 1483: **18a**; 1831: **18a**; 1832: **18a**; 2800: **18a**; 2941: **18a**; 3812: **18a**; 3922: **18a**; 4199: **18a**; 7701: *cf.* **18a**; 7727: **18a**; 8114: **18a** — *White* 2526: **18a**; NGF 8257: **18a**; NGF 10103: **18a**; NGF 10344: **18a**; NGF 10848: **18a** — *Whitford* 52: **16a**; 415: **5a**; 476: **16a**; 521: **16a**; 556: **33a**; 1135: **15**; 1316: **33a**; 1453: **16a**; FB 19802: **42** — *Whitmore* FRI 118: **54**; FRI 151: **38**; FRI 226: **11**; FRI 228: **33a**; FRI 328: **49**; FRI 382: **33a**; FRI 400: **37**; FRI 453: **33a**; FRI 630: **36a**; FRI 697: **40**; FRI 811: **37**; FRI 896: **29**; FRI 916: **11**; BSIP 1228: **18a**; BSIP 1351: **18a**; BSIP 1969: **18a**; BSIP 1976: **18a**; BSIP 2233: **18a**; BSIP 2247: **18a**; BSIP 2480: **18a**; BSIP 2726: **18a**; FRI 3367: **54**; FRI 3368: **36a**; FRI 3450: **36a**; FRI 3456: **14a**; FRI 3459: **33a**; FRI 3466: **54**; FRI 3826: **33a**; FRI 3830: **33a**; FRI 3840: **33c**; BSIP 3893: **18a**; BSIP 3941: **18a**; 3946: **28**; FRI 4009: **54**; FRI 4071: **39**; FRI 4076: **54**; FRI 4136: **33c**; FRI 4284: *cf.* **11**; FRI 4308: **28**; FRI 4410: **11**; FRI 4547: **39**; FRI 4760: **49**; FRI 8589: **33c**; FRI 8860: **33a**; FRI 8907: **33c**; FRI 12184: **39**; FRI 12317: **36a**; FRI 12705: **36a**; FRI 12818: *cf.* **51**; FRI 15205: **36a**; FRI 15390: **33c**; FRI 15876: **54**; FRI 15979: **54**; FRI 15992: **28**; NGF 16337: **18a**; FRI 20054: **33c**; FRI 20069: **40**; FRI 20073: **36a**; FRI 20074: **54**; FRI 20075: **52a**; FRI 20265: **33c**; FRI 20340: **39**; FRI 20497: **33c**; FRI 20571: **36b**; FRI 20664: *cf.* **11** — *Whitmore & Marfu'ah Wardani* 3157: **14a** — *Whitmore & Sidiyasa* 3259: **49**; 3319: **33a**; 3465: **17** — Whitmore's collectors BSIP 2856: **18a**; BSIP 2860: **18a**; BSIP 2916: **18a**; BSIP 2927: **18a**; BSIP 3607: **18a**; BSIP 5420: **18a**; BSIP 5463: **18a** — *Wiakabu et al.* LAE 73395: **35a** — *Widjaja* 1202: **7**; 1874: **51**; 2186: **10**; 2697: **33a** — *Wight* 2455/1: **21**; 2654: **33a**; Kew distr. no. 2655: **1**; Kew distr. no. 2655 /1: **21**; Kew distr. no. 2655/2: **47** — *Wigman Jr.* 5/8: **7** — *Wilkie* 94174: **52a** — *Williams* 490: **16a**; 940: **33a**; 974: **15**; 1139: **15**; 1140: **15**; 1141: **15**; 1372: **16a**; 2117: **33a**; 2719: **21**; 2912: **13**; 2926: **42**; 3008: **33a**; 3112: **33a** — *Wilson* 4072: **33b**; 11058: **33a** — *Winckel* 61: **32**; 93: **51**; 150: **51**; 170: **5a**; 175: **32**; 277: **51**; 311: **54**; 429: **51**; 813: **32**; 1507: **5a**; 1790: **51**; 1818: **54**;

INDEX

Lightning Source UK Ltd.
Milton Keynes UK
29 October 2009

145515UK00001B/63/A